Advanced Technologies of Water and Wastewater Treatment

Advanced Technologies of Water and Wastewater Treatment

Guest Editors

Athanasia K. Tolkou
George Z. Kyzas

MDPI

Basel • Beijing • Wuhan • Barcelona • Belgrade • Novi Sad • Cluj • Manchester

Guest Editors

Athanasia K. Tolkou
Department of Chemistry
Democritus University
of Thrace
Kavala
Greece

George Z. Kyzas
Department of Chemistry
Democritus University
of Thrace
Kavala
Greece

Editorial Office
MDPI AG
Grosspeteranlage 5
4052 Basel, Switzerland

This is a reprint of the Special Issue, published open access by the journal *Environments* (ISSN 2076-3298), freely accessible at: https://www.mdpi.com/journal/environments/special_issues/9YJ9178BY6.

For citation purposes, cite each article independently as indicated on the article page online and as indicated below:

Lastname, A.A.; Lastname, B.B. Article Title. *Journal Name* **Year**, *Volume Number*, Page Range.

ISBN 978-3-7258-5609-1 (Hbk)
ISBN 978-3-7258-5610-7 (PDF)
https://doi.org/10.3390/books978-3-7258-5610-7

Contents

About the Editors

Athanasia K. Tolkou

Athanasia K. Tolkou is a Chemist, graduated from the Aristotle University of Thessaloniki (AUTH), and holds a PhD in Chemistry in the field of wastewater treatment. She is now working at the Department of Chemistry, School of Science, Democritus University of Thrace (DUTH), Kavala, Greece, as an Assistant Professor. Her research interests include wastewater treatment, municipal, industrial, and hazardous wastewater treatment technologies (coagulation/flocculation, adsorption, ozonation, membrane bioreactors), recovery of nutrients (phosphates and nitrates) from wastewater, and removal of ions, heavy metals, and dyes from water and wastewater.

George Z. Kyzas

George Z. Kyzas is a Chemist and Full Professor in the Department of Chemistry at the Democritus University of Thrace (DUTh). He was born in Drama (Greece) and studied Chemistry at the Department of Chemistry at the Aristotle University of Thessaloniki (AUTh). He obtained his BSc, MSc, and PhD degrees from the same department that specializes in Chemical Technology. He is now working at the Department of Chemistry (International Hellenic University, Kavala, Greece) as the Head of the Department (since 2019). He is also the Head of the Hephaestus Laboratory of the School of Science (since 2023) and the Director of the MSc in Cosmetic Chemistry (since 2021). His research interests include the synthesis and characterization of various (majorly adsorbent) materials (inorganic, aluminates, polymers, graphene, activated carbons, agro-food residues, nanomaterials, CNTs, etc.) for environmental applications (wastewater treatment). At the age of 42 years old, Prof. Kyzas is included in the list of Highly Cited Researchers for 2022 (Thomson Reuters—Clarivate WoS) [0.1% Most Impactful Scientists in the World]. His name is also included in the list of World Top 2% Scientists for 2019, 2020, 2021, 2022, 2023, and 2024, which is compiled by Stanford University (USA) based on standardized citation indicators.

environments MDPI

Editorial

Advanced Technologies of Water and Wastewater Treatment

Athanasia K. Tolkou * and George Z. Kyzas *

Hephaestus Laboratory, School of Chemistry, Faculty of Sciences, Democritus University of Thrace, GR-65404 Kavala, Greece
* Correspondence: atolkou@chem.duth.gr (A.K.T.); kyzas@chem.duth.gr (G.Z.K.)

With more than 2 billion people worldwide suffering from water scarcity, clean water is one of the most important natural resources on earth, whereas wastewater, which corresponds to spent water, can be considered a valuable natural resource if treated and reused. The aim is for this *Environments* Special Issue on "Advanced Technologies of Water and Wastewater Treatment" to address the most pressing problems of natural water resource contamination, as well as wastewater treatment and reuse. We also aim to present innovative treatment options for aquatic environments.

The main topics of this Special Issue include water and wastewater treatment technologies that can address contaminants of emerging concern for the aquatic environment, such as membrane filtration, adsorption, coagulation, ion exchange, biological processes, ozonation, and advanced oxidation or hybrid processes. Moreover, submissions of papers that deal with the fate and removal of microplastics, heavy metals, pharmaceuticals, oxyanions, toxic dyes, and several other environmental pollutants of global concern, were invited.

Figure 1 illustrates statistical data on the articles published in this Special Issue. A total of seventeen (17) manuscripts were published, of which two (2) were *Reviews*, one (1) was a *Communication*, and fourteen (14) were *Articles* (Figure 1a). As can be seen in Figure 1b, the published articles covered a wide range of applications and topics, with a higher proportion focusing on wastewater treatment and the removal of heavy metals. Specifically, the published papers can be categorized as follows:

- Livestock waste management

In the study by El bied et al., 2024 [1], the impacts of different technologies and approaches were investigated, including the use of *Bacillus* biological additives (HIPO-PURÍN) and aeration using OXI-FUCH, on GHG (methane (CH_4), carbon dioxide (CO_2), and ammonia (NH_3)) emissions from stored pig slurry. By analyzing the results statistically, the authors demonstrated that the treatment with HIPO-PURÍN achieved significant reductions in CH_4 emissions by 67% and CO_2 emissions by 60%, which were increased to 99% and 87%, respectively, when combined with an aeration system (OXI-FUCH), compared with untreated slurry. Ammonia emissions were substantially reduced by 90% by using biological additives alone and by 76% when combined with aeration. The study was driven by the need to develop sustainable solutions for livestock waste management. In addition, Lamssali et al., 2024 [2], provided valuable insights into optimizing anerobic bioreactor processes for the efficient and environmentally friendly treatment of high-strength wastewaters from swine farms, paving the way to more effective and sustainable management of swine wastewater, which is crucial to meet environmental discharge standards and promoting eco-friendly agricultural practices.

- Drinking Water Treatment Systems

In the study by Aragon et al., 2024 [3], TiO_2-coated meltblown nonwoven fabrics that were prepared via Atomic Layer Deposition (ALD) were applied for the inactivation of *Escherichia coli* (*E. coli*). It was observed that in the presence of an ultraviolet light-emitting diode (UV-LED) light source (255 nm), 1.35 log *E. coli* inactivation was achieved.

- Water Reuse

An interesting review by Florides et al., 2024 [4], investigated the current status of water reuse practices in Europe, delving into various methods of water recovery. This review examined public perceptions and attitudes toward recycled water, offering insights into the societal outlook on this increasingly vital aspect of water management. According to the data presented, the current applications of water reuse in northern Europe are for environmental purposes (51%), while in southern Europe, it is mainly used for irrigation (44%), with Spain reusing the largest volume. In another review from de Campos and Soto, 2024 [5], the authors selected and reviewed articles published in the last six years involving the use of different constructed wetland systems (CWs) and their use alongside other technologies to treat different effluents and in which the quality of different effluents for reuse is discussed. From a total of 81 articles reviewed, 41 presented quantitative data on the quality of the treated effluents in relation to the requirements of reuse regulations in different countries. CWs can be used to treat gray and runoff water, as well as domestic and industrial effluents, with the purpose of reusing them.

- Wastewater Treatment and Reuse

Four (4) commercially available small decentralized technologies (DWWTs) with design capacities of 4–8 PE (population equivalent) were selected and operated with various wastewater compositions in Leipzig, Germany, as presented in the study by Rahman et al., 2024 [6]. The technologies were (i) a moving-bed biofilm reactor (MBBR), (ii) a sequencing batch reactor (SBR), (iii) a membrane bioreactor (MBR), and (iv) an aerated vertical-flow constructed wetland (AVFCW). This study clearly showed that the EU-certified small DWWT technologies can efficiently treat high-strength wastewater with concentrations beyond the EU testing ranges (i.e., BOD_5 > 500 mg/L; COD > 1000 mg/L; TSS > 700 mg/L) and are promising wastewater treatment alternatives for small communities in the rural parts of arid/semi-arid regions. Treating wastewater using small decentralized wastewater treatment (DWWT) technologies in combination with the safe reuse of the treated effluents for agricultural irrigational purposes can potentially create noticeable environmental and economic benefits in arid and semi-arid regions. Plant-wide models for optimizing the operation and maintenance of BTEX (Benzene, toluene, ethylbenzene, and xylenes)-contaminated wastewater treatment and reuse facilities that adhere to the IWA Good Modelling Practice Guidelines were used in the study by Bencsik et al., 2024 [7]. The model-based benchmarking tests of this study demonstrated the applicability of the presented concepts for plant-wide modeling by investigating the impacts of operational parameters of the activated sludge process on the removal efficiency of BTEX chemicals, alongside the demand for granular activated carbon usage in effluent polishing. Moreover, the removal characteristics of high-strength ammonium wastewater using A. faecalis No. 4 were studied and presented by Lee and Joo, 2023 [8], under several operating conditions, such as foam–recycle. Ammonium was first oxidized to hydroxylamine, nitrite, and nitrate under aerobic conditions by *A. faecalis No. 4*, providing intermediate substrates for subsequent denitrification. Finally, regarding wastewater treatment, a statistical characterization of full-scale thermophilic biological systems was used to inform process optimization by Collivignarelli et al., 2024 [9]. Advanced descriptive statistics (violin graphs) highlighted the treatment effectiveness of this process, which combines thermophilic biodegradation with an ultrafiltration treatment. The authors found that the loading mode of the reactor showed the presence of some peaks, which were uniformly distributed over the different years.

- Pharmaceuticals

Only one paper was submitted to this Special Issue focusing on the removal of pharmaceuticals. Skempi et al., 2023 [10], studied the heterogeneous activation of sodium persulfate (SPS) using nickel oxide/strontium carbonate ($NiO/SrCO_3$) for the degradation of sulfamethoxazole (SMX), a representative compound from the group of antibiotics. Their results showed that $NiO/SrCO_3$ exhibited high performance in the activation of SPS, leading to SMX elimination within brief time spans. Overall, the examined system showed promising results for the decomposition of SMX.

- Microplastics

Microplastics are one of the most persistent pollutants affecting the quality of the environment. Thus, in a study included in this Special Issue, conducted by Bule Možar et al., 2023 [11], five bacterial strains (*Bacillus cereus, Bacillus licheniformis, Bacillus subtilis, Delftia acidovorans*, and *Pseudomonas alcaligenes*) and five yeast strains (*Candida parapsilosis, Geotrichum candidum, Rhodotorula glutinis, Saccharomyces cerevisiae*, and *Trichosporon* sp.) were isolated from environmental samples. Among them, *Delftia acidovorans* was identified as the bacterium with the highest potential to biodegrade PS and PVC microplastics, while *Candida parapsilosis* showed the highest potential among the yeasts. The two microorganisms adapted better to the PVC MPs than to the PS MPs and used them as a carbon and energy source. A positive effect of increased cell density and a negative effect of high agitation speed on the biodegradation process were observed.

- Heavy Metal Removal

In their study, Kharel et al., 2023 [12], evaluated the strain of red microalgae *Galdieria sulphuraria CCMEE 5587.1* in a controlled laboratory environment for its ability to tolerate and remove two heavy metal (HM) ions, cadmium [Cd(II)] and lead [Pb(II)], in aqueous solutions as a single metal species. The study underscores the dual advantage of *G. sulphuraria CCMEE 5587.1*, making it a promising candidate for addressing heavy metal pollution in wastewater treatment processes. In addition, Altahaan and Dobslaw, 2024 [13], aimed to quantify the effect of heavy metal inputs resulting from the war on the water and sediment of the Tigris River within the urban area of Mosul, taking seasonal variations (test series from the winter and spring vs. summer and fall) into account and comparing measurement points upstream and downstream of Mosul, and to compare them with existing global limits. Their results demonstrated seasonally independent exceedances of the WHO limit values for Cd, Pb, Cr, and Ni in water and sediments. Moreover, Tolkou and Kyzas, 2023 [14], presented the application of Magnesium/Silica/Lanthamum@Activated Carbon (Mg-Si-La@AC) for the first time in the removal of As(III) from water, resulting in a residual As concentration <10 μg/L.

- Phosphorus Recovery from Wastewater

In the study by Vineyard et al., 2024 [15], BioWin modeling of calprex phosphorus recovery from wastewater was used to predict substantial struvite formation at the Nine Springs Wastewater Treatment plant (Madison, WI, USA), using BioWin 6.2 with modifications to model the acid digestion process. A full-scale CalPrex process was predicted to reduce struvite formation in the anaerobic digesters by 80% and total precipitate formation by 58%. Vineyard et al., 2024, also published another article [16] in this Special Issue, which focused on improving BioWin modeling of phosphorus solubilization in acid-phase digesters.

- Others

In the study by López-Patiño et al., 2023 [17], the main objective was to verify whether it is possible to use a general-pressure hydraulic network analysis program, such as EPANET, for the analysis of the hydraulic performance of siphonic roof drainage systems (SRDSs). SRDSs are an alternative to conventional roof drainage systems (CRDSs) due to the comparative benefits that they offer under fully primed conditions. It was concluded that EPANET is a software that can be used to analyze and, therefore, design siphonic roof drainage systems in buildings.

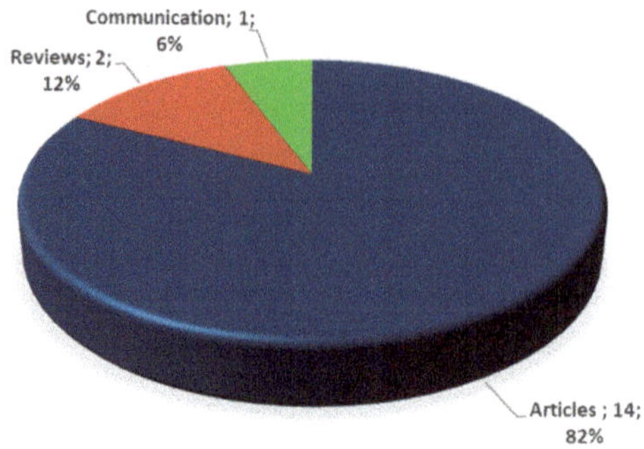

(a)

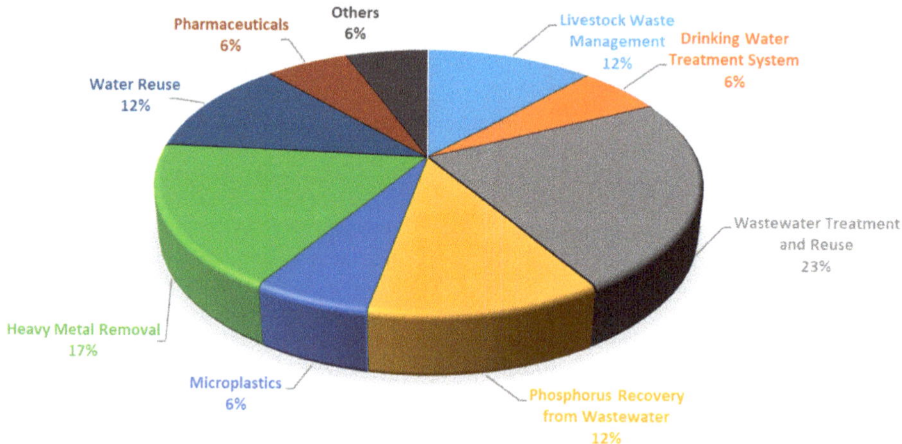

(b)

Figure 1. Statistical data on the published articles: (**a**) type of publication; (**b**) topics covered.

We would like to thank the authors for their contributions and their willingness to share innovative ideas and methods in this Special Issue. In addition, we would like to express our appreciation to the reviewers for the considerable amount of time they invested in providing accurate and fair manuscript evaluations. Finally, we would like to express our pleasure in working with the staff of the *Environments* Editorial Office on this productive and exciting publication.

Conflicts of Interest: The authors declare no conflicts of interest.

References

1. El bied, O.; Turbí, M.A.T.; Garrido, M.G.; Cano, Á.F.; Acosta, J.A. Reducing Methane, Carbon Dioxide, and Ammonia Emissions from Stored Pig Slurry Using Bacillus-Biological Additives and Aeration. *Environments* **2024**, *11*, 171. [CrossRef]
2. Lamssali, M.; Mantripragada, S.; Deng, D.; Zhang, L. Enhancing Swine Wastewater Treatment: A Sustainable and Systematic Approach through Optimized Chemical Oxygen Demand/Sulfate Mass Ratio in Attached-Growth Anaerobic Bioreactor. *Environments* **2024**, *11*, 162. [CrossRef]
3. Aragon, A.G.; Cárdenas Sánchez, J.A.; Zimeri, C.; Shim, E.; Fang, X.; Young, K.R.L. TiO_2-Coated Meltblown Nonwoven Fabrics Prepared via Atomic Layer Deposition for the Inactivation of E. Coli as a Model Photocatalytic Drinking Water Treatment System. *Environments* **2024**, *11*, 92. [CrossRef]
4. Florides, F.; Giannakoudi, M.; Ioannou, G.; Lazaridou, D.; Lamprinidou, E.; Loukoutos, N.; Spyridou, M.; Tosounidis, E.; Xanthopoulou, M.; Katsoyiannis, I.A. Water Reuse: A Comprehensive Review. *Environments* **2024**, *11*, 81. [CrossRef]
5. de Campos, S.X.; Soto, M. The Use of Constructed Wetlands to Treat Effluents for Water Reuse. *Environments* **2024**, *11*, 35. [CrossRef]

6. Rahman, K.Z.; Al Saadi, S.; Al Rawahi, M.; van Afferden, M.; Bernhard, K.; Friesen, J.; Müller, R.A. Small Decentralized Technologies for High-Strength Wastewater Treatment and Reuse in Arid and Semi-Arid Regions. *Environments* **2024**, *11*, 142. [CrossRef]
7. Bencsik, D.; Wadhawan, T.; Házi, F.; Karches, T. Plant-Wide Models for Optimizing the Operation and Maintenance of BTEX-Contaminated Wastewater Treatment and Reuse. *Environments* **2024**, *11*, 88. [CrossRef]
8. Lee, K.; Joo, H.S. The Effect of Foam-Recycle on Ammonium Removal by Aerobic Denitrification Using Alcaligenes Faecalis No. 4. *Environments* **2023**, *10*, 184. [CrossRef]
9. Collivignarelli, M.C.; Bellazzi, S.; Caccamo, F.M.; Sordi, M.; Crotti, B.; Abbà, A.; Baldi, M. Statistical Characterization of Full-Scale Thermophilic Biological Systems to Inform Process Optimization. *Environments* **2024**, *11*, 36. [CrossRef]
10. Skempi, D.J.; Kouvelis, K.; Petala, A.; Bampos, G.; Frontistis, Z. Heterogeneous Activation of Persulfate by Nickel Oxide/Strontium Carbonate Composite for Sulfamethoxazole Degradation in Water. *Environments* **2023**, *10*, 147. [CrossRef]
11. Bule Možar, K.; Miloloža, M.; Martinjak, V.; Cvetnić, M.; Ocelić Bulatović, V.; Mandić, V.; Bafti, A.; Ukić, Š.; Kučić Grgić, D.; Bolanča, T. Bacteria and Yeasts Isolated from the Environment in Biodegradation of PS and PVC Microplastics: Screening and Treatment Optimization. *Environments* **2023**, *10*, 207. [CrossRef]
12. Kharel, H.L.; Shrestha, I.; Tan, M.; Selvaratnam, T. Removal of Cadmium and Lead from Synthetic Wastewater Using Galdieria Sulphuraria. *Environments* **2023**, *10*, 174. [CrossRef]
13. Altahaan, Z.; Dobslaw, D. Assessment of the Impact of War on Concentrations of Pollutants and Heavy Metals and Their Seasonal Variations in Water and Sediments of the Tigris River in Mosul/Iraq. *Environments* **2024**, *11*, 10. [CrossRef]
14. Tolkou, A.K.; Kyzas, G.Z. Magnesium/Silica/Lanthanum@Activated Carbon for the Remediation of As(III) from Water. *Environments* **2023**, *10*, 171. [CrossRef]
15. Vineyard, D.; Karthikeyan, K.G.; Barak, P. BioWin Modeling of CalPrex Phosphorus Recovery from Wastewater Predicts Substantial Nuisance Struvite Reduction. *Environments* **2024**, *11*, 48. [CrossRef]
16. Vineyard, D.; Karthikeyan, K.G.; Davidson, C.; Barak, P. Improving BioWin Modeling of Phosphorus Solubilization in Acid-Phase Digesters. *Environments* **2024**, *11*, 31. [CrossRef]
17. López-Patiño, G.; Iglesias-Rey, P.L.; Martínez-Solano, F.J.; Fuertes-Miquel, V.S. Analysis of Siphonic Roof Drainage Systems with EPANET. *Environments* **2023**, *10*, 123. [CrossRef]

environments

MDPI

Article

Reducing Methane, Carbon Dioxide, and Ammonia Emissions from Stored Pig Slurry Using Bacillus-Biological Additives and Aeration

Oumaima El bied *, Martire Angélica Terrero Turbí, Melisa Gómez Garrido, Ángel Faz Cano and José Alberto Acosta

Sustainable Use, Management and Reclamation of Soil and Water Research Group, Agronomic Engineering Department, ETSIA, Universidad Politécnica de Cartagena, Paseo Alfonso XIII, 48, 30203 Cartagena, Spain
* Correspondence: oumaima.elbied@upct.es

Abstract: This study delves into the innovative application of a novel bacterial and enzyme mixture alone or combined with aeration in mitigating emissions from pig slurry storage and explores their impacts on the methane (CH_4), carbon dioxide (CO_2), and ammonia (NH_3) emissions from stored pig slurry. A dynamic chamber was used in this research to assess the efficacy of the treatments. Biological additives (HIPO-PURÍN) of specific microbial strains were tested (a mixture ofof *Bacillus subtilis, Bacillus megaterium, Bacillus licheniformis, Bacillus amyloliquefacien, and Bacillus thuringiensis*) alone and combined with an aeration system (OXI-FUCH). Controlled experiments simulated storage conditions, where emissions of ammonia, methane, and carbon dioxide were measured. By analyzing the results statistically, the treatment with HIPO-PURÍN demonstrated a significant reduction in CH_4 emissions by 67% and CO_2 emissions by 60% with the use of biological additives, which was increased to 99% and 87%, respectively, when combined with OXI-FUCH aeration, compared to untreated slurry. Ammonia emissions were substantially reduced by 90% with biological additives alone and by 76% when combined with aeration. The study was driven by the need to develop sustainable solutions for livestock waste management, particularly in reducing emissions from pig slurry. It introduces techniques that significantly lower greenhouse gases, aligning with circular economy goals and setting a new standard for sustainable agriculture. Furthermore, there is a need to validate that farmers can independently manage pig slurry using simple and effective treatments techniques with profound environmental benefits, encouraging broader adoption of climate-conscious practices.

Keywords: ammonia emissions; pig slurry; aeration; biological additives; dynamic chamber; aerobic treatment; bacillus bacteria; greenhouse gas emissions

1. Introduction

Intensive pig farming presents a major environmental challenge in numerous countries, including Spain. The substantial presence of meat production farms in the country requires an effective management of pig slurry [1]. The Region of Murcia ranks in fourth position in Spain in terms of pig population according to the 2023 census (2,485,375 pigs) [2], only surpassed by Aragon, Cataluña, and Castilla Leon; therefore, there is a considerable need for proper waste management. This high number of pigs results in an annual production of approximately 94.7608 Hm^3 of slurries, as reported by the Ministry of Agriculture, Food, and Environment 2023.

Pig slurry, a by-product of pig farming, consists of a complex mixture of water, nutrients (including nitrogen, phosphorus, and potassium), organic matter, and microorganisms, along with some traces of heavy metals [3]. The variability in pig slurry composition across farms is significant, influenced by factors such as housing systems, diet, climate, and farm management practices [4,5]. This pig slurry is recognized for its fertilizing properties, yet it presents considerable environmental concerns. This slurry serves as a valuable fertilizer but also poses significant environmental challenges [6].

According to the USDA Foreign Agricultural service 2023/2024 [7], the major pig-producing countries are China at 57.94 million metric tons (50% of global production), the European Union at 20.8 million metric tons (18%), and the United States at 12.39 million metric tons (11%), which generate millions of tons of pig slurry annually. Inadequate management of pig slurry can result in environmental issues such as nutrient runoff, superficial and subterranean water pollution, GHG and ammonia emissions, and soil degradation [8,9]. These challenges underscore the need for adept slurry management and creative strategies to ensure that agricultural use of pig slurry aligns with ecological preservation and public health protection.

Studies have identified three main waste streams that could be used to produce organic fertilizers: manure, sewage sludge, and food processing waste [10]. Among these three options, livestock manure represents the most significant waste stream, and pig manure is particularly associated with the highest environmental concerns regarding its safe disposal and management [11]. In Europe, the majority of produced manure is used as a fertilizer in agricultural fields without any treatment, either through spreading or grazing. The direct application to agricultural soil is among the most cost-effective choices for managing pig slurry; however, the feasibility of this alternative relies on having sufficient available agricultural surfaces and implementing sustainable application practices.

This practice of direct application of pig slurry to agricultural soil poses potential environmental risks, including nutrient imbalances, an overabundance of nutrients, and the presence of contaminants like metals [8,12]. Improper handling and application can lead to nutrient runoff, with the transfer of nitrogen from soil to surface water, often coupled with phosphorus transport, with this being the principal contributor to eutrophication [3]. Moreover, groundwater contamination is a significant concern, posing risks to drinking water sources.

Additionally, pig slurry releases greenhouse gases like methane and ammonia, contributing to climate change and air quality problems. Ammonia, nitrogen dioxide, and methane (NH_3, N_2O, and CH_4) are potent greenhouse gases [13]. These gases are primarily released during slurry storage and the composting of its solid fraction. Among them, ammonia plays a major role in acidification and eutrophication, while methane and nitrogen dioxide significantly impact climate change [3]. Unpleasant odors resulting from improper storage and spreading can affect both air quality and community well-being. The sector often grapples with neighborhood conflicts and olfactory nuisances, as documented in previous research [14]; therefore, balancing the benefits of pig slurry as a fertilizer with responsible management practices is essential to avoid negative consequences for soil quality, pathogen spread, and overall environmental health.

In recent decades, the negative environmental consequences of pig production and manure management have been controlled through stricter regulations on storage and spreading, the European Union has specific regulations regarding pig manure management through the Nitrate Directive 1991/676/EEC) [15] and the Water Framework Directive (2000/60/EC) [16]. Guidelines and regulations for mitigating the effects of livestock manure generally recommend applying nutrient limits that align closely with the actual needs of crops, with a primary focus on nitrogen.

As the intensive livestock sector faces increasing pressure to minimize the environmental impact of its operations, several methods have been evaluated as the best available technologies on the market for manure treatment, while improving the use of nutritional resources [17]. Numerous studies have been carried out to investigate the treatment of pig slurry, employing physical, chemical, and biological processes to address its environmental impact and optimize its management, such as precipitation, which involves adding chemicals to the slurry to precipitate nutrients, primarily phosphorus, into solid form. This technic is very effective in reducing phosphorus concentration, while it can be costly due to the chemical inputs, and furthermore generates additional solid waste that requires management [18,19]. Coagulation flocculation use coagulants and flocculants to aggregate suspended particles into larger clusters that can be easily removed, and this technique is efficient at removing suspended solids and some dissolved nutrients, and enhances subsequent treatment processes. The main limitations include the use of chemical additives, which can increase costs and the environmental footprint [20]. Adsorption utilizes materials like activated carbon or biochar to adsorb contaminants from the slurry. It is effective at removing a wide range of organic and inorganic pollutants and can improve effluent quality. However, adsorbent materials can be expensive and require periodic regeneration or replacement [21]. Acidification reduces pH levels in the slurry, inhibiting microbial activity and reducing ammonia emissions. This method is advantageous in reducing odor and ammonia volatilization and can improve nutrient availability for crops. The drawbacks include the need for acid inputs and the potential for soil acidification if not managed properly [22,23]. Solid–liquid separation consists in separating the solid and liquid fractions of the slurry, often using mechanical separators. This process reduces the volume of waste needing treatment, and the solids can be composted or used as a soil amendment. However, the liquid fraction still requires further treatment [24,25]. Aerobic biological nitrogen removal uses aerobic bacteria to convert ammonia into nitrogen gas through nitrification and denitrification. It is effective for nitrogen removal and reduces odor and pathogen levels. The main limitations are the energy-intensive nature due to aeration requirements and the need for careful control of environmental conditions [26]. Anaerobic digestion involves microorganisms breaking down organic matter in the absence of oxygen, producing biogas (methane) and digestate. This method produces renewable energy (biogas), reduces pathogen levels, and stabilizes organic matter. However, it requires a high initial investment for biogas plants and necessitates the management of biogas and digestate [27,28]. Selecting an appropriate treatment method relies on several factors, including the installation and maintenance expenses, desired treatment level, and the organic matter and nutrient removal [29].

Currently, there is no definitive solution for pig slurry treatment, and treatment plants employ diverse techniques based on their preferences. Many facilities adopt a combination of the popular methods previously mentioned. However, these procedures are neither technologically practical nor economically feasible. As a result, owners of pig farming facilities must pay waste management companies for handling their manure, and larger facilities must invest in their own treatment equipment to comply with existing regulations [30].

Recently, innovated biological additives have emerged on the market, introducing additional capabilities, including the enhancement of nitrogen removal through denitrification by introducing different bacteria, which have promise in mitigating the nitrogen excess in intensive livestock areas [31]. These additives are intended to modify certain properties of the slurry by either suppressing or enhancing specific microbiological processes [32]. The benefit of this approach lies in its accessibility and cost-effectiveness. Farmers can easily implement it by following the provided guidelines and instructions.

The goal of this research was to evaluate the effect of commercially available biological additives, which are expected to improve denitrification, when applied to slurry from a fattening pig farm [31]. The application of these biological additives, particularly HIPO-PURÍN bacteria formula, introduces an innovative methodology and an original bacteria formula in this research area, highlighting innovative contributions to scientific comprehension. Additionally, this research seeks to assist in decision-making, regarding whether to use the biological additives alone or in combination with the aeration system known as OXI-FUCH, which is a newly designed aeration technology that can be installed in pig slurry storage to compare the effectiveness of the bacteria when used on the same type of slurry under consistent conditions.

The combination of HIPO-PURÍN bacteria with the aeration system OXI-FUCH is intended to reduce greenhouse gases emissions into the atmosphere. The study had two primary objectives: the initial goal was to verify the efficacy of employing a dynamic chamber in the measurement of gaseous emissions originating from slurry storage, specifically NH_3, CH_4, and CO_2, and to show the efficacy and precision of utilizing a dynamic chamber as a measurement method. By validating this methodology, the researchers aimed to determine its reliability for future emission measurements. The second objective was to analyze the gaseous emissions (NH_3, CH_4, and CO_2) throughout a slurry storage period. The researchers sought to establish its reliability for future emission measurements, in three distinct modes: in raw pig slurry, incorporating a biological additive with aeration, and without aeration, pursuing the goal of analyzing the changes in gaseous emissions throughout the storage duration. This valuable information will enhance our understanding of the factors affecting emission variations and will aid in determining the optimal periods and durations for estimating gaseous emission factors in pig manure storage, during treatment with biological additives alone or combined with an aeration system.

2. Materials and Methods

2.1. Experimental Design

The experiment was conducted at a white pig farm in Fuente Alamo, a municipality in the Murcia region of southeastern Spain. It was conducted from 18 April to 22 June 2023. The pilot study aimed to replicate the conditions of swine manure storage tanks. A portable pool with a steel and PVC structure, purchased from Bestway®, España, Spain (ref. 84265026), made in China, with circular dimensions of Ø 457 × 122 cm was used Figure 1. The slurry was divided proportionately among the experimental pools, resulting in each pool being filled up with 13 m^3 of slurry at a depth of 1 m. The chosen volume was determined to meet the requirements of the VERA protocol [33]. To implement the study on a real-world scale, we had three pools, two were treated with the same quantity of the biological additives (HIPO-PURIN), one pool with OXI-FUCH aeration techniques and the other one without, while one pool remained untreated as a control. The slurry composition and gaseous emissions were continuously tracked throughout the experiment.

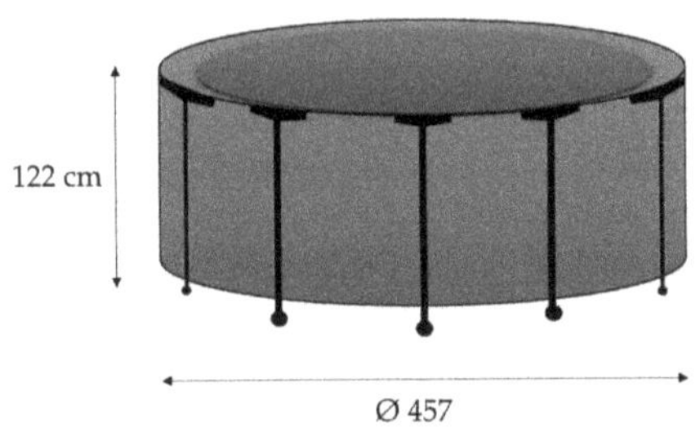

Figure 1. Bestway® Portable Pool (Ref. 84265026).

The experiment had three distinct phases. In the initial stage, no additives were introduced. During this phase, pig slurry was sampled with 3 replicates, and CO_2, CH_4, and NH_3 was measured at the same time as the replicates, in order to have initial values before starting the treatments. Subsequently, the biological additive was introduced in the other pools after the initial sampling and measurements. The concluding phase centered on greenhouse gas flux measurements, and pig slurry sampling was executed weekly. This entailed positioning the floating dynamic chamber over the pools and collecting the gas samples at $T_0 = 0$ min and $T_{30} = 30$ min post-chamber deployment, adhering to the guidelines outlined in the VERA protocol [33].

2.2. Biological Additive

The biological additive used in this experiment was obtained from the Sewervac company, Spain. Marketed under the commercial name HIPO-PURÍN, the product comprises different microorganisms, including *Bacillus subtilis, Bacillus megaterium, Bacillus licheniformis, Bacillus amyloliquefacien,* and *Bacillus thuringiensis,* which produce a range of enzymes such as proteases, cellulases, and lipases, the formula of HIPO-PURÍN also contains yucca extract.

Bacillus subtilis secretes proteases such as subtilisin and alkaline protease, which break down proteins into peptides and amino acids, it also produces cellulases and lipases, facilitating the hydrolysis of cellulose and lipids into simpler compounds [34]. *Bacillus megaterium* produces neutral and alkaline proteases, cellulases for cellulose degradation, and lipases for lipid hydrolysis [35]. *Bacillus licheniformis* is recognized for its alkaline protease and subtilisin activities, along with cellulases and lipases [36]. *Bacillus amyloliquefaciens* produces subtilisin and neutral proteases, cellulases, and lipases, which enable the breakdown of proteins, cellulose, and lipids [37]. *Bacillus thuringiensis,* in addition to its insecticidal properties, produces serine proteases, cellulases, and lipases that contribute to protein degradation, cellulose hydrolysis, and lipid breakdown [38].

The enzymes produced by these bacterial strains include cellulases (CMCase, avicelase, β-glucosidase, xylanase), lipases (Lip A, Lip B), amylases (α-amylase, β-amylase), and proteases (serine protease, threonine protease, cysteine protease, aspartic protease, metalloprotease). Some of the strains are facultative, allowing them to function in conditions where oxygen is limited. They can grow faster than anaerobic strains, thus avoiding the formation of sulfides and using ammonia for growth, which helps retain more nitrogen in manure and prevents its release into the atmosphere. Additionally, they stabilize pH, preventing it from rising to levels where ammonia becomes more volatile and can be released into the atmosphere.

In this study, the dosage was outlined as follows: for rehydration, a ratio of 1 kg per 100 L was recommended. The initial dose involved diluting 400 g of HIPO-PURÍN in 40 L of water, while the weekly dose entailed diluting 133.60 g of HIPO-PURÍN in 13 L of water. The cumulative dosage over a 6-week period was calculated as 2.00 kg, obtained by adding the initial and weekly doses.

For dose preparation, the product was deposited into a clean container, filled halfway with warm clean water (approximately 30 °C), stirred thoroughly, and allowed to rest for a minimum of 30 min before incorporating it to the slurry. The rehydration ratio recommended was 1 part product to at least 100 parts of water. The product is provided in bulk powder form and packaged in plastic buckets.

In the application of HIPO-PURÍN in this study, the initial dose was administered just before filling the slurry pool, in accordance with the predetermined dosage set by Sewervac technical department. The weekly or maintenance dose was applied around the study pool with a manual stirring, in order to incorporate the solution with the treated slurry. The application of the Sewervac product HIPO-PURÍN, adhered to the manufacturer's instructions and was uniformly distributed over the pig manure.

The objective of the biological additives was to allow an appropriate and easy slurry management by promoting liquefaction, achieving an effective method for controlling and reducing total nitrogen, greenhouse gases (GHG), and ammonia.

2.3. OXI-FUCH Aeration System

This research article delves into the optimization of wastewater treatment processes through the utilization of innovative technologies, biological additive alone or combined with aeration system. Focusing on the material characteristics employed, the study explored the implementation of an investment compressor and aeration modules, specifically employing microperforated upper pipes (OXI-FUCH) and weighted feeder pipes (Figure 1). The operational dynamics of the compressor, with 20 continuous hours of activity followed by 5 continuous hours of downtime, were meticulously examined. In this real-world experiment, the installation of the aeration system was under the purview of the OXI-FUCH technical department.

In this study, we explored the potential of enhancing HIPO-PURÍN with an aeration system composed of linear modules of pipes, known as OXI-FUCH modules Figure 2. These modules consist of two interconnected pipes: an upper micro-perforated pipe facilitating aeration by generating two columns of microbubbles, and a lower pipe filled with sand to serve as ballast for deposition at the reservoir bottom. The microbubbles thus produced traverse the slurry from bottom to surface at a slow and uniform pace, transferring oxygen through friction. This oxygen acts as an energy source for sustaining and promoting the growth of bacteria present in HIPO-PURÍN. Our methodology involved transforming the storage tank into a miniature treatment plant by implementing these OXI-FUCH modules. The design of these modules is crucial, as it prevents the merging of air flows, thereby preventing the formation of large bubbles. This design feature promotes the efficient transfer of oxygen, with microbubbles facilitating up to 6.6 times more oxygen transfer compared to large bubbles, due to their increased contact surface area with the water. The operation of the compressor in this study was structured with a continuous operation time of 20 h, followed by a continuous off-time of 5 h.

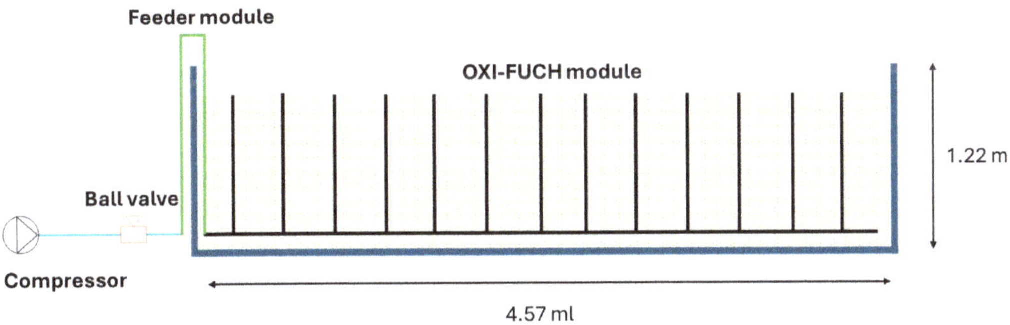

Figure 2. OXI-FUCH aeration system.

2.4. Dynamic Chamber

This research focused on the direct measurement of greenhouse gas emissions from pig slurry, utilizing a dynamic chamber as a primary methodology. The dynamic chamber, constructed from PVC as chosen in this study, was sealed to eliminate any gas exchange with the external environment. The PVC material was chosen because it is considered a non-adhesive material for the gases under investigation [33].

Complying with the VERA protocol recommendations, a specified area (0.564 m^2) of the pig slurry surface was covered, ensuring adherence to sampling standards. The chamber design included an inlet for ambient air entry and an outlet for the release of chamber air, promoting continuous gas exchange. Airflow was precisely regulated using a combination of a controlled, adjustable pump at the inlet and a suction device at the outlet. Additionally, an anemometer set to a speed of 0.2 m/s was used to monitor the flow at a speed of 0.2 m s^{-1}, as per VERA protocol guidelines [33], to monitor the airflow within the chamber. Gas sampling ports strategically positioned within the chamber facilitated representative air sampling, and the collected samples were analyzed using GASERA ONE for gas concentration measurements. The study adhered to a minimum sampling duration of 30 min per point, as recommended by the VERA protocol, considering the emission source and measurement objectives. Gaseous emissions were quantified as F (flux measured with the dynamic chamber) in kg (gas) ha^{-1} h^{-1}, calculated using the equation provided [33].

$$F = \frac{C_{out} - C_{in}}{Ab} Ai\ Vi$$

C_{out} and C_{in} represent the time-averaged gaseous concentrations of the gas (measured in kg m^{-3}) in the outlet and inlet air, respectively. The variables Ai, Vi, and Ab denote the cross-sectional area of the inlet (in m^2), the measured wind speed at the tunnel inlet (in m s^{-1}), and the source surface area covered by the tunnel canopy (in m^2).

Besides these parameters, environmental conditions such as temperature, humidity, and pressure within the chamber were routinely monitored. This thorough monitoring ensured the accurate documentation and incorporation of measurement conditions into the emission calculations.

By accurately quantifying emissions, the dynamic chamber played a pivotal role in comprehending the environmental effects of pig farming, especially pig slurry management. This understanding, in turn, can support the development of effective mitigation strategies to reduce greenhouse gas and ammonia emissions in the agricultural sector.

2.5. Gases Sampling Frequency

The aim of this study was to assess the seasonal variability in gas emissions during the storage of slurry with or without treatment. To achieve this, multiple measurements were taken, each with a duration of approximately four weeks and following a frequency of twice per week, as proposed by VERA PROTOCOL [33], conducted under varying ambient temperature conditions.

Specifically, the study focused on the measurement period of spring 2023, from 18 April to 22 June. Within each measurement period, 30 min average emission rates for each gas were computed based on air flow rates and gas concentrations obtained at 10 min intervals. Furthermore, detailed records were kept for several parameters, including manure management practices, slurry characteristics, and environmental conditions like slurry and ambient temperatures throughout the entire measurement period.

2.6. Liquid Pig Slurry Sampling and Analysis

Samples of the slurry were collected simultaneously for measurement of each of the studied gases (CH_4 and CO_2) and ammonia (NH_3). From three distinct points within the storage tank, slurry samples were collected from the surface to a depth of 20 cm, ensuring homogeneity through stirring before sampling. Subsequently, these samples were stored and transported at 4 °C to the laboratory for further examination.

All parameters were analyzed at the laboratory, except the pH and electrical conductivity (EC) that were measured in situ using HANNA multiparameter equipment with reference HI98194. The determination of Kjeldahl nitrogen (KN) content employed a modified Kjeldahl method, involving the digestion of 1 mL of pig slurry. Ammonium nitrogen (NH_4^+–N) was determined through steam distillation followed by titration with HCl 0.1 N, while total nitrogen (TN) encompassed both organic and inorganic forms, including Kjeldahl nitrogen, nitrite, and nitrates. Total phosphorus (TP) determination involved acidic hydrolysis and oxidation at 120 °C, followed by photometric analysis. Potassium (K^+) levels were measured using an atomic absorption spectrometer. Total suspended solids (TSSs) were assessed by filtering the sample through a pre-weighed standard glass-fiber filter, and the residue retained on the filter was dried and weighed using the 2440-D method (APHA–AWWA–WEF, 2012). Biochemical oxygen demand over five days (BOD_5) was determined using OXITOP WTW equipment and measured with a manometer (Darmstadt, Germany), while chemical oxygen demand (COD) was determined using photometric analysis of the chromium (III) concentration after 2 h of oxidation with potassium dichromate/sulfuric acid and silver sulfate at 148 °C (Macherey–Nagel GmbH & Co., KG, Nanocolor Test; ref. 985 028/29, Weilheim, Germany) according to German standard methods DIN 38 409-H41-1 and DIN ISO 15 705-H45 [39].

2.7. Statistical Analysis

The data were subjected to statistical analysis using one-way ANOVA, followed by Tukey's HSD post hoc test for pairwise comparisons of means (with a significance level set at $p < 0.05$). This analysis was performed using SPSS 24.0 software, to detect significant differences in greenhouse gas emissions and ammonia levels. Additionally, the parameters of pig slurry under both treatments, biological additives alone and in combination with aeration, were evaluated to draw conclusions for the study.

The statistical analysis was conducted using triple replicate sampling for both pig slurry characterization and gas measurements. The results from these replicates were averaged, and statistical analysis was applied to the mean values obtained from these triplicate experiments.

3. Results

3.1. Greenhouse Gas Emissions and Ammonia

Figures 3–5 provide a detailed overview of the carbon dioxide (CO_2), methane (CH_4), and ammonia (NH_3) emissions from both untreated and treated pig slurry samples over a period of six weeks. Notably, the treatment was applied immediately after the initial measurements, allowing for a thorough assessment of the additive's effects. These figures are instrumental in illustrating the emission patterns of the studied gases throughout the timeframe, enhancing our understanding of how the treatments impacted the slurry composition.

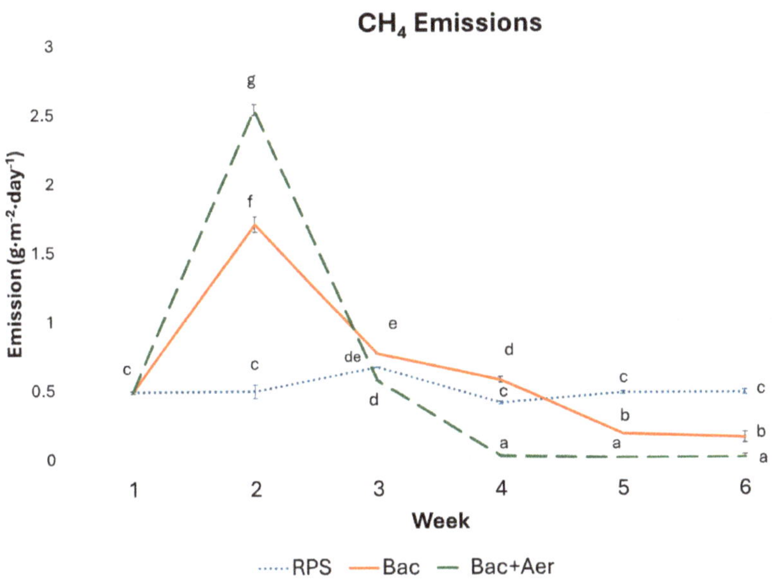

Figure 3. Methane (CH_4) progression of untreated and treated pig slurry (RPS: raw pig slurry, Bac: treated pig slurry with bacteria, Bac + Aer: treated pig slurry with bacteria and aeration). Different letters indicate significant differences between the sampling times and treatments ($p < 0.05$).

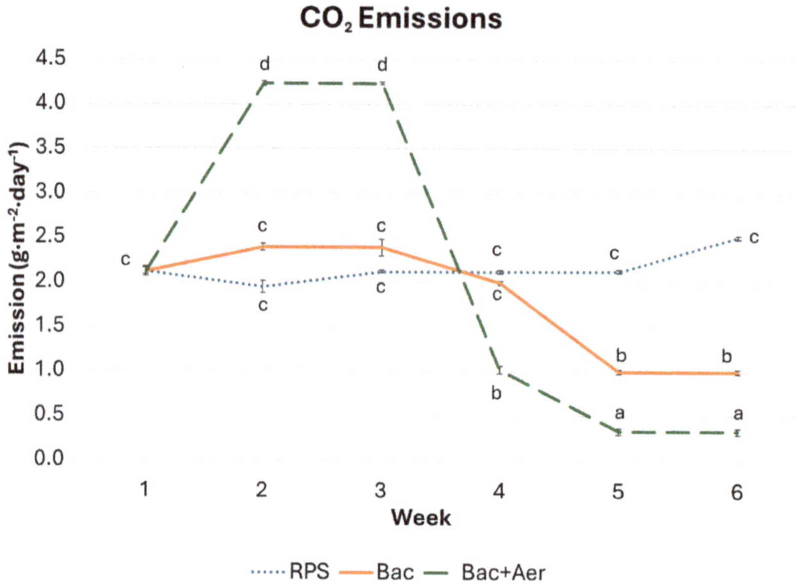

Figure 4. Carbon dioxide (CO_2) progression of untreated and treated pig slurry (RPS: raw pig slurry, Bac: treated pig slurry with bacteria, Bac + Aer: treated pig slurry with bacteria and aeration). Different letters indicate significant differences between the sampling times and treatments ($p < 0.05$).

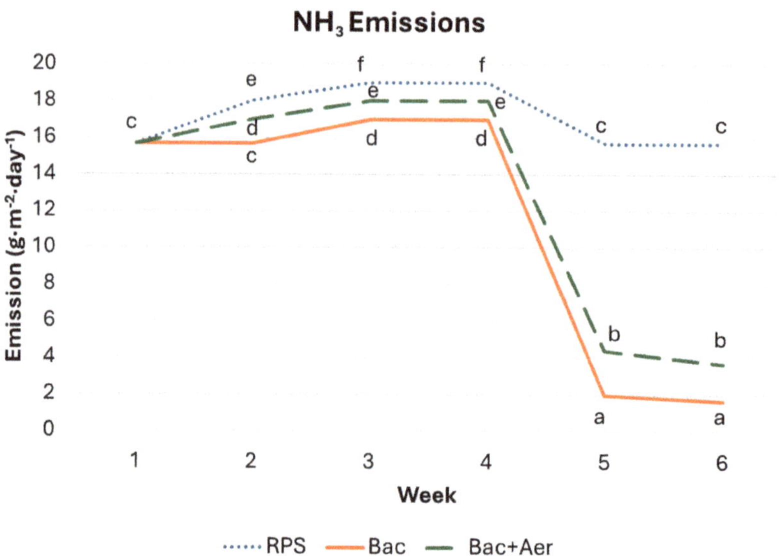

Figure 5. Ammonia (NH_3) progression of untreated and treated pig slurry.(RPS: raw pig slurry, Bac: treated pig slurry with bacteria, Bac + Aer: treated pig slurry with bacteria and aeration). Different letters indicate significant differences between the sampling times and treatments ($p < 0.05$).

3.2. Untreated and Treated Pig Slurry Analysis

Tables 1–3 present the physicochemical characteristics, macro-nutrients, and micronutrients of the pig slurry, respectively. This study investigated each parameter in both untreated and treated pig slurry, facilitating a thorough examination of the treatment evolution and reactions during the storage of slurry. The treatment was applied subsequent to the initial slurry sampling during the first week. Significant differences ($p < 0.05$) in the concentrations of various parameters such as DBO_5, COD, total nitrogen (TN), total suspended solids (TSSs), and phosphorus were observed between the untreated pig slurry (control) and the treated slurry with bacteria alone or coupled with aeration.

Table 1. Means and standard deviation values of the physicochemical parameters' progression in untreated and treated pig slurry.

Weeks		1	2	3	4	5	6
Tª (°C)	RPS	18.05 ± 0.49 a	17.92 ± 0.59 a	21.61 ± 0.15 c	19.94 ± 0.01 b	23.1 ± 0.14 d	22.96 ± 0.06 d
	BAC	17.36 ± 0.45 a	18.12 ± 0.04 b	23.79 ± 0.07 e	18.71 ± 0.02 c	22.06 ± 0.22 d	23.42 ± 0.8 e
	BAC + AER	18.79 ± 0.11 b	17.54 ± 0.13 a	23.53 ± 0.04 d	19.89 ± 0.01 c	24.06 ± 0.04 e	24.91 ± 0.51 e
pH	RPS	7.32 ± 0.01 a	7.68 ± 0.02 b	7.71 ± 0.02 b	7.71 ± 0.02 b	7.33 ± 0.01 a	7.35 ± 0.03 a
	BAC	7.32 ± 0.03 a	7.75 ± 0.06 b	7.73 ± 0.01 b	8.00 ± 0.02 d	7.96 ± 0.04 c	7.97 ± 0.08 c
	BAC + AER	7.32 ± 0.01 a	7.3 ± 0.01 a	7.69 ± 0.03 b	8.3 ± 0.2 d	8.26 ± 0.01 d	8.46 ± 0.01 d
EC (dS m^{-1})	RPS	22.09 ± 0.02 a	24.81 ± 0.12 c	25.74 ± 0.23 d	26.78 ± 0.01 e	23.55 ± 0.06 b	24.04 ± 0.09 b
	BAC	22.45 ± 0.56 bc	23.87 ± 0.02 c	24.56 ± 0.32 d	24.98 ± 0.1 d	19.38 ± 1.12 a	21.76 ± 1.15 b
	BAC + AER	23.54 ± 0.26 b	22.26 ± 0.67 b	22.38 ± 0.4 b	22.47 ± 0.02 b	16.16 ± 2.25 a	20.45 ± 1.63 a
TSSs (g L^{-1})	RPS	26.57 ± 1.25 d	11.67 ± 0.15 a	22.2 ± 0.66 c	22.4 ± 2.15 c	15.53 ± 1.19 b	14.5 ± 0.86 ab
	BAC	26.97 ± 1.55 d	26.43 ± 0.35 c	17.37 ± 0.71 ab	15.5 ± 0.46 a	15.3 ± 1.35 a	19.03 ± 2.36 b
	BAC + AER	26.77 ± 1.02 d	21.23 ± 1.59 d	17.93 ± 0.57 a	25.97 ± 0.35 c	21.03 ± 0.55 b	24.07 ± 2.07 c
COD (gL^{-1})	RPS	32.5 ± 0.5 e	21 ± 0.01 c	26 ± 0.02 d	17.5 ± 0.5 b	10.9 ± 0.2 a	10.75 ± 0.12 a
	BAC	32.5 ± 0.5 e	24 ± 1.2 d	17.5 ± 0.5 c	12.5 ± 0.5 b	9.4 ± 0.01 a	10.5 ± 0.92 a
	BAC + AER	32.5 ± 0.5 e	28 ± 1.9 c	26.5 ± 1.5 bc	22 ± 1.2 a	21.3 ± 0.2 a	19.85 ± 1.68 c
BOD_5 (g O_2L^{-1})	RPS	12 ± 0.36 c	11.68 ± 0.08 c	12.98 ± 0.03 d	9.41 ± 0.26 b	2.54 ± 0.09 a	2 ± 0.36 a
	BAC	9.94 ± 0.04 d	10.32 ± 0.54 d	7.09 ± 0.12 c	3.09 ± 0.56 b	2.01 ± 0.03 a	2.59 ± 0.23 a
	BAC + AER	11.9 ± 0.36 c	13.76 ± 0.65 e	10.62 ± 0.13 d	10.54 ± 0.14 d	6.92 ± 0.53b	3.77 ± 1,18 a

Note(s): EC: electrical conductivity, TSSs: total suspended solids; COD: chemical oxygen demand; BOD_5: biochemical oxygen demand; RPS: raw pig slurry; BAC: bacteria, BAC + AER: Bacteria with aeration. Data are the means of three replicates (SD). Different letters indicate significant differences between the treatment and the control (RPS). Significance: $p < 0.05$.

Table 2. Mean and standard deviation values of the macronutrient parameters' progression in untreated and treated pig slurry.

Weeks		1	2	3	4	5	6
N total (g L^{-1})	RPS	2.66 ± 0.11 c	2.33 ± 0.07 c	2.33 ± 0.06 c	2.7 ± 0.21 c	1.93 ± 0.06 b	2.11 ± 0.16 b
	BAC	2.66 ± 0.14 c	2.68 ± 0.04 c	2.56 ± 0.05 c	2.29 ± 0.08 c	1.87 ± 0.12 b	1.47 ± 0.22 a
	BAC + AER	2.63 ± 0.12 c	2.65 ± 0.37 c	2.67 ± 0.04 c	2.37 ± 0.13 c	1.72 ± 0.11 b	1.52 ± 0.06 a
Na^+ (mg L^{-1})	RPS	1504 ± 48 a	1724 ± 54.95 b	1923 ± 21.88 c	1987 ± 41 c	2060 ± 86.1 c	2324 ± 155 d
	BAC	1492 ± 40 a	1670 ± 41.11 b	1822 ± 21.06 c	2031 ± 25 d	1980 ± 64.01 d	2198 ± 122 e
	BAC + AER	1495 ± 51 a	1665 ± 48 b	1780 ± 75 bc	1979 ± 37 d	1866 ± 29 cd	2136 ± 128 e
K^+ (mg L^{-1})	RPS	2096 ± 31 a	2451 ± 51 b	2772 ± 52 c	2975 ± 57 d	3096 ±137 d	3441 ± 210 e
	BAC	2103 ± 52 a	2297 ± 64 b	2608 ± 56 c	2966 ± 37 e	2753 ± 11 d	3173 ± 219 f
	BAC + AER	2082 ± 78 a	2319 ± 79 b	2580 ± 125 c	2925 ± 51 de	2724 ± 42 cd	3091 ± 184 e
Ca_2^+ (mg L^{-1})	RPS	372 ± 49 bc	398 ± 11 c	430 ± 30 c	352 ± 2 bc	305 ± 11 ab	255 ± 22 a
	BAC	383 ± 41 cd	428.73 ± 16 d	348 ± 13 bc	302 ± 12 ab	240 ± 21 a	268 ± 20 a
	BAC + AER	365 ± 12 bc	411.8 ± 15 c	278 ± 43 a	296 ± 41 ab	301 ± 33 ab	299 ± 28 ab

Table 2. *Cont.*

Weeks		1	2	3	4	5	6
Mg_2^+ (mg L^{-1})	RPS	396 ± 22 c	224 ±15 a	248 ± 11 ab	272 ± 11 b	239 ± 13 ab	233 ± 1 ab
	BAC	348 ± 72 c	311 ± 11 bc	263 ± 26 ab	244 ± 6 ab	201 ± 10 a	227 ± 14 ab
	BAC + AER	341 ± 68 b	242 ±5 a	195 ± 6 a	193 ± 2 a	263 ± 19 ab	229 ± 34 a
P (mg L^{-1})	RPS	296 ± 46 b	139 ±14 ab	113 ± 6 ab	93 ±1.81 a	98 ±1.58 a	96.6 ± 2.05 a
	BAC	231 ± 97 b	70.2 ± 1.26 a	96 ±1.58 a	63.9 ± 1.58 a	64.8 ± 1.27 c	53.4 ± 1.43 c
	BAC + AER	225 ± 106 b	61.8 ± 2.04 a	82.7 ± 1.85 a	54.7 ± 0.73 c	59.4 ± 0.73 c	43.2 ± 1.59 c

Note(s): Data are the means of three replicates (SD). Different letters indicate significant differences between the treatment and the control (RPS). Significance: $p < 0.05$. RPS: raw pig slurry; BAC: bacteria, BAC + AER: Bacteria with aeration.

Table 3. Mean and standard deviation values of the micronutrient parameters' progression in untreated and treated pig slurry.

Weeks		1	2	3	4	5	6
Cu (mg L^{-1})	RPS	0.49 ± 0.05 a	0.82 ± 0.28 ab	1.45 ± 0.07 b	0.6 ± 0.15 a	1.41 ± 0.27 b	2.29 ± 0.58 c
	BAC	0.48 ± 0.05 a	0.76 ± 0.03 ab	1.36 ± 0.17 b	0.67 ± 0.19 ab	2.49 ± 0.18 c	2.44 ± 0.11 c
	BAC + AER	0.47 ± 0.07 ab	0.78 ± 0.01 b	1.49 ± 0.28 c	0.34 ± 0.06 a	1.43 ± 0.22 c	1.96 ± 0.37 d
Zn (mg L^{-1})	RPS	1.28 ± 0.06 a	2.12 ± 0.1 b	2.15 ± 0.14 b	2.3 ± 0.18 b	4.08 ± 0.67 c	5.28 ± 0.48 d
	BAC	1.39 ± 0.11 a	2.23 ± 0.09 b	2.53 ± 0.18 bc	2.98 ± 0.32 c	5.42 ± 0.48 d	5.17 ± 0.4 d
	BAC + AER	1.42 ± 0.05 a	2.45 ± 0.06 b	3.61 ± 0.31 cd	3.04 ± 0.23 bc	4.28 ± 0.64 de	4.52 ± 0.66 e
Fe (mg L^{-1})	RPS	2.8 ± 0.22 a	3.27 ± 0.28 ab	4.72 ± 0.18 de	3.69 ± 0.15 bc	4.2 ± 0.17 cd	5.46 ± 0.85 e
	BAC	3.31 ± 0.51 a	4.45 ± 0.26 c	4.7 ± 0.55 c	3.63 ± 0.28 b	5.38 ± 0.38 d	5.34 ± 0.23 d
	BAC + AER	3.36 ± 0.46 a	4.17 ± 0.11 a	6.62 ± 0.52 bc	5.53 ± 0.23 b	7.79 ± 0.76 cd	8.23 ± 0.52 d
Mn (mg L^{-1})	RPS	7.02 ± 0.61 c	4.68 ± 0.16 b	4.58 ± 0.15 b	4.21 ± 0.06 b	4.31 ± 0.54 b	4.91 ± 0.54 bc
	BAC	6.85 ± 0.57 c	5.71 ± 0.19 c	4.31 ± 0.54 b	2.75 ± 0.1 a	2.31 ± 0.06 a	2.61 ± 0.17 a
	BAC + AER	6.93 ± 1.18 d	4.43 ± 0.2 c	2.81 ± 0.15 ab	2.04 ± 0.07 a	3.41 ± 0.14 bc	1.98 ± 0.83 a

Note(s): Data are the means of three replicates (SD). Different letters indicate significant differences between the treatment and the control (RPS). Significance: $p < 0.05$. RPS: raw pig slurry; BAC: bacteria, BAC + AER: Bacteria with aeration.

4. Discussion

4.1. Methane (CH_4) Emissions

As illustrated in Figure 3, the weekly CH_4 emissions followed varied trajectories for each treatment in comparison to the control slurry. Using HIPO-PURÍN bacteria showed a notable reduction in CH_4 emission, significantly lower ($p < 0.05$) than the untreated pig slurry (0.455 g m^{-2} day^{-1} compared to 0.151 g m^{-2} day^{-1}, respectively). Combining the same bacteria with aeration resulted in improved outcomes compared to the untreated slurry (0.005 g m^{-2} day^{-1}).

The biological additives effectively reduced CH_4 emissions from the pig slurry (PS). Methane emission was observed in the control (RPS) from the third week onwards, exhibiting a significant increase relative to the first and second week, and proving to be consistent throughout the study period. The CH_4 emissions during the storage of raw pig slurry remained consistent, without showing important variations throughout the entire study duration.

After adding the biological additives, methane emissions increased in the second week, followed by a gradual decline in the subsequent weeks, until achieving a value of 0.151 g m^{-2} day^{-1}.

While microbial activity in slurry can be categorized into aerobic or anaerobic digestion, limited research has been reported on methane (CH_4) control strategies using different formulations of microbial stimulants in this field. The studies examining microbial stimulants drew inspiration from the observed impacts of commercial biological additives in reducing GHG and ammonia a farm scale. Numerous commercial products, comprising mixtures of bacteria, enzymes, and partially decomposed organic materials are utilized for slurry treatment [40]. Wheeler [41] noted a 46% decrease in CH_4 emissions upon the addition of biological additives; Bastami [42] demonstrated a 27% reduction in CH_4 emissions from cattle slurry stored in warm conditions and a 15% reduction in cold conditions; while in other studies, the addition of microbial additives did not cause a notable decrease in CH_4 emissions [43–45].

The variability in methane (CH_4) reduction can be attributed to several factors, with temperature playing a significant role in the efficiency of biological treatment [23,46]. Bastami [44] observed an effect of CH_4 reduction, contrasting with Amon and others [43–45]. This difference could potentially be due to variations in several factors such as the dosage of effective microorganisms (EM), microbial community composition, slurry characteristics, and environmental conditions across the studies [45].

In this study, in both treatment cases, the methane emission was increased in the second week after adding the biological additives, followed by a gradual decline in the following weeks. Methane production mainly occurred between the second and third weeks during the study. The introduction of biological additives can lead to changes in the microbial population within slurry. Initially, certain bacteria that produce methane may dominate, causing an increase in emissions. However, as the microbial community adjusts, the population of methane-consuming bacteria may increase, leading to a reduction in methane emissions over time. The pH can play a crucial role in CH_4 emissions, and according to Hamood and Ishak [47,48], the optimal pH range for methane production is usually between 6.5 and 7. Within this range, the microbial communities involved in methane production thrive, leading to increased methane emissions [49]. Aeration of slurry has been shown to elevate the pH to an alkaline range in various studies [45,50,51], and this phenomenon can explain the methane reduction during the aeration of pig slurry. Furthermore, an initial peak in CH4 was observed, followed by a rapid decline in the subsequent week [51], which can be attributed to the release of trapped CH_4 in the slurry as a result of supersaturation, coupled with the establishment of aerobic conditions in the following days upon aeration. These multiple phenomena can explain the peaks in methane emissions after adding the biological additives, followed by reduction in the same gas.

The bacteria in HIPO-PURÍN produce a range of enzymes such as proteases, cellulases, and lipases that are effective in breaking down crust layers and removing organic material present in excess feed, bedding, and animal waste products. Another enzyme produced by the bacteria is urease. This enzyme breaks down urea into a form that HIPO-PURÍN bacteria can readily utilize. It has a small amount of free enzymes that start the biodegradation process. Once the bacteria are established, they handle the enzyme production. They are able to produce biosurfactants. These are naturally occurring surfactants that help with the biodegradation process and also facilitate the suspension of crust layers. HIPO-PURÍN includes a buffering agent that helps neutralize the environment in which the microorganisms are intended to act, thereby reducing the acclimatization period. Additionally, it contains one crucial final ingredient, yucca extract. This component, a natural extract from the yucca plant, consistently mitigates ammonia levels in controlled settings. This can explain the methane reduction in the treatment pig slurry by the biological additive HIPO-PURIN.

Biological additives like Bacillus bacteria and yucca extract have the potential to reduce methane emissions from stored pig slurry through multifaceted mechanisms. Bacillus species, including *Bacillus subtilis, Bacillus megaterium, Bacillus licheni-formis, Bacillus amyloliquefaciens,* and *Bacillus thuringiensis* can disrupt methanogenic microbial pathways by outcompeting for substrates; enzymatically breaking down organic matter, by producing hydrolytic enzymes, such as cellulases (CMCase, avicelase, β-glucosidase, xylanase), lipases (Lip A, Lip B), amylases (α-amylase, β-amylase), and proteases (serine protease, threonine protease, cysteine protease, aspartic protease, metalloprotease) that help break down high molecular weight compounds [52–54]; and modulating the microbial community to favor non-methane producing species; therefore, when applied to pig slurry, these bacteria can potentially reduce methane emissions through several mechanisms, mainly indirect anaerobic digestion enhancement by competing with methanogens for the broken down organic matter, which indirectly reduce methanogenesis by utilizing some of the available carbon sources for their own metabolism [55,56]. Yucca extract complements this by reducing foam formation, potentially trapping methane, and exhibiting antimicrobial properties that can inhibit methanogenic activity [57].

In general, aeration of the slurry resulted in decreased CH_4 levels. However, significant spikes in CH_4 emissions occurred during aeration events, especially in the second week. This trend is consistent with findings from previous studies [45,58], highlighting the limited solubility of CH_4 in water, which causes it to form small bubbles within the slurry before escaping into the atmosphere. Despite infrequent aeration, the CH_4 production persisted in the early stages, due to the existence of easily degradable organic matter, and the sporadic aeration likely failed to fully eliminate the anaerobic conditions necessary for CH_4 generation. However, with prolonged aeration, easily accessible organic matter, such as volatile fatty acids underwent aerobic degradation [59], thereby reducing the potential for CH_4 generation.

Adding the HIPO-PURÍN bacteria combined with an aeration system created an environment that was less favorable for methane production [50]. The increased oxygen levels resulting from the aeration system further supported aerobic decomposition processes, which produced less methane compared to anaerobic conditions [50–61]. Overall, this combined approach helps to mitigate methane production by creating conditions that favor alternative pathways of organic matter degradation over methanogenesis, which explain the significant methane emissions ($p < 0.05$) between using the biological additive alone and combined with aeration.

4.2. Carbon Dioxide (CO_2) Emissions

As shown in Figure 4, the weekly CO_2 emissions followed varied trajectories for each treatment, particularly notable with aeration compared to the untreated control slurry. Adding HIPO-PURÍN bacteria led to a significant reduction in CO_2 emissions, significantly lower ($p < 0.05$) than the untreated raw pig slurry (2.51 g m^{-2} day^{-1} compared to 0.998 $g{\cdot}m^{-2}{\cdot}day^{-1}$, respectively). Combining the same bacteria with aeration led to better results (0.328 g m^{-2} day^{-1}), according to our research. For the control pig slurry, CO_2 emissions remained stable throughout the entire study period, with no significant variations observed.

During the second and third weeks following the introduction of aeration combined with biological additives to the pig slurry, there were notable peaks in CO_2 emissions. This period coincided with increased bubbling and agitation of the slurry surface due to aeration, suggesting heightened microbial activity. By the fourth week, a significant decline ($p < 0.05$) in CO_2 emissions was detected, and in the fifth and sixth weeks, emissions continued to decrease steadily.

Aeration of slurry combines the effects of mixing and oxygenation, as documented in other research [62,63]. The CO_2 is produced from organic matter by microorganisms during the aerobic composting process. During the initial aeration phase, CO_2 emissions escalate quickly and there may be a spike in CO_2 emissions [64]. This is because the increased oxygen availability stimulates microbial activity, leading to accelerated decomposition of organic matter and subsequent release of CO_2. As the aerobic decomposition process continues, CO_2 emissions may remain relatively high during the initial stages of aeration [26,50,51]. This phenomenon explains the observed peaks in CO_2 emissions during the second and third weeks after initiating aeration with biological additives, the introduction of aeration could also promote the release of CO_2 trapped within the organic matter. This initial surge in CO_2 emissions signified the increased metabolic activity of aerobic microorganisms, including the Bacillus species present in the biological additive, which thrive in the presence of oxygen. As these bacteria metabolize organic compounds, they release CO_2 as a by-product, contributing to the observed peaks in Figure 4.

However, as the aerobic decomposition process progressed and organic matter was gradually broken down, the availability of substrates for microbial metabolism decreased. Consequently, CO_2 emissions began to decline, leading to the notable decrease observed in the fourth week in Figure 4. Subsequently, during the fifth and sixth weeks, CO_2 emissions continued to decrease as the microbial population stabilized and the rate of organic matter decomposition diminished. This trend underscores the dynamic nature of microbial activity in response to changes in environmental conditions, ultimately influencing CO_2 emissions from pig slurry [26,65,66].

Overall, aeration can play a crucial role in managing CO_2 emissions from pig slurry by promoting aerobic decomposition and minimizing anaerobic processes that typically produce methane, a more potent greenhouse gas. Additionally, aeration can help improve the overall quality of the slurry by reducing odors and enhancing nutrient availability for agricultural use [67].

The reviewed studies indicated that biological additives and aeration can significantly reduce CO_2 emissions from pig manure. According to Kupper and Clemens [68,69], biological additives can reduce CO_2 emissions by 20–30%. The average CO_2% reduction in our study was 60%, which is consistent with the results (71%) reported by El bied [70].

Other studies [71,72] showed that aeration increases CO_2 emissions, because during aerobic treatment, organic matter is decomposed in the presence of oxygen, which leads to the formation of carbon dioxide, while Loyon [26] confirmed that aerobic treatment can indeed help reduce carbon dioxide emissions from pig manure over the long term when comparing the initial and final stages of pig slurry treatment with aeration after its stabilization. The overall reduction in greenhouse gas emissions was approximately 97.5%, primarily due to the complete aerobic decomposition of organic matter during the treatment process, resulting in minimal emissions post-treatment. In our study, the combination of both treatments provided a CO_2 emission reduction of up to 87%.

4.3. Ammonia (NH_3) Emissions

According to results shown in Figure 5, similar curves were observed for NH_3 during the first, third, and fourth weeks for the control and treated pig slurry. Following the initial treatment, the control and treated slurry reported significant increases ($p < 0.05$) during the three weeks previously indicated, and the values for the controlled slurry decreased in the fifth week then settled in a constant range, with the same behavior for the other treatments. NH_3 emissions significantly ($p < 0.05$) decreased when comparing control and pig slurry treated with bacteria (15.66 $mg \cdot m^{-2} \cdot day^{-1}$ compared to 1.667 $mg \cdot m^{-2} \cdot day^{-1}$, respectively). The same effects were seen when comparing the treated slurry with bacteria coupled with aeration (3.66 $mg \cdot m^{-2} \cdot day^{-1}$). In this study, we observed that NH_3 was less reduced under aerobic conditions compared to the treated slurry when only adding bacteria. The difference in the reduction in ammonia emissions between using the additives alone (90%) and combining them with aeration (76%) could be due to the specific mechanisms involved in each approach.

The average ammonia reduction in our study was 99% using only bacteria and 76% when coupled with the aeration system. Other studies [45,73] demonstrated that biological additives alone typically reduce ammonia emissions by only 10–25% and aeration alone reduces ammonia emissions by 35–50%, while El bied [70] reported that biological additives alone could reduce 77% of ammonia. Kupper [68] showed that aeration alone reduced ammonia emissions by 35–50%, while combining with biological additives could reduce ammonia emissions by 50–70%. Berg [74] confirmed that the combination of both treatments provided high reductions, with ammonia emissions reduced by up to 70%. Our findings on ammonia emissions with biological additive combined with aeration (76%) agree with the values reported by Berg and Kupper [68,74].

Once HIPO-PURÍN, previously rehydrated, is applied, it initiates a series of crucial processes. Firstly, urea is decomposed into two molecules of ammonia by the urease present in slurry and HIPO-PURÍN additive, making it available for bacterial utilization. According to Calvet [75], once urea is excreted, it undergoes a relatively rapid process that results in rapid urea saponification, leading to higher NH_3 concentrations [51], which explains the peaks during the first weeks after adding the treatment. As shown in Figure 5, there was a significant difference between the control and treated pig slurry in ammonia emissions during the second, third, and fourth weeks. To delay ammonia release until bacterial activation, the Yuca extract within the HIPO-PURÍN acts as a binding agent, allowing subsequent utilization of nitrogen and ammonia by microorganisms for growth and reproduction, with increased absorption leading to biomass conversion, which explains the small peaks in the treated slurry and the high peakks for the control.

Bacillus species are known for their potent enzymatic activity, which directly targets and metabolizes ammonia, resulting in significant reductions when added to pig slurry [76]. Bacillus bacteria can effectively reduce nitrogen and ammonia emissions from pig slurry through the nitrification–denitrification process [77]. When introduced into the slurry, Bacillus species facilitate both stages of this process and lead to reductions in the concentration of ammonia in the slurry, as ammonia is converted into less volatile nitrate forms. Subsequently, microorganisms utilize nitrogen and ammonia for growth and reproduction, with increased absorption of ammonia, leading to biomass conversion. Notably, HIPO-PURIN bacteria exhibit anaerobic functionality, even in suspension without oxygen. The establishment of an NH_3–NH_4^+ equilibrium, largely favoring non-volatile NH_4^+ due to intermittent aeration and microbial activity stabilizing pH, contributes to reduced ammonia volatilization. Furthermore, bacterial consumption of ammonia, particularly in the NH_4^+ form, significantly diminishes NH_3 emissions, highlighting the efficacy of HIPO-PURIN in mitigating atmospheric ammonia levels.

However, when combined with aeration, factors such as oxygen availability, microbial interactions, and the timing and duration of aeration may impact the efficiency of reducing ammonia [78]. Aeration may introduce conditions that alter the microbial community dynamics or affect the availability of nutrients, potentially influencing the performance of the biological additives in ammonia reduction [79,80]. This disruption could lead to reduced efficiency in lowering nitrogen and ammonia emissions, which explains the high reductions in both parameters in the treatments with and without aeration (76% and 90%, respectively). Therefore, while both strategies contributed to mitigating ammonia emissions, the observed difference could stem from the complex interplay of various factors influencing the microbial activity and ammonia metabolism within the system. Further investigation into optimizing the synergistic effects between the biological additives and aeration parameters may provide insights into improving their combined efficiency in reducing ammonia pollution from pig slurry, while also considering the additional benefit of aeration in methane (CH_4) reduction.

4.4. Compositions of the Treated and Control Raw Pig Slurry

The compositions of the slurry during the treatment process are detailed in Tables 1–3. Table 1 shows a notable contrast in electrical conductivity (EC) between the first and sixth weeks, with an increase from 22.09 $\mu S\ cm^{-1}$ to 24.04 $\mu S\ cm^{-1}$, a change deemed statistically significant ($p < 0.05$) for the control slurry. Adding bacteria alone or combined with aeration reduced this significantly (21.76 and 20.45 $\mu S\ cm^{-1}$, respectively).

Adding biological additives led to a significant change in pH between the first and sixth weeks, attributed to microbial degradation of organic matter, leading to the formation of carbonate and ammonium [78], consequently elevating the pH value from 7.32 to 7.97 and 8.22 for bacteria alone and combined with aeration, respectively. On the other hand, aeration increased the bulk pH over the medium term, likely due to the degradation of volatile fatty acids, as previously reported [23,59,81].

The duration of storage for pig slurry has a big impact the concentration and behavior of total suspended solids (TSSs). Over time, natural settling occurs, causing heavier solid particles to separate from the liquid phase, leading to the accumulation of TSSs at the tank's bottom [82], and this phenomenon explains the significant reduction in the TSSs in raw pig slurry (from 26.57 to 14.5 g L^{-1}) between the first and sixth weeks.

Adding bacteria, the TSSs evolved from 26.97 g L^{-1} to 19.03 g L^{-1} and 24.1 g L^{-1} when combined with aeration which showed a significant effect on TSSs removal relative to the initial values.

In the context of pig slurry management, microbial cultures, particularly Bacillus bacteria strains, play a crucial role in the breakdown of organic matter. These specialized bacteria are adept at solubilizing or converting solid particles into smaller components, thereby contributing to the reduction in total suspended solids (TSSs) within slurry [83,84].

Addition of bacteria combined with aeration exhibited a small yet significant effect ($p < 0.05$) on TSSs removal. Aerobic treatment methods hold promise for reducing total suspended solids (TSSs) in pig slurry [85], but there are challenges associated with agitation during the process, which can lead to a lower removal efficiency, which happened in this study. Agitation of pig slurry can disturb settled solids, causing them to become resuspended in the liquid phase and reducing the effectiveness of TSSs removal [86].

It is noteworthy that settling is not a treatment that removes suspended solids from the liquid phase, because when the tank is cleaned or emptied, those settled solids at the bottom remain present, requiring additional steps for their removal. Therefore, settling is considered a partial solid removal technique rather than a complete one, as it only separates solids physically but does not eliminate them entirely from the system.

The correlation between COD and TSSs changes is based on the levels of inorganic and organic solids in the mixture [87]. In pig slurry, there is typically a correlation between TSSs and COD caused by the presence of both organic and inorganic suspended solids, which collectively impact the COD content [9,22]. This study validated this observation, demonstrating that a decrease in suspended solids led to significantly reduced COD concentrations across all three tanks ($p < 0.05$).

Just as COD is correlated with TSSs, BOD_5 is similarly linked to the same parameter, with BOD_5 being linked to the organic content within suspended solids. The microorganisms responsible for BOD_5 decomposition can adhere to these solids [88,89]. Consequently, effective TSSs removal can result in decreased BOD_5 levels, due to the reduction in microbial habitat and organic matter accessible for biological degradation. Notably, a significant BOD_5 reduction was noted in all three cases ($p < 0.05$), clearly showing a correlation between the BOD_5 and TSSs removal efficiency.

The findings of this study indicated a correlation between the biological additives and total nitrogen removal, because the TN had the same behavior in both treatments, where was decreased from 2.66 g L^{-1} to 1.47g L^{-1} with biological additive alone and 1.52 g L^{-1} when combined with the OXI-FUCH aeration technique.

On the one hand, Bacillus bacteria can play a role in reducing total nitrogen levels in liquid pig slurry through their ability to break down organic matter [90]. According to Shao 2003 [91], the cellulose-degrading bacteria present in pig manure, especially *Bacillus subtilis*, demonstrated superior cellulose degradation capabilities. In liquid slurry systems, organic compounds are often abundant and can contribute to elevated nitrogen levels. Bacillus bacteria, known for their proficiency in organic matter decomposition, metabolize these compounds, releasing nitrogen in forms that are more accessible for uptake or for further transformation by other microorganisms [92]. By facilitating the breakdown of organic material, Bacillus bacteria contribute to the reduction in total nitrogen content in liquid pig slurry.

On the other hand, Bacillus thuringiensis bacteria demonstrated a noteworthy capacity to reduce total nitrogen levels by 55% when introduced to pig slurry. This significant finding parallels the pioneering research that identified Bacillus thuringiensis as a potent agent in nitrogen removal and shows its effectiveness in removing total nitrogen, which can achieve up to 82% reductions under specific conditions [93]. Similarly to the initial discovery, this study underscored bacteria's remarkable adaptability and efficacy in processing nitrogen compounds within pig slurry and revealed Bacillus thuringiensis' ability to transform nitrogen forms, resulting in a substantial decrease in total nitrogen content. In conclusion, the biological treatment yielded good results, achieving a 56% reduction in total nitrogen compared to the raw slurries.

This study demonstrates that biological additives, particularly when combined with aeration, can significantly improve the composition and quality of pig slurry by reducing key pollutants. The addition of bacteria and aeration significantly reduced electrical conductivity (EC) by 1.5% and 7.4%, respectively. Total suspended solids (TSSs) in the raw pig slurry showed a 45.4% decrease, with bacteria alone resulting in a 29.5% decrease and a 10.6% decrease when combined with aeration, affected by the absence of settling conditions. Total nitrogen (TN) was reduced by 44.7% with the addition of bacteria alone, and by 42.9% when combined with aeration.

In the final week of treatment, both treatments did not significantly affect ($p < 0.05$) Na^+, K^+, $Ca_2{}^+$, $Mg_2{}^+$, Cu, Zn, and Fe in comparison with the control slurry in the last week of treatment, while decreasing others such as P and Mn.

Bacillus bacteria are known for their ability to solubilize phosphorus, especially *Bacillus megaterium*, which means they can convert insoluble forms of phosphorus into soluble forms that are more readily available for uptake by microorganisms or plants [94]. When Bacillus bacteria are introduced to pig slurry, they may initiate processes such as phosphorus solubilization, where insoluble phosphorus compounds in the slurry are converted into soluble forms [95]. As a result, the concentration of phosphorus in the liquid portion of the slurry may decrease, as the solubilized phosphorus is utilized by the bacteria, because all bacterial species require several micronutrients, with phosphorus being particularly crucial for ATP (Adenosine Triphosphate) production. ATP conversion fuels cellular processes, with phosphorus in pig slurry typically existing in the form of phosphates ($PO_4{}^{3-}$) [96,97]. This explains the decrease in total phosphorus observed when comparing untreated pig slurry with pig slurry containing biological additives with and without aeration 96.6 mg L^{-1}, 53.4 mg L^{-1}, and 43.2 mg L^{-1}, respectively. Zhu et al. (2006) [59] investigated the impact of combining a biological additive with aeration on nutrient reduction in swine manure over a 15-day period. After just one day of aeration, all aerated treatments showed a 42% reduction in total soluble phosphorus and an increase in total insoluble phosphorus. Additionally, total Kjeldahl nitrogen decreased by about 40% in all treatments except the control.

As per the findings of Rongrong Wu [98], diverse biological approaches involving microorganisms have been identified for manganese (Mn) removal, encompassing biosorption, bioaccumulation, and biological oxidation. Therdkiattikul and Katsoyiannis [99,100] affirmed that biological oxidation employing microorganisms has emerged as a potential method for manganese removal from water sources. This observation elucidates the substantial contrast between untreated and treated pig slurry with bacteria. The former exhibited a concentration of 7 g L^{-1}, whereas the latter, treated with bacteria, displayed reduced concentrations of 2.61 g L^{-1} and 1.98 g L^{-1}, respectively.

The study results showed a significant decrease in Mn levels with biological additives. Untreated pig slurry had 4.91 g L^{-1} of Mn, while treated slurry had 2.61 g L^{-1} and 1.98 g L^{-1} without and with aeration, respectively. According to Rongrong [98], biological treatments that utilize microorganisms, such as biosorption, bioaccumulation, and biological oxidation, can remove Mn. Two other studies, by Therdkiattikul and Katsoyiannis [99,100], also supported that biological oxidation using microorganisms can used for Mn removal. Our study validated how adding biological additives to pig slurry can decrease Mn content.

Regarding cations, our findings align with those of previous studies [101], although Moral [102] reported higher concentrations of potassium (K). Several cations, including K, sodium (Na), copper (Cu), zinc (Zn), and manganese (Mn), are frequently added to diets as supplements to improve growth rates, even at concentrations surpassing physiological requirements [103,104]. For example, pigs are estimated to excrete roughly 66% of the sodium (Na) and 59% of the potassium (K) they ingest [105]. Additionally, Clemente et al. [106] observed elevated concentrations of Cu and Zn when analyzing the separated solid fraction of pig slurries. Slurry storage could have contributed to the observed trends of increased concentrations of potassium (K^+), sodium (Na^+), calcium (Ca^{2+}), magnesium (Mg^{2+}), and various metals Cu, Zn, and Fe alongside a decrease in total suspended solids (TSSs) after six weeks of treatment and storage [107]. Vaporization, or evaporation, of water from slurry can concentrate the remaining solutes, including ions and metals, leading to higher concentrations in the remaining liquid phase. This concentration effect can occur as water evaporates from the surface of the slurry, leaving behind a more concentrated solution of dissolved constituents [107]. Consequently, the observed changes in ion and TSSs concentrations can be influenced by both vaporization and the biological processes previously mentioned.

5. Conclusions

This study investigated the impact of different treatment approaches, including the use of Bacillus biological additives HIPO-PURÍN and aeration using OXI-FUCH, on GHG and ammonia emissions from stored pig manure.

Methane (CH_4) emissions displayed varied trajectories across treatments, with notable reductions observed when using HIPO-PURÍN bacteria alone or combined with aeration. While initial spikes in CH_4 emissions were noted following the introduction of biological additives, subsequent weeks saw a gradual decline, indicative of microbial community adjustments favoring methane-consuming bacteria. Aeration, although generally decreasing CH_4 levels, occasionally led to spikes due to trapped CH_4 release and accelerated aerobic decomposition.

Carbon dioxide (CO_2) emissions exhibited similar trends, with significant reductions noted with HIPO-PURÍN bacteria and aeration treatments compared to untreated slurry. These reductions were attributed to enhanced aerobic decomposition facilitated by increased oxygen availability.

Ammonia (NH_3) emissions, on the other hand, saw significant decreases with the addition of biological additives alone or combined with aeration, highlighting the efficacy of HIPO-PURÍN bacteria in mitigating NH_3 volatilization.

This investigation also revealed the significant impact of biological additives and aeration on the composition of stored pig slurry. The addition of bacteria, particularly Bacillus strains, led to pH elevations and a slight decrease in electrical conductivity. Moreover, biological treatments showed promising results in reducing chemical oxygen demand (COD), biological oxygen demand (BOD_5), total nitrogen (TN), phosphorus, and manganese levels.

Overall, our findings demonstrate the multifaceted effects of biological additives and aeration on GHG from pig slurry, underscoring their potential in reducing the environmental impacts associated with slurry management.

Future studies should focus on the direct application of biological additives and implementing aeration systems in pig slurry storage ponds with larger volumes. Investigating the long-term impact and feasibility of these treatments in real-world farm environments is essential to promote sustainable livestock waste management practices.

Author Contributions: Conceptualization, O.E.b. and J.A.A.; data curation, O.E.b. and J.A.A.; formal analysis, O.E.b. and M.A.T.T.; investigation, O.E.b., M.A.T.T., M.G.G. and J.A.A.; methodology, O.E.b. and M.A.T.T.; software, O.E.b.; supervision, Á.F.C. and J.A.A.; validation, O.E.b., Á.F.C. and J.A.A.; visualization, O.E.b., Á.F.C. and J.A.A.; writing—original draft, O.E.b. writing—review and editing, O.E.b. and J.A.A. All authors have read and agreed to the published version of the manuscript.

Funding: This research was funded by Ayuntamiento Fuente Alamo, Murcia, Spain. Project reference: CUE-AYTO FUENTE ÁLAMO 2020-T.

Data Availability Statement: The original contributions presented in the study are included in the article, further inquiries can be directed to the corresponding author.

Acknowledgments: We express our gratitude to the www.oxifuch.com (accessed on 23 May 2024) company for their invaluable support and cooperation, which significantly contributed to the success of this research endeavor.

Conflicts of Interest: The authors declare no conflicts of interest. The funders had no role in the design of this study; in the collection, analyses, or interpretation of the data; in the writing of the manuscript; or in the decision to publish the results.

References

1. Bayo, J.; Gómez-López, M.D.; Faz, A.; Caballero, A. Environmental Assessment of Pig Slurry Management after Local Characterization and Normalization. *J. Clean. Prod.* **2012**, *32*, 227–235. [CrossRef]
2. Ministerio de Agricultura, Pesca y Alimentación. *"Estadística de Ganado Porcino. Resultados" Edita: © Ministerio de Agricultura, Pesca y Alimentación*; NIPO: 003230290, Resultados de las Encuestas de Ganado Porcino; Ministerio de Agricultura, Pesca y Alimentación: Madrid, Spain, 2023.
3. Dennehy, C.; Lawlor, P.G.; Jiang, Y.; Gardiner, G.E.; Xie, S.; Nghiem, L.D.; Zhan, X. Greenhouse Gas Emissions from Different Pig Manure Management Techniques: A Critical Analysis. *Front. Environ. Sci. Eng.* **2017**, *11*, 11. [CrossRef]
4. Varma, V.S.; Parajuli, R.; Scott, E.; Canter, T.; Lim, T.T.; Popp, J.; Thoma, G. Dairy and Swine Manure Management—Challenges and Perspectives for Sustainable Treatment Technology. *Sci. Total Environ.* **2021**, *778*, 146319. [CrossRef]
5. Kerr, B.J.; Trabue, S.L.; Andersen, D.S.; Van Weelden, M.B.; Pepple, L.M. Dietary Composition and Particle Size Effects on Swine Manure Characteristics and Gas Emissions. *J. Environ. Qual.* **2020**, *49*, 1384–1395. [CrossRef] [PubMed]
6. Antezana, W.; De Blas, C.; García-Rebollar, P.; Rodríguez, C.; Beccaccia, A.; Ferrer, P.; Cerisuelo, A.; Moset, V.; Estellés, F.; Cambra-López, M.; et al. Composition, Potential Emissions and Agricultural Value of Pig Slurry from Spanish Commercial Farms. *Nutr. Cycl. Agroecosystems* **2016**, *104*, 159–173. [CrossRef]
7. U.S Foreign Agricultural Service. *Department of agriculture Production—Pork /Top Producing Countries*; U.S Foreign Agricultural Service: Washington, DC, USA, 2024.
8. Provolo, G.; Manuli, G.; Finzi, A.; Lucchini, G.; Riva, E.; Sacchi, G.A. Effect of Pig and Cattle Slurry Application on Heavy Metal Composition of Maize Grown on Different Soils. *Sustainability* **2018**, *10*, 2684. [CrossRef]
9. El Bied, O.; García-Valero, A.; Fechtali, T.; Faz, Á.; Acosta, J.A. Purification Performance of Filtration Process for Pig Slurry Using Marine Sands, Silty Loam Soils, Fly Ash and Zeolite. *Agronomy* **2021**, *11*, 1608. [CrossRef]
10. Chojnacka, K.; Moustakas, K.; Witek-Krowiak, A. Bio-Based Fertilizers: A Practical Approach towards Circular Economy. *Bioresour. Technol.* **2020**, *295*, 122223. [CrossRef]
11. Zhang, Y.; Jiang, Y.; Wang, S.; Wang, Z.; Liu, Y.; Hu, Z.; Zhan, X. Environmental Sustainability Assessment of Pig Manure Mono- and Co-Digestion and Dynamic Land Application of the Digestate. *Renew. Sustain. Energy Rev.* **2021**, *137*, 110476. [CrossRef]
12. Antoneli, V.; Mosele, A.C.; Bednarz, J.A.; Pulido-Fernández, M.; Lozano-Parra, J.; Keesstra, S.D.; Rodrigo-Comino, J. Effects of Applying Liquid Swine Manure on Soil Quality and Yield Production in Tropical Soybean Crops (Paraná, Brazil). *Sustainability* **2019**, *11*, 3898. [CrossRef]
13. Houghton, J.T.; Ding, Y.D.J.G.; Griggs, D.J. *Climate Change 2001: The Scientific Basis*; Cambridge University Press: Cambridge, UK, 2001.
14. Degré, A.; Verhève, D.; Debouche, C. Émissions Gazeuses En Élevage Porcin et Modes de Réduction: Revue Bibliographique. *Biotechnol. Agron. Soc. Environ.* **2001**, *5*, 135–143.
15. Directive 91/676/EEC. Concerning the Protection of Waters against Pollution Caused by Nitrates from Agricultural Sources. *Off. J. Eur. Communities L* **1991**, *375*, 1–8.
16. Water Framework Directive (2000/60/EC) Water Framework Directive (2000/60/EC). Available online: https://eur-lex.europa.eu/eli/dir/2000/60/oj (accessed on 23 May 2024).
17. Guía de Mejores Técnicas Disponibles del Sector Porcino 2017. pp. 1–146. Available online: https://www.mapa.gob.es/es/ganaderia/publicaciones/guiamtdssectorporcino_tcm30-105316.pdf (accessed on 23 May 2024).
18. Company, E.; Farrés, M.; Colprim, J.; Magrí, A. Exploring the Recovery of Potassium-Rich Struvite after a Nitrification-Denitrification Process in Pig Slurry Treatment. *Sci. Total Environ.* **2022**, *847*, 157574. [CrossRef] [PubMed]
19. Schott, C.; Cunha, J.R.; van der Weijden, R.D.; Buisman, C. Phosphorus Recovery from Pig Manure: Dissolution of Struvite and Formation of Calcium Phosphate Granules during Anaerobic Digestion with Calcium Addition. *Chem. Eng. J.* **2022**, *437*, 135406. [CrossRef]
20. El, O.; Kessler, M.; Ang, M.; Fechtali, T.; Cano, A.F.; Acosta, J.A. Turbidity and Chemical Oxygen Demand Reduction from Pig Slurry through a Coagulation Flocculation Process. *Agronomy* **2021**, *11*, 2158. [CrossRef]
21. Goldfarb, J.L.; Hubble, A.H.; Ma, Q.; Volpe, M.; Severini, G.; Andreottola, G.; Fiori, L. Valorization of Cow Manure via Hydrothermal Carbonization for Phosphorus Recovery and Adsorbents for Water Treatment. *J. Environ. Manag.* **2022**, *308*, 114561. [CrossRef] [PubMed]
22. El bied, O.; Turbí, M.A.T.; García-Valero, A.; Cano, Á.F.; Acosta, J.A. Mitigating Ammonia, Methane, and Carbon Dioxide Emissions from Stored Pig Slurry Using Chemical and Biological Additives. *Water* **2023**, *15*, 4185. [CrossRef]

23. Fangueiro, D.; Hjorth, M.; Gioelli, F. Acidification of Animal Slurry—A Review. *J. Environ. Manag.* **2015**, *149*, 46–56. [CrossRef] [PubMed]
24. Hjorth, M.; Christensen, K.V.; Christensen, M.L.; Sommer, S.G. Solid—Liquid Separation of Animal Slurry in Theory and Practice. A Review. *Agron. Sustain. Dev.* **2010**, *30*, 153–180. [CrossRef]
25. Møller, H.B.; Lund, I.; Sommer, S.G. Solid-Liquid Separation of Livestock Slurry: Efficiency and Cost. *Bioresour. Technol.* **2000**, *74*, 223–229. [CrossRef]
26. Loyon, L.; Guiziou, F.; Béline, F.; Peu, P. Gaseous Emissions (NH_3, N_2O, CH_4, CO_2) during Pig Slurry Biological Aerobic Treatment and Treatment by-Product Storages. *Int. Congr. Ser.* **2006**, *1293*, 299–302. [CrossRef]
27. Möller, K.; Müller, T. Effects of Anaerobic Digestion on Digestate Nutrient Availability and Crop Growth: A Review. *Eng. Life Sci.* **2012**, *12*, 242–257. [CrossRef]
28. Lymperatou, A.; Rasmussen, N.B.; Gavala, H.N.; Skiadas, I.V. Improving the Anaerobic Digestion of Swine Manure through an Optimized Ammonia Treatment: Process Performance, Digestate and Techno-Economic Aspects. *Energies* **2021**, *14*, 787. [CrossRef]
29. Petersen, S.O.; Sommer, S.G.; Beline, F.; Burton, C.; Dach, J.; Dourmad, J.Y.; Leip, A.; Misselbrook, T.H.; Nicholson, F.; Poulsen, H.D.; et al. Recycling of livestock manure in a whole-farm perspective. *Livest. Sci.* **2007**, *112*, 180–191. [CrossRef]
30. Rajasulochana, P.; Preethy, V. Comparison on Efficiency of Various Techniques in Treatment of Waste and Sewage Water—A Comprehensive Review. *Resour. Effic. Technol.* **2016**, *2*, 175–184. [CrossRef]
31. Provolo, G.; Finzi, A.; Perazzolo, F.; Mattachini, G.; Riva, E. Effect of a Biological Additive on Nitrogen Losses from Pig Slurry during Storage. *J. Environ. Qual.* **2016**, *45*, 1460–1465. [CrossRef]
32. Petersen, S.O.; Blanchard, M.; Chadwick, D.; Del Prado, A.; Edouard, N.; Mosquera, J.; Sommer, S.G. Manure Management for Greenhouse Gas Mitigation. *Animal* **2013**, *7*, 266–282. [CrossRef] [PubMed]
33. VERA. *Verification of Environmental Technologies for Agricultural Production Vera Test Protocol Covers and Other Mitigation Technologies for Reduction of Gaseous Emissions from Stored Manure*; VERA: Brondby, Denmark, 2018.
34. Akkale, C. Production, Optimization and Partial Characterization of Alkaline Protease from *Bacillus subtilis* spp. *subtilis* NRRL B-3384 and B-3387. *Hittite J. Sci. Eng.* **2023**, *10*, 135–144. [CrossRef]
35. Sekhon, A.; Dahiya, N.; Tiwari, R.; Hoondal, G. Production of Extracellular Lipase by Bacillus Megaterium AKG-1 in Submerged Fermentation. *Indian J. Biotechnol.* **2006**, *5*, 179–183.
36. Jellouli, K.; Ghorbel-Bellaaj, O.; Ayed, H.B.; Manni, L.; Agrebi, R.; Nasri, M. Alkaline-Protease from Bacillus Licheniformis MP1: Purification, Characterization and Potential Application as a Detergent Additive and for Shrimp Waste Deproteinization. *Process Biochem.* **2011**, *46*, 1248–1256. [CrossRef]
37. Guleria, S.; Walia, A.; Chauhan, A.; Shirkot, C.K. Molecular Characterization of Alkaline Protease of Bacillus Amyloliquefaciens SP1 Involved in Biocontrol of *Fusarium oxysporum*. *Int. J. Food Microbiol.* **2016**, *232*, 134–143. [CrossRef] [PubMed]
38. Lin, L.; Kan, X.; Yan, H.; Wang, D. Characterization of Extracellular Cellulose-Degrading Enzymes from *Bacillus thuringiensis* Strains. *Electron. J. Biotechnol.* **2012**, *15*, 2. [CrossRef]
39. Awwa, W. *APHA, Standard Methods for the Examination of Water and Wastewater, American Public Health Association*; American Public Health Association: Washington, DC, USA, 2005.
40. Nemali, K.; University, P. Ji-Qin Ni and Albert Heber-Purdue Agricultural and Biological Engineering. 2018. Available online: https://www.extension.purdue.edu/extmedia/ABE/ABE-161-W.pdf (accessed on 23 May 2024).
41. Wheeler, E.F.; Adviento-Borbe, M.A.A.; Brandt, R.C.; Topper, P.A.; Topper, D.A.; Elliot, H.A.; Graves, R.E.; Hristov, A.N.; Ishler, V.A.; Bruns, M.A. *Amendments for Short-and Medium-Term Mitigation of Odor Emissions from Dairy Manure*; American Society of Agricultural and Biological Engineers: St. Joseph, MI, USA, 2001.
42. Bastami, M.S.B.; Jones, D.L.; Chadwick, D.R. Reduction of Methane Emission during Slurry Storage by the Addition of Effective Microorganisms and Excessive Carbon Source from Brewing Sugar. *J. Environ. Qual.* **2016**, *45*, 2016–2022. [CrossRef] [PubMed]
43. Kavanagh, I.; Fenton, O.; Healy, M.G.; Burchill, W.; Lanigan, G.J.; Krol, D.J. Mitigating Ammonia and Greenhouse Gas Emissions from Stored Cattle Slurry Using Agricultural Waste, Commercially Available Products and a Chemical Acidifier. *J. Clean. Prod.* **2021**, *294*, 126251. [CrossRef]
44. Owusu-Twum, M.Y.; Polastre, A.; Subedi, R.; Santos, A.S.; Mendes Ferreira, L.M.; Coutinho, J.; Trindade, H. Gaseous Emissions and Modification of Slurry Composition during Storage and after Field Application: Effect of Slurry Additives and Mechanical Separation. *J. Environ. Manag.* **2017**, *200*, 416–422. [CrossRef]
45. Amon, B.; Kryvoruchko, V.; Amon, T.; Zechmeister-Boltenstern, S. Methane, Nitrous Oxide and Ammonia Emissions during Storage and after Application of Dairy Cattle Slurry and Influence of Slurry Treatment. *Agric. Ecosyst. Environ.* **2006**, *112*, 153–162. [CrossRef]
46. Husted, S.; Jensen, L.S.; Jørgensen, S.S. Reducing Ammonia Loss from Cattle Slurry by the Use of Acidifying Additives: The Role of the Buffer System. *J. Sci. Food Agric.* **1991**, *57*, 335–349. [CrossRef]
47. Hamood, A.; Mao, X.; Jaworski, A.J. Two-Stage Thermoacoustic Electricity Generator for Waste Heat Recovery. In *World Congress on Engineering*; International Association of Engineers: Hongkong, China, 2006.
48. Ishak, M.A.M.; Ani, A.Y.; Syed Ismail, S.N.A.; Ali, M.L.M.; Ahmad, R. Conversion of Biomass to Biofuels. In *Value-Chain of Biofuels: Fundamentals, Technology, and Standardization*; Elsevier: Amsterdam, The Netherlands, 2021; pp. 49–67, ISBN 9780128243886.
49. Okoro, O.V.; Sun, Z.; Birch, J. Meat Processing Waste as a Potential Feedstock for Biochemicals and Biofuels—A Review of Possible Conversion Technologies. *J. Clean. Prod.* **2017**, *142*, 1583–1608. [CrossRef]

50. Calvet, S.; Hunt, J.; Misselbrook, T.H. Low Frequency Aeration of Pig Slurry Affects Slurry Characteristics and Emissions of Greenhouse Gases and Ammonia. *Biosyst. Eng.* **2017**, *159*, 121–132. [CrossRef]
51. Mostafa, E.; Selders, A.; Buescher, W. Aeration of Pig Slurry Affects Ammonia and Greenhouse Gases Emissions. *Int. J. Environ. Sci. Technol.* **2019**, *16*, 7327–7338. [CrossRef]
52. Dettmer, A.; Coelho Cavalheiro, J.; Cavalli, E.; Misturini Rossi, D.; De Souza Gusatti, C.; Záchia Ayub, M.A.; Gutterres, M. Optimization of the Biotechnological Process for Hide Unhairing in Substitution of Toxic Sulfides. *Chem. Eng. Technol.* **2012**, *35*, 803–810. [CrossRef]
53. Franke-Whittle, I.H.; Confalonieri, A.; Insam, H.; Schlegelmilch, M.; Körner, I. Changes in the Microbial Communities during Co-Composting of Digestates. *Waste Manag.* **2014**, *34*, 632–641. [CrossRef] [PubMed]
54. Merlin Christy, P.; Gopinath, L.R.; Divya, D. A Review on Anaerobic Decomposition and Enhancement of Biogas Production through Enzymes and Microorganisms. *Renew. Sustain. Energy Rev.* **2014**, *34*, 167–173. [CrossRef]
55. Alam, M.J.; Islam, M.; Jeon, C.O.; Lee, K.C.; Kim, S.H.; Yang, C.J.; Kabir, M.E.; Lee, S.S. In Vitro Assessment of Probiotic Potential of Selected Bacteria Isolated from Pig Faeces with Potential Application of Odour Reduction. *Int. J. Vet. Sci. Med.* **2021**, *9*, 22–30. [CrossRef] [PubMed]
56. Prenafeta-Boldú, F.X.; Fernández, B.; Viñas, M.; Lizardo, R.; Brufau, J.; Owusu-Asiedu, A.; Walsh, M.C.; Awati, A. Effect of Bacillus Spp. Direct-Fed Microbial on Slurry Characteristics and Gaseous Emissions in Growing Pigs Fed with High Fibre-Based Diets. *Animal* **2017**, *11*, 209–218. [CrossRef] [PubMed]
57. Lee, S.R.; Han, J.K.; Choi, Y.J.; Nam, K. Reduction of Ammonia and Hydrogen Sulfide Emission from Swine Manure Using Aqueous Foams Amended with Microorganisms and Chemical Additives. *Clean* **2007**, *35*, 230–234. [CrossRef]
58. Blanes-Vidal, V.; Guàrdia, M.; Dai, X.R.; Nadimi, E.S. Emissions of NH_3, CO_2 and H_2S during Swine Wastewater Management: Characterization of Transient Emissions after Air-Liquid Interface Disturbances. *Atmos. Environ.* **2012**, *54*, 408–418. [CrossRef]
59. Zhang, Z.J.; Zhu, J. Characteristics of Solids, BOD5 and VFAs in Liquid Swine Manure Treated by Short-Term Low-Intensity Aeration for Long-Term Storage. *Bioresour. Technol.* **2006**, *97*, 140–149. [CrossRef]
60. Martinez, J.; Guiziou, F.; Peu, P.; Gueutier, V. Influence of Treatment Techniques for Pig Slurry on Methane Emissions during Subsequent Storage. *Biosyst. Eng.* **2003**, *85*, 347–354. [CrossRef]
61. Dai, X.R.; Blanes-Vidal, V. Emissions of Ammonia, Carbon Dioxide, and Hydrogen Sulfide from Swine Wastewater during and after Acidification Treatment: Effect of PH, Mixing and Aeration. *J. Environ. Manag.* **2013**, *115*, 147–154. [CrossRef]
62. Ni, J.-Q.; Hendriks, J.; Vinckier, C.; Coenegrachts, J. A New Concept. of Carbon. Dioxide Accelerated Ammonia Release from Liquid Manure in Pig House. *Environ. Int.* **2000**, *26*, 97–104. [CrossRef] [PubMed]
63. Petersen, V.; Markfoged, R.; Hafner, S.D.; Sommer, S.G. A New Slurry PH Model Accounting for Effects of Ammonia and Carbon Dioxide Volatilization on Solution Speciation. *Nutr. Cycl. Agroecosystems* **2014**, *100*, 189–204. [CrossRef]
64. Ma, S.; Xiong, J.; Cui, R.; Sun, X.; Han, L.; Xu, Y.; Kan, Z.; Gong, X.; Huang, G. Effects of Intermittent Aeration on Greenhouse Gas Emissions and Bacterial Community Succession during Large-Scale Membrane-Covered Aerobic Composting. *J. Clean. Prod.* **2020**, *266*, 121551. [CrossRef]
65. Loyon, L.; Guiziou, F.; Béline, F.; Peu, P. Gaseous Emissions (NH3, N2O, CH4, CO2) from A Biological Aerobic Treatment Unit of Pig Slurry. Available online: http://ramiran.uvlf.sk/doc04/Proceedings%2004/Loyon.pdf (accessed on 23 May 2024).
66. Marszałek, M.; Kowalski, Z.; Makara, A. Emission of Greenhouse Gases and Odorants from Pig Slurry—Effect on the Environment and Methods of Its Reduction. *Ecol. Chem. Eng. S* **2018**, *25*, 383–394. [CrossRef]
67. Heinonen-Tanski, H.; Leinonen, P.; Niskanen, E.M.; Mielonen, M.M.; Räsänen, H.; Valta, T.; Rinne, K.; Joki-Tokola, E. Aeration Improves the Hygiene of Cattle Slurry and the Quality of Grass Forage and Silage. *Acta Agric. Scand. B Soil Plant Sci.* **1998**, *48*, 212–221. [CrossRef]
68. Kupper, T.; Häni, C.; Neftel, A.; Kincaid, C.; Bühler, M.; Amon, B.; VanderZaag, A. Ammonia and Greenhouse Gas Emissions from Slurry Storage—A Review. *Agric. Ecosyst. Environ.* **2020**, *300*, 106963. [CrossRef]
69. Clemens, J.; Ahlgrimm, H.-J. Greenhouse Gases from Animal Husbandry: Mitigation Options. *Nutr. Cycl. Agroecosystems* **2001**, *60*, 287–300. [CrossRef]
70. El bied, O.; Angélica Terrero Turbí, M.; García-Valero, A.; Faz Cano, Á.; Alberto, J. Mitigating ammonia and greenhouse gas emissions from stored pig slurry using chemical additives and biochars. *Agronomy* **2023**, *15*, 2744.
71. Chai, C.; Zhang, D.; Yu, Y.; Feng, Y.; Wong, M.S. Carbon Footprint Analyses of Mainstream Wastewater Treatment Technologies under Different Sludge Treatment Scenarios in China. *Water* **2015**, *7*, 918–938. [CrossRef]
72. Ranieri, E.; D'Onghia, G.; Lopopolo, L.; Gikas, P.; Ranieri, F.; Gika, E.; Spagnolo, V.; Ranieri, A.C. Evaluation of Greenhouse Gas Emissions from Aerobic and Anaerobic Wastewater Treatment Plants in Southeast of Italy. *J. Environ. Manag.* **2023**, *337*, 117767. [CrossRef]
73. Guarino, M.; Fabbri, C.; Brambilla, M.; Valli, L.; Navarotto, P. Evaluation of simplified covering systems to reduce gaseous emissions from livestock manure storage. *Trans. ASABE* **2006**, *49*, 737–747. [CrossRef]
74. Berg, W.; Pazsiczki, I. Mitigation of Methane Emissions during Manure Storage. *Int. Congr. Ser.* **2006**, *1293*, 213–216. [CrossRef]
75. Calvet, S.; Estellés, F.; Cartanyà Ferré, J.; Babot Gaspa, D. *Guía Para La Minimización de Las Emisiones de Gases En Las Granjas Porcinas*; Edicions de la Universitat de Lleida: Lleida, Spain, 2020.
76. Shen, W.; Dai, H.; Gu, S.; Guo, F.; Li, T.; Rajasekar, A. Isolation and Application of PH- and Salt-Resistant *Bacillus* Strains to Reduce Ammonia Emission from Pig Manure during the Storage Period. *J. Appl. Microbiol.* **2023**, *134*, lxad134. [CrossRef] [PubMed]

77. Wang, Y.; Bi, L.; Liao, Y.; Lu, D.; Zhang, H.; Liao, X.; Liang, J.B.; Wu, Y. Influence and Characteristics of *Bacillus stearothermophilus* in Ammonia Reduction during Layer Manure Composting. *Ecotoxicol. Environ. Saf.* **2019**, *180*, 80–87. [CrossRef] [PubMed]
78. Li, G.; Yu, Y.; Li, X.; Jia, H.; Ma, X.; Opoku, P.A. Research Progress of Anaerobic Ammonium Oxidation (Anammox) Process Based on Integrated Fixed-Film Activated Sludge (IFAS). *Environ. Microbiol. Rep.* **2024**, *16*, e13235. [CrossRef] [PubMed]
79. Zhao, L.; Fu, G.; Zeng, A.; Cheng, B.; Song, Z.; Hu, Z. Effects of Different Aeration Strategies and Ammonia-Nitrogen Loads on Nitrification Performance and Microbial Community Succession of Mangrove Constructed Wetlands for Saline Wastewater Treatment. *Chemosphere* **2023**, *339*, 139685. [CrossRef] [PubMed]
80. Ma, T.; Zhan, Y.; Chen, W.; Xu, S.; Wang, Z.; Tao, Y.; Shi, X.; Sun, B.; Ding, G.; Li, J.; et al. Impact of Aeration Rate on Phosphorus Conversion and Bacterial Community Dynamics in Phosphorus-Enriched Composting. *Bioresour. Technol.* **2022**, *364*, 128016. [CrossRef] [PubMed]
81. Sørensen, P.; Eriksen, J. Effects of Slurry Acidification with Sulphuric Acid Combined with Aeration on the Turnover and Plant Availability of Nitrogen. *Agric. Ecosyst. Environ.* **2009**, *131*, 240–246. [CrossRef]
82. Zhu, K.; El-Din, M.G.; Moawad, A.K.; Bromley, D. Physical and Chemical Processes for Removing Suspended Solids and Phosphorus from Liquid Swine Manure. *Environ. Technol.* **2004**, *25*, 1177–1187. [CrossRef]
83. Paz Pérez-Sangrador, M.; Cristina León-Cófreces, M.; Acítores-Benavente, M.; Cruz García-González, M. Solids and Nutrient Removal from Flushed Swine Manure Using Polyacrylamides. *J. Environ. Manag.* **2012**, *93*, 67–70. [CrossRef]
84. Doyle, Y.; de la Noüe, J. Aerobic Treatment of Swine Manure: Physico-Chemical Aspects. *Biol. Wastes* **1987**, *22*, 187–208. [CrossRef]
85. O'Keefe, M.; McDonald, S.A.; Yan, M.; Davis, R.; Watts, P.; McGahan, E.; Tucker, R. Sludge Handling and Management. 2014, Project No. 2012/1029, Australian Pork Limited, Canberra, Australia, January 2014. Available online: https://extension.okstate.edu/fact-sheets/print-publications/pss/aerobic-treatment-system-pss-2275.pdf (accessed on 23 May 2024).
86. Abit, S.M.; William, J.; Graduate, J.; Associate, R. Oklahoma Cooperative Extension Service Extension Specialist for Soils under Non-Agricultural Uses. Available online: https://extension.okstate.edu/fact-sheets/aerobic-treatment-system.html (accessed on 23 May 2024).
87. Umweltbundesamt GmbH Final Report: Framework Service Contract. ENV.D2/FRA/2012/0013 Support. to the Implementation of the UWWTD: COD Substitution Scoping Study. 2017. Available online: https://circabc.europa.eu/sd/a/33c4c452-868e-4851-be02-56923cf75abb/COD_Study_Draft_Final_8Mar17_230517.pdf (accessed on 23 May 2024).
88. BOD; TSS. *The Biology and Troubleshooting of Facultative Lagoons*; Wiley: Hoboken, NJ, USA, 2015; pp. 195–198.
89. Alexander, R.T.; Cordat, E.; Chambrey, R.; Dimke, H.; Eladari, D. Acidosis and Urinary Calcium Excretion: Insights from Genetic Disorders. *J. Am. Soc. Nephrol.* **2016**, *27*, 3511–3520. [CrossRef] [PubMed]
90. Duan, M.; Zhang, Y.; Zhou, B.; Qin, Z.; Wu, J.; Wang, Q.; Yin, Y. Effects of *Bacillus subtilis* on Carbon Components and Microbial Functional Metabolism during Cow Manure–Straw Composting. *Bioresour. Technol.* **2020**, *303*, 122868. [CrossRef] [PubMed]
91. Shao, M.; Zhao, X.; Rehman, K.U.; Cai, M.; Zheng, L.; Huang, F.; Zhang, J. Synergistic Bioconversion of Organic Waste by Black Soldier Fly (*Hermetia illucens*) Larvae and Thermophilic Cellulose-Degrading Bacteria. *Front. Microbiol.* **2023**, *14*, 1288227. [CrossRef] [PubMed]
92. Hwang, O.; Yun, Y.M.; Trabue, S. Impact of *Bacillus subtilis* on Manure Solids, Odor, and Microbiome. *J. Environ. Manag.* **2023**, *333*, 117390. [CrossRef] [PubMed]
93. Xu, N.; Liao, M.; Liang, Y.; Guo, J.; Zhang, Y.; Xie, X.; Fan, Q.; Zhu, Y. Biological Nitrogen Removal Capability and Pathways Analysis of a Novel Low C/N Ratio Heterotrophic Nitrifying and Aerobic Denitrifying Bacterium (*Bacillus thuringiensis* Strain WXN-23). *Environ. Res.* **2021**, *195*, 110797. [CrossRef] [PubMed]
94. Saeid, A.; Prochownik, E.; Dobrowolska-Iwanek, J. Phosphorus Solubilization by *Bacillus* Species. *Molecules* **2018**, *23*, 2897. [CrossRef] [PubMed]
95. Wang, Z.; An, Y.; Chen, H.; Zhang, J.; Zhang, H.; Zhu, G.; Chen, J.; Li, W.; Wang, J.; Xu, H.J.; et al. Effects of Earthworms and Phosphate-Solubilizing Bacteria on Carbon Sequestration in Soils Amended with Manure and Slurry: A 4-Year Field Study. *Agronomy* **2022**, *12*, 2064. [CrossRef]
96. Zhang, Y.; Song, C.; Zhou, Z.; Cao, X.; Zhou, Y. Coupling between Nitrification and Denitrification as Well as Its Effect on Phosphorus Release in Sediments of Chinese Shallow Lakes. *Water* **2019**, *11*, 1809. [CrossRef]
97. Li, B.; Yang, Y.; Chen, J.; Wu, Z.; Liu, Y.; Xie, S. Nitrifying Activity and Ammonia-Oxidizing Microorganisms in a Constructed Wetland Treating Polluted Surface Water. *Sci. Total Environ.* **2018**, *628–629*, 310–318. [CrossRef]
98. Wu, R.; Yao, F.; Li, X.; Shi, C.; Zang, X.; Shu, X.; Liu, H.; Zhang, W. Manganese Pollution and Its Remediation: A Review of Biological Removal and Promising Combination Strategies. *Microorganisms* **2022**, *10*, 2411. [CrossRef] [PubMed]
99. Therdkiattikul, N.; Ratpukdi, T.; Kidkhunthod, P.; Chanlek, N.; Siripattanakul-Ratpukdi, S. Manganese-Contaminated Groundwater Treatment by Novel Bacterial Isolates: Kinetic Study and Mechanism Analysis Using Synchrotron-Based Techniques. *Sci. Rep.* **2020**, *10*, 13391. [CrossRef]
100. Katsoyiannis, I.A.; Zouboulis, A.I. Biological Treatment of Mn(II) and Fe(II) Containing Groundwater: Kinetic Considerations and Product Characterization. *Water Res.* **2004**, *38*, 1922–1932. [CrossRef]
101. Bloem, E.; Albihn, A.; Elving, J.; Hermann, L.; Lehmann, L.; Sarvi, M.; Schaaf, T.; Schick, J.; Turtola, E.; Ylivainio, K. Contamination of Organic Nutrient Sources with Potentially Toxic Elements, Antibiotics and Pathogen Microorganisms in Relation to P Fertilizer Potential and Treatment Options for the Production of Sustainable Fertilizers: A Review. *Sci. Total Environ.* **2017**, *607–608*, 225–242. [CrossRef]

102. Moral, R.; Perez-Murcia, M.D.; Perez-Espinosa, A.; Moreno-Caselles, J.; Paredes, C. Estimation of Nutrient Values of Pig Slurries in Southeast Spain Using Easily Determined Properties. *Waste Manag.* **2005**, *25*, 719–725. [CrossRef]
103. Dourmad, J.Y.; Jondreville, C. Impact of Nutrition on Nitrogen, Phosphorus, Cu and Zn in Pig Manure, and on Emissions of Ammonia and Odours. *Livest. Sci.* **2007**, *112*, 192–198. [CrossRef]
104. Kerkaert, H.R.; Woodworth, J.C.; Derouchey, J.M.; Dritz, S.S.; Tokach, M.D.; Goodband, R.D.; Manzke, N.E. Determining the Effects of Manganese Source and Level on Growth Performance and Carcass Characteristics of Growing-Finishing Pigs. *Transl. Anim. Sci.* **2021**, *5*, txab067. [CrossRef] [PubMed]
105. Hatfield, J.L.; Brumm, M.C.; Melvin, S.W. Chapter 4 Swine Manure Management. *Agric. Uses Munic. Anim. Ind. Byprod.* **1998**, *44*, 78–90.
106. Clemente, R.; Sáez-Tovar, J.A.; Bernal, M.P. Extractability, Distribution Among Different Particle Size Fractions, and Phytotoxicity of Cu and Zn in Composts Made with the Separated Solid Fraction of Pig Slurry. *Front. Sustain. Food Syst.* **2020**, *4*, 2. [CrossRef]
107. Choia, H.L. Changes of Physico-Chemical Properties of Pig Slurry During Storage. *Media Peternak.* **2011**, *34*, 179–183. [CrossRef]

environments

MDPI

Article

Enhancing Swine Wastewater Treatment: A Sustainable and Systematic Approach through Optimized Chemical Oxygen Demand/Sulfate Mass Ratio in Attached-Growth Anaerobic Bioreactor

Mehdi Lamssali [1], Shobha Mantripragada [2], Dongyang Deng [1,*] and Lifeng Zhang [2,*]

1 Department of Built Environment, College of Science and Technology, North Carolina A&T State University, Greensboro, NC 27411, USA

2 Department of Nanoengineering, Joint School of Nanoscience and Nanoengineering, North Carolina A&T State University, Greensboro, NC 27401, USA

* Correspondence: ddeng@ncat.edu (D.D.); lzhang@ncat.edu (L.Z.); Tel.: +1-336-285-3139 (D.D.); +1-336-285-2875 (L.Z.)

Abstract: The swine industry generates millions of gallons (thousands of cubic meters) of wastewater every day, posing significant environmental risk due to high concentrations of organics and nutrients. This study aims to investigate the effectiveness of attached-growth anaerobic bioreactors for treating swine wastewater by utilizing sulfate-reducing bacteria, focusing on the impact of chemical oxygen demand (COD)/sulfate mass ratios on organics degradation. A series of lab-scale anaerobic bioreactors were employed to treat swine wastewater for a 14-day period. The study evaluated changes in pH, acidity, alkalinity, COD, sulfate, and various nutrients along with total suspended solids (TSS) and volatile suspended solids (VSS) before and after treatment. At a COD/sulfate mass ratio of 2:1, the bioreactors achieved optimum removal efficiencies of 80% for TSS, 83% for VSS, 86–88% for COD, 82–87% for sulfate, 73% for sulfide, and 73% for sulfite. The nutrient removal efficiency was 67% for nitrate and 72% for nitrite. The acidity and alkalinity were effectively controlled, with alkalinity values reaching up to 2161 ± 92.5 mg/L and pH within the range of 7–7.24. The findings demonstrated that anaerobic bioreactor at a COD/sulfate mass ratio of 2:1 significantly enhanced the degradation of organic matter coupling with sulfate reduction in swine wastewater, providing an efficient and sustainable treatment method.

Keywords: swine wastewater; wastewater treatment; anaerobic bioreactor; sulfate-reducing bacteria; COD/sulfate mass ratio

1. Introduction

The swine industry, a vital component of agricultural output and economy, has experienced significant growth in recent years. Its expansion has led to a substantial increase in swine wastewater production, which has become a major environmental concern. In the United States, the swine industry generates about 245 million gallons (more than 925,000 cubic meters) of manure daily, containing high concentrations of organic matters and nutrients [1–3]. The release of untreated swine wastewater into water bodies can lead to oxygen depletion, eutrophication, and high turbidity, which causes fish kill and other ecological disturbances. Thus, it is crucial to effectively manage swine wastewater to mitigate its adverse effects on water quality and aquatic environment [4,5].

Swine wastewater is a complex mixture of organic matters, nutrients, microbes, solids, heavy metals, and antibiotics. These constituents can pose significant risks to both the environment and human health [6,7]. One effective technology for swine wastewater treatment is constructed wetlands, which utilize natural processes involving wetland vegetation, soils, and their associated microbial assemblages to treat wastewater [8]. These systems are cost-effective and require low maintenance. Advanced oxidation processes (AOPs) and membrane bioreactors (MBRs) are also employed to enhance the removal of pollutants and pathogens from swine wastewater [9,10]. AOPs use chemical oxidants to break down contaminants, while MBRs combine biological treatment with membrane filtration, offering high-quality effluent suitable for reuse. Integrating these technologies can significantly improve the efficiency of swine wastewater management, reduce greenhouse gas emissions, and promote resource recovery.

Among various technologies for wastewater treatment, anaerobic digestion stands out as a prominent method due to its effectiveness in reducing organic load and generating biogas as a renewable energy source. This process involves the microbial decomposition of organic matter in the absence of oxygen, leading to the production of methane-rich biogas and digestate, which can be used as a fertilizer [11]. A common feature of the anaerobic digestion method is the exploitation of sulfate-reducing bacteria (SRB) to render sulfate reduction to sulfide, alkalinity generation, and subsequent metal sulfide precipitation, and convert organics to bicarbonate ions instead of CO_2 and methane as opposed to traditional anerobic digestion, which is dominated by methanogens [12–14]. In anaerobic digestion system, both methanogens and SRB play critical roles in the degradation of organic matter. Methanogens produce methane as a byproduct of their metabolic processes, while SRB reduce sulfate to sulfide, utilizing organic compounds or hydrogen as electron donors. These two groups of microorganisms often compete for the same substrates. The chemical oxygen demand (COD)/sulfate mass ratio is a crucial parameter that determines which microbial community will dominate. When the COD/sulfate ratio is low, SRB tend to outcompete methanogens because they have a higher affinity for hydrogen and acetate, allowing them to effectively outcompete methanogens on these substrates [15–17]. Conversely, when the COD/sulfate ratio is high, methanogens are more likely to dominate, as the availability of organic substrates exceeds the sulfate reduction capacity.

Studies have demonstrated that SRB can effectively reduce the levels of organic pollutants, sulfates, and heavy metals in wastewater [18,19]. Along with COD/sulfate mass ratio, several factors influence the activity and efficiency of SRB in treating wastewater, including pH, temperature, and the availability of electron donors. Optimal pH levels for SRB activity typically range between 6.5 and 7.5, while temperatures between 25 °C and 37 °C are considered ideal [20,21]. Additionally, the presence of readily available organic substrates such as lactate, acetate, and ethanol can enhance the metabolic activity of SRB, leading to more efficient sulfate reduction [22]. There is an optimum COD/sulfate mass ratio for SRB in wastewater treatment to maximize treatment efficiency by ensuring sufficient organic substrate availability for SRB while maintaining an effective sulfate reduction and preventing inhibitory conditions. The optimum COD/sulfate ratio typically varies, depending on operational conditions and wastewater composition [23–25]. For example, sulfate removal efficiencies exceeding 90% were achieved in synthetic wastewater with sulfate concentrations up to 1965 mg/L and COD/sulfate ratios higher than 1.7 [24]. Despite their potential, the application of SRB in wastewater treatment also has challenges. The production of hydrogen sulfide, a toxic and odorous gas, necessitates careful management and control to prevent environmental and health hazards [26]. Deng and Lin have utilized ferrous chloride to take out sulfide generated from the sulfate reduction process and form ferrous sulfide precipitates to reduce the sulfide toxicity [27].

In spite of a number of reports in literature exploring SRB and COD/sulfate ratio for wastewater treatment, there is limited research in this regard for swine wastewater. Given the sulfate prevalence and the large amount of readily available organics in swine wastewater [28], a similar SRB mechanism may be effectively applied to swine wastewater treatment. The coupling of sulfate with organic degradation could reduce organics to bicarbonate, thus adding alkalinity to the system instead of producing greenhouse gases [28–30]. For the first time, this study aimed to investigate the effectiveness of treating swine wastewater using SRB in the anaerobic attached-growth bioreactor and explore the impact of COD/sulfate mass ratio on organics degradation and sulfate reduction. A series of lab-scale anaerobic bioreactors were employed and monitored to evaluate pH change, alkalinity production, and COD, sulfate, and nutrient reduction in swine wastewater after treatment. Our results revealed that at an optimum COD/sulfate ratio of 2:1, the bioreactors achieved significant contaminant and nutrient removal efficiencies for swine wastewater. This research demonstrated the great potential of optimizing COD/sulfate mass ratio in the attached-growth anaerobic bioreactors for swine wastewater treatment. Not only can such a treatment be environmentally friendly but also energy efficient because it does not require aeration for the microbial oxidation of organics in the swine wastewater. Additional benefits may include reduced CO_2 and methane production and emission, the removal of selected heavy metals (e.g., Cd, Cu, Hg, Pb and Zn) (if present) due to their low solubility with sulfide, and low biomass production. The findings from this study provided a holistic approach (as illustrated in Figure 1) to treat swine wastewater and contributed to the knowledge of treating high-strength agricultural wastewaters, where efficient reductions in organics and nutrients are required to meet environmental discharge standards.

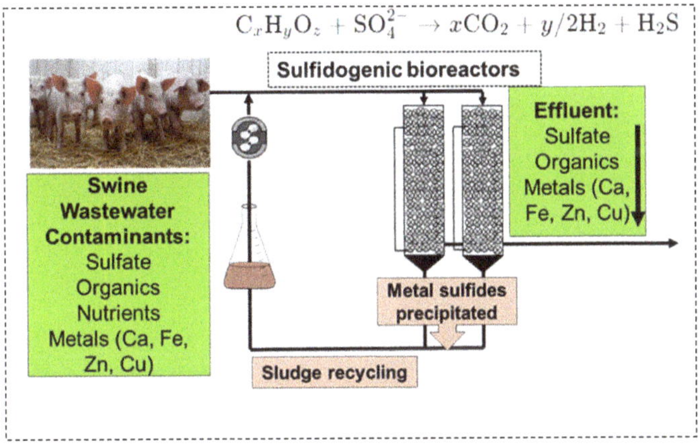

Figure 1. Schematic diagram of anaerobic attached-growth bioreactor system for swine wastewater treatment utilizing sulfate-reducing bacteria (SRB).

2. Material and Methods

2.1. Anaerobic Bioreactors Setup

Four lab-scale anaerobic bioreactors were established to investigate the removal of chemical oxygen demand (COD), sulfate, and nutrients from swine wastewater. The control bioreactor contained 2000 mL of sterile swine wastewater, collected from the swine farm operation facility at North Carolina Agricultural and Technical State University. In contrast, each of the three experimental bioreactors consisted of 1950 mL of swine wastewater and 50 mL of anaerobic digester sludge, sourced from a municipal wastewater treatment plant as the bacteria inoculum. Throughout the 14-day experimental period, all bioreactors were continuously stirred and anaerobic conditions were maintained via N_2 sparging.

Potassium hydrogen phthalate ($C_8H_5KO_4$) and sodium sulfate (Na_2SO_4), sourced from Fisher Scientific, were used to manipulate the COD/sulfate mass ratios in the bioreactors when necessary. The bioreactors were kept at a consistent temperature of 28 $\pm$ 1 °C [31,32]. To maintain anaerobic conditions, the airtight bioreactors were periodically sparged with syringe-filtered N_2 gas before and during operation.

Each bioreactor was filled with 50 pieces of Kaldnes K1 plastic media (Evolution Aqua Ltd., Wigan, UK) to support the growth of attached anaerobic microorganisms. The unique design of the K1 media provides an extensive surface area (approximately 500 m^2/m^3) for improving microbial colonization, enhancing biomass retention, increasing resistance to operation condition changes, and higher contact surface between the substrate and microorganisms [32,33].

2.2. Operation

To evaluate the effect of the COD/sulfate mass ratios on swine wastewater treatment, swine wastewater samples were collected from three separate wastewater collection trips, and the three bioreactors were individually operated based on the water collected from each trip. The results from each reactor represent distinct experimental runs (reactor 1 refers to trip 1, reactor 2 refers to trip 2, and reactor 3 refers to trip 3). The bioreactors were operated in batch mode at four different COD/sulfate mass ratios: 6:1, 4:1, 2:1, and 1:1. Each ratio was maintained to ensure consistent interactions between the microorganisms and the swine wastewater.

Operational conditions, such as temperature, were regularly monitored to maintain consistency. A daily 50 mL sample was collected from each bioreactor for analysis purposes. This sample volume was selected to ensure comprehensive data collection throughout the study while minimizing disturbance to the reactor environment. The samples were rigorously tested to measure pH, acidity, alkalinity, COD, sulfate, nutrient, and solid levels, providing insights into the performance of the bioreactors at each COD/sulfate mass ratio.

2.3. Measurement

Acidity and alkalinity were measured using an Auto titrator (Thermo Scientific Orion 950, Thermo Fisher Scientific, Waltham, MA, USA) following Standard Methods for the Examination of Water and Wastewater 2320 B and 2310 B, respectively [34]. pH was determined with a pre-calibrated pH meter (Mettler Toledo EasyPlus Easy pH Acid/Base Titrator). Total solids (TS), total suspended solids (TSS), total dissolved solids (TDS), and volatile suspended solids (VSS) were measured using Standard Methods for the Examination of Water and Wastewater [34]. COD concentrations were assessed using a closed reflux, colorimetric method with a Hach DR-3900 spectrophotometer [35]. Nutrient levels, including phosphorus (PO_4^{3-}), nitrogen-ammonia (NH_3-N), nitrite (NO_{2}-), nitrate (NO_{3}-), and various sulfur species including sulfate (SO_4^{2-}), sulfide (S^{2-}), and sulfite (SO_3^{2-}), were determined using a Hach (Loveland, CO, USA) DR-3900 spectrophotometer in accordance with Hach methods 8180, 10205, 10019, 8192, 8051, 8131, and 10308, respectively.

A scanning electron microscope (SEM, Auriga, Carl Zeiss, White Plains, NY, USA) was employed to examine microorganism morphology and accumulation on the K1 plastic media. For this purpose, one piece of K1 media was removed from each bioreactor at the beginning (day 0) and the end (day 14) of the experiment. The samples were then dried and prepared for SEM analysis to compare the surface morphology over time. To reduce surface charging during SEM, all samples were coated with gold–palladium [36].

3. Results and Discussion

3.1. Water Quality of Initial Swine Wastewater

The initial values of pH, acidity, alkalinity, nutrients, sulfate, and COD along with solids concentration of the collected swine wastewater are presented in Table 1.

Table 1. The initial water quality data of swine wastewater.

Parameter	Average ± SD
pH	7.1 ± 0.04
Alkalinity (mg/L)	1620.3 ± 9.4
Acidity (mg/L)	245 ± 6.2
Nitrite NO_2^-–N (mg/L)	0.14 ± 0.005
Nitrate NO_3^-–N (mg/L)	1.3 ± 0.10
Phosphorus PO_4^{3-} (mg/L)	284.6 ± 7.3
Nitrogen, Ammonia NH_3–N (mg/L)	255.4 ± 1.5
Sulfide S^{2-} (μg/L)	137 ± 2.7
Sulfite SO_3^{2-} (mg/L)	1.0 ± 0.004
Sulfate SO_4^{2-} (mg/L)	133.1 ± 1.9

Table 1. *Cont.*

Parameter	Average ± SD
COD (mg/L)	953.9 ± 0.7
TSS (mg/L)	1195 ± 8.5
TDS (mg/L)	2011.8 ± 5.1
VSS (mg/L)	804.5 ± 3.3

3.2. Swine Wastewater Treatment

3.2.1. Solids Concentration (TSS, VSS, and TDS)

The efficacy of anaerobic bioreactor treatment is strongly influenced by the COD/sulfate mass ratios, as evidenced by the variable removal rates of TSS and VSS in this study (Figure 2). At a COD/sulfate ratio of 6:1, the average TSS and VSS removal efficiencies were 29% and 25%, respectively. The TSS and VSS removal efficiencies increased to 48% and 37%, respectively, at a COD/sulfate ratio of 4:1. The removal efficiencies further improved to 80% and 83%, respectively, achieving their highest levels at a COD-to-sulfate ratio of 2:1, indicating an optimum mass ratio for solids management. TSS, encompassing both organic and inorganic particles like metal hydroxides, decreased significantly as the majority of organic matter was degraded and metals were precipitated with hydroxide ions that were generated from the hydrolysis of produced bicarbonate ions. Consequently, this reduction in organics (including biomass) also led to a decrease in VSS. There was a noticeable decline in the average TSS and VSS removal efficiencies to 65% and 48%, respectively, at a COD/sulfate ratio of 1:1 compared to 2:1. The decreased solids removal efficiencies at ratios other than 2:1 can be attributed to an imbalance between sulfate and COD. An excess of sulfate relative to COD or an excess of organics can hinder the optimal growth of SRB, resulting in reduced organic degradation and metal precipitation [37,38].

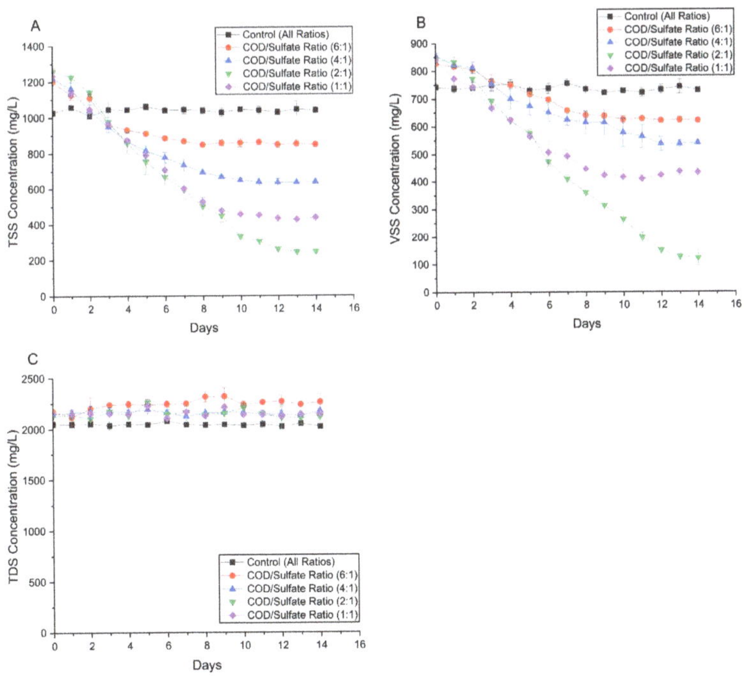

Figure 2. Profiles of total suspended solids (TSS) (**A**), volatile suspended solids (VSS) (**B**), and total dissolved solids (TDS) (**C**) in the bioreactors at various COD/sulfate ratios across the treatment period of 14 days.

While the removal efficiencies of TSS and VSS showed significant variation upon the change in COD/sulfate ratios, TDS remained relatively constant across all the bioreactors with negligible removal efficiencies observed (Figure 2). This observation suggests that TDS components, likely comprising inorganic salts, minerals and other small particles, are less amenable to the biological treatment processes presented in this study. The observed stability of TDS concentrations across different COD/sulfate ratios is consistent with the principle that anaerobic processes predominantly degrade biodegradable organic and inorganic matter, exerting minimal influence on dissolved inorganic substances [39].

Furthermore, the control reactor showed no substantial degradation of TSS, TDS, or VSS, irrespective of the COD/sulfate ratios. This indicated the observed removal of solids in the bioreactors was primarily a result of biological degradation, rather than physical or chemical pathways. The lack of degradation within the control reactor reinforced the importance of biological activity to achieve a reduction in solids in swine wastewater and the necessity of maintaining an appropriate balance of bacteria, sulfate, nutrients, and electron acceptors to stimulate the microbial consortia responsible for the anerobic bioreactor treatment process [40].

3.2.2. pH, Acidity, and Alkalinity

In the context of anaerobic bioreactor, pH, acidity, and alkalinity stand as important parameters that can profoundly impact system performance and stability. SRB thrive in slightly acidic to neutral pH environments, typically between pH 6.0 and 8.0. Outside this range, their metabolic activities can be inhibited [16]. The pH, acidity, and alkalinity of all the bioreactors under different COD/sulfate conditions were monitored in this research (Figure 3).

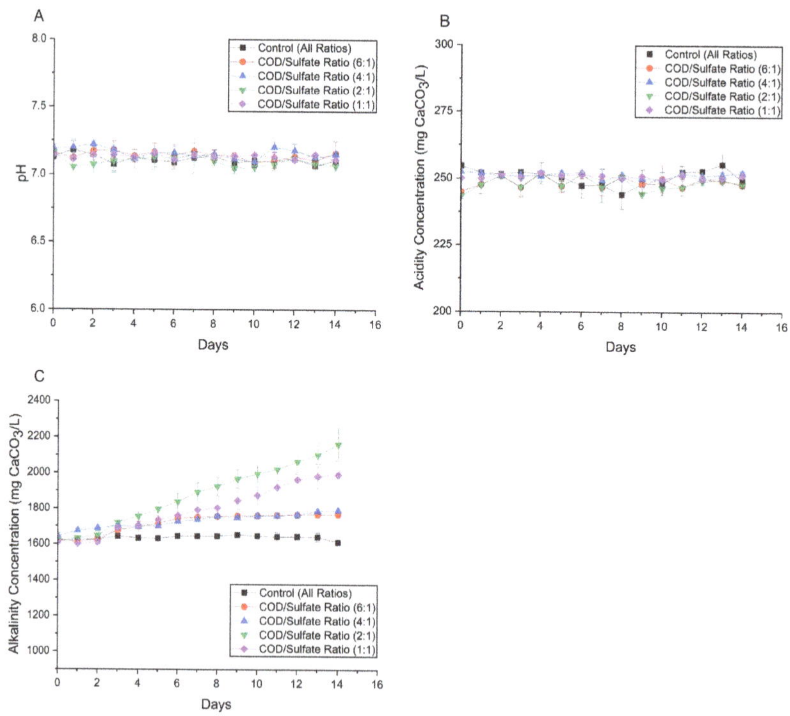

Figure 3. Profiles of pH (**A**), acidity (**B**), and alkalinity (**C**) in the bioreactors at various chemical oxygen demand (COD)/sulfate ratios across the treatment period of 14 days.

Regardless of the COD/sulfate ratios employed, the pH in the bioreactors remained within the range of 7–7.24. It is reported that this pH range supports the stability and activity of the enzymes involved in sulfate reduction, promoting efficient microbial growth and function [41,42]. The pH in our bioreactors indicated that SRB can effectively carry out their metabolic processes, including the reduction of sulfate to sulfide.

The consistent acidity results, irrespective of the COD/sulfate ratios, further validated the robustness of the sulfate reduction, dominating the anaerobic bioreactor environment [43,44]. While the control reactors exhibited mostly stable alkalinity profiles, bioreactors R1–R3 showed consistent increases in alkalinity, particularly at COD/sulfate ratios of 2:1 and 1:1 (Figure 3). These rises in alkalinity peaked at 2161.1 ± 92.5 mg/L and 2349.1 ± 17 mg/L for the respective ratios, which can be attributed to the SRB's metabolic processes. When SRB reduce sulfate (SO_4^{2-}) to sulfide (S^{2-}), they consume hydrogen ions (H^+) in the process. This consumption of H^+ ions increases the pH and contributes to alkalinity. Additionally, the production of bicarbonate (HCO_3^-) as a byproduct of SRB metabolism also adds to the alkalinity of the environment [13,45].

Overall, the bioreactors exhibited a stable and efficient treatment performance, as indicated by consistent pH and acidity values as well as an alkalinity increase, and reached the optimum level at a COD/sulfate ratio of 2:1.

3.2.3. COD and Sulfate Removal

In this study, the variation in the COD/sulfate mass ratios presented significant changes in the removal efficiencies of COD and sulfate. At a COD/sulfate ratio of 6:1, the removal efficiencies for both COD and sulfate were around 60% in bioreactors R1–R3 (Figure 4). As the ratio increased to 4:1, there was a marked improvement in the bioreactors' removal efficiencies of COD and sulfate with both exceeding 70%. The zenith of bioreactor performance was observed at a 2:1 COD/sulfate ratio. Both COD and sulfate removal efficiencies ranged from 81% to 86% across reactors R1–R3. A shift to the 1:1 ratio showed a slightly lower but still significant COD and sulfate removal efficiency in the mid-to-high range (70%).

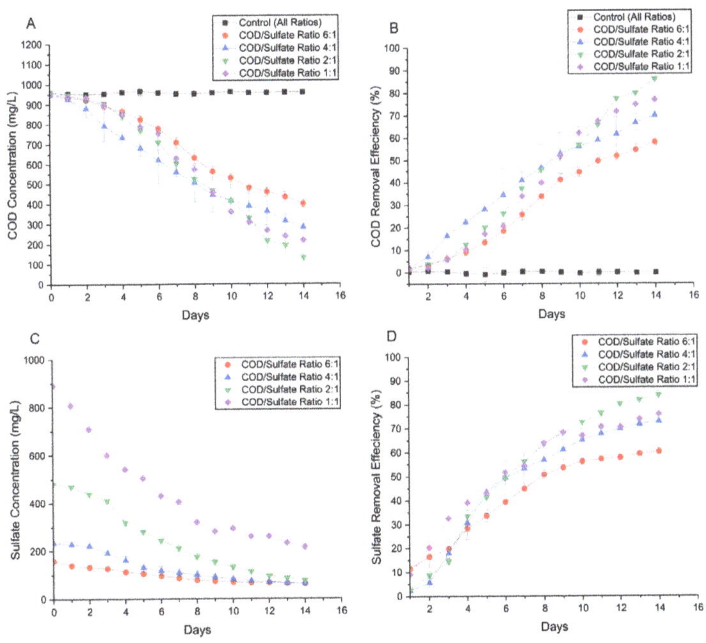

Figure 4. Profiles of COD and sulfate concentrations (**A**,**C**) and their removal efficiencies (**B**,**D**) in the bioreactors at various COD/sulfate ratios across the treatment period of 14 days.

It is reported that the COD/sulfate mass ratio significantly influences the activity of SRB [13]. Omil et al. reported that a COD/sulfate ratio of 2:1 resulted in the highest sulfate reduction efficiency and COD removal rates in anaerobic bioreactors treating industrial wastewater [18]. Lens et al. found that a COD/sulfate ratio between 1.8:1 and 2.2:1 provided optimal conditions for SRB activity, leading to effective sulfate reduction and organic matter degradation [46]. In another study, Stefanie et al. suggested that a COD/sulfate ratio of approximately 1.5:1 to 2:1 supported balanced microbial growth and efficient sulfate reduction in sulfate-reducing bioreactors [47]. Our results are consistent with previous studies showing that when the COD/sulfate ratio was maintained at 2, sulfate reduction processes were maximized, and the production of methane was substantially minimized [48,49]. This ratio represented the balance between available organic matter (COD) and sulfate, which SRB used as an electron acceptor in their metabolic processes. This optimal ratio ensured that SRB had sufficient organic substrate to thrive and effectively reduce sulfate. At this ratio, the sulfate presence was sufficient to serve as the primary electron acceptor, and SRB could efficiently utilize the available organic matter. This reduced the substrates available for methanogens, thereby minimizing methane production. The amount of sulfate that was reduced in our processing also proved the dominance of SRB.

3.2.4. Nutrient Degradation and Reduced Sulfur Species

At the COD/sulfate ratio of 6:1, across the bioreactors R1–R3, the average removal efficiencies for sulfide, sulfite, nitrate, and nitrite reached 39%, 43%, 38%, and 52%, respectively (Figure 5). With the COD/sulfate ratio of 4:1, the bioreactors exhibited enhanced nutrient removal. The observed efficiencies for sulfide, sulfite, nitrate, and nitrite were 46%, 55%, 53%, and 62%, respectively. When the COD/sulfate ratio was adjusted to 2:1, significant removal efficiencies were noted; i.e., removal efficiencies of sulfide and sulfite both reached 73%, while nitrate and nitrite degradation reached 67% and 72%, respectively. Interestingly, at the COD/sulfate of 1:1, while the sulfide and sulfite removal efficiencies remained high at 65% and 66%, nitrate and nitrite removal efficiencies declined to 57% and 46%, respectively.

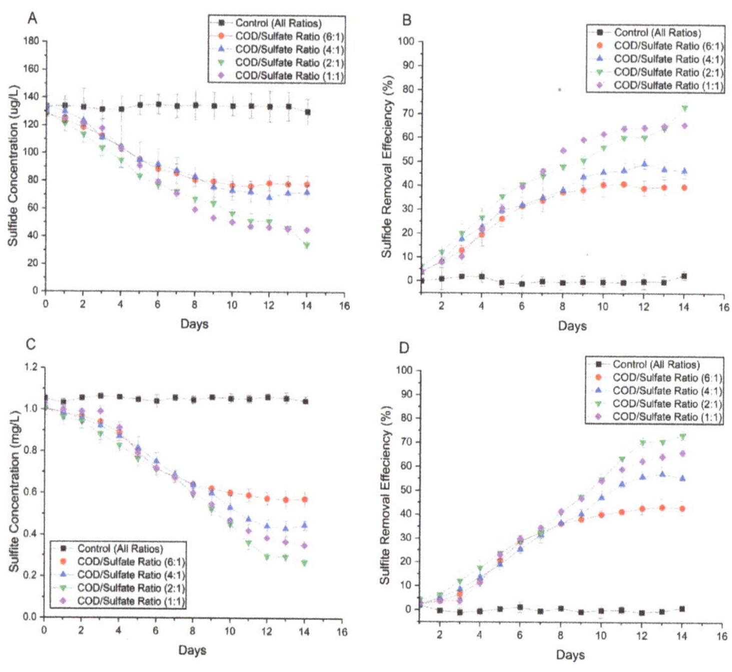

Figure 5. *Cont.*

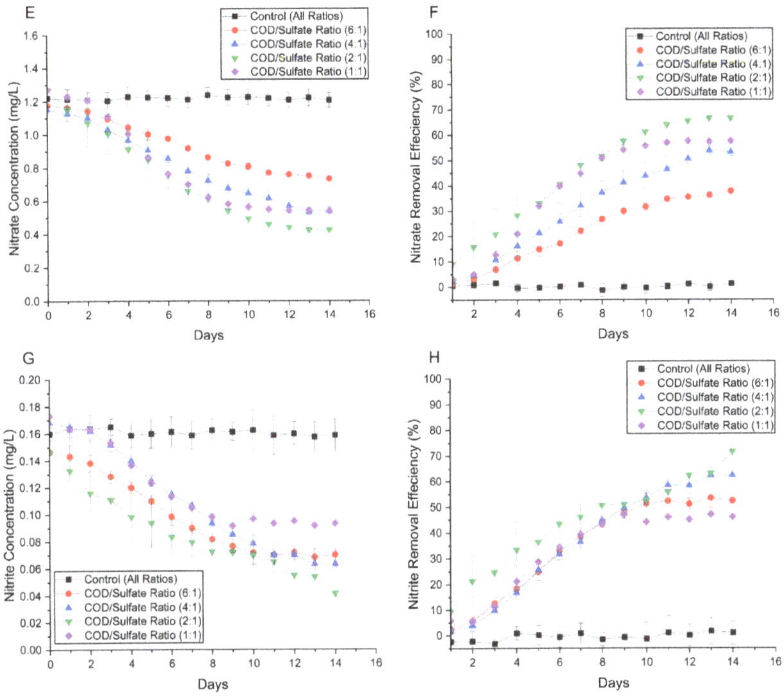

Figure 5. Profiles of sulfide, sulfite, nitrate, nitrite concentrations (**A**,**C**,**E**,**G**) and their removal efficiencies (**B**,**D**,**F**,**H**) in the bioreactors at various COD/sulfate ratios across the treatment period of 14 days.

The reduction of nitrate, nitrite, sulfide, and sulfite in swine wastewater could be attributed to the complex action of the overall microbial community in the bioreactors. It is known that the reduction of nitrate and nitrite in an anaerobic bioreactor with dominant SRB is a complex process, which can be influenced by microbial competition, environmental factors, and the metabolic versatility of the microbial community [16]. Nitrate and nitrite reduction can occur through the activities of denitrifying bacteria, nitrate-reducing SRB, and even SRB under certain conditions [50]. The coexistence of these microbial populations is possible and often observed, contributing to the overall stability and efficiency of anaerobic wastewater treatment systems. Sulfide and sulfite reduction can be a result of sulfate reduction process. In anaerobic digestion reactors, both sulfite and sulfide can be reduced due to the presence of SRB [42]. These bacteria utilize organic compounds as electron donors to reduce sulfate to sulfide, while sulfite, being an intermediate in the sulfate reduction pathway, can also be directly reduced by these bacteria under anaerobic conditions [16]. The simultaneous reduction of sulfite and sulfide occurs as part of the complex biochemical processes within the reactor, contributing to the breakdown of organic matter and the production of hydrogen sulfide (H_2S). Hydrogen sulfide remaining in the treated wastewater can be precipitated through the addition of ferrous chloride ($FeCl_2$) to form ferrous sulfide (FeS) to eliminate H_2S toxicity in the water [27].

Across all COD/sulfate ratios, phosphorus and nitrogen-ammonia showed limited degradation (Figure 6). The phosphorous removal hovered around 3.8–7.7%, and the removal of nitrogen-ammonia was between 5.8 and 10.9%. Despite the evident relationship between COD/sulfate ratio and sulfur-based or nitrogenous compounds, phosphorous and nitrogen-ammonia compounds remained largely unaltered within the studied range of COD/sulfate ratios. Such observations are consistent with previous reports [51,52], in which limited removal efficiencies for phosphorous and nitrogen-ammonia were observed during the wastewater treatment processes. These findings suggested that phosphorous and nitrogen-ammonia degradation depended on different microbial or physicochemical processes, which were not present in our bioreactors, and thus, alternative post-treatment methods may be required for the effective removal of these contaminants [53–55].

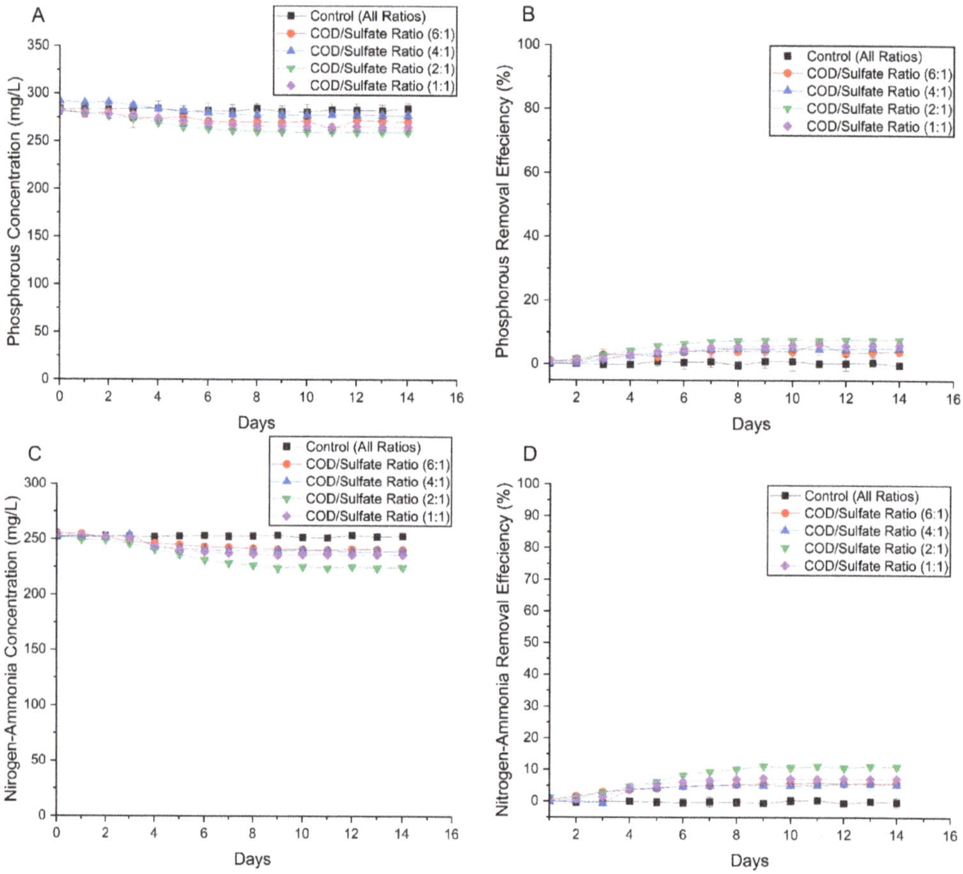

Figure 6. Profiles of phosphorous and N-ammonia concentrations (**A**,**C**) and their removal efficiencies (**B**,**D**) in the bioreactors at various COD/sulfate ratios across the treatment period of 14 days.

3.3. Solid Characterization by Scanning Electron Microscopy (SEM)

SEM was conducted to examine the change in the bioreactor sludge before and after treatment at different COD/sulfate ratios in order to obtain the microbial evolution information for a deeper understanding of the COD/sulfate mass ratios effect (Figure 7).

At a COD/sulfate ratio of 6:1, there was no significant presence of anaerobic bacteria on the sludge. At a COD/sulfate ratio of 4:1 and 1:1, there was only a marginal formation of anaerobic bacteria. This is consistent with the lower removal efficiencies as observed for TSS and VSS at these COD/sulfate ratios. The limited bacterial presence may be indicative of an environment not fully conducive to the proliferation of anaerobic bacterial colonies, particularly those of SRB and acidogenic bacteria, which are crucial for initiating the degradation process in anaerobic systems [56–58].

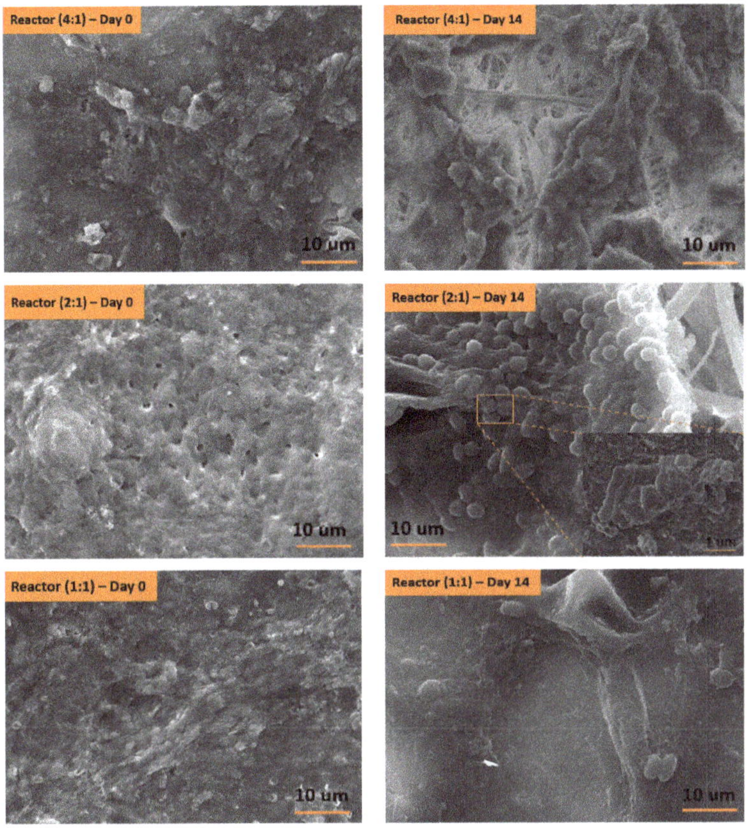

Figure 7. SEM images of sludge surface before (day 0) and after (day 14) bioreactor treatment at different COD/sulfate ratios (showed in bracket).

Conversely, at COD/sulfate ratio of 2:1, the SEM image of the sludge surface demonstrated a robust presence of anaerobic bacteria, a substantial biofilm development (Figure 7: reactor (2:1)–Day 14). Biofilm formation is a hallmark of healthy anaerobic bacterial activity and is known to enhance the degradation of multiple waste components [59–61]. The rod-shaped bacteria observed were consistent with known species such as *Clostridium spp.* and *Desulfovibrio spp.*, which are recognized for their roles in acidogenesis and sulfate reduction, respectively [57,62–65].

These SEM results are consistent with the removal efficiencies of TSS and VSS, suggesting that the COD/sulfate ratio of 2:1 is an optimal environment for the growth of sulfate reducers. The correlation between the microbial presence from the SEM examination and the anaerobic treatment efficiency underscores the need for tailoring the COD/sulfate balance to support the microbial community that is necessary for efficient waste degradation.

4. Conclusions

This study demonstrated the efficacy of treating swine wastewater using anaerobic bioreactors, with emphasis on the crucial role of the chemical oxygen demand (COD)/sulfate mass ratio in the process of treatment. By employing a COD/sulfate ratio of 2:1, the bioreactors achieved significant contaminant removal efficiencies: 86–88% for COD, 82–87% for sulfate, 80% for total suspended solids (TSS), and 83% for volatile suspended solids (VSS). Additionally, the following nutrient removal efficiencies were recorded: 73% for sulfide, 73% for sulfite, 67% for nitrate, and 72% for nitrite. These results highlighted that an optimal COD/sulfate ratio could enhance overall swine wastewater treatment efficiency by promoting a balanced microbial ecosystem, fostering synergistic interactions between sulfate-reducing bacteria (SRB) and methanogens. Scanning electron microscopy (SEM) further validated these findings by showing substantial biofilm development and active anaerobic bacterial presence at the optimal COD/sulfate ratio.

Overall, this research provided valuable insights into optimizing anerobic bioreactor processes for efficient and environmentally friendly treatment of high-strength wastewaters from swine farms, paving the way for more effective and sustainable management of swine wastewater, which is crucial to meet environmental discharge standards and promoting eco-friendly agricultural practices.

Supplementary Materials: The following supporting information can be downloaded at: https://www.mdpi.com/article/10.3390/environments11080162/s1.

Author Contributions: Conceptualization: D.D. and L.Z.; data curation: M.L.; formal analysis: M.L. and S.M.; funding acquisition and resources: D.D. and L.Z.; investigation and validation: M.L. and S.M.; methodology: D.D. and M.L.; project administration and supervision: D.D. and L.Z.; writing—original draft: M.L. and S.M.; writing—review and editing: L.Z. and D.D. All authors have read and agreed to the published version of the manuscript.

Funding: This research received no external funding.

Data Availability Statement: The original contributions presented in the study are included in the article/Supplementary Material, further inquiries can be directed to the corresponding authors.

Conflicts of Interest: The authors declare no conflicts of interest.

References

1. Venslauskas, K.; Navickas, K.; Rubežius, M.; Tilvikienė, V.; Supronienė, S.; Doyeni, M.O.; Barčauskaitė, K.; Bakšinskaitė, A.; Bunevičienė, K. Environmental impact assessment of sustainable pig farm via management of nutrient and co-product flows in the farm. *Agronomy* **2022**, *12*, 760. [CrossRef]
2. Sharara, M.; Kim, D.; Sadaka, S.; Thoma, G. Consequential life cycle assessment of swine manure management within a thermal gasification scenario. *Energies* **2019**, *12*, 4081. [CrossRef]
3. Deviney, A.; Classen, J.; Bruce, J.; Sharara, M. Sustainable swine manure management: A tale of two agreements. *Sustainability* **2020**, *13*, 15. [CrossRef]
4. Burkholder, J.; Libra, B.; Weyer, P.; Heathcote, S.; Kolpin, D.; Thorne, P.S.; Wichman, M. Impacts of waste from concentrated animal feeding operations on water quality. *Environ. Health Perspect.* **2007**, *115*, 308–312. [CrossRef]
5. Varma, V.S.; Parajuli, R.; Scott, E.; Canter, T.; Lim, T.T.; Popp, J.; Thoma, G. Dairy and swine manure management—Challenges and perspectives for sustainable treatment technology. *Sci. Total Environ.* **2021**, *778*, 146319. [CrossRef] [PubMed]
6. Cole, D.; Todd, L.; Wing, S. Concentrated swine feeding operations and public health: A review of occupational and community health effects. *Environ. Health Perspect.* **2000**, *108*, 685–699. [CrossRef] [PubMed]
7. Rosa, D.M.; Sampaio, S.C.; Pereira, P.A.; Mauli, M.M.; Reis, R.R.D. Swine wastewater: Impacts on soil, plant, and leachate. *Eng. Agrícola* **2017**, *37*, 928–939. [CrossRef]
8. Vymazal, J. Constructed wetlands for wastewater treatment: Five decades of experience. *Environ. Sci. Technol.* **2011**, *45*, 61–69. [CrossRef]
9. Oturan, M.A.; Aaron, J.-J. Advanced oxidation processes in water/wastewater treatment: Principles and applications. A review. *Crit. Rev. Environ. Sci. Technol.* **2014**, *44*, 2577–2641. [CrossRef]
10. Judd, S. *The MBR Book: Principles and Applications of Membrane Bioreactors for Water and Wastewater Treatment*; Elsevier: Amsterdam, The Netherlands, 2010.
11. Lourinho, G.; Rodrigues, L.; Brito, P. Recent advances on anaerobic digestion of swine wastewater. *Int. J. Environ. Sci. Technol.* **2020**, *17*, 4917–4938. [CrossRef]
12. Hao, O.J.; Chen, J.M.; Huang, L.; Buglass, R.L. Sulfate-reducing bacteria. *Crit. Rev. Environ. Sci. Technol.* **1996**, *26*, 155–187. [CrossRef]
13. Zhang, Z.; Zhang, C.; Yang, Y.; Zhang, Z.; Tang, Y.; Su, P.; Lin, Z. A review of sulfate-reducing bacteria: Metabolism, influencing factors and application in wastewater treatment. *J. Clean. Prod.* **2022**, *376*, 134109. [CrossRef]
14. Bhattacharya, S.; Uberoi, V.; Dronamraju, M. Interaction between acetate fed sulfate reducers and methanogens. *Water Res.* **1996**, *30*, 2239–2246. [CrossRef]
15. Lens, P.; Kuenen, J.G. The biological sulfur cycle: Novel opportunities for environmental biotechnology. *Water Sci. Technol.* **2001**, *44*, 57–66. [CrossRef] [PubMed]

16. Muyzer, G.; Stams, A.J. The ecology and biotechnology of sulphate-reducing bacteria. *Nat. Rev. Microbiol.* **2008**, *6*, 441–454. [CrossRef] [PubMed]
17. Zinder, S.H. Physiological ecology of methanogens. In *Methanogenesis: Ecology, Physiology, Biochemistry & Genetics*; Ferry, J.G., Ed.; Springer: Boston, MA, USA, 1993; pp. 128–206.
18. Omil, F.; Lens, P.; Visser, A.; Pol, L.H.; Lettinga, G. Long-term competition between sulfate reducing and methanogenic bacteria in UASB reactors treating volatile fatty acids. *Biotechnol. Bioeng.* **1998**, *57*, 676–685. [CrossRef]
19. Xu, Y.-N.; Chen, Y. Advances in heavy metal removal by sulfate-reducing bacteria. *Water Sci. Technol.* **2020**, *81*, 1797–1827. [CrossRef] [PubMed]
20. Kushkevych, I. Isolation and purification of sulfate-reducing bacteria. In *Microorganisms*; Blumenberg, M., Shaaban, M., Elgaml, A., Eds.; IntechOpen: London, UK, 2020.
21. Lens, P.; Visser, A.; Janssen, A.; Pol, L.H.; Lettinga, G. Biotechnological treatment of sulfate-rich wastewaters. *Crit. Rev. Environ. Sci. Technol.* **1998**, *28*, 41–88. [CrossRef]
22. Rabus, R.; Hansen, T.A.; Widdel, F. Dissimilatory sulfate-and sulfur-reducing prokaryotes. *Prokaryotes* **2006**, *2*, 659–768.
23. Li, J.; Li, A.; Li, Y.; Cai, M.; Luo, G.; Wu, Y.; Tian, Y.; Xing, L.; Zhang, Q. PICRUSt2 functionally predicts organic compounds degradation and sulfate reduction pathways in an acidogenic bioreactor. *Front. Environ. Sci. Eng.* **2022**, *16*, 47. [CrossRef]
24. Damianovic, M.H.R.Z.; Foresti, E. Anaerobic degradation of synthetic wastewaters at different levels of sulfate and COD/sulfate ratios in horizontal-flow anaerobic reactors (HAIB). *Environ. Eng. Sci.* **2007**, *24*, 383–393. [CrossRef]
25. Barbosa, L.d.P.; Bertolino, S.; Freitas, P.; Oliveira, V.; Pina, P.D.; Leão, V.; Teixeira, M. Effects of different COD/sulfate ratios on the growth of metal tolerant sulfate reducing bacteria (SRB). *Adv. Mater. Res.* **2009**, *71*, 569–572. [CrossRef]
26. Kushkevych, I.; Dordević, D.; Vítězová, M. Toxicity of hydrogen sulfide toward sulfate-reducing bacteria *Desulfovibrio piger* Vib-7. *Arch. Microbiol.* **2019**, *201*, 389–397. [CrossRef] [PubMed]
27. Deng, D.; Lin, L.-S. Continuous sulfidogenic wastewater treatment with iron sulfide sludge oxidation and recycle. *Water Res.* **2017**, *114*, 210–217. [CrossRef] [PubMed]
28. Villamar, C.A.; Cañuta, T.; Belmonte, M.; Vidal, G. Characterization of swine wastewater by toxicity identification evaluation methodology (TIE). *Water Air Soil Pollut.* **2012**, *223*, 363–369. [CrossRef]
29. Deng, D.; Weidhaas, J.L.; Lin, L.-S. Kinetics and microbial ecology of batch sulfidogenic bioreactors for co-treatment of municipal wastewater and acid mine drainage. *J. Hazard. Mater.* **2016**, *305*, 200–208. [CrossRef]
30. Deng, D.; Lin, L.-S. Two-stage combined treatment of acid mine drainage and municipal wastewater. *Water Sci. Technol.* **2013**, *67*, 1000–1007. [CrossRef] [PubMed]
31. Barton, L.L.; Hamilton, W.A. *Sulphate-Reducing Bacteria: Environmental and Engineered Systems*; Cambridge University Press: Cambridge, UK, 2007.
32. Hu, B.; Wheatley, A.; Ishtchenko, V.; Huddersman, K. Performance linked to residence time distribution by a novel wool-based bioreactor for tertiary sewage treatment. *Appl. Microbiol. Biotechnol.* **2012**, *94*, 817–828. [CrossRef] [PubMed]
33. Cordone, L.; Carlson, C.; Plaehn, W.; Shangraw, T.; Wilmoth, D. Case Study and Retrospective: Aerobic Fixed Film Biological Treatment Process for 1, 4-Dioxane at the Lowry Landfill Superfund Site. *Remediat. J.* **2016**, *27*, 159–172. [CrossRef]
34. Rice, E.W.; Bridgewater, L.; Association, A.P.H. *Standard Methods for the Examination of Water and Wastewater*; American Public Health Association: Washington, DC, USA, 2012; Volume 10.
35. *ASTM D1252*; Standard Test Methods for Chemical Oxygen Demand (Dichromate Oxygen Demand) of Water. ASTM International: West Conshohocken, PA, USA, 2012.
36. Mohammed, A.; Abdullah, A. Scanning electron microscopy (SEM): A review. In Proceedings of the 2018 International Conference on Hydraulics and Pneumatics—HERVEX, Băile Govora, Romania, 7–9 November 2018; pp. 7–9.
37. Omil, F.; Lens, P.; Pol, L.H.; Lettinga, G. Effect of upward velocity and sulphide concentration on volatile fatty acid degradation in a sulphidogenic granular sludge reactor. *Process Biochem.* **1996**, *31*, 699–710. [CrossRef]
38. Rinzema, A.; Boone, M.; van Knippenberg, K.; Lettinga, G. Bactericidal effect of long chain fatty acids in anaerobic digestion. *Water Environ. Res.* **1994**, *66*, 40–49. [CrossRef]
39. Ahmed, M.; Lin, L.-S. Ferric reduction in organic matter oxidation and its applicability for anaerobic wastewater treatment: A review and future aspects. *Rev. Environ. Sci. BioTechnol.* **2017**, *16*, 273–287. [CrossRef]
40. Shrestha, S.; Pandey, R.; Aryal, N.; Lohani, S.P. Recent advances in co-digestion conjugates for anaerobic digestion of food waste. *J. Environ. Manag.* **2023**, *345*, 118785. [CrossRef] [PubMed]
41. Gerardi, M.H. *The Microbiology of Anaerobic Digesters*; John Wiley & Sons: Hoboken, NJ, USA, 2003.
42. Postgate, J.R. *The Sulphate-Reducing Bacteria*; CUP Archive: Cambridge, UK, 1979.
43. Appels, L.; Baeyens, J.; Degrève, J.; Dewil, R. Principles and potential of the anaerobic digestion of waste-activated sludge. *Prog. Energy Combust. Sci.* **2008**, *34*, 755–781. [CrossRef]
44. Bella, K.; Rao, P.V. Anaerobic digestion of dairy wastewater: Effect of different parameters and co-digestion options—A review. *Biomass Convers. Biorefinery* **2023**, *13*, 2527–2552. [CrossRef]
45. Diao, C.; Ye, W.; Yan, J.; Hao, T.; Huang, L.; Chen, Y.; Long, J.; Xiao, T.; Zhang, H. Application of microbial sulfate-reduction process for sulfate-laden wastewater treatment: A review. *J. Water Process Eng.* **2023**, *52*, 103537. [CrossRef]
46. Lens, P.; Pol, L.H. *Environmental Technologies to Treat Sulfur Pollution*; IWA publishing: London, UK, 2000.

47. Stefanie, J.O.E.; Visser, A.; Pol, L.W.H.; Stams, A.J. Sulfate reduction in methanogenic bioreactors. *FEMS Microbiol. Rev.* **1994**, *15*, 119–136.
48. O'Flaherty, V.; Mahony, T.; O'Kennedy, R.; Colleran, E. Effect of pH on growth kinetics and sulphide toxicity thresholds of a range of methanogenic, syntrophic and sulphate-reducing bacteria. *Process Biochem.* **1998**, *33*, 555–569. [CrossRef]
49. Spanjers, H.; Weijma, J.; Abusam, A. Modelling the competition between sulphate reducers and methanogens in a thermophilic methanol-fed bioreactor. *Water Sci. Technol.* **2002**, *45*, 93–98. [CrossRef]
50. Reis, M.; Almeida, J.; Lemos, P.; Carrondo, M. Effect of hydrogen sulfide on growth of sulfate reducing bacteria. *Biotechnol. Bioeng.* **1992**, *40*, 593–600. [CrossRef] [PubMed]
51. Poh, P.; Gouwanda, D.; Mohan, Y.; Gopalai, A.; Tan, H. Optimization of wastewater anaerobic digestion using mechanistic and meta-heuristic methods: Current limitations and future opportunities. *Water Conserv. Sci. Eng.* **2016**, *1*, 1–20. [CrossRef]
52. Zieliński, M.; Kazimierowicz, J.; Dębowski, M. Advantages and limitations of anaerobic wastewater treatment—Technological basics, development directions, and technological innovations. *Energies* **2022**, *16*, 83. [CrossRef]
53. Rahimi, S.; Modin, O.; Mijakovic, I. Technologies for biological removal and recovery of nitrogen from wastewater. *Biotechnol. Adv.* **2020**, *43*, 107570. [CrossRef]
54. Di Costanzo, N.; Cesaro, A.; Di Capua, F.; Esposito, G. Exploiting the nutrient potential of anaerobically digested sewage sludge: A review. *Energies* **2021**, *14*, 8149. [CrossRef]
55. Mao, Y.; Xiong, R.; Gao, X.; Jiang, L.; Peng, Y.; Xue, Y. Analysis of the status and improvement of microalgal phosphorus removal from municipal wastewater. *Processes* **2021**, *9*, 1486. [CrossRef]
56. Dar, S.A.; Kleerebezem, R.; Stams, A.J.; Kuenen, J.G.; Muyzer, G. Competition and coexistence of sulfate-reducing bacteria, acetogens and methanogens in a lab-scale anaerobic bioreactor as affected by changing substrate to sulfate ratio. *Appl. Microbiol. Biotechnol.* **2008**, *78*, 1045–1055. [CrossRef]
57. Saia, F.T.; Souza, T.S.; Duarte, R.T.D.; Pozzi, E.; Fonseca, D.; Foresti, E. Microbial community in a pilot-scale bioreactor promoting anaerobic digestion and sulfur-driven denitrification for domestic sewage treatment. *Bioprocess Biosyst. Eng.* **2016**, *39*, 341–352. [CrossRef] [PubMed]
58. Harirchi, S.; Wainaina, S.; Sar, T.; Nojoumi, S.A.; Parchami, M.; Parchami, M.; Varjani, S.; Khanal, S.K.; Wong, J.; Awasthi, M.K. Microbiological insights into anaerobic digestion for biogas, hydrogen or volatile fatty acids (VFAs): A review. *Bioengineered* **2022**, *13*, 6521–6557. [CrossRef] [PubMed]
59. Nielsen, P.H. Biofilm dynamics and kinetics during high-rate sulfate reduction under anaerobic conditions. *Appl. Environ. Microbiol.* **1987**, *53*, 27–32. [CrossRef] [PubMed]
60. Gagliano, M.; Ismail, S.B.; Stams, A.J.; Plugge, C.M.; Temmink, H.; Van Lier, J. Biofilm formation and granule properties in anaerobic digestion at high salinity. *Water Res.* **2017**, *121*, 61–71. [CrossRef] [PubMed]
61. Cayetano, R.D.A.; Kim, G.-B.; Park, J.; Yang, Y.-H.; Jeon, B.-H.; Jang, M.; Kim, S.-H. Biofilm formation as a method of improved treatment during anaerobic digestion of organic matter for biogas recovery. *Bioresour. Technol.* **2022**, *344*, 126309. [CrossRef] [PubMed]
62. Sarti, A.; Pozzi, E.; Chinalia, F.A.; Ono, A.; Foresti, E. Microbial processes and bacterial populations associated to anaerobic treatment of sulfate-rich wastewater. *Process Biochem.* **2010**, *45*, 164–170. [CrossRef]
63. Nguyen, L.N.; Nguyen, A.Q.; Nghiem, L.D. Microbial community in anaerobic digestion system: Progression in microbial ecology. *Water Wastewater Treat. Technol.* **2019**, 331–355.
64. Nakasaki, K.; Nguyen, K.K.; Ballesteros Jr, F.C.; Maekawa, T.; Koyama, M. Characterizing the microbial community involved in anaerobic digestion of lipid-rich wastewater to produce methane gas. *Anaerobe* **2020**, *61*, 102082. [CrossRef] [PubMed]
65. El Houari, A.; Ranchou-Peyruse, M.; Ranchou-Peyruse, A.; Bennisse, R.; Bouterfas, R.; Goni Urriza, M.S.; Qatibi, A.-I.; Guyoneaud, R. Microbial Communities and Sulfate-Reducing Microorganisms Abundance and Diversity in Municipal Anaerobic Sewage Sludge Digesters from a Wastewater Treatment Plant (Marrakech, Morocco). *Processes* **2020**, *8*, 1284. [CrossRef]

environments

Communication

TiO_2-Coated Meltblown Nonwoven Fabrics Prepared via Atomic Layer Deposition for the Inactivation of *E. coli* as a Model Photocatalytic Drinking Water Treatment System

Alexander G. Aragon [1,2], Jaime A. Cárdenas Sánchez [2,3], Carlos Zimeri [4], Eunkyoung Shim [4], Xiaomeng Fang [4] and Kyana R. L. Young [2,*]

[1] Department of Chemistry, Wake Forest University, 1834 Wake Forest Road, Winston-Salem, NC 27109, USA; aaragon@forsythtech.edu
[2] Department of Engineering, Wake Forest University, 455 Vine Street, Winston-Salem, NC 27101, USA; cardja20@wfu.edu
[3] Department of Biology, Wake Forest University, 1834 Wake Forest Road, Winston-Salem, NC 27109, USA
[4] Department of Textile Engineering, Chemistry and Science, North Carolina State University, 1020 Main Campus Drive, Raleigh, NC 27606, USA; cjzimeri@ncsu.edu (C.Z.); eshim@ncsu.edu (E.S.); xfang3@ncsu.edu (X.F.)
* Correspondence: youngk@wfu.edu

Abstract: The controlled manufacturing of semiconductor photocatalysts is crucial to their development for drinking water treatment. In this study, TiO_2-coated meltblown nonwoven fabrics prepared via Atomic Layer Deposition (ALD) are applied for the inactivation of *Escherichia coli* (*E. coli*). It is observed that in the presence of an ultraviolet light-emitting diode (UV-LED) light source (255 nm), 1.35 log *E. coli* inactivation is achieved. However, exposure to catalyst-coated fabrics in addition to the light source resulted in >4 log *E. coli* inactivation, suggesting a much higher rate of hydroxyl radical formation on the surface, leading to cell death.

Keywords: antimicrobial; nonwovens; drinking water; atomic layer deposition; photocatalysis

1. Introduction

Water is an essential natural resource and is humanity's greatest need for the sustenance of life for continued health, wellbeing and development [1,2]. Unfortunately, 20% of the world's population has no access to clean drinking water, and this remains an issue due to the increased global population in recent years [3,4]. Many types of water contaminants exist, such as chemical waste, heavy metals, and biological agents such as disease-causing microbes. In most cases where clean water is inaccessible, the greatest concern is disease, such as cholera, typhus, polio, tuberculosis, diphtheria and diarrhea [5]. One such microbe that signals issues with water quality is *Escherichia coli* (*E. coli*). *E. coli* serves as an indicator microbe to verify the presence of fecal indicator organisms, and thus serves as a good model for testing disinfection methods. It is crucial to develop drinking water treatment methods that inactivate these organisms in both a safe and sustainable way, and utilizing *E. coli* as a model allows for ease of testing.

Drinking water disinfection methods, including chlorination, ozonation, radiation, advanced filtration and the use of ultraviolet (UV) light, are common technologies employed in drinking water treatment [2,6]. However, there are several key drawbacks in using these conventional methods, such as the formation of disinfection byproducts that are potentially toxic/carcinogenic, high operational and maintenance costs, the development of antibiotic drug resistance over time, and a low disinfection efficiency [2,3,5,7,8]. These methods are also met with implementation challenges, and thus alternative methods must be sought to achieve truly sustainable, low-cost, and minimally toxic water treatment methods. Photocatalytic disinfection remains an innovative and promising option for water purification that removes some of these challenges [3,9]. Given the minimal operational cost and highly sustainable nature of photocatalysis, the development of catalysts utilized for this type of disinfection is a current area of research with high interest [4].

Traditional photocatalysis utilizes semiconductor materials, such as metal oxides [10–12], perovskite oxides [13,14], and metal chalcogenides [10,15], that are synthesized chemically with the use of potentially toxic reagents. These catalysts are typically also produced in bulk powders or thin films, where surface area and reproducibility concerns arise in terms of large-scale water treatment usage. Therefore, it is crucial to develop methods to produce semiconductor photocatalysts with a high uniformity, minimal toxicity, and high surface area for industrial use. ALD provides strong levels of controllability and uniformity to produce such semiconductor photocatalysts, including TiO_2, which is utilized in this study. To date, metal oxides have been heavily studied as antibacterial agents prepared via traditional chemical methods [16,17], and the preparation of TiO_2 and ZnO via ALD shows promise for the large-scale production of photocatalysts for drinking water treatment.

ALD is a powerful tool for the atomically precise design and synthesis of catalytic materials. It is key to developing new technologies to produce valuable products from resources including fossil fuel, biomass, carbon dioxide and water. Large industries, such as energy, pharmaceuticals, food processing, and environmental remediation, all use ALD because of the improved uniformity and control it provides for creating advanced systems. TiO_2 is a very common material for these applications in electronics and pharmaceuticals, such as bacterial inactivation [18]. ALD has been shown to be effective at controlling metal and metal oxide active sites and improving catalytic activity, selectivity, and longevity. To date, there have been a few key applications of ALD in catalysis, including the photocatalytic degradation of organic pollutants and production of carbon dioxide, to name a few [19,20]. In this work, the utilization of TiO_2 deposited via ALD saw much greater efficiency gains and stronger stability due to the increased attachment on the surface of the substrates used. This highlights the strong potential of ALD as a method to produce effective and stable catalysts, but more work is needed to improve the cost-effectiveness of the production and understand the mechanisms by which they operate.

In this study, nano-TiO_2 was attached to the surface of meltblown nonwoven fabrics via ALD methods to prepare a photocatalyst for the inactivation of *E. coli* in water. Textiles have been shown to also be useful for antimicrobial treatment [21,22] and thus can be seen as a strong substrate choice for the preparation of photocatalytic disinfection technologies based on ALD. Due to the near-zero waste of raw materials and environmental pollution, ALD proves to be a high-value synthetic process for the development of antibacterial textiles [22,23]. The overall structural properties of thin films on meltblown nonwoven fabrics are discussed to understand the deposition process and confirm the uniform synthesis of TiO_2 on the surface. The photocatalytic activity of catalyst-coated meltblown nonwoven fabrics is discussed and compared to a UV-LED light source alone. Lastly, the overall stability of catalyst films is discussed with respect to reusability.

2. Materials and Methods

2.1. Bacterial Stock Preparation

E. coli was used as the model bacterial microbial contaminant to assess the inactivation efficacy of the ALD photocatalyst on the meltblown nonwoven fabric. *E. coli* BA-1883 (American Type Culture Collection, Manassas, VA, USA) was aseptically transferred to Nutrient Broth (Difco™, Franklin Lakes, NJ, USA) and placed in an incubator at 37 °C for 5 h. The bacterial growth was then centrifuged at 2000 rpm for 2 min, with the supernatant being disposed and fresh phosphate buffered saline (PBS) added to resuspend the pellet. The pellet and PBS mixture were vortexed and centrifuged at 2000 rpm for 2 min; the supernatant was disposed, and fresh PBS was added, bringing the *E. coli* stock concentration to 10^6 colony-forming units (cfu) per milliliter.

2.2. Preparation of Meltblown Nonwoven Fabrics

The meltblown nonwoven fabrics used in this study were produced using a pilot-scale Reicofil®® (Reifenhauser, Troisdorf, Germany) meltblown line equipped with a traditional Exxon-style die (die orifice diameter = 400 μm, die orifice density = 35 holes/inch, setback = 1.2 mm, total die width = 1.3 m). Processing parameters were varied to control the meltblown fabric structures and are summarized in Table S1, with detailed structural characteristics shown in Table S2 (Supplementary Materials).

2.3. Synthesis of TiO_2 Films

The synthesis of TiO_2 films was done via ALD methods utilizing an Arradiance GEMSTAR-6 benchtop chamber equipped with a furnace. Fabric samples along with a Si reference wafer were placed into the chamber, and the sequential depositions of Ti (titanium isopropoxide) and O (deionized water, Milli-Q, Randolph, MA, USA) were done chemically. The deposition conditions included a chamber temperature of 150 °C with a Ti pulse time of 100 ms followed by a N_2 purge of 23 s and with an O pulse time of 1000 ms followed by another N_2 purge of 28 s to complete a cycle (growth rate = 0.03 nm/cycle). The full growth of the films was measured via XRR experiments utilizing a Bruker D8 Discover X-ray diffractometer equipped with Cu kα radiation and a Goebel mirror.

2.4. Photocatalytic Studies

2.4.1. Ultraviolet Light-Emitting Diode Irradiation

For disinfection, a UV-LED (PearlBeam™, AquiSense Technologies, Erlanger, KY, USA) was used as a light source. Detection of irradiance was conducted utilizing the ILT-5000 radiometer, equipped with an SED 270/QT detector (International Light Technologies, Peabody, MA, USA). For irradiance of the water sample, a reflection factor = 0.975, petri factor = 0.894, water factor = 0.981, and divergence factor = 0.975 were used to correct the irradiance values. The irradiance at the water surface is 0.041 mW/cm^2.

2.4.2. Experimental Conditions

An aliquot of 1.5 mL of the *E. coli* stock was added to 13.5 mL of Milli-Q water (25 °C, 18.2 Ω) in a glass 60 × 15 mm quartz Petri dish and placed on a magnetic stir plate to gently mix at 250 rpm. The Petri dish was placed under a collimated beam for 60 s, where the UV-LED irradiated the sample at λ = 255 nm, selected for its proximity to the peak absorbance (λ = 254 nm) responsible for DNA damage. Simultaneously, the meltblown nonwoven fabric was added to the Petri dish, and the UV-LED was powered on. At the conclusion of irradiation, the meltblown nonwoven fabric was then removed from the Petri dish, and the solution was processed using the membrane filtration method (Standard Methods 9222B); the filter was placed on m-Endo Agar LES (Remel™, Lenexa, KS, USA) in the incubator for 24 h at 37 °C and then transferred to the EC Medium w/MUG (Remel™, Lenexa, KS, USA) for 4 h at 44 °C. After incubation, fluorescent colonies were quantified using the plate count method. The control groups were Milli-Q (negative) and bacteria stock solution (positive). The experimental groups were meltblown nonwoven fabric, meltblown nonwoven fabric deposited with photocatalyst, meltblown nonwoven fabric deposited with photocatalyst and UV-LED-irradiated, Laprotex™, Laprotex™ deposited with photocatalyst, Laprotex™ deposited with photocatalyst and UV-LED-irradiated, photocatalyst, photocatalyst irradiated with UV-LED, and UV-LED.

2.4.3. Inactivation Kinetics

The *log inactivation* for the *E. coli* was determined via Equation (1), where N_0 is the initial concentration of cfu/mL with no irradiation and N_f is the concentration of cfu/mL after UV-LED irradiation:

$$log\frac{N_f}{N_0} = log\ inactivation \tag{1}$$

A comparison of log inactivation was conducted to assess the inactivation efficacy of the experimental conditions. Data for TiO_2 were collected in triplicate.

3. Results and Discussion

Thin films of TiO_2 were deposited on the polypropylene-based nonwovens via ALD, and a full cycle of ALD is pictured in Scheme 1.

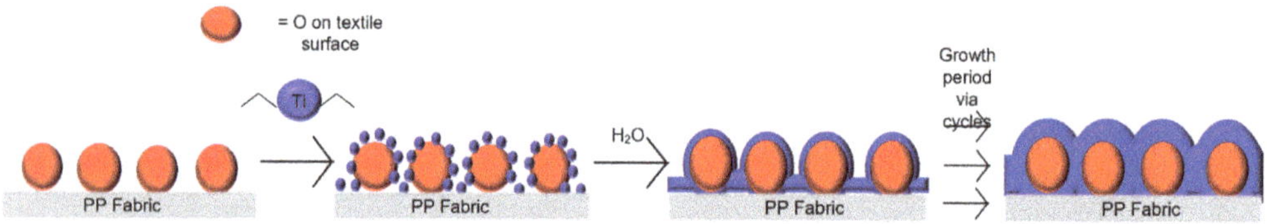

Scheme 1. ALD deposition cycle on nonwoven fabric surface.

The Ti precursor first reacts with hydroxyl groups on the surface of the meltblown nonwoven fabrics, followed by a reaction between this intermediate product and the water precursor to form a layer of TiO_2 on the surface. The thickness of these films can be adjusted via the number of ALD cycles run, but for this work, an average film thickness of ~12 nm was utilized for the TiO_2 films, as determined via XRR experiments shown in Figure S1 (Supplementary Materials). In addition to the determination of the film thickness, which gives an idea of the load of TiO_2, a more exact quantification of the TiO_2 concentration via TGA analysis should be done in the future to fully understand the impact of the TiO_2 load on the disinfection efficiency of the coated fabrics.

The morphology of ALD-treated nonwovens compared to a pristine nonwoven fiber is shown in Figure 1a,b and was obtained via SEM imaging.

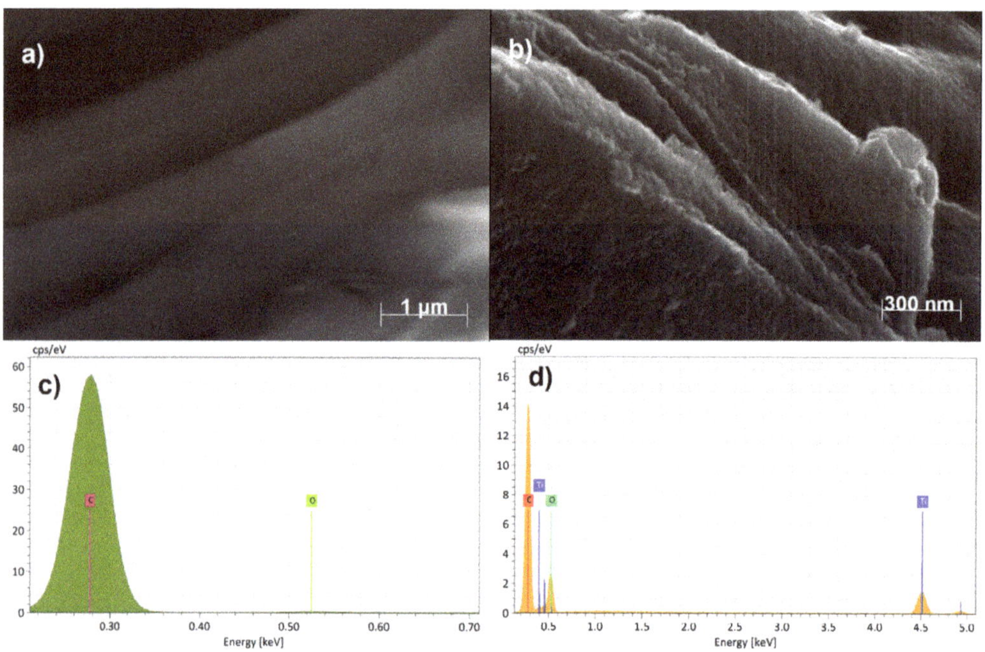

Figure 1. (**a**) SEM image of pristine fabric; (**b**) SEM image of TiO_2-coated fabric; (**c**) EDS spectrum of pristine fabric; (**d**) EDS spectrum of TiO_2-coated fabric.

A rough surface of metal oxide coating is achieved after the ALD process when comparing the pristine fiber to TiO_2-coated nonwovens, respectively. The SEM images were collected in triplicate over a series of three meltblown nonwoven fabric samples, and the TiO_2 films were found to be amorphous, which is consistent with other literature reports of low-temperature ALD TiO_2 films [24,25]. The EDS spectra shown in Figure 1c,d, collected over the series of meltblown fabrics, paired with line scan data shown in Figure S2 (Supplementary Materials) suggest that Ti is present on the surface of the meltblown nonwoven fabrics after ALD and is present in the same regions as O on the surface. The morphology and elemental composition of ALD films deposited on the commercially available tested textile are shown in Figure S3 (Supplementary Materials).

To confirm the successful deposition of TiO_2 on the surface of the meltblown nonwoven fabrics, both X-ray photoelectron spectroscopy (XPS) and IR spectroscopy were utilized to probe the chemical structure of the pristine and ALD-treated meltblown nonwoven fabrics.

The IR spectrum of the treated and pristine nonwovens is shown in Figure 2a. The peaks at 2900, 1350, and 1053 cm^{-1} represent the stretch vibrations of C-H, O-H, and C=C bonds, respectively. After ALD treatment, a broad peak is present near 3500 cm^{-1}, which represents the Ti-OH stretch vibration, while the broad feature observed below 1000 cm^{-1} is the Ti-O-Ti stretch and suggests deposition of TiO_2 on the surface of the meltblown nonwoven fabric. The C 1s and O 1s characteristic peaks appear at 284 and 532 eV, respectively, for the pristine nonwoven fabric, as shown in the XPS survey spectrum in Figure 2b. When compared to the TiO_2-coated meltblown nonwoven fabrics, the presence of the characteristic Ti 2p peak at 458 eV further suggests the successful deposition of TiO_2 on the surface.

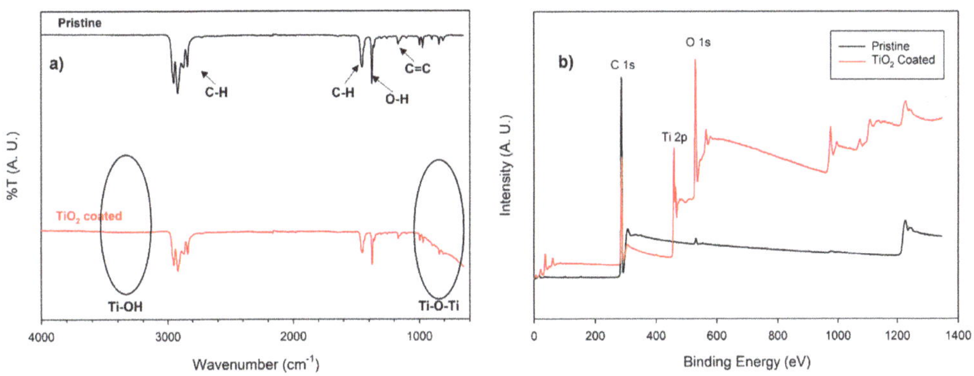

Figure 2. (**a**) IR spectrum of pristine and ALDtreated meltblown nonwoven fabrics; (**b**) XPS survey spectrum of pristine and ALD-treated meltblown nonwoven fabrics.

In photocatalytic systems, the application of external light provides the energy to move electrons from the valence band to the conduction band; this movement of electrons allows the production of powerful oxidants (hydroxyl radicals) in a bulk water system. The oxidant causes damage to the DNA of waterborne microbes, thus rendering them inactive and unable to reproduce. The use of UV-LED as an external light source was analyzed as a singular treatment mechanism and compared to light-irradiated photocatalysts for the disinfection of bulk water samples. The samples tested were at pH 7 and temperature = 25 °C, and a collimated beam of UV-LED at λ = 255 nm (intensity = 0.041 mW/cm^2) was utilized to assess the efficacy of the *E. coli* inactivation of TiO_2-coated fabrics. An overview of the photoreactor setup is shown in Figure 3a.

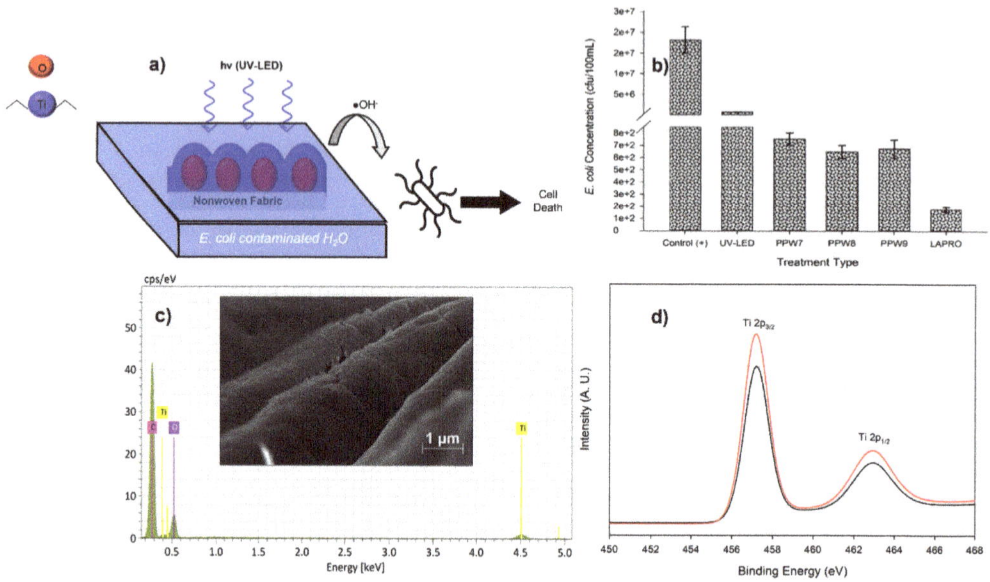

Figure 3. (**a**) Overview of photoreactor setup used for photocatalytic experiments (Conditions: pH = 7, temperature = 25 °C, λ = 255 nm, dose = 2.46 mJ/cm^2); (**b**) E. coli concentration after exposure to UV-LED alone or TiO_2-coated meltblown nonwoven with UV-LED irradiation (n = 3); (**c**) EDS spectrum and SEM image (inset) of TiO_2-coated meltblown nonwoven fabric after catalysis; (**d**) XPS spectrum for the Ti 2p electron before (black) and after (red) catalysis for TiO_2-coated nonwoven fabric.

When exclusively irradiating the bulk water with UV-LED, UV dose = 2.46 mJ/cm^2 (0.041 mW/cm^2 $\times$ 60 s), 1.35 $\pm$ 0.0422 log inactivation of *E. coli* was achieved (Table 1).

Table 1. Comparison of log inactivation for experimental types of *E. coli* inactivation.

Fabric Type	Catalyst	External Light	Log Inactivation
PPW7	TiO_2	UV-LED	4.40 $\pm$ 0.043
PPW8	TiO_2	UV-LED	4.40 $\pm$ 0.032
PPW9	TiO_2	UV-LED	4.40 $\pm$ 0.041
Laprotex™	TiO_2	UV-LED	5.00 $\pm$ 0.031
-	-	UV-LED	1.35 $\pm$ 0.0422

For experiments performed when the photocatalyst was deposited on the meltblown nonwoven fabric, inactivation of *E. coli* was evident (Figure 3b). The PPW7, PPW8, and PPW9 deposited with TiO_2 and a UV dose = 2.46 mJ/cm^2 achieved a minimum of 4.40 $\pm$ 0.032 log inactivation of *E. coli*. The significant improvement in log *E. coli* inactivation for the coated meltblown nonwovens suggests a high potential for the utilization of this technology as a multifunctional drinking water treatment option. For the comparison of the treatment efficacy between fabric types and TiO_2, the bacteria inactivation was impacted by the commercial Laprotex™ fabric ($p < 0.001$) compared to the insignificant difference when using the PPW fabrics ($p = 0.260$). These findings suggest the strong potential shown by the coated meltblown nonwoven fabrics for drinking water treatment, and further work is needed to test their capability under different irradiation sources, such as UVA-LED, which is more cost-effective than UVC-LED. Another factor that can significantly impact the photocatalytic efficiency is the exact load of TiO_2 on the meltblown nonwoven fabric, and future studies would also utilize TGA analyses to determine this. Though the mechanism of photocatalytic inactivation is not yet fully understood, it is believed that through the creation of hydroxyl and superoxide radicals on the surface of the catalyst in solution, direct oxidative damage of the DNA of *E. coli* is achieved, following previous literature reports [26]. Density Functional Theory calculations are needed to understand the band structure, differential charge density, and density of states for the catalyst-coated nonwoven fabrics to probe the movement of photogenerated radical oxidants on the surface.

The stability of the as-synthesized TiO_2 thin film catalysts was probed to determine the reusability of the coated meltblown nonwoven fabrics for larger scale use. Figure 3c details the morphology and elemental composition of the TiO_2-coated nonwovens after catalysis. The overall film morphology remains relativity unchanged after catalysis. As evidenced by the EDS spectrum, the TiO_2 films remain relatively unchanged given the presence of the Ti in the spectra. To further examine the change in catalyst films during catalysis, XPS was utilized to probe the changes in the chemical structure of TiO_2 and is shown in Figure 3d. The spectrum for the Ti 2p electron shows no change in the characteristic peaks of Ti (IV) for the TiO_2-treated nonwoven fabric, which suggests a strong stability and reusability of the TiO_2 catalyst. Detailed atomic percentages as determined via XPS are shown in the Supplementary Materials (Table S3). Although there is not much change in the films, as evidenced through the post-catalysis EDS and XPS spectra, work is still needed to fully examine the reusability of the coated fabrics through multiple cycles and in large scale drinking water treatment systems.

4. Conclusions

Uniform TiO_2 films were deposited on meltblown nonwovens via ALD and applied as a photocatalyst for the inactivation of *E. coli* in bulk water. It was found that the photocatalyst-coated meltblown nonwovens outperformed the UV-LED light source utilized, as evidenced by a near 4-fold increase in *E. coli* log inactivation for all fabrics tested with TiO_2. In addition, a significant difference of *E. coli* log inactivation between the Laprotex™ and PPW fabrics coated with TiO_2 was observed. This suggests that the bacterial inactivation was affected by the presence of the Laprotex™ fabric. Photocatalyst-coated films were shown to have great stability given the minimal change in the structural morphology and elemental composition of the coated fabrics after catalysis. The minimal change in the Ti 2p XPS spectrum suggests that TiO_2 films are completely intact after catalysis. The utilization of ALD for the preparation of catalysts shows a strong potential for the development of sustainable, low-environmental-impact drinking water treatment technologies in the next few years, and more work is needed to understand the reusability and efficacy under different irradiation sources of this technology and to improve its cost-effectiveness. Furthermore, the determination of the distinct hydroxyl radical groups produced in the photocatalytic process will allow for the continued exploration of photocatalysts suitable for this application.

Supplementary Materials: The following supporting information can be downloaded at: https://www.mdpi.com/article/10.3390/environments11050092/s1, Figure S1: Fitted XRR data for metal oxide ALD film on Si reference; Figure S2: EDS line scan data of TiO_2 coated fabrics; Figure S3: ALD films on commercial textile; Table S1: Processing conditions of meltblown nonwoven fabrics production; Table S2: Meltblown fabrics structural characteristics; Table S3: Atomic weight percentages as determined via XPS for TiO_2 coated nonwovens before and after catalysis.

Author Contributions: Conceptualization, A.G.A. and K.R.L.Y.; methodology, A.G.A., J.A.C.S. and C.Z.; validation, A.G.A. and K.R.L.Y.; formal analysis, A.G.A., C.Z. and K.R.L.Y.; investigation, A.G.A., J.A.C.S. and C.Z.; resources, E.S., X.F. and K.R.L.Y.; writing—original draft preparation, A.G.A.; writing—review and editing, E.S., X.F. and K.R.L.Y.; visualization, A.G.A. and K.R.L.Y.; supervision, E.S., X.F. and K.R.L.Y.; project administration, K.R.L.Y.; funding acquisition, K.R.L.Y. All authors have read and agreed to the published version of the manuscript.

Funding: The Office of Research and Sponsored Programs at Wake Forest University provided generous financial support for the research detailed here within (CPG 1066).

Data Availability Statement: The original contributions presented in the study are included in the article and Supplementary Materials, further inquiries can be directed to the corresponding author.

Acknowledgments: The authors thank Olubunmi Ayodele and Shobha Mantripragada from the Joint School of Nanoscience and Nanoengineering, University of North Carolina for providing XPS data.

Conflicts of Interest: The authors declare no conflicts of interest.

References

1. Odonkor, S.T.; Ampofo, J.K. *Escherichia coli* as an Indicator of Bacteriological Quality of Water: An Overview. *Microbiol. Res.* **2013**, *4*, e2. [CrossRef]
2. Vietro, N.D.; Tursi, A.; Beneduci, A.; Chidichimo, F.; Milella, A.; Fracassi, F.; Chatzisymeon, E.; Chidichimo, G. Photocatalytic Inactivation of *Escherichia coli* Bacteria in Water Using Low Pressure Plasma Deposited TiO_2 Cellulose Fabric. *Photochem. Photobiol. Sci.* **2019**, *18*, 2248–2258. [CrossRef] [PubMed]
3. Park, J.; Kettleson, E.; An, W.-J.; Tang, Y.J.; Biswas, P. Inactivation of *E. coli* in Water Using Photocatalytic, Nanostructured Films Synthesized by Aerosol Routes. *Catalysts* **2013**, *3*, 247–260. [CrossRef]
4. Ajiboye, T.O.; Babalola, S.O.; Onwudiwe, D.C. Photocatalytic Inactivation as a Method of Elimination of *E. coli* from Drinking Water. *Appl. Sci.* **2021**, *11*, 1313. [CrossRef]
5. Khani, M.; Amin, N.A.S.; Hosseini, S.N.; Heidarrezaei, M. Kinetics Study of the Photocatalytic Inactivation of *Escherichia coli*. *IJNBM* **2016**, *6*, 139. [CrossRef]
6. Ribeiro, M.A.; Cruz, J.M.; Montagnolli, R.N.; Bidoia, E.D.; Lopes, P.R.M. Photocatalytic and Photoelectrochemical Inactivation of *Escherichia coli* and *Staphylococcus aureus*. *Water Supply* **2014**, *15*, 107–113. [CrossRef]
7. Dunlop, P.S.M.; Ciavola, M.; Rizzo, L.; McDowell, D.A.; Byrne, J.A. Effect of Photocatalysis on the Transfer of Antibiotic Resistance Genes in Urban Wastewater. *Catal. Today* **2015**, *240*, 55–60. [CrossRef]
8. Friedman, N.D.; Temkin, E.; Carmeli, Y. The Negative Impact of Antibiotic Resistance. *Clin. Microbiol. Infect.* **2016**, *22*, 416–422. [CrossRef]
9. Benabbou, A.K.; Derriche, Z.; Felix, C.; Lejeune, P.; Guillard, C. Photocatalytic Inactivation of *Escherischia Coli*: Effect of Concentration of TiO_2 and Microorganism, Nature, and Intensity of UV Irradiation. *Appl. Catal. B Environ.* **2007**, *76*, 257–263. [CrossRef]
10. Aragon, A.G.; Kierulf-Vieira, W.; Łęcki, T.; Zarębska, K.; Widera-Kalinowska, J.; Skompska, M. Synthesis and Application of N-Doped TiO_2/CdS/Poly(1,8-Diaminocarbazole) Composite for Photocatalytic Degradation of 4-Chlorophenol under Visible Light. *Electrochim. Acta* **2019**, *314*, 73–80. [CrossRef]
11. Badhe, R.A.; Ansari, A.; Garje, S.S. Study of Optical Properties of TiO_2 Nanoparticles and CdS@TiO_2 Nanocomposites and Their Use for Photocatalytic Degradation of Rhodamine B under Natural Light Irradiation. *Bull. Mater. Sci.* **2021**, *44*, 11. [CrossRef]
12. Madkour, M.; Allam, O.G.; Abdel Nazeer, A.; Amin, M.O.; Al-Hetlani, E. CeO_2-Based Nanoheterostructures with p–n and n–n Heterojunction Arrangements for Enhancing the Solar-Driven Photodegradation of Rhodamine 6G Dye. *J. Mater. Sci. Mater. Electron.* **2019**, *30*, 10857–10866. [CrossRef]
13. He, Z.; Sun, C.; Yang, S.; Ding, Y.; He, H.; Wang, Z. Photocatalytic Degradation of Rhodamine B by Bi_2WO_6 with Electron Accepting Agent under Microwave Irradiation: Mechanism and Pathway. *J. Hazard. Mater.* **2009**, *162*, 1477–1486. [CrossRef]
14. Olagunju, M.O.; Zahran, E.M.; Reed, J.M.; Zeynaloo, E.; Shukla, D.; Cohn, J.L.; Surnar, B.; Dhar, S.; Bachas, L.G.; Knecht, M.R. Halide Effects in $BiVO_4/BiO_X$ Heterostructures Decorated with Pd Nanoparticles for Photocatalytic Degradation of Rhodamine B as a Model Organic Pollutant. *ACS Appl. Nano Mater.* **2021**, *4*, 3262–3272. [CrossRef]
15. Bessekhouad, Y.; Chaoui, N.; Trzpit, M.; Ghazzal, N.; Robert, D.; Weber, J.V. UV–Vis versus Visible Degradation of Acid Orange II in a Coupled CdS/TiO_2 Semiconductors Suspension. *J. Photochem. Photobiol. A Chem.* **2006**, *183*, 218–224. [CrossRef]
16. Choi, H.; Stathatos, E.; Dionysiou, D.D. Photocatalytic TiO_2 Films and Membranes for the Development of Efficient Wastewater Treatment and Reuse Systems. *Desalination* **2007**, *202*, 199–206. [CrossRef]
17. Baghdadi, A.M.; Saddiq, A.A.; Aissa, A.; Algamal, Y.; Khalil, N.M. Structural Refinement and Antimicrobial Activity of Aluminum Oxide Nanoparticles. *J. Ceram. Soc. Jpn.* **2022**, *130*, 257–263. [CrossRef]
18. Akyildiz, H.I.; Diler, S.; Islam, S. Evaluation of TiO_2 and ZnO Atomic Layer Deposition Coated Polyamide 66 Fabrics for Photocatalytic Activity and Antibacterial Applications. *J. Vac. Sci. Technol. A* **2021**, *39*, 022405. [CrossRef]
19. Pham, K.; Pelisset, S.; Kinnunen, N.; Karvinen, P.; Hakala, T.K.; Saarinen, J.J. Controlled Photocatalytic Activity of TiO_2 Inverse Opal Structures with Atomic Layer Deposited (ALD) Metal Oxide Thin Films. *Mater. Chem. Phys.* **2022**, *277*, 125533. [CrossRef]
20. Islam, S.; Akyildiz, H.I. Atomic Layer Deposition of TiO_2 Thin Films on Glass Fibers for Enhanced Photocatalytic Activity. *J. Mater. Sci. Mater. Electron.* **2022**, *33*, 18002–18013. [CrossRef]
21. Li, L.; Yu, P.; Li, Y.; Wu, X.; Li, W.; Cheng, X. A Facile Approach to Fabricating Antibacterial Textile with High Efficiency and Compact Process. *Adv. Mater. Interfaces* **2021**, *8*, 2101197. [CrossRef]
22. Popescu, M.C.; Ungureanu, C.; Buse, E.; Nastase, F.; Tucureanu, V.; Suchea, M.; Draga, S.; Popescu, M.A. Antibacterial Efficiency of Cellulose-Based Fibers Covered with ZnO and Al_2O_3 by Atomic Layer Deposition. *Appl. Surf. Sci.* **2019**, *481*, 1287–1298. [CrossRef]

23. Wang, Z.; Zhang, L.; Liu, Z.; Sang, L.; Yang, L.; Chen, Q. The Antibacterial Polyamide 6-ZnO Hierarchical Nanofibers Fabricated by Atomic Layer Deposition and Hydrothermal Growth. *Nanoscale Res. Lett.* **2017**, *12*, 421. [CrossRef] [PubMed]
24. Nam, T.; Kim, J.M.; Kim, M.K.; Kim, H.; Kim, W.H. Low-Temperature Atomic Layer Deposition of TiO_2, Al_2O_3, and ZnO Thin Films. *J. Korean Phys. Soc.* **2011**, *59*, 452–457. [CrossRef]
25. Aghaee, M.; Maydannik, P.S.; Johansson, P.; Kuusipalo, J.; Creatore, M.; Homola, T.; Cameron, D.C. Low Temperature Temporal and Spatial Atomic Layer Deposition of TiO_2 Films. *J. Vac. Sci. Technol. A* **2015**, *33*, 041512. [CrossRef]
26. Wu, P.; Imlay, J.A.; Shang, J.K. Mechanism of *Escherichia coli* Inactivation on Palladium-Modified Nitrogen-Doped Titanium Dioxide. *Biomaterials* **2010**, *31*, 7526–7533. [CrossRef]

environments

MDPI

Review

Water Reuse: A Comprehensive Review

Fivos Florides, Maria Giannakoudi, Giorgos Ioannou, Despoina Lazaridou, Elissavet Lamprinidou, Nikolaos Loukoutos , Maria Spyridou, Eleftherios Tosounidis , Maria Xanthopoulou and Ioannis A. Katsoyiannis *

Laboratory of Chemical and Environmental Technology, Department of Chemistry, Aristotle University of Thessaloniki, P.O. Box 116, 54124 Thessaloniki, Greece; floridesfivos00@gmail.com (F.F.)
* Correspondence: katsogia@chem.auth.gr

Abstract: Water scarcity has emerged as a pressing global concern, driven by population growth, urbanization, and climate change. As freshwater resources dwindle, the imperative for water reuse becomes increasingly apparent. Reusing water presents a sustainable solution to mitigate scarcity, offering a way to maximize the efficiency of available resources. This review delves into the multifaceted landscape of water consumption and reuse, aiming to provide a comprehensive analysis and understanding of this critical issue. It explores the diverse implications of unregulated water consumption, spanning from its impacts on household routines to its profound influence on economic activities. Additionally, it scrutinizes the legislative framework surrounding water usage, shedding light on the policies and regulations in place. Furthermore, the review investigates the current status of water reuse practices in Europe, delving into various methods of water recovery. Finally, it examines public perceptions and attitudes toward recycled water, offering insights into the societal outlook on this increasingly vital aspect of water management.

Keywords: water; reuse; circular economy

1. Introduction

In many cases, the term "blue planet" is used when someone refers to Earth. The name is not a coincidence since 75% of it is covered by water. Nevertheless, the overwhelming percentage of it (97.5%) is non-potable, leaving only 2.5% for use and consumption. Despite that, humankind consumes 10 billion tons of water worldwide daily. Already for the first 10 days of 2024, 104 billion tons of water have been consumed [1]. According to the TheWorldCounts database, if the world continues its rampant water consumption at the same rate, the remaining usable water reserves will dry up within 16 years. Therefore, it is prudent to analyze this thoughtless expenditure of water resources and find the main culprits in search of the answer to why our planet is becoming increasingly grey [1].

As far as Europe is concerned, every year an average of 244,000 hm^3 (where 1 hm^3 = 106 m^3) of water is consumed for economic activities. Much of this (11.6%) is consumed by the citizens of the European Union in a domestic context, making proper and easily accessible information, public awareness, and a more generalized, environmentally friendly use of water the only way to a better future [2].

This review aims to define and analyze many aspects concerning water consumption and water reuse to have a clearer and multilayered understanding of the matter. Aspects studied include the variety of implications of unchecked water consumption both within the household and for economic activities, the legislative framework around this issue, the status quo of water reuse in Europe, water recovery methods, and finally, information on the outlook of the general public when it comes to recycled water.

2. The Impacts of Reckless Water Consumption

The way water resources were managed in the past coupled with the inevitable consequences of water scarcity, led to a necessary shift towards a new and different way of managing the existing resources [3]. The—constantly increasing—global demands cannot be met by the available water resources. The result is their gradual depletion. With the crucial role water has in the economy and human activities, a myriad of problems is created.

More specifically, it was shown that water withdrawal, for agricultural, industrial, and urban use, experienced a minimum fivefold increase between 1990 and 2000. Additionally, it is reported that there will be a 55% augmentation up until 2050. Agriculture, in particular, accounts for 70% of global water consumption, with the percentage varying by region. As for Europe, southern countries are more vulnerable, due to the agricultural activities being mainly concentrated there [4].

In America, for example, the areas experiencing significant water scarcity are those where rapid growth of population, expansion of irrigated land, and consequently increased water demand are evident. However, in areas where agriculture was not as developed, this phenomenon is still restricted. Simultaneously, serious issues arose in India, where uncontrolled and unregulated industrial development, combined with overpopulation, led to the deposition of vast amounts of untreated waste in the environment, negatively impacting water quality and public health as well [5]. Apart from these two examples, a plethora of other countries are experiencing similar problems. According to the "Water Security Risk Index", 18 countries are faced with significant water shortage issues, with 15 of them being located in the Middle East [5].

International observations already indicate that approximately 1/10 of the population lacks access to basic drinking water services. With the global population increasing and a lack of initiatives to address the crisis of available water resources, it is evident that the number of people deprived of the fundamental right to water will rise substantially in the upcoming years [6]. This statement is also supported by Figure 1, where it is indicated that the number of urban citizens who encounter water shortage—seasonal or long-term—will continue to increase [7]. Furthermore, it has been revealed that the use of renewable water sources (renewed through the water cycle, for example through groundwater from rainfall) is persistently decreasing. Iceland which holds the primary position in Europe, has halved its usage from 1961 to 2019. In general, if water withdrawal rates are significantly higher than renewal rates, sources will be depleted [8].

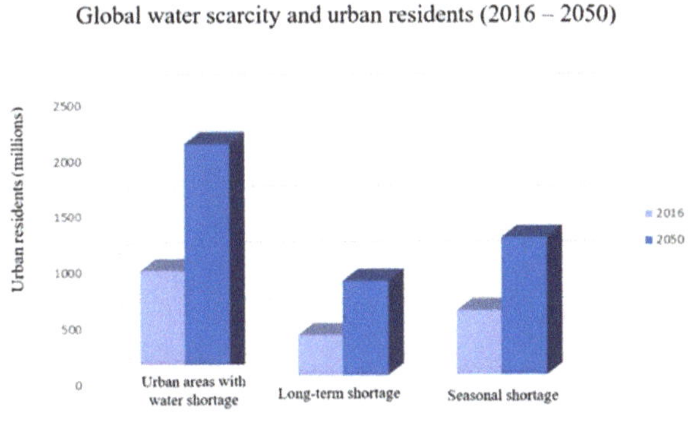

Figure 1. Urban citizens that live under water shortage conditions (2016–prediction 2050) [7].

In summary, chronic water scarcity has been studied as a multidimensional problem. As it was mentioned earlier, this global issue is becoming increasingly severe due to increasing consumer demands, rapid population growth, urbanization, agricultural development, improper utilization of available water, and ultimately, climate change [5]. Therefore, a shift towards water reuse, more specifically a transition towards a circular economy, is deemed necessary.

2.1. The Role of Industry

In Europe, about 215,000 hm^3 of water is used for a multitude of industrial processes [9]. Admittedly, more than half of the quantity (140.00 hm^3) returns to the environment, but without a guarantee for its quality, as most industrial processes pollute the water they use with several substances, which are often dangerous for humans and the environment. In fact, according to Eurostat data, only 60% of the water used at an industrial level is treated before it ends up in the environment [2]. It is no coincidence that many countries of the European Union, such as Greece, Spain, and Portugal already encounter significant drought issues during the summer while the state is unable to manage this new order of things. The problem is not only found in southern Europe but also in countries such as Germany and England, which are experiencing unprecedented levels of water scarcity. In a general context, places with high agricultural activity are the most affected areas when it comes to water stress [2].

This fact is not at all coincidental, as shown in Figure 2, the consumption of water in the agricultural industry amounts to a percentage equal to 41% of the consumption for economic activities [9]. This main source of consumption will continue to exist for many decades to come due to the ever-increasing consumption and increasing need for cultivatable lands, especially in the plains of southern Europe. Therefore, water in this industry is used for several processes, such as the irrigation of crops, the use of it for herbicides and insecticides, and temperature control of the fruits (i.e., heating in the winter months and cooling during summer months) [9]. At the same time, its exploitation is accompanied by several pollutants, drastically reducing its quality. More specifically, these pollutants include the various toxic substances contained in insecticides and herbicides, as well as the various pollutants leaching from fertilizers [10].

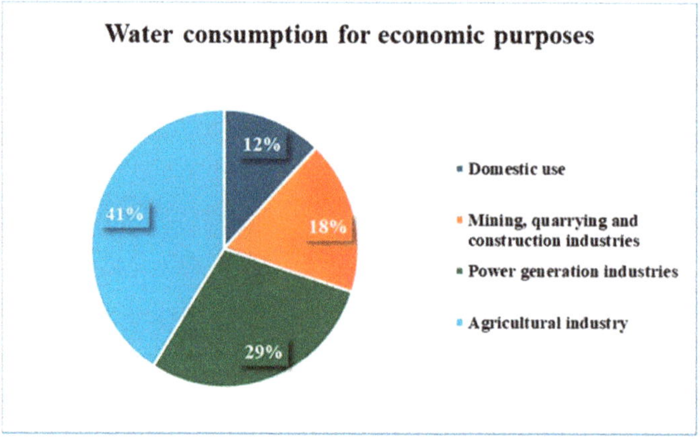

Figure 2. Percentage of total water consumed for economic purposes by industry in Europe [9].

Moving to the next target, the energy production industry is the next to be in the crosshair. Accounting for 29% (Figure 2) of total consumption, water in this industry is primarily used for temperature control of various combustion and nuclear power reactors [9]. In this case, there may not necessarily be significant pollution by substances; nevertheless, the quality of the water is significantly reduced due to the so-called thermal pollution. In this industry, the temperature of the water deviates from the normal limits, causing a significant decrease in the amount of dissolved oxygen in it, due to a decrease in the solubility of the gas with the increase in temperature, negatively affecting fauna and flora after its deposition in the environment [11].

Finally, the mining, quarrying, construction, and manufacturing industries are worth mentioning. Combined, these industries account for 18% (Figure 2) of the total water consumed for economic activities [9]. This percentage may seem small in relation to the industries that have been mentioned before; nevertheless, the keyword in this example is pollution. With the mining and quarrying industry utilizing water in a multitude of processes such as the processing of various minerals, the suppression and limitation of dust, the transport of mud, etc., combined with the very limited treatment of the waste it produces, it is no coincidence that it is a major culprit in terms of water consumption [12]. At the same time, in the context of construction, water consumption is primarily found in processes such as washing machines, preparing, and curing raw concrete, and dust suppression. In this industry, the treatment before final disposal is even more limited, and as a result, the polluted water ends up in the environment contaminated with heavy metals, a high amount of inorganic solids, and derivatives of the various chemicals used [13].

2.2. The Role of Domestic Use

Of course, the industry is not the only one responsible for the uncontrollable expenditure of water resources. As seen in Figure 2, the citizens of the European Union consume about 12% of the total water, i.e., an average of 29,280 hm^3 of fresh water [9]. To put this volume of water in perspective, this amount is roughly equivalent to 7756 Olympic-sized swimming pools. This quantity is divided into several processes that take place in a household, such as water for drinking, food preparation, bathrooms and showers, dishwashers and clothes washers, toilet cisterns, yard irrigation, swimming pools, cleaning, etc. Pollutants that significantly reduce water quality due to the above activities include detergents, pharmaceuticals, chemicals, soaps, large amounts of organic matter, and many others. It may seem that these processes do not consume large amounts of water individually; however, combined, and given that the population of Europe is 746 million, it is not difficult to imagine how this excessive water consumption is built up, stone by stone [14,15].

3. Legislative Frameworks

For a long time, the European Union (EU) has been dealing with problems related to the protection of available water resources by passing legislation limiting reckless waste exploitation but also monitoring the effective treatment of wastewater [16]. The first directive established by the EU related to water reuse is the Drinking Water Quality Directive (DIRECTIVE 80/778/EC) [17]. This directive mainly refers to the implementation of regulations concerning the presence of toxic chemicals and harmful micro-organisms in drinking water. Furthermore, it establishes guidelines when it comes to the determination of physical, chemical, and biological parameters for the various uses of water, especially drinking water. Several pieces of legislation followed Directive 80/778/EC. The various pieces of legislation are presented in the following timeline (Figure 3).

The EU Council Directive (91/271/EC) refers to measures that have to be implemented on a smaller scale, within each community and it aims for more thorough urban wastewater [18]. According to the directive, the secondary treatment of wastewater must be implemented in order to avoid the negative effects that the disposal of untreated wastewater will have on the environment. It also raises the issue of controlling, by general regulations, the channeling of industrial waste into sewerage networks and the disposal of sewage and sewage sludge from municipal sewage treatment plants. More specifically, Article 12 of the directive states that treated wastewater should be reused whenever it is necessary, and disposal methods should minimize adverse effects on the environment [18].

In 1998 and according to Directive 98/83/EC, limits related to the quality of water consumed by humans were established [19]. In 2000, the European Council (EC) enacted the directive (2000/60/EC), which established the water policy framework. The purpose of this directive is to lay down regulations for the protection of inland surface, transitional, coastal, and underground water [20]. The implementation of this framework contributes to many environmental fronts such as ensuring an adequate supply of surface and underground water of good quality, minimizing the pollution load of groundwater, protecting territorial and marine water, and at the same time achieving the objectives of relevant international agreements [20]. The above directive was amended in 2013 according to directive 2013/39/EU of the European Parliament and the Council [21]. The directive adopted in 2006 (2006/118/EC) sets conditions for the protection of groundwater from pollution and degradation. This directive sets limits and quality standards for the chemical and biological status of groundwater as well as for the pumping systems used to transport water [22].

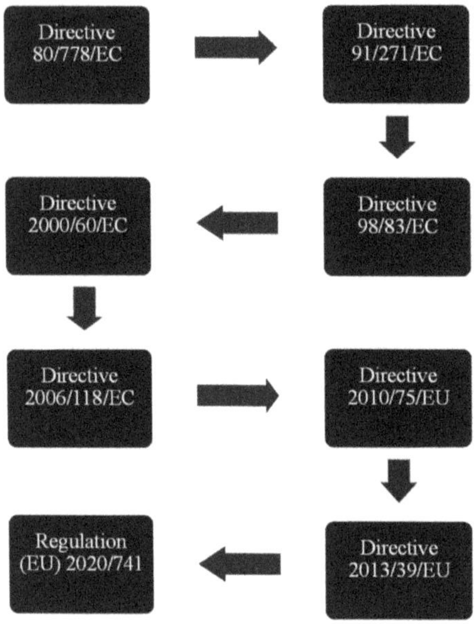

Figure 3. Timeline of European guidelines for water reuse.

Water is the most important factor affecting the sustainability of human life and the planet. Every day, excessive volumes of waste are produced, degrading the quality and quantity of available drinking water [23]. To avoid the significant effects of water degradation, wastewater reuse should be carried out through specific technologies [16]. According to the report published in 2016 by the European Council, only 2.5% of water in Europe is reused [24].

The European Union 2020 enacted Regulation (EU) 2020/741, which sets certain limits connected to minimum requirements for water reuse [25]. The above regulation defines the minimum requirements for water quality in order to preserve water and limit its expenditure. The purpose of the regulation is to guarantee that the reused water is non-harmful for agricultural irrigation, thereby safeguarding the environment, human and animal health and promoting the circular economy, enhancing adaptation to climate change while at the same time helping to implement the objectives of Directive 2000/60/EC, addressing the phenomenon of water scarcity and the consequent pressure on water resources [25].

When it comes to the production, supply, and use of reclaimed water, the qualified authority shall ensure the establishment of a risk management plan for water reuse. Furthermore, the authority is responsible for issuing permits and for ensuring compliance with the regulation. Depending on many factors like geographical position and climate, a Member State according to the regulation is able to decide whether water reuse is required [25].

According to Annex 1 and Section 1 of Regulation 2020/741, reclaimed water can be used in several ways. Water reuse can find application in agricultural irrigation and can be applied in many crops such as edible plants that are consumed raw, edible plants that are processed, and non-edible plants. Furthermore, the water should be reused, always with the protection of the environment, in industrial facilities and in recreational areas [25]. More extensively, for industrial waste and for its recovery and reuse, Directive 2010/75/EU specifies the required prevention and control of pollution from industrial emissions. According to Annex VI, Part 5 of the above directive, the maximum value of total suspended solids for industrial wastewater discharges is set at 45 mg/L [26].

Depending on their quality, the waters are divided into four categories. Table 1 shows the quality categories of reclaimed water, the permitted uses, the indicative technological purpose, and their minimum quality requirements.

Table 1. Reclaimed water quality categories, irrigation method, indicative technological target, and minimum reclaimed water quality requirements [18,25].

Reclaimed Water Quality Class	Indicative Technological Target			Quality Requirements			
	Crop Category	Irrigation Method	Treatment	*E. coli* (Number/100 mL)	BOD5 (mg/L)	TSS (mg/L)	Turbidity (NTU)
A	Crops of edible plants that are eaten raw in direct contact with reclaimed water and root crops that are eaten raw.	All irrigation methods	Secondary treatment, filtration, and disinfection	≤10	≤10	≤10	≤5
B	Crops of edible plants are eaten raw and the edible part does not come into direct contact with the reclaimed water. Edible crops that are processed and non-edible crops. Crops used to feed dairy or meat-producing animals are also included.	All irrigation methods	Secondary treatment and disinfection	≤100	≤25	≤35	-
C	Crops of edible plants are eaten raw and the edible part does not come into direct contact with the reclaimed water. Edible crops that are processed and non-edible crops. Crops used to feed dairy or meat-producing animals are also included.	Drip irrigation or other irrigation method that avoids direct contact with the edible part of the crop	Secondary treatment and disinfection	≤1000	≤25	≤35	-
D	Industrial and energy crops and seed crops.	All irrigation methods	Secondary treatment and disinfection	≤10,000	≤25	≤35	-

Circular Economy Model

Thus far, the predominant economic model applied was the linear economy model. A model like that considers used water as waste and operates on the principle of "extract, produce, dispose", as shown in Figure 4a. It exploits water resources, which are later treated as waste, obstructing their reuse even when possible. Essentially, it is an economic system that considers the product's life shorter and thus limited. This conservative model is deemed to be unsustainable in the long term [27].

Gradually, the transition towards a circular economy model was perceived as the only sustainable path for the environment and the economy. Water naturally exhibits a circular nature through its cycle. Therefore, the application of a circular model that incorporates policies about water usage is feasible, and it provides solutions and promotes ecological water management practices [27].

(**a**) (**b**)

Figure 4. (**a**) Linear economy model; (**b**) simplified presentation of closed-loop water management principles.

Before diving more into sustainable water use through this model, a definition should be provided. A circular model is defined as a techno-economic model that advocates for the conservation and enhancement of natural resources and for their increased efficiency. The term "circular economy" became more widely known in the 1960s and 1970s when a redesign of the model of product mobility was needed. However, the exact time of its emergence is not clear. The organization "Ellen MacArthur" played a significant role in promoting and popularizing the concept [28].

This model ought to be applied in the context of water management. A simplified version of its principles is shown in Figure 4b. More specifically, compared to the linear model, it is based on nine strategies: "Rethink, Avoid, Reduce, Replace, Reuse, Recycle, Cascade, Store, and Recover". The first strategy involves reorganizing the way water is used to achieve a more circular approach. The next three strategies focus on reducing the water quantities used. Moreover, the strategies "Reuse, Recycle, Cascade" promote sustainable resource management through continuous reuse. Finally, the last two strategies concern the maintenance and preservation of used water in tanks for future recovery [29]. This results in a system that promotes sustainable development and protects water resources.

For a complete and uninterrupted transition, from the conservative linear model to the circular water management model, certain strategic tools, known as "roadmaps", play a crucial role. They are used by governments to describe the current situation that must change, to set and prioritize goals for the future, and to define a strategy with which the set goals can be achieved. Due to the radical change that must be applied, "roadmaps" should include the idea of water reuse [30]. However, it is observed that many countries that use "roadmaps" for a circular economy do not include the need for water reuse [4].

Therefore, the implementation of a circular economy model—with the additional help of roadmaps—that clearly incorporates water reuse can provide numerous economic, social, and environmental benefits. Among these advantages are the conservation of water resources, increased water availability, its sustainable use, the maximization of the water's value, as well as processes in which it is used, reduced environmental pollution, and an upgrade in the quality of life [4].

4. Water Reuse in Europe

Current wastewater treatment methods allow water of sufficient quality to be recovered for most uses. Therefore, the quality of water used for irrigation must meet certain limits, but without necessarily reaching the quality of drinking water. This way, higher-quality water can be stored or used for domestic applications. In several cases around the world—after appropriate treatment—reclaimed water has been used to provide ecological flows or environmental volumes that consequently meet many requirements set by relevant regulations. Water can be reclaimed for industrial, agricultural, or recreational uses [31].

The use of reclaimed water in agriculture reduces fertilizer consumption since the nutrients contained in these waters can be absorbed into the soil, resulting in a general improvement in crop production. It also provides alternatives for the discharge of treated wastewater in areas where discharge is difficult, such as sites located on public property [4,31].

Despite the advanced technology and the proven benefits of reusing wastewater, there is currently no international norm. There are global reference recommendations—particularly for agricultural use—from the World Health Organization (WHO) published in 1989, with subsequent revisions and extensions [31].

In northern Europe, 51% of reclaimed water is used for environmental purposes (enhancing existing water sources, creating wetlands, etc.), while in southern Europe, it is mainly used for irrigation (44%). Spain, Italy, Cyprus, and Greece are pioneers in the use of reclaimed water in agriculture. Cyprus uses 76% (in 2013) of treated wastewater as irrigation water. In Spain, in 2016, 10% of wastewater was recovered and 61% of this was used in agriculture. Researchers said that with current water reuse applications in the EU, water stress reduction is projected to be only 1% by 2030, although the potential had been estimated at 14%. In Israel, water reuse covers 50% of irrigation water needs, monitors less than a dozen parameters but defines the required level of treatment and crop/product type, and irrigation technique. Italy—on the other hand—sets parameter limits, some of which are stricter than those for drinking water [4].

Spain, Italy, Germany, Greece, and France were the five countries with the highest water reuse volume for the year 2000, and Spain, Italy, Bulgaria, Turkey, Germany, France, Portugal, Greece, Poland, and Belgium are the ten countries projected to have the highest reuse volume in 2050 in Europe. Spain appears to have the leading position in Europe with a projection that it will maintain this position over time. Some other countries, such as Greece, while currently occupying a high position in the ranking, do not seem likely to maintain their position in 2050 [32].

Spain has the highest water reuse volume in Europe and ranks 10th in the world. It shows great variations in the distribution of water resources across its territory and is experiencing water shortages in the southeast [32]. Reclaimed water in Spain is mainly used in agriculture, with more than 40% of all reclaimed wastewater being used for irrigation of pastures, agriculture, woody crops, ornamental plants, nurseries and forages, and products of human consumption in fresh water, among others. Second to agricultural use, irrigation of parks and recreational areas comes second (36% of total reuse), with industry using 10%. It is important to note that, ultimately, reclaimed water makes up 5.4% of total water use in Spain, but this percentage can exceed 25% in certain regions [31].

In Italy, the reuse of wastewater for agriculture (and more recently for landscape irrigation) is the most promising option among the various reuse applications [33]. However, regulated reuse of urban wastewater in agriculture is not yet developed in most Italian regions. The improvement of existing cases of agricultural reuse and the development of irrigation practices requires a reform of the current legislation. More realistic regulations should be established based on the findings of recent research work and experience from the uncontrolled reuse, which is common in the inland areas of southern Italy. The growing interest in wastewater reuse is mainly focused on the control of water body pollution in the northern and central regions, while in the southern regions reuse could have a significant impact on alleviating shortages in water supply, especially in the agricultural sector [33].

In Greece, due to the limited water supply and the high demand for freshwater—especially during the hot season—it is necessary to exploit every possible water resource [34]. Crop irrigation is one of the most intensive consumers of water, especially during the summer. Despite this need, only a few wastewater reuse projects have been implemented, and most of them are pilot projects for crop or landscape irrigation. The most important projects currently in operation are those in Thessaloniki, Chalkida, Malia, Livadia, Amfissa, Kallikratia, and Hersonissos. One of the most important wastewater reuse projects is that of Thessaloniki, in northern Greece. In this project, treated wastewater is mixed with fresh water in a ratio of 1:5, mainly because of the high salinity of the wastewater. Part of the total WWTP of "EELTH" effluent, 165,000 m^3/day is mixed with water from the Axios River for the irrigation of about 2500 hectares of rice, corn, alfalfa, sugar beet, and cotton. Several other projects are currently being planned, such as those in Heraklion, Agios Nikolaos, and various Aegean islands [34].

It should be mentioned that there are several cases of indirect reuse of wastewater from wastewater treatment plants in central Greece. However, the potential for reuse in Greece is limited, since the reuse of wastewater generated by the WWTP in Psyttaleia, Athens, which covers about half of the Greek population, is not economically feasible due to the location of the treatment plant [34].

5. Recovery Methods

The first historical reference to the reuse of water is recorded in 3000 BC in Crete, where the Minoan civilization used its wastewater for agricultural irrigation. The reuse of wastewater finds application even today throughout the world [3]. According to the European Union, 40% of the wastewater produced could be recycled [35]. When the reuse of wastewater is required, then an additional treatment should be carried out to reduce the risks and to ensure the required quality. The main need for extensive wastewater treatment is to remove pathogenic microorganisms and chemical contaminants [32].

Each recovery technology has its characteristics, and it is usually necessary to use two or more technologies to achieve the required water quality levels. The selection of the appropriate process must consider a variety of factors such as the quality and quantity of water in terms of water recovery, the final quality required for the corresponding use, the economic cost, and the environmental impact [32].

Restoration technologies can be classified as conventional and extended technologies. Conventional ones are characterized by the need for large amounts of energy and minimal space. On the other hand, extensive technologies require a large area of land due to the fact that they are based on natural processes, and as a result, the processes occur at natural rates, the energy requirements are very low, and the technologies require low levels of operation and maintenance [32].

One of the conventional recovery technologies and the most widespread is water recycling with membrane processes and more specifically with reverse osmosis. According to new wastewater management research, there are two terms: minimum liquid discharge (MLD) and zero liquid discharge (ZLD) [36]. The MLD concept uses less energy-intensive processes compared to ZLD but can exhibit a high rate of water recycling. Using membrane technologies, MLD can reach efficiencies of up to 80–95%. The ZLD process refers to the complementation of MLD technology with technologies based on thermal thickening, evaporation, and crystallization. The above processes are very energy-intensive, but they can treat wastewater with a high load concentration and reach an efficiency rate almost equal to 100% [16,37]. Membrane technology offers numerous benefits such as high-quality treated water, compact and durable construction, cost-effective operation, and maintenance [38]. Membrane processes such as reverse osmosis (RO) and nanofiltration (NF) are highly effective in removing contaminants from water, including bacteria, viruses, dissolved salts, and organic compounds [39]. This ensures that recycled water meets quality standards for safe reuse. They can be tailored to specific water quality requirements, allowing for flexible treatment options depending on the intended reuse application. Different membrane types and configurations can be utilized to achieve desired water quality standards. While membrane processes require initial investment, they can offer long-term cost savings by reducing the need for freshwater intake and wastewater disposal [38]. The cost of membrane technology has been decreasing over time due to advancements.

Other recovery technologies are through physicochemical systems (coagulation–flocculation), membrane technologies (ultrafiltration), rotating biological contactors, and disinfection technologies (UV radiation, ClO_2, etc.). The above technologies belong to the category of conventional technologies. Extended technologies include stabilization ponds, constructed wetlands, and infiltration–filtration systems [32]. Each recovery technology has its characteristics, and it is usually wise to use more than one technology to achieve the ideal quality performance. The choice of technology must be made based on the quantity and quality of the wastewater to be recovered, the required final quality thereof, the economic cost, and the environmental impact. The directive (2010/75/EU) defines the term Best

Available Technique (BAT). This term is applied to the selection of the most appropriate water recovery technique for the specific purpose of its use [26,32].

6. Barriers to Water Reuse

The reuse of purified water leaving a decentralized wastewater treatment plant is mainly applied to the irrigation of agricultural crops—in a proportion exceeding 50% of the total reused water—but also in industry, firefighting, recreational purposes, recharging groundwater resources, water supply for daily domestic use (e.g., flushing), or even for drinking water. However, while in many countries a large proportion of urban wastewater is treated appropriately, very little of the treated water is reused [4,40]. The causes of this finding are technical, economic, institutional (regulatory), and social barriers [4,41].

Regarding the first category, there are a few conditions that need to be met for the treatment of wastewater to be considered adequate and they are related to concentration limits for certain constituents in the effluent. These are inorganic constituents, such as nitrogen (N), sulfur (S), phosphorus (P), chlorine (Cl), and their compounds or toxic heavy metals, such as copper (Cu), lead (Pb), boron (B), arsenic (As) and zinc (Zn), organic pollutants, i.e., compounds such as lipids, proteins, carbohydrates, surfactants, phenols, pharmaceuticals, and pesticides, which are categorized as persistent organic pollutants, but also several other constituents. Other characteristics of wastewater that need to be addressed are pH and pathogenic microorganisms, which are responsible for causing diseases [3]. The combination of primary and secondary treatment is not sufficient to produce water suitable for irrigation, so the tertiary treatment should be included in the design of the plant. Such processes include membrane filtration (ultrafiltration, membrane bioreactors), advanced oxidation methods (chlorination, ozonation, UV radiation), and adsorption with materials that possess high specific surface areas such as granular activated carbon [4,40]. Nevertheless, while these processes achieve high purification efficiencies so that the effluent stream can even be used as potable water, they significantly increase costs. It is expected, however, that advances in the know-how, performance, and operation of the treatment processes as well as the training of the personnel will help to overcome the current obstacles [42].

Most municipal wastewater ends up in large, decentralized plants located at low altitudes so that it can reach them by gravity and thus lower the energy costs for pumping. Obviously, the reclaimed water cannot be distributed either through the sewage network or the water supply and irrigation network. Consequently, its transfer to the crops requires not only operating costs but also the construction of a new distribution network that must cope with the huge distances between the arable land and the wastewater treatment plant, cross-connection issues with other networks, and proper management of the water distribution [4,42,43].

It has been argued that the institutional barrier may have an even more catalytic effect than the aforementioned category. Insufficient information and, possibly, awareness among policymakers contribute to the ineffective legislative infrastructure. Water quality is vital for the good quality of crops and therefore the health of consumers. Water quality criteria must therefore be strict, without at the same time making it impossible to achieve a circular economy. The examples of Israel and Italy are illustrative, where the former has more flexible legislation, also related to the type of crops, while the latter has limited reuse due to strict regulatory provisions. Legislation should keep pace with research progress which is constantly being applied to existing processes. In turn, applications feed the research sector with valuable data, thus creating a constructive perpetual cycle. However, there is often disagreement between stakeholders (e.g., agriculture and wastewater treatment), which discourages public opinion [4,44].

European regulation 2020/741, which sets limits for Total Suspended Solids (TSS), turbidity, Biological Oxygen Demand (BOD), and Escherichia coli bacteria, is an attempt to establish a common policy to ensure the quality of water for irrigation and to inform citizens. Nonetheless, the number of EU Member States opposed to the reuse of reclaimed water has not been reduced to zero [4,25,44]. In the US, water reuse is gaining momentum due to alarming data on the availability of conventional water resources. Interestingly, the regulations between different states—even those that are adjacent—show great disparity. Thus, it is suggested that the need for reuse—expressed through more lenient provisions—is a very localized problem that usually stems from the low availability of conventional water resources. Finally, common to all states is that the statutory treatment requirements show a gradation depending on the use of the reclaimed water [44].

It is a fact that water reuse is not widely accepted by society. One reason for this is the lack of public awareness of this unprecedented practice of the circular economy model. Consumers' concerns relate to the use, cost, quality (odors, color), and lack of confidence in the proper functioning of waste treatment plants [41,42]. For their part, farmers are wary of its use for reasons related to its stable levels of quality, risks regarding the quality of crops and land, and the commercial value of the products produced. Potential financial facilitation in the supply of reclaimed water and the fact that its high content of nutrients such as nitrogen and phosphorus reduce the need for fertilizers are two reasons that may draw their attention to it. To consolidate water reuse, consumer awareness should go hand in hand with the production of goods from crops irrigated with that kind of water. The contribution of the academic community can play an important role here [4].

7. Acceptance

The following paragraphs analyze research carried out in America, Europe, and Greece. The purpose of these questionnaires was to record and draw conclusions regarding the acceptance of the human population regarding the inclusion of reclaimed water in their daily life.

7.1. Acceptance of Water Recycling in the USA

The goal of this research is to explore the factors that affect underlying public reluctance toward using reclaimed water. The initial goal was to ascertain, through a more qualitative method, the elements that individuals self-report as being significant in their decision-making regarding water recycling. The second goal was to employ a multivariate technique to assess the degree of correlation between expected dislike toward recycled water and willingness to use it, while also accounting for many theoretically significant individual difference characteristics [45].

Using questions developed on a five-point Likert scale ranging from 1 "strongly disagree" to 5 "strongly agree", participants expressed [46]:

(1) Their willingness to drink different water alternatives.
(2) Their trust in different public/private/government/non-profit organizations related to drinking water.
(3) Their willingness to implement water reuse for indoor/outdoor activities.
(4) Willingness to pay for drinking water/wastewater reuse.

Numerous crucial elements for gaining public acceptance have been discovered by the substantial literature on water recycling [47,48]. Specifically, it has been shown that unpleasant emotional responses—also referred to in the literature as the "yuck factor"—have a significant influence on how society as a whole and individuals make decisions regarding water recycling [46]. When adjusting for other factors like environmental and risk attitudes, trust, social pressure to support the scheme, and perceived control, Nancarrow et al. [49] demonstrated that emotions, in this case, reports of general negative or positive feelings toward a water reuse scheme, predicted willingness to use recycled water at the individual level. According to Haddad et al. [50], there was a strong correlation between the readiness to utilize recycled water and psychological characteristics such as sensitivity to concepts of disgust and contagion. In a more recent study, after adjusting for individual differences and political party identification, negative emotional responses toward water recycling were specifically linked to degrees of pathogen disgust sensitivity, or how quickly one becomes disgusted by stimuli connected to infections [51].

The impact of expected distaste on the desire to utilize recycled water for higher contact uses (drinking), even after adjusting for other significant variables (demographic, experiential, attitudinal, and ideological), remains to be determined. It has been discovered that certain demographic traits, such as gender, age, and educational attainment, predict the perceived risk connected to water recycling. Research focusing on this topic has generally shown that older people, women, and people with lower levels of education believe that recycling water poses a greater risk. It should be highlighted, although, that results of demographic associations vary throughout research and situations [47]. In numerous research, practical considerations like price and the quality of the water have also been shown to be significant [52].

Personal beliefs and attitudes seem to be important as well. In certain situations, acceptance is predicted by environmental awareness and pro-environmental sentiments [45]. Numerous studies have revealed that the general public believes that water recycling is harmful to public health and beneficial to the environment [53]. Therefore, depending on how the general public conceptualizes water recycling, prosocial ideals may lead to contradicting outcomes. The acceptance of water recycling may be influenced by ideological factors, such as moral reasons. In many groups, the perceived religious purity of the water recycling technique may hold significance [45]. Previous experience could also be significant; consuming tap water has been linked to less emotional distress when considering drinking recycled water [51].

The heterogeneity brought about by various water reuse techniques adds to the confusion surrounding the criteria that matter most when making judgments about water recycling. One of the best indicators of attitudes toward water recycling is the extent of "bodily contact" with reclaimed water. Numerous studies have shown that "low contact" usage, such as flushing the toilet, is more acceptable than "high contact" uses, like drinking or taking a shower [46]. Furthermore, depending on the type of water being treated (e.g., grey versus black water; indirect versus direct potable reuse), the perceived relationship between recycled water and bodily products (e.g., feces) may differ between water reuse methods. Crucially, discomfort with contamination is strongly predicted by the idea of bodily products [54], especially when there is a lot of physical interaction. It is challenging to compare research examining attitudes toward specific water reuse methods since the degree of bodily contact and perceived association with body products frequently vary from context to context. This makes it challenging to identify what factors are most crucial in boosting acceptability. From a policy perspective, it is especially crucial to increase acceptance of high-contact uses, such as drinking recycled water, as restricting recycled water's use to low-contact applications might either severely limit its potential applications or need costly pipe upgrading [46].

In the literature on water reuse acceptance, there is debate over whether it is beneficial to concentrate on the "yuck factor" [45]. This framing, according to Russell and Lux [55] is restrictive because it presents the general population as emotional and irrational in contrast to more "rational" expert viewpoints. Studies that show that disparities in opinion between the general public and scientific specialists are not as great or unbridgeable when communities are involved in governance lend credence to this claim. Furthermore, according to Russell and Lux [55], the "yuck factor" narrative suggests that public opinion is set and that responses to disgust are unavoidable. Rather, they contend that the key to raising acceptance is a strategy centered on fostering public conversation and trust. Additional research on water recycling has indicated that institutional and cultural, value-based elements are crucial for winning over the public, which lends support to this strategy [56,57]. The extent to which these two strategies might actually be complementary rather than mutually incompatible is unknown.

7.2. Social Acceptance in Europe Regarding the Reuse of Treated Water in Specific Applications

In this section, an attempt is made to outline the level of social acceptance of the reuse of treated wastewater and gray water in applications such as agricultural irrigation and domestic (toilet flush, garden irrigation) or recreational (park, golf course irrigation) use. Survey data for regions of Europe are analyzed. It is important to mention that although the most recent European directive (2020/741) from the European Commission sets the minimum requirements for the implementation of water reuse systems, nevertheless, many EU states have not made decisions to implement the directive due to factors such as differentiation of the legal system depending on the region, water management problems and/or social acceptance [58].

A total of 54 farmers/growers in Reggio-Emilia in Emilia-Romagna (Italy) participated in a related survey on the social acceptance of water reuse applied to cropland irrigation [59]. At that time, the local Wastewater Treatment Plant conducted a water reuse project that not all respondents were aware of. However, over half had a positive view of the water quality in the area [59]. Overall, 93% of the farmers had a positive view of the quality of treated water. The perceived risk (e.g., for health, environmental, etc.) of water reuse cultivation emerged as a catalytic factor influencing acceptance.

Additionally, certain crop types tended to be more accepted than others, such as growing alfalfa as opposed to tomatoes. Some farmers (46%) recognized the advantages of water reuse, especially those who were informed about the reuse project [59]. The research confirms the relationship between low educational levels and the perception of fewer advantages and therefore less acceptance. In summary, research reveals that reduced risk perception is associated with increased acceptance without necessarily identifying specific benefits. At the same time, it sets the framework of water scarcity and climate change around the issue, but also covers the future needs of water users [59].

In France, the reuse of treated wastewater is not a common topic of discussion, nor is it covered by the media [60]. However, 70% of consumers of agricultural products in the Pic Saint-Loup region in the south of France who took part in a related survey would agree with its application in the irrigation of products such as fruit and vegetables and 90% with the irrigation of green areas (e.g., parks) (Figure 5). Research confirms that consumer acceptance increases by providing "neutral" information about the issue. Furthermore, it is confirmed that factors such as awareness of environmental issues and perceived health risks play an important role in shaping acceptance [60]. It is recommended to initiate intensive, targeted information campaigns and to deepen studies, including taste tests (as a significant minority show some inconsistencies between attitudes, food purchase intentions, disgust, and perceived risks).

Evidence from Mesa-Pérez et al. [61] collected from eight European Union countries (Belgium, Bulgaria, France, Germany, Greece, Spain, Italy, Portugal) identify (a) the high cost of reclaimed water for farmers and (b) the societal distaste for products irrigated with recycled water as key factors in the applicability of reuse systems and as a cornerstone of the EU strategy to promote the use of reclaimed water in agriculture [61].

The example of domestic gray water treatment and reuse systems comes from a group of households in the Sant Cugat del Vallès District in Barcelona, Spain [62]. In this case, too, the most critical factor affecting acceptance emerged as the perceived health risk, which most users (>90%) rated positively or neutrally. However, the reuse systems implemented did not perform optimally, resulting in dissatisfaction among many users. In many cases, the dissatisfaction was outweighed by the positive consequences for the environment. The cost of the systems in this case did not play a particularly decisive role in acceptance, in contrast to the level of information that was negatively evaluated. The estimated level of acceptance was 72% of experienced users (Figure 5) [62].

Students in Germany who had no previous experience with reuse systems showed acceptance especially for applications such as watering plants and the garden, flushing the toilet, or washing the car [63]. Showering and drinking water received the least acceptance. The perceived health risk was judged to be a determining factor, and higher educational level contributed to a greater likelihood of acceptance [63]. In addition, knowledge of possible unplanned reuse led to a higher level of acceptance of planned reuse. Factors shaping acceptance included vocabulary and level of information about recycled water, as well as emotional factors (e.g., disgust). The research concludes that applying quality standards similar to those of drinking water and the educational preparation for outreach education could lead undecideds to acceptance.

A study by Lyach et al., 2023 [64], correlates water savings methods with the reuse of treated greywater in Czech households (Figure 5). Although the majority of respondents practiced saving methods, only 17% viewed reuse positively. Households would accept water reuse in toilet flushing, car washing, room cleaning, and lawn watering, but not in activities involving contact with water [64]. Hygiene emerged as the most important factor influencing acceptance, while comfort and practical utility were also considered important. Lack of understanding of water purification technology as well as trust are factors influencing acceptance. The "eww" effect for treated gray water is, also, strong among the Czech public.

In summary, from the above research, we conclude that the acceptance of the reuse of treated water (grey water, wastewater) is directly influenced by similar factors regardless of the region of Europe for which the study is conducted. At the same time, complementary factors are being considered depending on the specific applications of water reuse. The role of demographic characteristics (age, gender, level of education, etc.) is not confirmed in all studies. Primarily, the most consistent catalyst factor that appears to have a dominant effect on the level of acceptance is perceived health risk. Secondarily, other factors that are considered important are environmental awareness, disgust, the information provided, and cost.

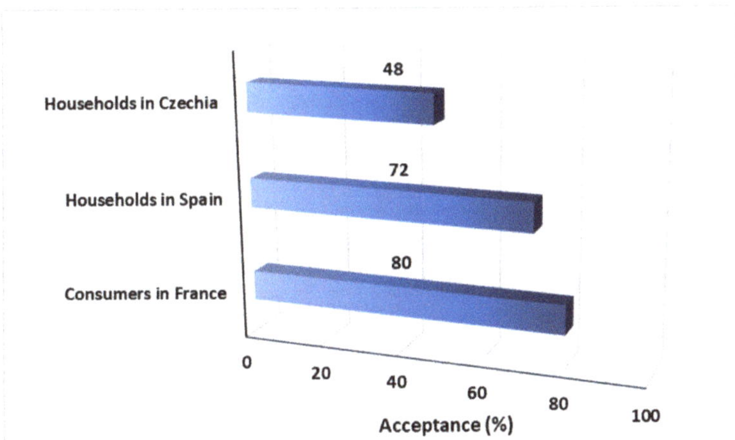

Figure 5. Acceptance of water reuse in different applications [60,62,64].

7.3. Acceptance of Water Reuse in Greece

Water scarcity and droughts are likely to deteriorate further due to increasing human activity and observed climate change. Over the last few years, drought phenomena have increased significantly in number and intensity in regions of the European Union, affecting approximately 11% of the population. Greece (especially its southern part and islands) is one of the Mediterranean countries that is expected to suffer severe water scarcity problems in the upcoming years [65].

Water originating from the treatment of mainly municipal waste has already emerged as a reliable source for various activities and uses. In Greece, more than 492 Wastewater Treatment Plants (WWTPs) produce wastewater with high reuse potential. Greece is a country that implements water reuse, mainly for crop irrigation. Recycled water is primarily used in semiarid areas where the phenomenon of water scarcity is more intense. The most important water reuse projects that have been implemented are located in the area of Thessaloniki, Chalkida, and Hersonissos, in Crete [66–68]. Specifically, according to data from the Hellenic Association of Municipal Water Supply and Sewerage Companies, until March 2023, approximately 10.95 hm^3/yr of recycled water was used in Greece, of which 1.16% was allocated for agricultural irrigation. However, the majority of treated wastewater is still discharged into the sea [65].

While water reuse is an issue of high national priority, several factors are responsible for the limited use of recycled water in Greece to date [65]. The acceptance of the use of recycled water by society is emerging as a critical parameter for the effective implementation of a water reuse system. Numerous international studies have been conducted to examine the factors that influence the greater acceptance of this practice [66]. Most of the research on the acceptance of the use of recycled water in Greece has been conducted in the form of questionnaires or personal interviews, the majority of which has focused on the region of Crete, where the use of recycled water for agricultural irrigation is widespread. Corresponding surveys for other regions of Greece, such as Chalkida, Thessaly, and Thessaloniki, are limited. Respondents included people of different ages, educational levels, social backgrounds, and incomes, and in several cases farmers who had already implemented the use of recycled water in the irrigation of their crops [66].

Based on the results, Greek society adopts a positive attitude towards the use of recycled water for various applications. However, the attitudes of citizens show significant changes when they take into account the origin and possible uses of recycled water. Specifically, the vast majority of respondents stated that they would not choose to use recycled water as drinking water [66]. Its use as irrigation water is more acceptable in crops that undergo some kind of treatment before consumption, such as olive trees, in contrast to vegetable crops, such as tomatoes, which can be consumed without further processing [69]. At the same time, the use of recycled water in activities that are not directly or indirectly related to food (e.g., urban parks, sports fields, firefighting) is more acceptable than all the above applications [65,67,70].

Acceptance of the use of recycled water can be examined from both the perspective of consumers and farmers. From the consumer's perspective, acceptance is affected by environmental awareness, education, income levels, and age. Farmers who face water scarcity problems and apply environmentally friendly practices to cultivate the land are more likely to adopt recycled water for irrigation. Consumers are generally willing to incorporate recycled water into their food purchasing habits, which will also contribute to the wider acceptance of farmers' use of recycled water to irrigate their crops. Thus, a significant change in the social attitude towards the use of recycled water is likely [65,66,69].

The impact of targeted educational and informational programs on increasing public acceptance of the use of recycled water is worth mentioning. Lack of information and education is a proven factor influencing peoples' judgment [67]. The skepticism of consumers, as well as farmers, about water reuse, is mainly due to concerns about health risks. They consider that recycled water can be a source of toxic substances and pathogenic microorganisms and that its use will lead to a deterioration of their health, opinions that are largely attributed to insufficient knowledge and information about the methods of wastewater treatment and water reuse [66,67].

Furthermore, citizens' trust in public authorities plays a major role in the acceptance of recycled water use by people. The overall conclusion is a pervasive doubt about the ability of the authorities to enforce regulations for the welfare of consumers, which further intensifies citizens' skepticism towards water reuse [66,67]. Nevertheless, there is a remarkable social acceptance of the use of recycled water, which encourages the expansion of its applications.

The pricing of recycled water is a special parameter influencing its acceptance by society. The majority of consumers and farmers claim that they are willing to pay a lower price for recycled water than for fresh water. In general, economic factors such as the price of fresh water and income are important in the social acceptance of water reuse [66,69,71]. Particularly encouraging is the research of Petousi et al. [66], from which it appears that a significant percentage of farmers in the region of Hersonissos, Crete, are even willing to invest in new irrigation systems for recycled water.

In summary, despite the challenges, using recycled water is a valuable resource. Greek society demonstrates a positive attitude towards its reuse, mainly in agriculture. The acceptance of using recycled water is multi-factorial. Various surveys reveal diverse views of the community. Citizens express health concerns and doubts about the authorities' ability to enforce the necessary regulations. However, there is a general social acceptance of the use of recycled water, which is enhanced by factors such as environmental awareness and appropriate information. Food producers and consumers are willing to incorporate recycled water into their daily practices and habits, indicating a potential shift in social attitudes, which is believed to be further fueled by citizens' need for fresh water.

8. Conclusions

In this paper, the main goal was to review water reuse through literature on different levels. The reason why water reuse has become so important is none other than the rising water scarcity levels, which are projected to intensify even more shortly. Water withdrawal for agricultural, industrial, and urban use has been one of the main factors that contribute to the water scarcity phenomenon, with agriculture being the main consumer. Industry is also a major consumer and is accompanied by a large production of waste which needs further treatment. The EU has been publishing directives and regulations on water reuse for the last 40 years, the latest being regulation (EU) 2020/741, which sets certain limits connected to minimum requirements for water reuse. The current applications of water reuse in northern Europe are for environmental purposes (51%), while in southern Europe, it is mainly used for irrigation (44%), with Spain reusing the largest volume. However, while a large proportion of urban wastewater is treated appropriately, very little of the treated water is reused due to various barriers that include technical, economic, institutional–regulatory, and social barriers. An extensive review of studies involving the acceptance of water reuse was carried out to identify the reasons why water is not being reused to its full potential. The results varied because of several factors, such as the region, the reuse method (e.g., irrigation, indoor/outdoor activities, drinking water), and due to economic, social, and emotional factors.

Author Contributions: Conceptualization, I.A.K.; writing—original draft preparation, F.F., M.G., G.I., D.L., E.L., N.L., M.S. and E.T.; writing—review and editing, M.X.; supervision, I.A.K. All authors have read and agreed to the published version of the manuscript.

Funding: This research received no external funding.

Data Availability Statement: Not applicable.

Conflicts of Interest: The authors declare no conflicts of interest.

References

1. Counts, T.W. Watch Our World's Social and Environmental Challenges. Available online: https://www.theworldcounts.com/ (accessed on 12 February 2024).
2. Water Statistics. Available online: https://ec.europa.eu/eurostat/statistics-explained/index.php?title=Water_statistics (accessed on 12 February 2024).
3. Asano, T.; Burton, F.; Leverenz, H.; Tsuchihashi, R.; Tchobanoglous, G. *Water Reuse: Issues, Technologies, and Applications*, 1st ed.; McGraw-Hill Education: New York, NY, USA, 2007; ISBN 978-0-07-145927-3.
4. Mannina, G.; Gulhan, H.; Ni, B.-J. Water Reuse from Wastewater Treatment: The Transition towards Circular Economy in the Water Sector. *Bioresour. Technol.* **2022**, *363*, 127951. [CrossRef] [PubMed]
5. Shevah, Y. Water Scarcity, Water Reuse, and Environmental Safety. *Pure Appl. Chem.* **2014**, *86*, 1205–1214. [CrossRef]
6. Torretta, V.; Katsoyiannis, I.; Collivignarelli, M.C.; Bertanza, G.; Xanthopoulou, M. Water Reuse as a Secure Pathway to Deal with Water Scarcity. *MATEC Web Conf.* **2020**, *305*, 00090. [CrossRef]
7. He, C.; Liu, Z.; Wu, J.; Pan, X.; Fang, Z.; Li, J.; Bryan, B.A. Future Global Urban Water Scarcity and Potential Solutions. *Nat. Commun.* **2021**, *12*, 4667. [CrossRef]
8. Water Use and Stress—Our World in Data. Available online: https://ourworldindata.org/water-use-stress (accessed on 12 February 2024).

9. *Water Use in Europe—Quantity and Quality Face Big Challenges*; European Environment Agency: Copenhagen, Denmark, 2018.
10. Mateo-Sagasta, J.; Marjani, S.; Turral, H. *Water Pollution from Agriculture: A Global Review. Executive Summary*; Food and Agriculture Organization of the United Nations (FAO): Rome, Italy; International Water Management Institute (IWMI): Colombo, Sri Lanka, 2017.
11. Verones, F.; Hanafiah, M.M.; Pfister, S.; Huijbregts, M.A.J.; Pelletier, G.J.; Koehler, A. Characterization Factors for Thermal Pollution in Freshwater Aquatic Environments. *Environ. Sci. Technol.* **2010**, *44*, 9364–9369. [CrossRef] [PubMed]
12. Fernández Scagliusi, M.d.l.Á. El uso de agua en la minería. En particular, la huella hídrica y la huella del agua, dos indicadores fundamentales (Water use in mining. In particular, the two kinds of water footprint, two key indicators). *Rev. Catalana Dret Ambient.* **2021**, *12*, 1–37. [CrossRef]
13. Zhou, Q.; Yang, N.; Li, Y.; Ren, B.; Ding, X.; Bian, H.; Yao, X. Total Concentrations and Sources of Heavy Metal Pollution in Global River and Lake Water Bodies from 1972 to 2017. *Glob. Ecol. Conserv.* **2020**, *22*, e00925. [CrossRef]
14. Lazaridis, I. The Evolutionary History of Human Populations in Europe. *Curr. Opin. Genet. Dev.* **2018**, *53*, 21–27. [CrossRef]
15. Shan, Y.; Yang, L.; Perren, K.; Zhang, Y. Household Water Consumption: Insight from a Survey in Greece and Poland. *Procedia Eng.* **2015**, *119*, 1409–1418. [CrossRef]
16. Procházková, M.; Touš, M.; Horňák, D.; Miklas, V.; Vondra, M.; Máša, V. Industrial Wastewater in the Context of European Union Water Reuse Legislation and Goals. *J. Clean. Prod.* **2023**, *426*, 139037. [CrossRef]
17. *Council Directive 80/778/EEC of 15 July 1980 Relating to the Quality of Water Intended for Human Consumption*; 1980; Volume 229. EUR-Lex. Available online: https://eur-lex.europa.eu/legal-content/EN/TXT/?uri=celex:31980L0778 (accessed on 12 February 2024).
18. *Council Directive 91/271/EEC of 21 May 1991 Concerning Urban Waste-Water Treatment*; 1991; Volume 135. EUR-Lex. Available online: https://eur-lex.europa.eu/legal-content/EN/TXT/?uri=celex:31991L0271 (accessed on 12 February 2024).
19. *Council Directive 98/83/EC of 3 November 1998 on the Quality of Water Intended for Human Consumption*; 1998; Volume 330. EUR-Lex. Available online: https://eur-lex.europa.eu/legal-content/EN/TXT/?uri=celex:31998L0083 (accessed on 12 February 2024).
20. *Directive 2000/60/EC of the European Parliament and of the Council of 23 October 2000 Establishing a Framework for Community Action in the Field of Water Policy*; 2000; Volume 327. EUR-Lex. Available online: https://eur-lex.europa.eu/eli/dir/2000/60/oj (accessed on 12 February 2024).
21. *Directive 2013/39/EU of the European Parliament and of the Council of 12 August 2013 Amending Directives 2000/60/EC and 2008/105/EC as Regards Priority Substances in the Field of Water Policy Text with EEA Relevance*; 2013; Volume 226. EUR-Lex. Available online: https://eur-lex.europa.eu/legal-content/EN/TXT/?uri=uriserv:OJ.L_.2013.226.01.0001.01.ENG&toc=OJ:L:2013:226:FULL (accessed on 12 February 2024).
22. Directive 2006/118/EC of the European Parliament and of the Council of 12 December 2006 on the Protection of Groundwater against Pollution and Deterioration. Available online: https://www.legislation.gov.uk/eudr/2006/118/contents (accessed on 13 February 2024).
23. Ahmed, J.; Thakur, A.; Goyal, A. Industrial Wastewater and Its Toxic Effects. In *Biological Treatment of Industrial Wastewater*; Shah, M.P., Ed.; The Royal Society of Chemistry: London, UK, 2021. [CrossRef]
24. *EU-Level Instruments on Water Reuse: Final Report to Support the Commission's Impact Assessment*; 2016. Publications Office of the European Union. Available online: https://op.europa.eu/en/publication-detail/-/publication/b4b562f5-9ad0-11e6-868c-01aa75ed71a1/language-en (accessed on 12 February 2024).
25. *Regulation (EU) 2020/741 of the European Parliament and of the Council of 25 May 2020 on Minimum Requirements for Water Reuse (Text with EEA Relevance)*; 2020; Volume 177. EUR-Lex. Available online: https://eur-lex.europa.eu/legal-content/EN/TXT/?uri=CELEX:32020R0741 (accessed on 12 February 2024).
26. Directive 2010/75/EU of the European Parliament and of the Council of 24 November 2010 on Industrial Emissions (Integrated Pollution Prevention and Control). *Off. J. Eur. Union* **2010**, *334*, 17–119.
27. Vinayagam, V.; Sikarwar, D.; Das, S.; Pugazhendhi, A. Envisioning the Innovative Approaches to Achieve Circular Economy in the Water and Wastewater Sector. *Environ. Res.* **2024**, *241*, 117663. [CrossRef] [PubMed]
28. Winans, K.; Kendall, A.; Deng, H. The History and Current Applications of the Circular Economy Concept. *Renew. Sustain. Energy Rev.* **2017**, *68*, 825–833. [CrossRef]
29. Morseletto, P.; Mooren, C.E.; Munaretto, S. Circular Economy of Water: Definition, Strategies and Challenges. *Circ. Econ. Sustain.* **2022**, *2*, 1463–1477. [CrossRef]
30. McDowall, W. Technology Roadmaps for Transition Management: The Case of Hydrogen Energy. *Technol. Forecast. Soc. Chang.* **2012**, *79*, 530–542. [CrossRef]
31. Jodar-Abellan, A.; López-Ortiz, M.I.; Melgarejo-Moreno, J. Wastewater Treatment and Water Reuse in Spain. Current Situation and Perspectives. *Water* **2019**, *11*, 1551. [CrossRef]
32. Joint Research Centre (European Commission); Gawlik, B.M.; Alcalde-Sanz, L. *Water Reuse in Europe: Relevant Guidelines, Needs for and Barriers to Innovation*; Publications Office of the European Union: Luxembourg, 2014; ISBN 978-92-79-44399-2.
33. Barbagallo, S.; Cirelli, G.L.; Indelicato, S. Wastewater Reuse in Italy. *Water Sci. Technol.* **2001**, *43*, 43–50. [CrossRef] [PubMed]
34. Ilias, A.; Panoras, A.; Angelakis, A. Wastewater Recycling in Greece: The Case of Thessaloniki. *Sustainability* **2014**, *6*, 2876–2892. [CrossRef]

35. Water Assessments for 2012—European Year of Water—European Environment Agency. Available online: https://www.eea.europa.eu/themes/water/european-waters/water-quality-and-water-assessment/water-assessments-2012 (accessed on 13 February 2024).
36. Panagopoulos, A.; Giannika, V. Decarbonized and Circular Brine Management/Valorization for Water & Valuable Resource Recovery via Minimal/Zero Liquid Discharge (MLD/ZLD) Strategies. *J. Environ. Manag.* **2022**, *324*, 116239. [CrossRef]
37. Panagopoulos, A.; Haralambous, K.-J. Minimal Liquid Discharge (MLD) and Zero Liquid Discharge (ZLD) Strategies for Wastewater Management and Resource Recovery—Analysis, Challenges and Prospects. *J. Environ. Chem. Eng.* **2020**, *8*, 104418. [CrossRef]
38. Khanzada, N.K.; Al-Juboori, R.A.; Khatri, M.; Ahmed, F.E.; Ibrahim, Y.; Hilal, N. Sustainability in Membrane Technology: Membrane Recycling and Fabrication Using Recycled Waste. *Membranes* **2024**, *14*, 52. [CrossRef] [PubMed]
39. Zhong, X.; Li, R.; Wang, Z.; Wang, W.; Yu, D. Eco-Fabrication of Antibacterial Nanofibrous Membrane with High Moisture Permeability from Wasted Wool Fabrics. *Waste Manag.* **2020**, *102*, 404–411. [CrossRef] [PubMed]
40. Dolgen, D.; Alpaslan, M.N. Reuse of Wastewater from the Circular Economy (CE) Perspective. In *A Sustainable Green Future: Perspectives on Energy, Economy, Industry, Cities and Environment*; Oncel, S.S., Ed.; Springer International Publishing: Cham, Switzerland, 2023; pp. 385–408, ISBN 978-3-031-24942-6.
41. Voulvoulis, N. Water Reuse from a Circular Economy Perspective and Potential Risks from an Unregulated Approach. *Curr. Opin. Environ. Sci. Health* **2018**, *2*, 32–45. [CrossRef]
42. Poudel, S.; Shrestha, A.; Kandel, N.; Adhikari, S.; Paudel, S.R. A Review of Reclaimed Water Reuse for Irrigation in South Asian Countries. *ACS EST Water* **2023**, *3*, 3790–3806. [CrossRef]
43. Jjemba, P.; Johnson, W.; Bukhari, Z.; LeChevallier, M. Review of the Leading Challenges in Maintaining Reclaimed Water Quality during Storage and Distribution. *J. Water Reuse Desalination* **2014**, *4*, 209. [CrossRef]
44. Hastie, A.G.; Otrubina, V.V.; Stillwell, A.S. Lack of Clarity Around Policies, Data Management, and Infrastructure May Hinder the Efficient Use of Reclaimed Water Resources in the United States. *ACS EST Water* **2022**, *2*, 2289–2296. [CrossRef]
45. Wester, J.; Timpano, K.R.; Çek, D.; Broad, K. The Psychology of Recycled Water: Factors Predicting Disgust and Willingness to Use. *Water Resour. Res.* **2016**, *52*, 3212–3226. [CrossRef]
46. Pathiranage, W.B.; Bray, L.; Jones, K.; Redwine, N.; Saralvarez, J.; D'Alessio, M. Perception and Acceptance towards Water Reuse in the Southeast United States: A Public Survey. *Sci. Total Environ.* **2024**, *908*, 168224. [CrossRef]
47. Mankad, A.; Tapsuwan, S. Review of Socio-Economic Drivers of Community Acceptance and Adoption of Decentralised Water Systems. *J. Environ. Manag.* **2011**, *92*, 380–391. [CrossRef]
48. Po, M.; Kaercher, J.; Nancarrow, B. *Literature Review of Factors Influencing Public Perceptions of Water Reuse. Australian Water Conservation and Reuse Research Program*; CSIRO Land and Water: Perth, Australia, 2004.
49. Nancarrow, B.E.; Leviston, Z.; Po, M.; Porter, N.B.; Tucker, D.I. What Drives Communities' Decisions and Behaviours in the Reuse of Wastewater. *Water Sci. Technol.* **2008**, *57*, 485–491. [CrossRef]
50. Haddad, B.M. *The Psychology of Water Reclamation and Reuse: Survey Findings and Research Road Map*; WateReuse Foundation: Alexandria, VA, USA, 2009; ISBN 978-1-934183-21-2.
51. Wester, J.; Timpano, K.R.; Çek, D.; Lieberman, D.; Fieldstone, S.C.; Broad, K. Psychological and Social Factors Associated with Wastewater Reuse Emotional Discomfort. *J. Environ. Psychol.* **2015**, *42*, 16–23. [CrossRef]
52. Hurlimann, A.; Hemphill, E.; McKay, J.; Geursen, G. Establishing Components of Community Satisfaction with Recycled Water Use through a Structural Equation Model. *J. Environ. Manag.* **2008**, *88*, 1221–1232. [CrossRef]
53. Dolnicar, S.; Hurlimann, A.; Grün, B. What Affects Public Acceptance of Recycled and Desalinated Water? *Water Res.* **2011**, *45*, 933–943. [CrossRef]
54. Olatunji, B.; Williams, N.; Tolin, D.; Abramowitz, J.; Sawchuk, C.; Lohr, J.; Elwood, L. The Disgust Scale: Item Analysis, Factor Structure, and Suggestions for Refinement. *Psychol. Assess.* **2007**, *19*, 281–297. [CrossRef]
55. Russell, S.; Lux, C.; Hampton, G. Beyond "Information": Integrating Consultation and Education for Water Recycling Initiatives. *Soc. Nat. Resour.* **2009**, *22*, 56–65. [CrossRef]
56. Ormerod, K.J.; Scott, C.A. Drinking Wastewater: Public Trust in Potable Reuse. *Sci. Technol. Hum. Values* **2013**, *38*, 351–373. [CrossRef]
57. Ross, V.L.; Fielding, K.S.; Louis, W.R. Social Trust, Risk Perceptions and Public Acceptance of Recycled Water: Testing a Social-Psychological Model. *J. Environ. Manag.* **2014**, *137*, 61–68. [CrossRef]
58. Ramm, K.; Smol, M. Water Reuse—Analysis of the Possibility of Using Reclaimed Water Depending on the Quality Class in the European Countries. *Sustainability* **2023**, *15*, 12781. [CrossRef]
59. Michetti, M.; Raggi, M.; Guerra, E.; Viaggi, D. Interpreting Farmers' Perceptions of Risks and Benefits Concerning Wastewater Reuse for Irrigation: A Case Study in Emilia-Romagna (Italy). *Water* **2019**, *11*, 108. [CrossRef]
60. Garin, P.; Montginoul, M.; Noury, B. Waste Water Reuse in France—Social Perception of an Unfamiliar Practice. *Water Supply* **2020**, *21*, 1913–1926. [CrossRef]
61. Mesa-Pérez, E.; Berbel, J. Analysis of Barriers and Opportunities for Reclaimed Wastewater Use for Agriculture in Europe. *Water* **2020**, *12*, 2308. [CrossRef]
62. Domènech, L.; Saurí, D. Socio-Technical Transitions in Water Scarcity Contexts: Public Acceptance of Greywater Reuse Technologies in the Metropolitan Area of Barcelona. *Resour. Conserv. Recycl.* **2010**, *55*, 53–62. [CrossRef]

63. Schmid, S.; Bogner, F.X. What Germany's University Beginners Think about Water Reuse. *Water* **2018**, *10*, 731. [CrossRef]
64. Lyach, R.; Remr, J. Motivations of Households towards Conserving Water and Using Purified Water in Czechia. *Sustainability* **2023**, *15*, 2202. [CrossRef]
65. Angelakis, A.N.; Tzanakakis, V.A.; Capodaglio, A.G.; Dercas, N. A Critical Review of Water Reuse: Lessons from Prehistoric Greece for Present and Future Challenges. *Water* **2023**, *15*, 2385. [CrossRef]
66. Petousi, I.; Fountoulakis, M.; Stentiford, E.; Manios, T. Farmers' Experience, Concerns and Perspectives in Using Reclaimed Water for Irrigation in a Semi-Arid Region of Crete, Greece. *Irrig. Drain.* **2015**, *64*, 647–654. [CrossRef]
67. Kantanoleon, N.; Zampetakis, L.; Manios, T. Public Perspective towards Wastewater Reuse in a Medium Size, Seaside, Mediterranean City: A Pilot Survey. *Resour. Conserv. Recycl.* **2007**, *50*, 282–292. [CrossRef]
68. Paranychianakis, N.; Salgot, M.; Snyder, S.; Angelakis, A. Water Reuse in EU-States: Necessity for Uniform Criteria to Mitigate Human and Environmental Risks. *Crit. Rev. Environ. Sci. Technol.* **2015**, *45*, 1409–1468. [CrossRef]
69. Menegaki, A.N.; Hanley, N.; Tsagarakis, K.P. The Social Acceptability and Valuation of Recycled Water in Crete: A Study of Consumers' and Farmers' Attitudes. *Ecol. Econ.* **2007**, *62*, 7–18. [CrossRef]
70. Bakopoulou, S.; Katsavou, I.; Polyzos, S.; Kungolos, A. *Using Recycled Water for Agricultural Purposes in the Thessaly Region, Greece: A Primary Investigation of Citizens Opinions*; WIT Press: Southampton, UK, 2008; Volume 109, p. 878, ISBN 978-1-84564-113-9.
71. Tzanakakis, V.; Angelakis, A.; Paranychianakis, N.; Dialynas, Y.; Tchobanoglous, G. Challenges and Opportunities for Sustainable Management of Water Resources in the Island of Crete, Greece. *Water* **2020**, *12*, 1538. [CrossRef]

environments MDPI

Review

The Use of Constructed Wetlands to Treat Effluents for Water Reuse

Sandro Xavier de Campos [1] and Manuel Soto [2,*]

1 Research Group on Environmental and Sanitary Analytical Chemistry (QAAS), State University of Ponta Grossa (UEPG), Av. General Carlos Cavalcanti, Ponta Grossa 84030-900, PR, Brazil; campos@uepg.br
2 Department of Chemistry, Faculty of Sciences & CICA, University of A Coruña, E-15008 A Coruña, Galiza, Spain
* Correspondence: m.soto@udc.es; Tel.: +34-647-38-77-69

Abstract: Constructed wetland systems (CWs) are technologies based on natural processes for pollutant removal and have been more and more accepted in the treatment of domestic and industrial wastewater. This study selected and reviewed articles published in the last six years involving the use of different CW conceptions and their association with other technologies to treat different effluents and evaluated the quality of the effluents for reuse. From a total of 81 articles reviewed, 41 presented quantitative data on the quality of the treated effluent in relation to the requirements of the reuse regulations in different countries of the world. CWs can be used to treat gray water and runoff water, as well as domestic and industrial effluents with the purpose of reusing them. While studies on the removal of new chemical and biological substances have increased, challenges are associated with the optimization of CWs to improve the removal of pathogens and new contaminants that have appeared more recently. The potential for the improved removal of those pollutants lies in the association of CWs with conventional and advanced technologies in new configurations. We concluded that studies related to the reuse of effluents using CWs are in constant evolution, with experiments at different scales. The perspectives are promising since CWs are an economic, environmentally friendly, and efficient technology to help in the mitigation of water scarcity problems imposed by climate changes.

Keywords: constructed wetlands; wastewater; runoff; effluent quality; effluent reuse

1. Introduction

Urban and industrial wastewater discharges remain a major source of pollution worldwide. Urban runoff, stormwater overflows, and untreated sewage discharges are increasingly important sources of pollution. Population growth and the change in annual rainfall patterns associated with climate change make it increasingly difficult to meet the growing demand for recreational, industrial, agricultural, and domestic purposes. Thus, the regeneration of used water with the aim of giving it a second use is increasingly imperative [1,2].

Solutions based on nature, such as constructed wetlands (CWs), offer high possibilities for the sustainable use of water, facilitating its treatment and reuse in situ, as well as contributions to adaptation to climate change through the use and promotion of vegetation, both in urban and rural areas [3,4]. The circular economy criteria and objectives require opting for technologies and configurations that allow the recovery of nutrients and other resources contained in wastewater while allowing the reuse or recycling of the water itself for different uses. CWs offer very interesting benefits regarding both sustainability and circularity [5].

The reuse options are conditioned by the destination that will be given to the treated or pre-treated wastewater. For a few reuse destinations, a pretreatment may be sufficient, while CW effluents can be reused in a greater number of applications by achieving efficiencies similar to or higher than those of conventional treatments in the removal of organic pollutants, while enabling a partial or advanced removal of nutrients, pathogens and persistent pollutants [5,6]. The latter depends on the configuration, type, and operating conditions of the CWs. Pathogenic pollution is one of the factors that limits many of the potential uses of reclaimed water, so it must also be a factor to be considered.

The general objective of this article was the review of the recent scientific literature addressing the quality of the effluents of CWs that treat various types of wastewater, from runoff and gray water to urban and industrial wastewater. A few reviews have been published on the potential of CWs for effluent reuse, either referring to specific geographical areas (Europe; Morocco) or to certain types of effluents, especially urban runoff and sewer overflows [1,4,6], emphasizing the need for a more general review. Specifically, this review aimed to analyze the requirements established in the legislation for the different reuse destinations and the quality of the effluents obtained in CWs. For this, a total of 81 studies published in the last six years, which addressed this objective, were reviewed. Of these, a total of 41 articles had quantitative data on the specific objective of determining the quality of the treated effluent in relation to the requirements of the reuse regulations in the different countries of the world where the studies were carried out. Quantitative data from the last set of articles were presented in tables to facilitate an analysis and comparison, and to allow further discussion. Thus, this work contributes to an update regarding the potential use of CWs as a technology to be adopted or not in the reuse of these different effluents in the context of water scarcity caused by the effects of climate change and the growing requirements of recently adopted reuse regulations. The evaluation of the quality of treated effluents for different uses requires, first of all, the analysis of the regulations that have been established in this respect in recent years. This aspect, as well as a brief description of CWs technology, are addressed in the following sections of this introduction.

1.1. Policies for the Regeneration and Reuse of Wastewater

Starting in the state of California in 1918, countries around the world and international organizations (e.g., World Health Organization—WHO) created their regulations and guidelines to incentivize water reuse practices [7]. In the last two decades, the reuse of different effluents started to be planned in the management of hydric resources in many countries worldwide. Therefore, many regulations in different regions of the planet seek to define the main standards that must be monitored so that the reuse can meet the target proposed without causing environmental problems or presenting risks to the population's health.

According to Santos et al. [8], water reuse regulations advanced in relation to the effects of climate change, pollution, and increased consumption due to the global population growth, which required government planning to incorporate alternative sources in the existing hydric matrices. This type of planning action can only become effective with proper normative instruments that seek the institutionalization of that practice. Not surprisingly, the initial actions were observed in regions that experienced water scarcity such as California (United States of America—USA), Victoria (Australia), China, and Mediterranean countries. However, different regions took different paths due to their specificities and socio-economic development factors. China, for example, is recognized worldwide for their substantial knowledge about water reuse in agriculture, while California and Australia opted for advancing in the adoption of potable water standards. In addition, several regions and global agencies started their own regulation processes with flexible norms and specific objectives such as those proposed by the European Union (EU) and Food and Agriculture Organization (FAO), which aimed at wastewater reuse for irrigation. With the appearance of new scientific and technological advances, countries and their states started to propose their norms with more specific parameters and broadened the possibilities of urban, environmental, and industrial reuse. Table 1 seeks to present in general terms some regulations available in different regions of the world. The socio-economic conditions of developed and developing regions and with greater or lesser water scarcity were taken into account.

Table 1. Applications and parameters for effluent reuse in different regions of the world.

Country	USA [1]	China [2]	EU [3]	Victoria [4]	Spain [5]
Examples of reuse applications	Food crops for human consumption consumed raw [a] Urban reuse: unrestricted [b] Impoundments: unrestricted [c] Indirect potable reuse: Groundwater [d]	Recreational use in rivers and ponds [a] Urban use—Road cleaning [b] Urban use—Municipal landscape [c] Industrial use—Boiler water [d]	Class A—All food crops for direct consumption [a] contact with reclaimed water, consumed raw Class B—Crops for indirect consumption [b] Class C—Crops for indirect consumption drip irrigation [c] Class D- Industrial, energy, and seeded crops [d]	Agricultural food production, consumed raw [a] Domestic garden watering, including [b] Industrial: example, wash-down water [c] Urban (non-potable) controlled public access [d]	Urban use—Irrigation of private gardens [a] Agricultural uses—crops to be eaten raw [b]rage Industrial use—leaning water in the food industry [c] Recreational use—Golf course irrigation [d]
pH	6.0–9.0 [a,b,c]; 6.5–8.5 [d]	6.0–9.0 [a,b,c], 6.5–8.5 [d]	-	6.0–9.0 [a,b,c,d]	-
BOD (mg/L)	≤10 [a,b,c,d]	6 [a]; 15 [b]; 20 [c]; 10 [d]	≤10 [a,b,c,d]	≤10 [a,b]; ≤20 [c,d]	-
Turbidity (NTU)	≤2 [a,b,c,d]	5 [a]; 10 [b,c]; 3 [d]	≤5 [a,b,c,d]	≤2 [a,b]	2 [a]; 15 [c]; 10 [b,d]
TSS	-	-	≤10 [a,b,c,d]	5 [a,b], 35 [c], 30 [d]	10 [a]; 20 [d]; 35 [c];
Total Coliforms (CFU/100 mL)	-	30 [a,b,c,d]	-	-	-
Fecal Coliforms (CFU/100 mL)	Not detectable	5000 [a] 20,000 [d]	≤10 [a], ≤100 [b], ≤1000 [c], ≤10,000 [d]	≤100 [c]; ≤1000 [d]	0 [a]; 100 [b]; 1000 [c]; 200 [d]
Legionella spp.: CFU/L	-	-	<1000 [a,b,c,d]	-	100 [a,c,d]; 1000 [b]
Helminth eggs (egg/L)	-	-	≤1 [a,b,c,d]	-	0.1 [a,b,c,d]
Chlorine Residue (mg/L)	1 [a,b,c,d]	0.05 [a,d]; 1 [b,c]	-	-	-
Color	-	30 [a,b,c,d]	-	-	-
NH_3 -N (mg/L)	-	5 [a]; 10 [b,d]; 20 [c]	-	-	-
NT (mg/L)	-	15 [c]	-	-	-
PT (mg/L)	-	1 [a,b]	-	-	-

Table 1. *Cont.*

Country	Egypt [6]	FAO [7]	Portugal [8]	Cyprus [9]	Greece [10]
Examples of reuse applications	Group A—All types of grass and flowers [a] Group B—All kinds of vegetables manufactured [b] Grupo C—Addition to spray irrigation is not used [c] Group D—Crops for the production of biodiesel [d]	Class A—Irrigation of crops to be eaten uncooked [a] Class B—Irrigation of cereal crops and fodder [b] Class C—Exposure of workers does not occur [c]	Class A—Irrigation of vegetables to be eaten raw [a] Class B—Public parks, gardens, and sport fields [b] Class C—Vegetables to be eaten cooked [c] Class D—Crops to be used as raw material [d]	Crops for human consumption; raw [a] Fodder crops [b] Industrial crops [c] Areas of limited public access [d]	Public access is not expected; industrial crops [a] Industrial use: Ex. Cooling water [b] Unrestricted irrigation crops [c] Urban uses: cemeteries, golf courses, and public parks [d]
pH	-	$6.5–8.4^{a,b,c}$	-	$6.5–8.5^{a,b,c,d}$	-
BOD (mg/L)	15^{a}, 30^{b}, 80^{c}, 350^{d}	$\leq 10^{a}$; $\leq 30^{b}$; $<30^{c}$;	$\leq 10^{a}$ $\leq 25^{b,c,d}$	$\leq 10^{a,b}$ $\leq 70^{c,d}$	$\leq 10^{b,c,d}$
Turbidity (NTU)	5^{a}	$\leq 2^{a}$	$\leq 5^{a,b,c,d}$	-	-
TSS	15^{a}, 30^{b}, 50^{c}, 300^{d}	$\leq 30^{b,c}$	$\leq 10^{a}$, $\leq 35^{b,c,d}$	$\leq 10^{a,b}$, $\leq 30^{c,d}$	$\leq 10^{b,c}$ $\leq 2^{d}$
Total Coliforms (CFU/100 mL)	-	-	-	$\leq 5^{a}$, $\leq 50^{b}$, $\leq 200^{c}$, $\leq 1000^{d}$	-
Fecal Coliforms (CFU/100 mL)	-	$\leq 14^{a}$, $\leq 200^{b,c}$	$\leq 10^{a}$, $\leq 100^{b}$, $\leq 1000^{c}$, $\leq 10{,}000^{d}$	-	$\leq 200^{a}$, $\leq 5^{b,c,d}$
Legionella spp.: CFU/L	-	-	$<1000^{a,b,c,d}$	-	-
Helminth eggs (egg/L)	-	-	$\leq 1^{a,b,c,d}$	-	-
Chlorine Residue (mg/L)	-	-	-	-	-
Color	-	-	-	-	-
NH_3 -N (mg/L)	-	-	-	-	-
NT (mg/L)	-	-	-	$15^{a,b,c,d}$	-
PT (mg/L)	30	-	$5^{a,b,c,d}$	2^{a}, $10^{b,c,d}$	$\leq 15^{a}$, $\leq 2^{b,c,d}$

Available at (all links accessed on 3 July 2023): [1] https://www.epa.gov/sites/default/files/2019-08/documents/2012-guidelines-water-reuse.pdf; [2] http://www.reclaimedwater.net/; [3] https://eur-lex.europa.eu/legal-content/EN/TXT/?uri=CELEX:52018PC0337; [4] https://www.epa.vic.gov.au/about-epa/publications/1910-2; [5] https://www.boe.es/eli/es/rd/2007/12/07/1620/con; [6] https://www.eib.org/attachments/registers/156263822.pdf; [7] https://www.fao.org/3/t0551e/t0551e00.htm#Contents; [8] https://diariodarepublica.pt/dr/detalhe/decreto-lei/119-2019-124097549; [9] https://www.moa.gov.cy/moa/environment/environmentnew.nsf/page17_en/page17_en?OpenDocument, [10] https://ypen.gov.gr/. [a,b,c,d] Class of reuse, when not cited unspecified values.

In most of the guidelines, total and fecal coliforms are used as indicator microorganisms for pathogenic microbial quality, while turbidity, BOD, and TSS are used as physicochemical quality parameters (Table 1). Another piece of evidence is that China keeps one of the most complete regulations in the world regarding the requirements and objectives of water reuse, which contributes to its position in the ranking of users of reused water, reaching around 7.1 billion m^3/year, followed by the USA, where 6.63 billion m^3/year are reused [8]. With the aim of advancing and expanding water reuse, both CW and conventional effluent treatment technologies have been undergoing changes in their configurations, to adapt to increasingly demanding reuse regulations, which almost always requires new innovative approaches.

1.2. Constructed Wetlands

Constructed wetlands (CWs) have been used to treat effluents since the 1960s [9]. Aiming at reproducing the mechanisms involved in several natural processes, CWs are designed selectively with the purpose of obtaining treatment conditions that result in the best quality possible for the final effluent. CWs are treatment systems that include vegetation and suitable bed substrates (i.e., the solid filter medium on which the vegetation is rooted) and flow structures, in addition to the use of a wide variety of microbial flora, which altogether play a crucial role in the removal of pollutants.

Regarding their hydraulic characteristics, CWs can be classified as surface and subsurface flow [9]. Surface flow CWs (SCWs) are very similar to natural wetland zones; however, with a wastewater volume on the substrate, they are not very deep. SCWs are units where wastewater flows over the surface of the substrate, thus forming a water-free layer, which increases oxygen availability and photodegradation processes. Some examples of SCWs are the free-floating and submerged macrophytes, macrophytes with floating leaves, emergent macrophytes, emergent macrophytes floating mats, and the use of trees. The subsurface flow CWs are classified as horizontal (HFCW) and vertical (VFCW), according to the flow direction. In these subsurface flow systems, wastewater flows below the surface of the filter medium, passing through it. VFCW might present a water-saturated or unsaturated substrate, a descending or ascending flow, or the fill and drain system, which is also known as a reciprocating or tidal flow [9]. While in HFCWs and permanently saturated VFCWs an anaerobic or anoxic environment dominates, unsaturated and fill and drain VFCWS can achieve a good oxygenation of the environment and ensure predominant aerobic conditions. In one of the most common arrangements for vertical unsaturated flow systems, the application of sewage occurs intermittently and uniformly on the surface, where it percolates into the filter medium. After passing the effluent, the filter pores are naturally occupied with air, facilitating the maintenance of aerobic conditions in the environment for the oxidative processes that occur in nitrification and also for the oxygenation of organic matter. Unless otherwise noted, in this paper we use the term VFCW to refer to systems with unsaturated conditions.

Many studies have investigated hybrid systems that associate SCW, HFCW, and VFCW (specially unsaturated VFCW) in different arrangements taking advantage of the characteristics of each type of CW to create a more efficient combined system, mainly aiming at the removal of nitrogen, phosphorus, and emergent contaminants. Thus, studies on both single CWs and hybrid CWs by themselves or associated with other conventional or advanced treatment systems have increased noticeably. This occurs because CWs present the advantages of construction and maintenance low cost, being a widely employed technology in developed countries as well as in developing countries.

The literature shows that CWs can treat different effluents with removal results reaching 90% for the main parameters associated with solids and organic matter, obtaining effluents that meet most of the requirements for the different types of reuse [10–15]. However, when a more efficient removal of nutrients (N, P) and pathogens is required, hybrid and combined systems are the most recommended [16–18]. In this way, there are also many studies reporting the possibility of associating CWs with advanced oxidative processes for the removal of emergent contaminants [12–14], even if many of these harmful substances have not been included in the parameters for effluent reuse yet (Table 1).

2. Materials and Methods

This article presents a selection of published studies related to the use of CWs and their advancements aiming to obtain reuse water. The Scopus database was surveyed using the search expressions of constructed wetlands, reuse, quality, and wastewater (all connected by the logical operator "AND"). The expression quality was used aiming to limit the search to studies that presented results or discussions related to the characteristics of effluents on its suitability for reuse, that is to say, in relation to different physical, chemical, and biological parameters. We also sought to relate the quality obtained after the treatment using CWs with reuse regulations. By including the word 'quality', we observed a reduction in the number of studies found from the beginning of the specific publications in 1994 from 621 to 274. Seeking a representative sample of the experiments carried out using CWs and taking into consideration the current effluent reuse criteria, we selected 81 articles published in the last six years (a period from 2018 to 2023).

This paper provides a description of the experiments using CWs considering their main objectives, novel results obtained, and conclusions. In addition, experiments detailing the type of CW, plants, substrates used, removals obtained, and reuse proposals were organized in tables for better visualization. Quantitative data were extracted directly from these articles, when originally available. On other occasions, they were calculated by the authors of this review from other information in the referred article. Articles with results unrelated to regulations and parameters required for water reuse were not included in the tables, so 41 articles were included in the tables. However, some of those articles were included in this report because of their interest in justifying the importance of using CW in the treatment of certain types of effluents and the potential interest for water reuse.

Among the studies published that focused on water reuse, the main focus on the use of CWs as the only treatment or integrated to other technologies was related to runoff, industrial, and domestic water. Due to the specificities of the studies, we decided to dedicate part of the review to the use of CWs in the treatment of domestic gray water and group these studies with the results of the runoff water treatment. Thus, the paper was divided into topics covering studies of CWs dedicated to the reuse of gray water and runoff and studies related to the treatment and reuse of other domestic and industrial effluents.

3. Results and Discussion

3.1. *Treatment of Gray Water and Runoff Using CWs for Water Reuse*

Gray water is produced from sinks, laundry, and showers and might represent up to 75% of the total domestic wastewater, with approximately 250–300 L generated per person per day in developed countries and 100–120 L in developing countries [2]. Gray water presents a lower concentration of organic matter, nutrients, and pathogens than black water and, consequently, is easier to be treated [15,16].

Considering the diversity of effluents generated and the possibility of proposing decentralized treatment systems, CWs stand out as a suitable technology for the treatment of gray water and runoff aiming at their reuse. Regarding urban runoff, CWs also contribute to the removal of highly toxic contaminants such as heavy metals that might present risk in the reuse of the treated effluent. Due to their landscaping characteristics, for using ornamental plants, experiments are carried out seeking to use CWs in urban spaces, thus enabling the effluent reuse in situ, which is allied to energy production [1,18].

3.1.1. Treatment of Gray Water in CWs

Lakho et al. [15] investigated a system that combined VFCW, filtration, and disinfection for the treatment of gray water from a restaurant in Belgium. The VFCW was built and operated for six months. After going through an activated charcoal system, ultrafiltration, reverse osmosis, ionic exchange membrane, remineralization, and ultraviolet disinfection, the effluent reached a level of potability in accordance with the Belgian potable water regulation and was reused in the restaurant [15]. Another innovation proposed for the use of CW in the treatment of gray water from a restaurant was the operation of a pilot system combining a constructed wetland microbial fuel cell (CWMFC) and a biological filter (FB) for the continuous treatment and recycling of hand washing water [18]. The CWMFC system reached a full bacterial removal for an *E. coli* influent load of 4 log and 432 mg L^{-1} chemical oxygen demand (COD). The final effluent quality met the South-African standards for noble reuse and was also able to generate 4.33 mW m^{-3}-treated effluent. However, those authors [18] reported the need for further studies related to pathogen removal.

Kotsia et al. [19] investigated a VFCW pilot system conceived as a treatment garden, in which they used ornamental plants to treat synthetic gray water aiming to improve the aesthetics and acceptability of the system. A high organic matter removal efficiency was observed; on the other hand, the total phosphorus (TP) removal reduced gradually from 100% during the first year of operation to 15% throughout the second year. Their results showed that *Pittosporum tobira* and *Hedera helix* can grow in VFCW treating gray water without any changes in their physical aspects. The biochemical oxygen demand (BOD) and total suspended solids' (TSSs) final concentration in the effluent was below 10 mg L^{-1} and met the Australian criterion for reuse in toilet flush, except for pathogen quantity. Thus, the use of a simple disinfection system such as the visible ultraviolet (UV/Vis) or chlorination is recommended for the elimination of pathogens [19].

Due to the difficulty of finding large areas needed for the construction of CWs for gray water treatment in urban environments, a solution would be the use of containers for VFCWs. An installation like that might provide the treatment of gray water and their reuse as irrigation water for urban façades, provided that it is followed by a disinfection treatment to comply with irrigation standards set by the WHO [20].

Another great challenge in the use of CWs to treat gray water is the presence of excess personal care products (PCPs) found in water from toilets and showers. Ren et al. [21] developed studies on a pilot scale associating a membrane bioreactor (MBR) and VFCW for gray water PCPs. The removal of PCPs in the MBR was mainly through the adsorption and biodegradation of activated sludge. PCPs were removed >80% in the CW system, through plant uptake and biodegradation, among other mechanisms. The VFCW presented lower removal efficiency in winter since several plants died, which led to a continuous decrease in the number of microorganisms adhered to the roots of the plants and reduction in the oxygen transport capability. Despite that, the final effluent met Chinese reuse requirements regarding organic matter and solids.

In addition to analyzing the efficiency of the removal of the main contaminants, studies have given more and more importance to the phytotoxic effects that gray water effluents reused for the irrigation of different crops might cause. An analysis carried out on gray water effluents from washing machines and kitchen sinks treated using biological minireactors and HFCW showed that tomatoes irrigated with that water did not suffer any negative effects in relation to their growth, photosynthetic activity, hydric state, osmotic potential, or productivity. Moreover, their results showed that treated gray water did not affect soil salinity and even improved plant height. Despite the positive results, those authors recommended further studies to monitor the long-term effect on the soil and health of those consuming the tomatoes [22].

Among the contaminants recently studied, research involving CWs in the treatment of gray water have shown certain concerns with the removal of antibiotic resistant bacteria. Results from the treatment of laundry water using a VFCW showed that an increase ranging between 36.34% and 40.79% in the bacteria resistant to ciproflaxin and ceftriaxone might occur. Another important observation was the strong association of a lack of the degradation of ciproflaxin and ceftriaxone with a lack of the removal of surfactants, which required the use of a disinfection method before the use of the treated water. To guarantee an efficient removal of these surfactants in CWs, an efficient COD removal is also important. In addition, the highest correlation between COD and LAS removal was obtained regarding an organic surface loading rate, suggesting that the area plays a more important role than the volume of the system. *Pseudomonas* spp. predominated in the degradation of such substances [23].

Other studies reported that to meet higher quality requirements for the noble reuse of gray water effluents, the implementation of an pilot advanced system of disinfection is always required [24,25]. With this purpose, an HFCW and ultraviolet-visible (UV/Vis) disinfection were used to treat gray water from the washbasins of a primary school in Morocco. After the treatment, the effluent produced was used to irrigate lawn areas favoring plant growth. With the UV/Vis disinfection in 50 mWs/cm^2, the effluent met the unrestricted irrigation requirements set by the WHO, Morocco, and California (USA), which is one of the strictest guidelines for fecal coliforms (23 CFU/100 mL) [25].

It can be observed in recently published works on the use of CW in gray water treatment that there is a predominance of the use of unsaturated VFCW. This preference may be due to the smaller area required to install these units due to more efficient oxidative processes that also allow for the efficient removal of ammonia.

3.1.2. Treatment of Runoff Water in CWs

Studies related to the treatment of runoff water have shown that those effluents showed a concentration of heavy metals and that the CWs might provide a good efficiency in the removal of such substances. An HFCW-SCW hybrid system built on a pilot scale was used to treat runoff water from the parking lot of a retail shop in Eastern Sicily, Italy [26]. The hybrid system showed good efficiency in the removal of heavy metals, mainly lead (Pb), zinc (Zn), and copper (Cu). The removal efficiency for *Clostridium perfringens* was observed in the HFCW unit. Algi growth occurred in the SCW unit, which reduced the efficiency of the TSS, BOD, and COD removal, compromising the water quality. Preliminary results suggested the reliability of the technology in the treatment of runoff water for urban reuse according to Italian parameters [26].

In another study, Tuttolomondo et al. [27] investigated an VFCW on a pilot scale for the treatment of the first discharge runoff water and verified the effect of such water on the reuse to irrigate pepper and rosemary. Their results showed a good removal of organic matter, nutrients, and heavy metals, such as nickel and chrome. The *Escherichia coli* (*E. coli*) concentrations were low in the effluent and levels were always below 100 CFU/100 mL, reaching the values required for agricultural use. Regarding the plants irrigated, both showed positive results, mainly in relation to metal removal [2].

3.1.3. Main Characteristics of CW Systems Used for the Treatment of Gray Water and Runoff Aiming for Water Reuse

Table 2 summarizes the main characteristics of the studies involving the treatment of gray water and runoff using CWs. Many of these studies were carried out on a pilot or field scale, but some small-scale studies developed in a laboratory were also included. Out of the 10 systems included in Table 2, 3 investigated runoff water and 7 investigated gray water. As for the type of technology, most were simple VFCW (seven cases), while there was one simple HFCW and two hybrid CWs. The vegetable species used were quite diversified, both defined as single-crop or polycrop farming, as well as parallel comparative studies, so that the 10 systems included in Table 2 used 17 different plant species. Out of those, only *Phragmites australis* appeared in three reports and *Typha latifolia* in two reports; the remaining 14 species were found only once in the reports. Gravel was the most frequently used substrate to prepare the CW beds. It could be used alone or in a combination with sand or other materials such as zeolite. There were also reports of the use of volcanic origin substrates, clay, or bioceramics. The HLR of the VFCW systems ranged between 10 and 130 mm/d, with a 63 mm/d (data number, n = 6) average. When comparing VFCW with other types of CW, they present a higher capacity; for this reason, they can operate greater hydraulic and organic loads. However, the reduced appearance of the other technologies in the reports in Table 2 do not allow for a deeper comparison.

Table 2. Main characteristics of systems involving CWs in the treatment of gray water and runoff.

System Number, Type, and Size	Plants	Bed Substrate	HRT, d (HLR, mm d^{-1})	Removal (% or LU (Pathogens)) COD (BOD), TSS, Others [c,d,e,f,g,h,i,j,k]	TN, TP, Others [L]	*E. coli*, TC, EC	Reuse	Country
1. HFCW, 12.5 m^2 [a]	*Typha latifolia*	Gravel	6.25 (96)	89 (87), na, 88 [c], 84 [k]	42, 50, 84 [L]	na, na, na	Irrigation of green spaces	Australia
2. HFCW, 1.0 m^2 [a]	*Equisetum giganteum*	Gravel	3.57 (196)	38 (na), na, 35 [k]	na, na	na, na, na	Discharge into soil	Brazil
3. VFCW, 46.80 m^2 [b]	*Phragmites australis; Arundo donax L.*	Silica quartz river gravel	7 (130)	65–69 (75–83), 65.9 [h], 66.7 [i]	na, na	0.94, na, na	Agricultural irrigation; discharge into soil	Italy
4. VFCW, 0.03 m^2 [a]	*Cyperus papyrus*	Clay aggregate	0.09 (768)	99 (na), na	na, na	4.0, na, na	Potable	South Africa
5. VFCW, 0.45 m^2 [a]	*Pittosporum tobira; Hedera Helix; Polygala myrtifolia*	Gravel and sand	na (74–110)	96 (99), 94	na, na	na, 2.2, na	Toilet flushing and washing machines	Australia
6. VFCW, 2.5 m^2 [a]	*Phragmites australis*	Zeolite; lava sand; Rhine sand	na (18–80)	96–98 (85), na	na, 83.4	na, na, na	Irrigation	Germany
7. VFCW, 60 m^2 [a]	-	Lava rock layer	na (50–80)	84 (97), 92,	42, 24	na, na, na	Potable	Belgium
8. VFCW, 0.5 m^2 [a]	*Phragmites australis and Acorus calamus*	Soil Volcanics; Pebble; quartz Sand; Bioceramics	na (80)	73.5 (80), 90 [c], 90 [j]	na, 87, 90 [L]	na, na, na	Reuse non-potable, irrigation	China
9. HCW (HFCW + SCW), 6.75 + 3.5 m^2 [b]	*Canna indica; Typhia Latifolia*	Volcanic gravel	4 (na)	47 (50), na, 60–63 [d], 9- 33 [e], 6–39 [f], 53–90 [g], 30–74 [h], 61–91 [i]	30, 40	1.5, na, na	Irrigation and toilet flushing	Italy
10a. HCW (VFCW + HFCW + HFCW), 0.3 + 0.8 + 2.0 m^2 [a]	*A. gayanus*	granitic gravel	na (75)	90.4 (95.5), 96.8	na, na	2, 2, 2	Irrigation	Africa's Sahel
10b HCW (VFCW + HFCW + HFCW), 0.3 + 0.8 + 2.0 m^2 a	*C. zizanioides*	granitic gravel	na (75)	93.9 (97.5), 98.5	na, na	3, 2, 2	Irrigation	Africa's Sahel
10c HCW (VFCW + HFCW + HFCW), 0.3 + 0.8 + 2.0 m^2 a	*unplanted*	granitic gravel	na (75)	88.7 (94.9), 96.0	na, na	2, 2, 1	Irrigation	Africa's Sahel

References (System number): 1. [25]; 2. [23]; 3. [27]; 4. [18]; 5. [19]; 6. [20]; 7. [17]; 8. [21]; 9. [26]; 10. [24]. Remarks: [a] Grey Water. [b] Stormwater. [c] Turbidity. [d] Cr. [e] Fe. [f] Ni. [g] Pb. [h] Cu. [i] Zn, [j] Personal care products, [k] Surfactants, [L] $N-NH_3$. *TC, total coliforms, EC: Enterococci.* HCW: hybrid constructed wetland. LU: log units. na: not available.

The VFCW COD removal resulted in 82 ± 18% on average (n = 7). The general BOD_5 and TSS removals were similar or slightly higher than that of COD (85% and 84% on average, with n = 6 and n = 3, for BOD_5 and TSS, respectively). Regarding the removal of nutrients and pathogens, only 4 out of the 10 studies surveyed presented information about these contaminants. Thus, the small number of systems and the variability of factors did not allow us to obtain indicative removal values or the appropriate operating conditions for efficient treatment, highlighting that more research is still needed. Finally, irrigation was the major water destination, reported in 6 out of the 10 studies, followed by use for toilet flushing in 3 reports. Other reported destinations included washing machines and potable water uses.

3.2. *Treatment of Domestic Wastewater and Industrial Effluents by CWs for Water Reuse*

Despite having been initially designed to treat domestic wastewater on a small scale, recently, CWs have been used to treat effluents from agriculture and dairy industries, winery, tannery, paper, and pulp, among others. Regarding industrial effluents and their reuse, CWs contribute to a system that can treat water containing high organic loads and different contaminants, while producing a small amount of sludge [28,29]. The efficiency of the removal of the main contaminants of domestic and municipal wastewater in CWs is comparable to that of modern technologies of wastewater treatment, such as the activated sludge process (ASP), sequence batch reactor (SBR), mobile bed biofilm reactor (MBBR), upflow anaerobic sludge blanket (UASB), etc. CWs can remove BOD (85–90%), COD (65–80%), TSS (90–95%), TN (65–80%), and TP (35–50%) [30]. Furthermore, if CWs are used in their hybrid configuration and are associated with disinfection systems, they can be highly efficient in removing pathogenic microorganisms. Thus, CWs can contribute the purpose of domestic and industrial wastewater recovery aiming at reusing it, and present the benefits of low maintenance and operational costs, robustness, and efficiency in the removal of several emergent substances that are introduced in the environment by anthropic action every year [31,32].

The use of CWs results in a more acceptable landscape for the population that live around the sewage treatment stations, turning the spaces into gardens, which transform a gray landscape into a green one with gains for the local flora and fauna [33]. In addition, there is evidence that CWs can be used on a real scale by aiming to recover water in places of water shortage such as the Gaza Strip [34].

Therefore, in the last few years, many studies have focused on the use of CWs for the treatment of domestic wastewater aiming at its reuse, mainly as a post-treatment to conventional technologies, to provide greater efficiency and ecological benefits [35].

3.2.1. Efficiency of CWs in the Reuse of Domestic Effluents

Landscape Integration of CWs and Aesthetic Gains from the Use of Ornamental Plants

Studies have been developed showing that CWs operated in large-scale facilities present efficiency in the treatment of effluents and result in ecological and aesthetic gains due to the use of ornamental plants. De Anda et al. [36] studied a unit designed to treat domestic sewage in a large research center in Spain for 2 years. The unit included a septic tank, anaerobic filter, HFCW, and chlorine disinfection. The system also allowed the production of *Agapanthus africanus* as an ornamental plant. Chlorination was required to meet the Spanish guidelines regarding fecal coliform removal for the irrigation of green areas. The water was reused to irrigate grass close to the research center, and the HFCW provided an environment where several species of birds, lizards, butterflies, and bees could live.

A study carried out by Kaushal et al. [37] evaluated three large-scale SCWs used for the treatment of effluents from urban, rural, and industrial facilities for 1.5 years. They investigated the removal of *E. coli*, Enterococci, and total coliforms for agricultural reuse in India. Their results showed that although the microbiological removal was over 70%, it was not possible to meet the requirements for agricultural reuse in India regarding those parameters. Thus, it was necessary to promote a disinfection process. On the other hand, they contributed to the wildlife habitat and improved the aesthetics of previously degraded territories.

Sandoval–Herazo et al. [38] evaluated the process of removing wastewater pollutants on an HFCW microcosm scale using different ornamental plants and substrates obtained from recyclable material. In that experiment, the ornamental plant *Lavandula* sp. was not able to adapt and died 45 days after sowing without producing flowers; the *Spathiphyllum wallisii* produced 12 flowers, while the *Zantedeschia aethiopica* produced 10 flowers. Their results revealed that the use of substrates originated from PET bottles is a viable alternative to be implemented in CWs. The plants *Spathiphyllum wallisii* and *Zantedeschia aethiopica* contributed noticeably to the removal of wastewater pollutants, resulting in good quality for type C agricultural reuse, pursuant to the EU Commission Norms.

HFCWs were used to treat effluents in three resorts located in Thailand. Those HFCWs were designed to receive effluent from a septic tank and decanters. The project included different types of plants aiming to promote a decorative aspect, since it was built in a high circulation area. Despite the effluent having reached Thai standards for reuse in buildings, a low removal of fecal coliforms was observed due to the temperature reaching (30 ± 5 °C), and a disinfection process had to be added [39].

Dell'osbel et al. [40] evaluated the performance of a pilot hybrid system combining VFCW and HFCW in the treatment of urban effluents. A primary screening treatment and secondary biodigester treatment were employed. The use of five ornamental plants was investigated seeking to aggregate the landscaping potential and better acceptance of the treatment station. The system presented a good capability of removing nutrients from the hybrid system. Pursuant to the Brazilian regulations, the results obtained showed that the final effluent could be used to wash cars and applied to other uses in which the users had direct contact with the water and in which the operator might be exposed to spray aspiration. Achieving the established effluent quality for these uses required recirculation as well as post-treatment disinfection [40].

The Challenge of Eliminating Pathogenic Microorganisms

As verified in previous reports, one of the greatest challenges in the use of CWs was the removal of pathogenic microorganisms, which usually required the use of disinfection to meet the quality requirements for reuse. Thus, a great part of the studies on the domestic effluent treatment using CWs in the last few years has been directed towards this purpose.

Quartaroli et al. [41] compared the use of calcium hypochlorite and sodium hypochlorite as disinfectant agents in a laboratory HFCW system. They carried out batch disinfection tests, using three hypochlorite dosages (5, 10, and 15 mg L^{-1}) and the contact times ranged between 5 and 50 min. All disinfection dosages used showed microorganism final results below 103 CFU/100 mL, meeting the requirements set by FAO for water reuse in agriculture. Trihalomethane was found after disinfection tests; there was also a possibility of the formation of other potentially harmful chlorination byproducts such as haloacetic acids, haloacetonitrile, haloketone, and trichloronitromethane. Therefore, a higher-quality effluent regarding COD removal can be obtained by associating CWs with other treatments to reduce the risk of effluent contamination by those byproducts.

Ali et al. [42] compared the quality of effluents obtained from full-scale HFCW and VFCW as a post-treatment to anaerobic reactors aiming at pathogen removal. Both systems resulted in over 90% removal of the bacterial population. With a tertiary system involving a SCW, a 50% increase was observed in the removal of pathogenic microorganisms, thus enabling the use of such a treated effluent in irrigation.

With a similar objective, Russo et al. [43] used a HFCW as the tertiary treatment of a large-scale sewage treatment station in Italy. Their objective was to obtain a suitable effluent to be used in agriculture, considering the regulations in force in Italy and the European Union. Although CWs have shown a high efficiency in *E. coli* reduction, the results are not enough to meet the limits for the reuse of wastewater in agriculture set by the Italian regulations (10 CFU/100 mL with a maximum value accepted of 100 CFU/100 mL) and were higher than the EU water quality value as required for classes A, B, C, or D. The combination of CW treatments with disinfection by UV treatment was investigated by the same authors [43], who obtained highly effective results with a complete removal of *E. coli*, somatic coliphages, and *C. perfringens* spores. Conversely, low efficiency was observed for enterococci. Thus, although *E. coli* removal from the effluent after the UV treatment met the Italian requirement for reuse, the total removal of enterococci was not possible. Even if neither the Italian regulations nor the EU have set a limit for enterococci, the risk associated with their environmental dispersion is hard to estimate; thus, their removal is needed to obtain better water quality with the least harmful impact to human's health and the environment [43].

Stefanakis et al. [44] verified that one possibility of removing coliforms and enterococci in CW systems involving domestic effluent treatment would be the system aeration. This procedure was implemented in a real-scale VFCW. The effluent was generated from sedimentation tanks and slow biological filters. The number of fecal coliforms, *E. coli*, and enterococci after the aerated VFCW treatment phase met the WHO guidelines for the reuse of wastewater in agriculture and eliminated the need for a final disinfection. A higher efficiency of aerated VFCW in the removal of microbiological contamination when compared to passive CW systems was reported for the first time, which might have implications in the selection of processes and CW technology for reuse. This also shows that aeration might be a new and efficient treatment scheme to be employed in new treatment stations or to update existing ones, seeking to improve pathogen removal. It has been suggested that high concentrations of dissolved oxygen likely alter the characteristics of the microbial consortium, including the development of groups that feed on pathogens [44].

Gonzales-Gustavson et al. [45] investigated the removal of different types of highly pathogenic viruses from an effluent originated in a large-scale domestic sewage treatment station, which included the coagulation, flocculation, and low-pressure UV disinfection phases or SCW as a tertiary treatment. The system served 112,000 inhabitants in Northeastern Spain and the objective of that study was to reuse that water to irrigate vegetables. The SCW was more efficient in reducing virus concentrations compared to the conventional post-treatment, although the effluent showed more variable virus concentrations, probably due to the variability of conditions in the CW. Their results showed that the viral load found in the final effluent did not allow its use in lettuce irrigation according to WHO recommendations. The authors also indicate that the CW land surfaces that would be required to achieve the effluent quality proposed by the WHO would make this option unfeasible in practice for large flows.

Gonzalez-Flo et al. [46] evaluated the performance and quality of water obtained using a combined full-scale system involving an HFCW and chlorine disinfection in Granollers, Barcelona, Spain. The whole system produced over 100 $m^3 \cdot d^{-1}$ reuse water for the irrigation of gardens and streets, and sewage network cleaning. The effluent met the water reuse requirements set by Catalonia for pH, electrical conductivity, TSS, *E. coli*, *Legionella* eggs, and nematoids in 90% of the analyses.

Studies developed in India by Thalla et al. [47] compared the efficiency of real HFCW and VFCW systems allied to disinfection in the treatment of a domestic wastewater with a concentration of 500 mg COD L^{-1} and 300 mg BOD L^{-1}. The VFCW showed better efficiency in the general removal than the HFCW. After chlorination disinfection, the water could be reused according to the American Environmental Protection Agency (EPA) in landscaping, impoundments, building, and industrial reuse such as tower cooling and recirculation, as well as environmental reuse in groundwater recharge [47].

Otter et al. [48] reported a study combining a VFCW with a chlorine generation pilot system in loco. The VFCW received domestic effluent originated from the treatment with activated sludge and from ponds in a Spanish city. The VFCW reduced the chlorine demand in 85%. During the effluent passage through the VFCW, increased conductivity and chlorine concentration were observed due to the planted vegetations' high evapotranspiration rates. The system was considered an alternative of efficient disinfection in decentralized applications of effluent reuse in remote locations with limited access to the electricity network and with restricted requirements for pathogen indicators [48].

Therefore, the disinfection stage associated with CWs is extremely important with a view to using the recovered water for some applications. To this end, chlorination is proposed (the intervention most commonly used) for reuse in urban environments [36,47,48]. For agricultural reuse, care must be taken with residual concentrations, as the presence of chlorine in irrigation water is responsible for contamination of the soil, causing toxicity mainly in the leaves of irrigated crops [49,50].

Use of Microalgae in SCW to Improve the Effluent Quality

When systems including SCW are used, recent studies indicate perspectives to improve the effluent quality based on the use of microalgae. Yehia et al. [51] investigated the treatment of a mixture of domestic effluent with agricultural and industrial material using four pilot SCW units in the presence and absence of microalgae in Northern Egypt (Delta). Their results showed that the best BOD and COD removal was achieved using Chlorella reaching 88% and 84% removal efficiency, respectively. Those authors also reported that Azolla was the microalgae providing the best removal of effluents with the highest TN concentration, while Spirulina was considered to be the most efficient microalgae in the removal of metals such as aluminum (Al), iron (Fe), and manganese (Mn). When compared with other plants used in SCW, microalgae present the advantage of fast growth and ability to absorb nutrients, and good tolerance to temperature changes, in addition to the potential economic benefit of the biomass of the harvested algae. The effluent BOD in all units was below 15 mg L^{-1} for most of the year. This allowed the application of the treated effluent in the irrigation of agricultural products and green landscapes in education and recreation facilities, according to Egyptian regulations.

Some reports indicated that the conventional SCW without microalgae did not meet the requirements for unrestricted irrigation [51]. The combined use of VFCW with microalgae was investigated in the post-treatment of anaerobic reactor effluents [52]. According to the Brazilian regulation ABNT 13969/97, the treated effluent could be reused to wash floors and pavements and irrigate gardens and landscapes. In systems involving large-scale (35,000 m^2) SCW without microalgae, results revealed that there was an increase in the sulfate concentration ($SO_4{}^{-2}$) [53]. Such an increase might be due to the denitrifying bacteria activity, since chemolithoautotrophic bacteria reduce nitrate, while the S-oxidant bacteria oxidize sulfide to return $SO_4{}^{-2}$ during denitrification. This bacterial activity might explain the high NO^{-3} removal and the $SO_4{}^{-2}$ release in CW, which requires post-treatment so that the effluent can be reused [53].

CWs as Post-Treatment of Domestic Effluents

CWs also appear as a good proposal in the post-treatment of domestic effluents in systems involving ponds, UASB reactors, and Imhof tanks, among other technologies. Ergaieg et al. [54] used two HFCW as a post-treatment for maturation ponds in a real-scale tertiary treatment of domestic effluents in Tunisia. The proposal aimed to improve the quality of the water used by farmers in that region regarding the removal of BOD and pathogens. One of the problems found by those authors was the risk of clogging the system due to higher hydraulic loads observed in some moments, which despite meeting the requirements of the Tunisian regulation for reuse as irrigation water, was not in compliance with the WHO regulations regarding coliform removal rates.

Omidinia-Anarkoli et al. [55] studied a pilot-scale hybrid CW (HFCW + VFCW) in the post-treatment of domestic effluents from stabilization ponds. The performance of two substrates and the effect of the presence and absence of plants were compared. When gravel was used as substrate, the VFCW showed maximum BOD removal efficiency. Phosphate removal ($PO_4{}^{-3}$) showed seasonal dependency, and the highest values were observed in hot seasons. Their study also showed that VFCW as a post-treatment for ponds is a solution to obtain effluent for several reuse applications in developing countries that face hydric crisis, such as Iran. The fecal coliform concentrations in the effluent in cold periods tended to be over the maximum standard, while in hot periods they met the reuse requirements [55].

Pinelli et al. [56] studied the use of full-scale SCW in the treatment of effluents originating from irrigated rural areas mixed with the municipal sewage, and compared this treatment with facultative ponds. The system comprised a decanter as a pre-treatment followed by the SCW and solar disinfection. Their results were satisfactory and reached the quality standards imposed by the Egyptian regulations for water reuse.

Gabr [57] described the design of a full-scale treatment system composed of an Imhoff tank and the HFCW post-treatment to obtain water to be reused in agriculture in Dakhla Oasis in the Egypt's Western desert. The installation was designed to serve 5000 people. Darvishmotevalli [58] investigated a treatment involving Imhoff tanks and HFCW aiming at domestic sewage reuse. Despite reaching a removal efficiency over 99%, for intestinal nematode parasites eggs and protozoan cysts, the treatment did not manage to produce suitable effluent for agricultural reuse according to the EU guidelines, due to pathogen concentration, thus requiring an advanced disinfection unit.

A study comparing the VFCW use and its absence as a post-treatment to UASB reactors verified that when the technologies were used separately, they did not achieve water quality to meet the Indian requirements for reuse. However, the UASB reactor removal capacity allied to the VFCW post-treatment reached up to 98% removal regarding organic matter parameters and could be reused [59].

In terms of combining CWs with anaerobic reactors, several applications have gained significant interest [42,52]. Biogas produced from anaerobic reactors could be converted into electricity and used to operate an aerated CW, favoring better performance in the removal of conventional and emerging contaminants. It should be noted that some risks also existed in some integrated processes and deserve attention. For example, although the formation of free chlorine is a fundamental step in the disinfection process adopted, the simultaneous existence of ammonia and free chlorine could produce chloramines which are considered very toxic substances in the environment and for aquatic life [60,61]. Furthermore, some other possible by-products, such as chlorate and perchlorate, are also toxic to humans or deteriorate in the post-treatment process of CW effluents. Therefore, a considerable risk assessment must be carried out in order to select the optimal integrated system in terms of the type of wastewater and the intended reuse purpose.

Hybrid CW Configurations and Combined Technologies Facing Emerging Pollutants Removal

When seeking to use CWs to obtain reuse water without employing other technologies, the best alternative has been found in hybrid configurations. Hybrid CWs have the advantage of combining aerobic, anoxic, and anaerobic environments and related treatment processes. The results show the potential for the long-term operation of hybrid CWs for the treatment of domestic–industrial-mixed wastewater. Studies have shown that when VFCW + HFCW is used to treat domestic wastewater, effluents reached reuse conditions in gardening, agriculture, and cleaning with sanitary purposes [62]. However, even when employing hybrid systems including VFCW + VFCW + HFCW, the salt removal difficulty still remains with increased electrical conductivity in the effluent when the hybrid system is used in a hot climate due to the excess evapotranspiration. The advantage of use in such climate conditions is that the VFCW does not require a resting time, and the effluent can meet quality requirements for use, being classified as Classes C and D according to the EU (irrigation of indirect consumption crops and in drip irrigation), which also results in a reduction in construction and operation costs [63].

Another possibility of optimizing processes involving CWs, reducing the time needed to remove contaminants in VFCW, is the use of biochar from the plant used in the treatment unit. When using biochar obtained from *Phragmits australis*, a significant reduction was observed in fecal coliforms, COD and BOD, and the metals Cu, Mn, Cd, and Al in only 72 h [64]. Those results indicated that the wastewater quality was highly improved by the biochar treatment. However, it was still not enough to reach reuse levels, with stricter requirements regarding the removal of the main physical, chemical, and biological parameters. Moreover, contaminant desorption mechanisms for this type of biochar must be considered in further investigations [64].

In more recent works involving the use of CWs, one of the emerging concerns, due to the changes in the characteristics of the effluents generated by the global population, is related to the removal of emergent contaminants and antibiotic-resistant microorganisms. Miladenov et al. [65] evaluated the removal of emergent contaminants at a trace level by a real system consisting of an aerobic reactor, anaerobic filter, and an HFCW as a tertiary treatment. Those authors investigated some facilities in the USA and South Africa, where the effluent was reused. Only 11 out of 111 compounds were found in the final effluent. However, these included chemical products with harmful effects to human health and to the environment. In addition, new compounds generated during the treatment were found whose toxicity is unknown. Despite being limited to the three systems studied only, those new results showed the need for future monitoring of organic chemical product traces for the reuse of effluents obtained from CWs.

Chen et al. [66] studied the use of a pilot hybrid CW (VFCW + HFCW + SCW) associated with membrane bioreactors, advanced oxidative processes, and UV/Vis disinfection in the removal of antibiotic-resistant genes. They verified that the CW system favored the removal of the existing bacteria (Transposon tnpA, insertion sequence IS91, and integron intI1), but was not enough to eliminate risks. They concluded that to guarantee safety in water reuse, the disinfection system efficiency must be improved.

Cherif et al. [67] proposed a pilot system involving a sequence of treatments to produce noble reuse water with enough quality to be used in fish farming. The system included a membrane biofilter (MBR), a VFCW, sand and activated coal filters, nanofiltration, and reverse osmosis desalinization. Such a combined system enabled water reuse in fish production with increased productivity. However, despite all phases employed, the low removal of emergent contaminants, such as benzotriazole and clarithromycin, showed that the system must be improved, which might include the use of coagulation and flocculation.

The application of hybrid CWs to various wastewaters has demonstrated that the combination could improve pollutant removal efficiency. Hybrid CWs could cover the limitation of each CW. For example, by combining HFCW for VFCW, the desirable condition for the nitrification–denitrification process would be created due to the aerobic, anoxic, and anaerobic environment. Some operational and design parameters such as HLR, bed material, system configuration (number of beds; system layout), influent pollutant concentrations and effluent recirculation can affect the performance of hybrid CWs. It is interesting to note that hybrid CWs are effective in removing organic matter (BOD; COD) and suspended solids, while in terms of pathogens and nutrient removal like N and P components, removal efficiencies depend on system properties and operating conditions.

Main Characteristics of CW Systems Used for the Treatment of Domestic Effluents Intended for Water Reuse

Further details of the studies published and their main characteristics can be observed in Table 3. A great variation is observed in treatment scales ranging between 15,000 m^2 and 0.06 m^2, with the microcosm-scale CWs. Among the 30 plants mentioned in the studies, the growing use of ornamental plants and microalgae has been observed. However, the most used plants were *Phragmites australis* (9) and *Typha latifolia* (5). The technologies appearing in the 23 studies cited in Table 3 are VFCW (7), SCW (6), hybrid CW (7), HFCW (5), and microcosm-scale CW (1). The HRT ranged from hours to over 11 days, with an average of 4.7 ± 5.3 d (n = 20). On the other hand, the HLR ranged from 26 to 1080 mm/d, with the higher HLR in a system involving aerated VFCW.

Table 3. Main characteristics of systems involving CWs in the treatment of domestic effluents.

System Number, Type, Size	Plants	Bed Substrate	HRT, d (HLR, mm d^{-1})	Removal (% or LU (Pathogens)) COD (BOD), TSS, Others [c,d,e,f,g]	TN, TP, Others [h,i]	*E. coli*, TC, EC	Reuse	Country
1. HFCW, 336 m^2 [a]	*Agapanthus africanus*	Volcanic porous rock (tezontle)	11.75 (22.3)	>95 (>95), na	na, na	0.94, 1.09, na	Garden irrigation	Mexico
2. SCW, na [a]	*Typha latifolia; Canna indica*	Gravel and sand	2.5–4 (na)	na, na, na	na, na	0.68, 0.87, 0.90	Garden irrigation	India
3. CW microcosms 0.06 m^2 [a]	*Lavandula sp., Spathiphyllum wallisii, Zantedeschia ethiopica*	Red volcanic gravel (RVG); polyethylene terephthalate (PET)	3 (na)	na (63–68), na	35–50, 35–38	0.41, na, na	Garden irrigation	México
4. SCW, 9.4 m^2 [a]	*Canna indica L.; Bird of Paradise; Sagittaria lancifolia; Lchinodorus*	Sand; Gravel	5 (na)	46–65 (51–76), 48–78, 57–82 [c]	30–48, 40–58	na, na, na	Non-body contact Non-sensitive receiving zones	India
5. HCW (SCW+ VFCW + HFCW), 1.17 + 1.17 + 1.17 m^2 [b]	*Canna x generalis, Equisetum sp., Chrysopogon zizanioides, Hymenachne grumosa and Cyperus papyros nanus*	Bricks of broken clay; gravel	24 (na)	77 (84), na, 99.7 [c]	93.8, 94.0, 93.8 [i]	na, na, na	Car washing and other uses with direct contact	Brazil
6. HFCW, na [b]	*Chrysopogon zizanioides*	Polystyrene, recycled; gravel; thick sand.	na(na)	85 (na), 70	89, na, 81 [i]	na, na, na	Agriculture irrigation	Brazil
7. SCW, 20 m^2 [a]	*Phragmites australis, Sparganium erectum.*	na	5 (na)	41 (54), na	na, na, 67 [j]	na, na, na	Agriculture irrigation	Lebanon
8. VFCW, 0.48 m^2 [a]	*Phragmits australis*	Sand, gravel, biochar	na (200)	71.9 (64.3), na	na, na	0.46, 0.17, na	Agriculture irrigation	Egypt
9a. HCW (VFCW + SCW), 166.8 + 167.4 m^2 [b]	*Typha latifolia, phragmites australis, Vetiver grass, Eichhornia crassipe Centella asiatica. pistia stratiotes*	Gravel; aggregate crush	2 (11.9)	73.6 (76.2), 82	37, na	1.31, na, na	Environmental	Pakistan
9b. HCW.(HFCW + SCW) [b], 123.7 + 107.9 m^2 [b]			1.3 (19.0)	71.5 (72.5), 91	47, na	1.24, na, na	Environmental	Pakistan
10a. HCW (HFCW + VFCW), 9 + 2.25 m^2 [a]	*Phragmites australis*	Gravel	4.1 (10.8)	68.2 (75.7), 84.4	na, na, 19.7 [i]	1.24, na, na	Agriculture irrigation	Iran
10b. HCW (HFCW + VFCW), 9 + 2.25 m^2 [a]	*Phragmites australis*	Pumice	4.1 (7.8)	64.1 (71.6), 85.1	na, na, 4.4 [i]	1.18, na, na	Agriculture irrigation	Iran
11. HCW (HFCW + VFCW + SCW), 360,000–420,000 m^2 [a]	*Calamus, reed water onion*	Gravel	2 (54–62)	na (na), na, 37 [d]	na, na	na, na, na	Agriculture irrigation	China
12. VFCW, 22 m^2 [a]	*Juncus; Phramitis*	na	na (69.3)	44.2 (na), na	na, na, 33.6 [j]	na, na, na	Agriculture irrigation	Tunisia
13. HCW (VFCW + HFCW), 0.78 + 0.78 m^2 [a]	*Canna Indica; Calibanus hookeri*	Gravel; sand; soil	1 (550)	89.9 (92.7), 85.5	na, 88.8, 99.1 [i]	na, na, na	Garden, irrigation, flushing in toilet	India
14. HCW (VFCW + VFCW + HFCW), 72 + 48 +72 m^2 [b]	*Phragmites australis and Typha latifolia*	Silex, granite, or river gravel	na (23.4)	90.7 (99.6), 98.3	80.9, na, 95.6 [h], 90.7 [i]	4.8, na, na	Agriculture irrigation	Senegal
15. VFCW, 0.72 m^2 [b]	*Pistia Stratiotes and Phragmites karka*	Crushed stone Sand, soil	3- 7 (39.7)	90 (>98), 92	89, 80, 70 [i]	na, na, na	Agriculture irrigation	India

Table 3. *Cont.*

System Number, Type, Size	Plants	Bed Substrate	HRT, d (HLR, mm d^{-1})	Removal (% or LU (Pathogens)) COD (BOD), TSS, Others [c,d,e,f,g]	TN, TP, Others [h,i]	*E. coli*, TC, EC	Reuse	Country
16. Aerated saturated VFCW, 1.160 m^2 [a]	*Typha latifolia*	Gravel	0.31 (1078)	22.3 (75.5), 5.6	na, na 89.1 [h], 0.0 [i],	1.29, nd, 1.31	Environmental	United Kingdom
17. SCW, 10,000 m^2 [a]	*Phragmites australis; Typha latifolia*	na	na (300)	na (na), na	na, na	2.10, 2.89, na	Agriculture irrigation	Spain
18. SCW 550 m^2 [a]	*Phragmites australis*	na	5 (na)	30–96 [e]; 85–88 [f]; 1–93 [g]	na, na	na, na, na	Garden irrigation; street sewerage cleaning,	Spain
19a. VFCW 0.7 m^2 [b]	*Pennisetum pedicellatum; Cyperus rotundus*	Gravel; sand	1 (357)	66 (90), na	na, na, 84.5 [h], 90 [i]	na, na, na	Landscape, construction, industrial	India
19b. HFCW; 1.45 m^2 [b]	*Pennisetum pedicellatum; Cyperus rotundus*	Gravel; soil	1 (204)	63 (80), na	na, na, 67 [h], 85 [i]	na, na, na	Landscape, construction, industrial	India
20a. SCW, 45 m^2	*Reeds*	na	2 (300)	79 (81), na	60, 69	1.64, 1.82, na	Agriculture irrigation	Egypt
20b. SCW, 45 m^2	*Chlorella*	na	2 (300)	84 (88), na	76, 80	2.32, 2.36, na	Agriculture irrigation	Egypt
20c. SCW, 45 m^2	*Spirulina*	na	2 (300)	82 (87), na	75, 80	2.29, 2.32, na	Agriculture irrigation	Egypt
20d. SCW, 45 m^2	*Azolla*	na	2 (300)	81 (83), na	80, 70	1.86, 2.30, na	Agriculture irrigation	Egypt
21. HCW (ATS + VFCW + VFCW), 0.51 + 0.51 + 0.51 m^2 [b]	*Hymenachne grumosa*	Basaltic gravel; basaltic crushed stone	21 (6.6)	72 (na), 48, 98 [c]	70.1, na, 99.9 [i]	1.26, 5.84, na	Garden irrigation, floor, sidewalk washing	Brazil
22. HFCW, 1321 + 1100 m^2 [a]	*Phragmites australis*	Medium gravel	0.53 (362)	49.4 (54.2), 56.9	56.4, 49.2	1.07, na, na	Agriculture irrigation	Tunisia

References (System number): 1. [36]; 2. [37]; 3. [38]; 4. [39]; 5. [40]; 6. [41]; 7. [53]; 8. [64]; 9. [42]; 10. [55]; 11. [66]; 12. [67]; 13. [62]; 14. [63]; 15. [59]; 16. [44]; 17. [45]; 18. [46]; 19. [47]; 20. [51]; 21. [52]; 22. [54]. Remarks: [a] Municipal Industry wastewater. [b] University wastewater. [c] Turbidity. [d] Antibiotic resistance gene. [e] Pharmaceuticals. [f] Personal care products. [g] Pesticides. [h] N−NH_3. [i] P-PO_4. *TC, total coliforms, EC: Enterococci.* [j] Helminth eggs. ATS: Algae turf scrubber. LU: log units. na: not available.

The BOD removal varied between 71.87% (n = 4) in systems involving SCW, 75.17% (n = 4) in HFCW, 79.86 (n = 6) in VFCW, and 82.9% (n = 4) in hybrid systems, which resulted in the achievement of reuse levels for this parameter that appears in all regulations. Out of the 23 studies, 12 cited studies included results on the removal of pathogens, and although some reached 5.87 Log Units (rarely 99.9% removal), many did not reach the limits imposed by reuse regulations and required additional disinfection. Regarding reuse, most systems targeted agriculture irrigation (13). Those were followed by the irrigation of gardens and environmental discharge (3), while car wash, street cleaning, toilet flush, and industrial use were also cited (1).

The reviewed literature showed that CWS and their different configurations are being explored in their potential of improving landscaping aspects of domestic sewage treatment stations. The CW use as post-treatment for conventional technologies enables improvements in the quality of the effluents generated that favor their reuse.

3.2.2. Efficiency of CWs in the Reuse of Industrial Effluents

The studies published reported that CWs are employed in the treatment of effluents of industries such as car manufacturing, oil, tanning and batteries, winery, cooling, brewery, and petrochemical, aiming at the reuse of wastewater. CWs were observed to be quite efficient in the removal of contaminants present in industrial effluents such as heavy metals and toxic organic compounds.

A treatment proposal for the reuse of effluents with heavy metals in a large automobile manufacturer in Italy was developed on a pilot scale using a hybrid CW VF + HFCW [68]. Wastewater collected from industrial and civil buildings was subjected to physical–chemical and biological pre-treatments prior to the hybrid CW system. The authors concluded that the CW systems implemented were highly efficient in the removal of TSS and heavy metals such as Fe (97.9% removal) and Zn (92.9% removal). However, electrical conductivity, alkalinity, and calcium did not meet the requirements for reuse according to the EPA regulations.

Al-Khafaji et al. [69] reported the use of a pilot SCW with duckweeds aiming to remove heavy metals from batteries and tannery industries. Their results revealed efficiency in the removal of Cd, Cr, Ni, and Pb. The duckweed plant is a poor Cd accumulator; however, it is a hyperaccumulator of Pb, Cr, and Ni. Applying an HRT of 3 d, the system removed low percentages of Cd (32.3%), Cr (44.9%), but high percentages of Ni (74.1%), Pb (79.1%), and Zn (92.9%). The effluent final quality, in relation to these parameters, enables its reuse in agriculture. Javeed [66] evaluated the potential of real SCW systems built to treat effluents from tannery and metallurgical industries, mainly regarding the removal of heavy metals from heavily loaded effluents. The removal of metals in each type of effluent proceeded quite similarly, with respect to retention time. For both treatment lines, the mean Ni, Cu, Cr, and Zn removal values were 32%, 48%, 48%, and 57% at 20 days of TRH, respectively, and increased to 88%, 85%, 95% and 94% at 60 d HRT. Seeking to verify the toxicity of the effluent treated, phytotoxicity tests were carried out. Thus, the effluent used in the irrigation of *Abelmoscus esculentus* L. showed a hyperaccumulation trend. However, those values were within the limit accepted by the WHO for the accumulation of metals by plants [70].

There are studies on the use of CW in the treatment of industrial effluents with high organic load such as those from oil, wine, and beer production. Due to the fact that effluents originated from the olive oil production present up to 4000 times more COD and high phenol concentrations, pre-treatment was applied in open tanks for sedimentation or even pH correction. The removal rates of COD, TSS, TKN, and phenol in the pilot VF + SCW hybrid system reached 54.1%, 52.0%, 44.4%, and 60.1%, respectively. On the other hand, the pilot SCW system (with enhanced pre-sedimentation and pH correction) achieved COD, TSS, TKN, and phenol removals of 49.4%, 72.0%, 26.9%, and 51.1%, respectively. Even if a hybrid CW with VFCW + HFCW was used, it was only possible to produce effluent for agricultural reuse in crops that did not require contact, that is, further treatment or lower surface loading rates would be required for more restrict uses [71].

In studies on wine cellars, the use of hybrid CW was investigated regarding agricultural use standards [72]. The treatment system was subjected to a pre-treatment phase (coarse sieving), followed by a Imhoff tank and an equalization tank. The hybrid CW presented a total area of around 230 m^2 and consisted of a VF + HF + SCW bed, producing water for reuse in suitable conditions for food crops to be consumed raw with edible parts produced above the ground, as described by the EU regulation [73]. When treating effluents from a brewery, a system made of an HFCW in two stages was used to produce tomato irrigation water [69]. The HFCW system was used in the post-treatment of an UASB reactor effluent. Those authors observed that the effluent going through the UASB and HFCW treatments reached the irrigation water standard as required by the Ethiopian environment protection agency, showing a high tomato yield capability [74].

Recent studies have also demonstrated the use of CWs in the treatment of effluents to help the production of water to be reused in cooling systems and petrochemical industries. Wagner et al. [70] investigated the use of a pilot hybrid CW (HF + VF + SCW) in the pre-treatment of a membrane system for the reuse of industrial synthetic effluents for tower cooling. The study also verified the removal of phenolic contaminants. Their results showed that the HFCW removed PO_4^{-3}, TSS, and TOC as a result of adsorption and filtration. Regarding VFCWs, they outstood in the removal of phenolic compounds resulting from biodegradation. The SCW did not contribute to the removal of substances and even increased the effluent salinity. However, they provided some options of water storage and habitat for the aesthetic improvement of the unit benefitting the local flora and fauna. Another study developed by the same research group [75] sought the reuse of water from a cooling tower of a system made of green and gray technologies. The treatment plant showed a configuration that included a SCW as a pre-treatment, followed by a system composed of VF + HFCW. After leaving the hybrid CW, the effluent was subjected to nanofiltration, electrochemical oxidation, and reverse osmosis. In addition, they studied the removal of corrosion inhibitors. The pre-treatment using hybrid CWs before the nanofiltration resulted in the removal of phosphate, nitrate, and benzotriazole, but increased TOC and electrical conductivity was also observed. Thus, for the water reuse in cooling towers, the use of reverse osmosis was required [75].

Jain et al. [76] proposed a new pilot CW configuration with a flow deflector (HFCW-FD) for the treatment of wastewater from petrochemical industries and their reuse. Their results revealed that the unit enabled a flow path that was nine times longer, and the use of a filter with a variable depth reduced the total area requirement, and served as a polishing unit. The HFCW-FD made the effluent quality suitable for reuse in irrigation, industrial reuse, and other environmental uses. Thus, the system proposed required less space and maintenance than conventional CWs and managed to produce good quality water for reuse [76].

CWs have been increasingly applied in treating various industrial wastewaters with specific characteristics. Considering the specific characteristics of various industrial effluents with low or high pH, soil materials with a neutralizing ability but low cost for potential use in full projects should be selected. For the use of CW in the treatment of huge industrial effluent flows, the main drawback is the large area that would be required. However, this problem could be minimized by associating it with the economic production of plant biomass and its energy recovery.

Main Characteristics of CW Systems Treating Industrial Effluents Intended for Water Reuse

The main difference in relation to studies on domestic effluents regards the removal of heavy metals originating from industrial activities. The studies listed in Table 4 showed Cr removal ranging between 30.4% and 95.5%, 32.4% for Cd, from 20.3 to 93.2% for Ni, 79.1% for Pb, 24.3–97.0% for Cu, 35.8–95.6% for Zn in SCW systems, and 98.0% for Fe and 92.9 for Zn in VFCW systems. Other examples of removal were reported for phenols, that is, 87.2% in HFCW systems, and from 8% to 20% in the VFCW system. The hybrid CW systems showed great efficacy in the removal of nutrients, obtaining values between 67% and 80% for TN and 95% to 100% for TP. COD removals varied from 20% up to 100% in one HFCW system.

Table 4. Main characteristics of systems involving CWs in the treatment of industrial effluents.

System Number, Type, Size	Plants	Bed Substrate	HRT, d (HLR, mm d^{-1})	Removal (% or LU (Pathogens)) COD (BOD), TSS, Others [h,i,j k,l,m,n,o,p,q,r,s,]	TN, TP, Others [t]	*E. coli*, TC, EC Others [u,v]	Reuse	Country
1a. VFCW, 1 m^2 [a]	*Juncus maritimus Lam*	Gravel, sand	na (25)	20 (na), 83.7, 84.6 [n], 96.1 [p], 15.9 [o]	na, na, 95 [t]	na, na, na	Industrial	Italy
1b. VFCW, 1 m^2 [a]	*Typha latifolia*	Gravel, sand	na (25)	20 (na), 98.9, 73.2 [n], 92.8 [p], 9.6 [o]	na, na, 95 [t]	na, na, na	Industrial	Italy
1c. HFCW, 1m^2 [a]	*Cyperus papyrus*	Gravel	4 (50)	60 (na), 99.2, 79.1 [n], 98.1 [p], 29.3 [o]	na, na, 95[99]	na, na, na	Industrial	Italy
2a. HCW (VFCW + SCW), 0.52 + 2.89 m^2 [b]	*Phragmites australis, Typha latifolia*	Carbonate material, gravel, sand, clay	>14 (7.6)	54.1 (na), 52.0, 60.1 [q]	44.4, na	na, na, na	Irrigation	Greece
2b. SCW, 2.89 m^2 [b]	*Typha latifolia*	Clay	14 (7.1)	49.9 (na), 72.0, 51.1 [q]	26.9, na	na, na, na	Irrigation	Greece
3. SCW, 0.3 m^2 [c]	*Lemna minor*	na	3	na, (na), na, 44.9 [i], 32.3 [j], 74.5 [k], 79.1 [l], 92.9 [n]	na, na	na, na, na	Irrigation	Iraq
4. HCW (VFCW + HFCW + SCW), 140 + 60 + 30 m^2 [d]	*Phragmites australis, Cyperus papyrus canna indica,*	Gravel, sand, coarse vulcanic	>9 (13)	81 (78), 69	56, 38, 57 [t]	1.3, 1.6, 0, 1.8	Irrigation	Italy
5. HCW (VFCW + HFCW + SCW), 1.37 + 2.10 + 2.10 m^2 [e]	*Phragmites australis.*	Gravel, sand	5–7 (29)	80–100 (na), na, 60 [r], 80 [s]	80–100, 80–100	na, na, na	Industrial	Netherlands
6. HFCW, 11.4 m^2 [f]	*Cyperus alternifolius, Typha latifolia*	Clay rock	20 (na)	74, (na), 67	66, 61	na, na, na	Irrigation	Ethiopia,
7. SCW, 0.92 m^2 [e]	*Hemarthria compressa*	Alluvial stones	(20–60)	70 (na), 98.3, 35.8–95.6 [n], 30.6–95.5 [i], 24.3–97.1 [m], 20.3–93.2 [k]	na, na	na, na, na	Irrigation	Pakistan
8. HCW (VFCW + HFCW + SCW), 1.37 + 2.10 + 2.10 m^2 [e]	*Phragmites australis*	Gravel, sand	5 (na)	67 (na), na, 90.6 [r]	67, na, 95 [u]	na, na, na	Industrial	Netherlands
9. HFCW, 0.6 m^2 [g]	*Chrysopogon zizanioides*	Gravel, sand, soil	4 (16)	66 (73), 78, 93 [h], 86 [q]	na, na	na, na, na	Irrigation, Industrial	India

References (System number): 1. [68]; 2. [71]; 3. [69]; 4. [72]; 5. [74]; 6. [73]; 7. [70]; 8. [75]; 9. [76]. Remarks: Industries – [a] Automotive, [b] Olive, [c] Tanning and batteries, [d] Winery, [e] Cooling, [f] Brewery, [g] Petrochemical, [h] Turbidity, [i] Cr, [j] Cd, [k] Ni. [l] Pb, [m] Cu, [n] Zn, [o] Cl^-, [p] Fe, [q] Phenol, [r] benzotriazole, [s] acid benzoic [t] N-NH_3, [u] PO4-P, TC, total coliforms, EC: Enterococci, LU: log units. [v] Bacteria indicators, na: not available.

The most widely used substrates were gravel and sand (n = 6). However, clay rock, alluvial stones, carbonate material, and coarse volcanic materials were also mentioned. As for the treatment of domestic sewage, *Phragmites australis* and *Typha latifolia* were the most used plants. Due to the type of industries investigated, with little microbiological contamination, only two reports were found with removal results for those contaminants. Irrigation (n = 6) and industrial (n = 3) were the reuse purposes proposed for the treated effluents.

4. Final Considerations

The advantages and disadvantages of CWs are well known and discussed in numerous reviews [9,12,30]. Low maintenance requirements, good treatment efficiency, and economic feasibility make CWs preferable to other wastewater treatment systems in certain situations. These advantages are manifested in a particular and more pronounced way in the case of decentralized treatment, in areas with less developed sewage network, or when there is a lack of qualified labor. However, CWs are also widely used in many highly developed countries, due to their advantages in sustainability. Multiple removal mechanisms occur in CWs, such as microbial degradation, absorption on the bed substrate, and phytoremediation, among others. Several mechanisms or all of them work simultaneously, resulting in an efficient treatment process [52,68,76]. The high flexibility of CWs allows the implementation of different modifications and intensifications, such as filter media with different properties, different flow regimes, macrophyte species, a combination of different types of CWs (hybrid CWs) or with other treatment systems, effluent recirculation, artificial aeration, etc. This provides them with a great potential to adapt the treatment to the goals of reusing the effluent [13,77,78].

The main drawback of CWs is the high land area required, which can vary from 1 to 8 m^2/hab.·equivalent for a complete secondary and tertiary treatment [9,79], while in the case of using CW as post-treatment, the footprint will be lower. This results in high hydraulic retention times or low hydraulic loading rates for optimum performance. However, in practice very variable values are applied, in the range of 0.3–20 d HRT and 6–300 mm d-1 HLR, excluding some extreme values, as can be observed in Tables 2–4. These wide ranges are the result of very different factors, such as the role of CWs in the global treatment system, the characteristics of the waste water to be treated or the final quality objectives required for reuse.

A drawback common to other technologies is the incomplete degradation of recalcitrant organic matter and other recalcitrant pollutants. This may make it necessary to combine it with other treatment technologies, depending on the objectives of the reuse, as highlighted in the previous sections of this review. However, the great flexibility of CW systems also allows for installations to be adapted to the quality required for the reuse of the effluent, which is highly variable according to the destination (Table 1). For example, nutrient removal is not necessary in most cases of irrigation and agricultural reuse. In this way, the elimination of pathogens, even if it can be high, depending on the type and operating conditions of the CWs, is usually the limiting factor of the effluent quality and the main reason for the need of an additional disinfection post-treatment.

Reports published in the last few years addressing the use of CWs in treatment systems aiming at the reuse of different effluents showed that the studies are promising and that for most parameters defined by the regulations, CWs can achieve the required removal level. Regarding runoff water treatment, studies are mainly directed to the removal of metals and the verification of toxicity in the effluent reuse for the irrigation of different plants. As for gray water, CWs might enable the treatment in loco using ornamental plants to promote better acceptance of sewage reuse, and possibilities of the application of such water in nobler uses when associated with disinfection systems.

Although the removal of pathogens from domestic and municipal wastewater might reach up to 99.9% (3.0 LU), or higher (Table 3), removals were usually in the range from 0.5 to 2.3 LU. Of the total of articles reviewed for CWs treating domestic wastewater, 12 studies contained information on the elimination of pathogenic agents, using various indicators, especially TC, FC, and *E. coli*, and less frequently, enterococci, fecal streptococci, nematode eggs, and viruses. TC and FC were measured simultaneously in six treatment lines. The average removals in these six facilities were 1.68 ± 0.88 LU for TC and 1.58 ± 0.75 LU of FC, observing a good direct correlation between these two parameters (R^2 = 0.94, $p < 0.01$). On the other hand, and considering that *E. coli* is considered an indicator parameter of fecal coliforms, we considered together the data for both indicators (6 studies for *E. coli* and 11 for FC). The elimination of pathogens for this set of facilities (n = 16, either FC or *E. coli*, excluding an outlier of 4.8 LU of FC) was in the range of 0.4–2.3 LU, with an average of 1.28 ± 0.57 LU. Observed pathogen removals were lower than those reported by López et al. [80]. In a review of about 60 systems including CW, these authors found TC and FC removals in the range of 1.2–3.9 UL, with small differences depending on the type of CW, although they were higher in the following order: VFCW < HFCW < HCW [80].

In our revision, pathogen removal was not related to HRT or HLR, which can be explained by the diversity of CW typologies, construction parameters, and operating conditions, in addition to the potential presence of short-circuiting that may cause the actual TRH to be lower than the design TRH [81]. Pathogen removal mechanisms that occur in CWs include antibiotic and antibacterial action, predation, filtration, sedimentation, photodegradation, absorption into the bed substrate, and phytoremediation, among others [80,81]. López et al. [80] reported that the main factors affecting the removal of TC and FC in CWs are the size of the support medium, pH conditions, and the organic matter removal rate. However, there are other factors that influence pathogen reduction, such as design, hydraulic regime, oxygen, environmental conditions, seasonal fluctuation, and the presence and type of vegetation [80,81]. Thus, pathogen removal in CW can be highly unpredictable.

At low influent pathogen concentrations, removals can be expected to be lower [81]. This indicates the difficulty of achieving the very demanding levels of effluent quality for some water reuses in terms of pathogens (Table 1) through the use of CW. In this way, many systems do not manage to meet the requirements imposed by reuse regulations, requiring disinfection using conventional systems such as chlorination or UV/Vis treatment to increase their efficiency. However, the formation of trialomethane due to chlorine use was observed, and the removal of enterococci using ultraviolet radiation was low. Aeration, hybrid configurations, and microalgae contribute to the achievement of more restrictive reuse levels, but the direct consumption crop irrigation standards have not been achieved yet.

The removal of emergent contaminants has been investigated in the CW studies, and even after an association with advanced oxidative processes, it was not possible to remove some types of antibiotic resistant bacteria, corrosion inhibitors, and medicines; therefore, further studies on reuse water monitoring are still required. In the industrial sector, for example, food and car industries, CWs showed efficiency in the treatment of effluents. Most studies have focused on the reduction in heavy metals and phenols from these industrial effluents.

The most frequently used technology involved the use of VFCW due to the possibility of operating with a higher hydraulic load, thus reducing the space needed for its implementation. The most used plants in the treatment of different effluents are *Phragmites australis* and *Typha latifolia*, while sand and gravel are the main substrates employed. When investigating the reuse, although a potability level has been achieved with the integration of CWs to other processes, the most cited possibility of use after the treatments is irrigation.

This systematic review of the recent literature on the treatment of different types of wastewater in CWs mainly included pilot-scale and field studies conducted in 22 countries worldwide. This included the treatment of various types of wastewater, from runoff and gray water to urban and industrial wastewater. Quantitative data on the quality of the obtained effluent were evaluated in comparison with the requirements of the respective regulations for the reuse of the treated effluent. Thus, this review extended the scope of other previous ones that were restricted to specific geographic areas or specific types of wastewater. This review verified that the future of the use of CWs aiming at the reuse of effluents resides in the perspective of their association with conventional and advanced systems of treatment in new facilities to be projected. In addition, studies aimed at the removal of different emerging contaminants should be developed, which can both favor the inclusion of these substances in the reuse regulations and be required by that probable inclusion in the near future. The use of CWs generally represents an aesthetic improvement in the construction of sewage treatment plants. In this regard, it is derived as a general conclusion from the reviewed articles that the use of CWs shows a great potential to increase the acceptance of the reuse of treated effluents by the local population.

Author Contributions: Writing, review, and editing, S.X.d.C.; Supervision, review, and editing, M.S. All authors have read and agreed to the published version of the manuscript.

Funding: This research is part of the project TED2021-132667B-I00 funded by MCIN/AEI/10.13039/501100011033 and by the European Union "NextGenerationEU"/PRTR. It was also funded CNPQ—National Council for Scientific and Technological Development of the Ministry of Science, Technology, and Innovation of Brazil for the funding granted (Project 402540/2022).

Data Availability Statement: The data that support the findings of this study are available from the corresponding author (M. Soto), upon reasonable request.

Acknowledgments: Sandro Xavier de Campos thanks the University of A Coruña for the research stay there and the CNPq for the scholarships granted for this stay.

Conflicts of Interest: The authors declare no conflicts of interest.

References

1. Nan, X.; Lavrnić, S.; Toscano, A. Potential of constructed wetland treatment systems for agricultural wastewater reuse under the EU framework. *J. Environ. Manag.* **2020**, *275*, 111219. [CrossRef]
2. Sun, H.; Zhang, H.; Zou, X.; Li, R.; Liu, Y. Water reclamation and reuse. *Water Environ. Res.* **2019**, *91*, 1080–1090. [CrossRef]
3. Addo-Bankas, O.; Zhao, Y.; Vymazal, J.; Yuan, Y.; Fu, J.; Wei, T. Green walls: A form of constructed wetland in green buildings. *Ecol. Eng.* **2021**, *169*, 106321. [CrossRef]
4. Biswal, B.K.; Balasubramanian, R. Constructed Wetlands for Reclamation and Reuse of Wastewater and Urban Stormwater: A Review. *Front. Environ. Sci.* **2022**, *10*, 836289. [CrossRef]
5. Masi, F.; Rizzo, A.; Regelsberger, M. The role of constructed wetlands in a new circular economy, resource oriented, and ecosystem services paradigm. *J. Environ. Manag.* **2018**, *216*, 275–284. [CrossRef] [PubMed]
6. Hdidou, M.; Necibi, M.C.; Labille, J.; El Hajjaji, S.; Dhiba, D.; Chehbouni, A.; Roche, N. Potential Use of Constructed Wetland Systems for Rural Sanitation and Wastewater Reuse in Agriculture in the Moroccan Context. *Energies* **2022**, *15*, 156. [CrossRef]
7. Shoushtarian, F.; Negahban-Azar, M. Worldwide Regulations and Guidelines for Agricultural Water Reuse: A Critical Review. *Water* **2020**, *12*, 971. [CrossRef]
8. Santos, A.S.P.; Pachawo, V.; Melo, M.C.; Vieira, J.M.P. Progress on legal and practical aspects on water reuse with emphasis on drinking water—An overview. *Water Supply* **2022**, *22*, 3000–3014. [CrossRef]
9. Vymazal, J. The Historical Development of Constructed Wetlands for Wastewater Treatment. *Land* **2022**, *11*, 174. [CrossRef]
10. Shukla, R.; Gupta, D.; Singh, G.; Mishra, V.K. Performance of horizontal flow constructed wetland for secondary treatment of domestic wastewater in a remote tribal area of Central India. *Sustain. Environ. Res.* **2021**, *31*, 1–10. [CrossRef]
11. Torrijos, V.; Ruiz, I.; Soto, M. Effect of step-feeding on the performance of lab-scale columns simulating vertical flow-horizontal flow constructed wetlands. *Environ. Sci. Pollut. Res.* **2017**, *24*, 22649–22662. [CrossRef]
12. Sánchez, M.; Ruiz, I.; Soto, M. The Potential of Constructed Wetland Systems and Photodegradation Processes for the Removal of Emerging Contaminants—A Review. *Environments* **2022**, *9*, 116. [CrossRef]
13. Sánchez, M.; Fernández, M.I.; Ruiz, I.; Canle, M.; Soto, M. Combining Constructed Wetlands and UV Photolysis for the Advanced Removal of Organic Matter, Nitrogen, and Emerging Pollutants from Wastewater. *Environments* **2023**, *10*, 35. [CrossRef]
14. Spiniello, I.; De Carluccio, M.; Castiglione, S.; Amineva, E.; Kostryukova, N.; Cicatelli, A.; Rizzo, L.; Guarino, F. Landfill leachate treatment by a combination of a multiple plant hybrid constructed wetland system with a solar photoFenton process in a raceway pond reactor. *J. Environ. Manag.* **2023**, *331*, 117211. [CrossRef]
15. Lakho, F.H.; Le, H.Q.; Mattheeuws, F.; Igodt, W.; Depuydt, V.; Desloover, J.; Rousseau, D.P.L.; Van Hulle, S.H. Life cycle assessment of two decentralized water treatment systems combining a constructed wetland and a membrane based drinking water production system. *Resour. Conserv. Recycl.* **2022**, *178*, 106104. [CrossRef]
16. Collivignarelli, M.C.; Miino, M.C.; Gomez, F.H.; Torretta, V.; Rada, E.C.; Sorlini, S. Horizontal Flow Constructed Wetland for Greywater Treatment and Reuse: An Experimental Case. *Int. J. Environ. Res. Public Health* **2020**, *17*, 2317. [CrossRef]

17. Lakho, F.H.; Le, H.Q.; Mattheeuws, F.; Igodt, W.; Depuydt, V.; Desloover, J.; Rousseau, D.P.; Van Hulle, S.W. Decentralized grey and black water reuse by combining a vertical flow constructed wetland and membrane based potable water system: Full scale demonstration. *J. Environ. Chem. Eng.* **2021**, *9*, 104688. [CrossRef]
18. Bolton, C.R.; Randall, D.G. Development of an integrated wetland microbial fuel cell and sand filtration system for greywater treatment. *J. Environ. Chem. Eng.* **2019**, *7*, 103249. [CrossRef]
19. Kotsia, D.; Deligianni, A.; Fyllas, N.M.; Stasinakis, A.S.; Fountoulakis, M.S. Converting treatment wetlands into "treatment gardens": Use of ornamental plants for greywater treatment. *Sci. Total Environ.* **2020**, *744*, 140889. [CrossRef] [PubMed]
20. Morandi, C.; Schreiner, G.; Moosmann, P.; Steinmetz, H. Elevated Vertical-Flow Constructed Wetlands for Light Greywater Treatment. *Water* **2021**, *13*, 2510. [CrossRef]
21. Ren, X.; Zhang, M.; Wang, H.; Dai, X.; Chen, H. Removal of personal care products in greywater using membrane bioreactor and constructed wetland methods. *Sci. Total Environ.* **2021**, *797*, 148773. [CrossRef] [PubMed]
22. Hajlaoui, H.; Akrimi, R.; Sayehi, S.; Hachicha, S. Usage of treated greywater as an alternative irrigation source for tomatoes cultivation. *Water Environ. J.* **2022**, *36*, 484–493. [CrossRef]
23. Junior, R.M.; Passoni, C.M.; Santos, F.M.; Bernardes, F.S.; Filho, F.J.C.M.; Paulo, P.L. Assessment of Surfactant Removal Capacity and Microbial Community Diversity in a Greywater-Treating Constructed Wetland. *Resources* **2023**, *12*, 38. [CrossRef]
24. Compaoré, C.O.T.; Maiga, Y.; Nikiéma, M.; Mien, O.; Nagalo, I.; Panandtigri, H.T.; Mihelcic, J.R.; Ouattara, A.S. Constructed wetland technology for the treatment and reuse of urban household greywater under conditions of Africa's Sahel region. *Water Supply* **2023**, *23*, 2505–2516. [CrossRef]
25. Laaffat, J.; Aziz, F.; Ouazzani, N.; Mandi, L. Biotechnological approach of greywater treatment and reuse for landscape irrigation in small communities. *Saudi J. Biol. Sci.* **2017**, *26*, 83–90. [CrossRef] [PubMed]
26. Ventura, D.; Barbagallo, S.; Consoli, S.; Ferrante, M.; Milani, M.; Licciardello, F.; Cirelli, G.L. On the performance of a pilot hybrid constructed wetland for stormwater recovery in Mediterranean climate. *Water Sci. Technol.* **2019**, *79*, 1051–1059. [CrossRef] [PubMed]
27. Tuttolomondo, T.; Virga, G.; Licata, M.; Leto, C.; La Bella, S. Constructed Wetlands as Sustainable Technology for the Treatment and Reuse of the First-Flush Stormwater in Agriculture—A Case Study in Sicily (Italy). *Water* **2020**, *12*, 2542. [CrossRef]
28. Zhang, N.; Gao, F.; Cheng, S.; Xie, H.; Hu, Z.; Zhang, J.; Liang, S. Mn oxides enhanced pyrene removal with both rhizosphere and non-rhizosphere microorganisms in subsurface flow constructed wetlands. *Chemosphere* **2022**, *307*, 135821. [CrossRef]
29. A Mohammed, N.; Ismail, Z.Z. Green sustainable technology for biotreatment of actual dairy wastewater in constructed wetland. *J. Chem. Technol. Biotechnol.* **2021**, *96*, 1197–1204. [CrossRef]
30. Shukla, A.; Parde, D.; Gupta, V.; Vijay, R.; Kumar, R. A review on effective design processes of constructed wetlands. *Int. J. Environ. Sci. Technol.* **2022**, *19*, 12749–12774. [CrossRef]
31. Yan, C.; Huang, J.; Lin, X.; Wang, Y.; Cao, C.; Qian, X. Performance of constructed wetlands with different water level for treating graphene oxide wastewater: Characteristics of plants and microorganisms. *J. Environ. Manag.* **2023**, *334*, 117432. [CrossRef]
32. Wang, J.; Yu, X.; Lin, H.; Wang, J.; Chen, L.; Ding, Y.; Feng, S.; Zhang, J.; Ye, B.; Kan, X.; et al. The efficiency of full-scale subsurface constructed wetlands with high hydraulic loading rates in removing pharmaceutical and personal care products from secondary effluent. *J. Hazard. Mater.* **2023**, *451*, 131095. [CrossRef]
33. Feng, Y.; Wu, Y.; Wei, B.; Zhu, H.; Xu, Y. Characteristics and applications of hybrid constructed wetlands for low-polluted water: Cased in Bagong River of the Yellow River Watershed, China. *Ecol. Eng.* **2022**, *182*, 6718. [CrossRef]
34. Pennellini, S.; Awere, E.; Kakavand, N.; Bonoli, A. Assessment of secondary wastewater treatment technologies for agricultural reuse in Rafah, Gaza Strip: Application of evidential reasoning method. *Clean. Eng. Technol.* **2023**, *13*, 100611. [CrossRef]
35. Agaton, C.B.; Guila, P.M.C. Ecosystem Services Valuation of Constructed Wetland as a Nature-Based Solution to Wastewater Treatment. *Earth* **2023**, *4*, 78–92. [CrossRef]
36. De Anda, J.; López-López, A.; Villegas-García, E.; Valdivia-Aviña, K. High-Strength Domestic Wastewater Treatment and Reuse with Onsite Passive Methods. *Water* **2018**, *10*, 99. [CrossRef]
37. Kaushal, M.; Patil, M.D.; Wani, S.P. Potency of constructed wetlands for deportation of pathogens index from rural, urban and industrial wastewater. *Int. J. Environ. Sci. Technol.* **2018**, *15*, 637–648. [CrossRef]
38. Sandoval-Herazo, L.C.; Alvarado-Lassman, A.A.; Marín-Muñiz, J.L.; Méndez-Contreras, J.M.; Zamora-Castro, S.A. Effects of the Use of Ornamental Plants and Different Substrates in the Removal of Wastewater Pollutants through Microcosms of Constructed Wetlands. *Sustainability* **2018**, *10*, 1594. [CrossRef]
39. Liamlaem, W.; Benjawan, L.; Polprasert, C. Sustainable wastewater management technology for tourism in Thailand: Case and experimental studies. *Water Sci. Technol.* **2019**, *79*, 1977–1984. [CrossRef] [PubMed]
40. Dell'Osbel, N.; Colares, G.S.; Oliveira, G.A.; Rodrigues, L.R.; da Silva, F.P.; Rodriguez, A.L.; López, D.A.; Lutterbeck, C.A.; Silveira, E.O.; Kist, L.T.; et al. Hybrid constructed wetlands for the treatment of urban wastewaters: Increased nutrient removal and landscape potential. *Ecol. Eng.* **2020**, *158*, 106072. [CrossRef]
41. Quartaroli, L.; Cardoso, B.H.; Ribeiro, G.d.P.; da Silva, G.H.R. Wastewater Chlorination for Reuse, an Alternative for Small Communities. *Water Environ. Res.* **2018**, *90*, 2100–2105. [CrossRef] [PubMed]
42. Ali, M.; Rousseau, D.P.; Ahmed, S. A full-scale comparison of two hybrid constructed wetlands treating domestic wastewater in Pakistan. *J. Environ. Manag.* **2018**, *210*, 349–358. [CrossRef] [PubMed]

43. Russo, N.; Marzo, A.; Randazzo, C.; Caggia, C.; Toscano, A.; Cirelli, G.L. Constructed wetlands combined with disinfection systems for removal of urban wastewater contaminants. *Sci. Total Environ.* **2019**, *656*, 558–566. [CrossRef] [PubMed]
44. Stefanakis, A.; Bardiau, M.; Trajano, D.; Couceiro, F.; Williams, J.; Taylor, H. Presence of bacteria and bacteriophages in full-scale trickling filters and an aerated constructed wetland. *Sci. Total Environ.* **2019**, *659*, 1135–1145. [CrossRef] [PubMed]
45. Gonzales-Gustavson, E.; Rusiñol, M.; Medema, G.; Calvo, M.; Girones, R. Quantitative risk assessment of norovirus and adenovirus for the use of reclaimed water to irrigate lettuce in Catalonia. *Water Res.* **2019**, *153*, 91–99. [CrossRef] [PubMed]
46. Gonzalez-Flo, E.; Romero, X.; García, J. Nature based-solutions for water reuse: 20 years of performance evaluation of a full-scale constructed wetland system. *Ecol. Eng.* **2023**, *188*, 106876. [CrossRef]
47. Thalla, A.K.; Devatha, C.P.; Anagh, K.; Sony, E. Performance evaluation of horizontal and vertical flow constructed wetlands as tertiary treatment option for secondary effluents. *Appl. Water Sci.* **2019**, *9*, 147. [CrossRef]
48. Otter, P.; Hertel, S.; Ansari, J.; Lara, E.; Cano, R.; Arias, C.; Gregersen, P.; Grischek, T.; Benz, F.; Goldmaier, A.; et al. Disinfection for decentralized wastewater reuse in rural areas through wetlands and solar driven onsite chlorination. *Sci. Total Environ.* **2020**, *721*, 137595. [CrossRef]
49. Caldeira, D.C.D.; Silva, C.M.; Zanuncio, A.J.V.; Correa Filho, J.R.R. Fertirrigation with nanofiltration retentate from thermomechanical pulp mill effluents. *Ind. Crop. Prod.* **2023**, *199*, 116713. [CrossRef]
50. McGehee, C.; Raudales, R.E. Irrigation Sources with Chlorine Drinking Water Standard Limits Cause Phytotoxicity on 'Rex' Lettuce Grown in Hydroponic Systems. *HortTechnology* **2023**, *33*, 125–130. [CrossRef]
51. Yehia, S.; El-Saadi, A.; Galal, M.M. Application of algae to free surface wetlands for effluent reuse. *Water Environ. J.* **2020**, *35*, 748–758. [CrossRef]
52. Souza, S.B.; Celente, G.d.S.; Colares, G.S.; Machado, L.; Lobo, E.A. Algae turf scrubber and vertical constructed wetlands combined system for decentralized secondary wastewater treatment. *Environ. Sci. Pollut. Res.* **2019**, *26*, 9931–9937. [CrossRef]
53. Karam, F.; Haddad, R.; Amacha, N.; Charanek, W.; Harmand, J. Assessment of the Impacts of Phyto-Remediation on Water Quality of the Litani River by Means of Two Wetland Plants (*Sparganium erectum* and *Phragmites australis*). *Water* **2023**, *15*, 4. [CrossRef]
54. Ergaieg, T.; Miled, B. Full-scale hybrid constructed wetlands monitoring for decentralized tertiary treatment of municipal wastewater. *Arab. J. Geosci.* **2021**, *14*, 1407–1417. [CrossRef]
55. Omidinia-Anarkoli, T.; Shayannejad, M. Improving the quality of stabilization pond effluents using hybrid constructed wetlands. *Sci. Total Environ.* **2021**, *801*, 149615. [CrossRef] [PubMed]
56. Pinelli, D.; Zanaroli, G.; Rashed, A.A.; Oertlé, E.; Wardenaar, T.; Mancini, M.; Vettore, D.; Fiorentino, C.; Frascari, D. Comparative Preliminary Evaluation of 2 In-stream Water Treatment Technologies for the Agricultural Reuse of Drainage Water in the Nile Delta. *Integr. Environ. Assess. Manag.* **2020**, *16*, 920–933. [CrossRef] [PubMed]
57. Gabr, M.E. Design methodology for sewage water treatment system comprised of Imhoff 's tank and a subsurface horizontal flow constructed wetland: A case study Dakhla Oasis, Egypt. *J. Environ. Sci. Health Part A* **2022**, *57*, 52–64. [CrossRef] [PubMed]
58. Darvishmotevalli, M.; Moradnia, M.; Asgari, A.; Noorisepehr, M.; Mohammadi, H. Reduction of pathogenic microorganisms in an Imhoff tank-constructed wetland system. *Desalination Water Treat.* **2019**, *154*, 283–288. [CrossRef]
59. Kumar, S.; Pratap, B.; Dubey, D.; Dutta, V. Integration of Constructed Wetland Microcosms with Available Wastewater Treatment Technologies for the Polishing of Domestic Wastewater and Their Potential Reuses. *Int. J. Environ. Res.* **2022**, *16*, 99. [CrossRef]
60. Bernstein, A.; Siebner, H.; Kaufman, A.G.; Gross, A. Onsite Chlorination of Greywater in a Vertical Flow Constructed Wetland—Significance of Trihalomethane Formation. *Water* **2021**, *13*, 903. [CrossRef]
61. Hong, N.; Li, Y.; Liu, J.; Yang, M.; Liu, A. A snapshot on trihalomethanes formation in urban stormwater: Implications for its adequacy as an alternative water resource. *J. Environ. Chem. Eng.* **2022**, *10*, 107180. [CrossRef]
62. Singh, K.K.; Vaishya, R.C. Municipal Wastewater Treatment uses Vertical Flow Followed by Horizontal Flow in a Two-Stage Hybrid-Constructed Wetland Planted with Calibanus hookeri and Canna indica (*Cannaceae*). *Water Air Soil Pollut.* **2022**, *12*, 510. [CrossRef]
63. Torrens, A.; de la Varga, D.; Ndiaye, A.K.; Folch, M.; Coly, A. Innovative Multistage Constructed Wetland for Municipal Wastewater Treatment and Reuse for Agriculture in Senegal. *Water* **2020**, *12*, 3139. [CrossRef]
64. Asaad, A.A.; El-Hawary, A.M.; Abbas, M.H.H.; Mohamed, I.; Abdelhafez, A.A.; Bassouny, M.A. Reclamation of wastewater in wetlands using reed plants and biochar. *Sci. Rep.* **2022**, *12*, 19516. [CrossRef] [PubMed]
65. Mladenov, N.; Dodder, N.G.; Steinberg, L.; Richardot, W.; Johnson, J.; Martincigh, B.S.; Buckley, C.; Lawrence, T.; Hoh, E. Persistence and removal of trace organic compounds in centralized and decentralized wastewater treatment systems. *Chemosphere* **2021**, *286*, 131621. [CrossRef]
66. Chen, J.; Liu, C.; Teng, Y.; Zhao, S.; Chen, H. The combined effect of an integrated reclaimed water system on the reduction of antibiotic resistome. *Sci. Total Environ.* **2022**, *838*, 156426. [CrossRef]
67. Cherif, H.; Risse, H.; Abda, M.; Benmansour, I.; Roth, J.; Elfil, H. Nanofiltration as an efficient tertiary wastewater treatment for reuse in the aquaponic system in Tunisia. *J. Water Process. Eng.* **2023**, *52*, 103530. [CrossRef]
68. Riggio, V.A.; Ruffino, B.; Campo, G.; Comino, E.; Comoglio, C.; Zanetti, M. Constructed wetlands for the reuse of industrial wastewater: A case-study. *J. Clean. Prod.* **2018**, *171*, 723–732. [CrossRef]
69. Al-Khafaji, M.S.; Al-Ani, F.H.; Ibrahim, A.F. Removal of Some Heavy Metals from Industrial Wastewater by Lemmna Minor. *KSCE J. Civ. Eng.* **2017**, *22*, 1077–1082. [CrossRef]

70. Javeed, F.; Nazir, A.; Bareen, F.E.; Shafiq, M.; Scholz, M. Industrial water treatment within a wetland planted with Hemarthria compressa and subsequent effluent reuse to grow Abelmoschus esculentus. *J. Water Process. Eng.* **2022**, *45*, 102511. [CrossRef]
71. Gikas, G.D.; Tsakmakis, I.D.; A Tsihrintzis, V. Hybrid natural systems for treatment of olive mill wastewater. *J. Chem. Technol. Biotechnol.* **2017**, *93*, 800–809. [CrossRef]
72. Milani, M.; Consoli, S.; Marzo, A.; Pino, A.; Randazzo, C.; Barbagallo, S.; Cirelli, G.L. Treatment of Winery Wastewater with a Multistage Constructed Wetland System for Irrigation Reuse. *Water* **2020**, *12*, 1260. [CrossRef]
73. Alayu, E.; Leta, S. Evaluation of irrigation suitability potential of brewery effluent post treated in a pilot horizontal subsurface flow constructed wetland system: Implications for sustainable urban agriculture. *Heliyon* **2021**, *7*, e07129. [CrossRef]
74. Wagner, T.V.; de Wilde, V.; Willemsen, B.; Mutaqin, M.; Putri, G.; Opdam, J.; Parsons, J.R.; Rijnaarts, H.H.; de Voogt, P.; Langenhoff, A.A. Pilot-scale hybrid constructed wetlands for the treatment of cooling tower water prior to its desalination and reuse. *J. Environ. Manag.* **2020**, *271*, 110972. [CrossRef] [PubMed]
75. Wagner, T.; Saha, P.; Bruning, H.; Rijnaarts, H. Lowering the fresh water footprint of cooling towers: A treatment-train for the reuse of discharged water consisting of constructed wetlands, nanofiltration, electrochemical oxidation and reverse osmosis. *J. Clean. Prod.* **2022**, *364*, 132667. [CrossRef]
76. Jain, M.; Majumder, A.; Gupta, A.K.; Ghosal, P.S. Application of a new baffled horizontal flow constructed wetland-filter unit (BHFCW-FU) for treatment and reuse of petrochemical industry wastewater. *J. Environ. Manag.* **2023**, *325*, 116443. [CrossRef] [PubMed]
77. Pascual, A.; Pena, R.; Gómez-Cuervo, S.; de la Varga, D.; Alvarez, J.A.; Soto, M.; Arias, C.A. Nature based solutions for winery wastewater valorisation. *Ecol. Eng.* **2021**, *169*, 106311. [CrossRef]
78. Pascual, A.; Álvarez, J.; de la Varga, D.; Arias, C.; Van Oirschot, D.; Kilian, R.; Soto, M. Horizontal flow aerated constructed wetlands for municipal wastewater treatment: The influence of bed depth. *Sci. Total Environ.* **2023**, *169*, 106311. [CrossRef] [PubMed]
79. Sánchez, M.; Ruiz, I.; Soto, M. Sustainable wastewater treatment using a new combined hybrid digester—Constructed wetland system. *J. Environ. Chem. Eng.* **2023**, *11*, 110861. [CrossRef]
80. López, D.; Leiva, A.M.; Arismendi, W.; Vidal, G. Influence of design and operational parameters on the pathogens reduction in constructed wetland under the climate change scenario. *Rev. Environ. Sci. Bio/Technol.* **2019**, *18*, 101–125. [CrossRef]
81. Shingare, R.P.; Thawale, P.R.; Raghunathan, K.; Mishra, A.; Kumar, S. Constructed wetland for wastewater reuse: Role and efficiency in removing enteric pathogens. *J. Environ. Manag.* **2019**, *246*, 444–461. [CrossRef] [PubMed]

environments

MDPI

Article

Small Decentralized Technologies for High-Strength Wastewater Treatment and Reuse in Arid and Semi-Arid Regions

Khaja Zillur Rahman [1,*], Shamsa Al Saadi [2], Mohamed Al Rawahi [3], Manfred van Afferden [1], Katy Bernhard [1], Jan Friesen [1] and Roland A. Müller [1]

1 Department of Systemic Environmental Biotechnology, Helmholtz Centre for Environmental Research—UFZ, Permoserstrasse 15, 04318 Leipzig, Germany; manfred.afferden@ufz.de (M.v.A.); katy.bernhard@ufz.de (K.B.); jan.friesen@ufz.de (J.F.); roland.mueller@ufz.de (R.A.M.)

2 Institute of Advanced Technology Integration (IATI), Ministry of Higher Education, Research and Innovation (MoHERI), P.O. Box 92, Muscat 123, Oman; shamsa.alsaadi@moheri.gov.om

3 Directorate General of In-Country-Value (ICV), Department of Research, Development and Innovation, Oman Investment Authority, P.O. Box 188, Muscat 100, Oman; mohamed.alrawahi@oia.gov.om

* Correspondence: khaja.rahman@ufz.de

Abstract: Rural and semi-urban areas in arid/semi-arid regions are facing severe water scarcity and a series of environmental challenges nowadays, specifically due to rapid urbanization and economic development, climate change, population growth, increasing water demand, influxes of refugees caused by war and regional political conflict, etc. To solve the emerging problems, the safe reuse of treated wastewater in agriculture can provide an additional water resource for countries with high water scarcity. The aim of this study was to investigate the treatment performance and effectiveness of small decentralized wastewater treatment (DWWT) technologies treating high-strength wastewater with concentrations far beyond the European Union testing ranges of parameters such as five-day biochemical oxygen demand (BOD_5 > 500 mg/L), chemical oxygen demand (COD > 1000 mg/L), or total suspended solids (TSS > 700 mg/L). Four (4) commercially available DWWT technologies with a design capacity of 4–8 PE (population equivalent) were selected and operated with various wastewater compositions in Leipzig, Germany. The technologies were (i) the moving bed biofilm reactor (MBBR), (ii) the sequencing batch reactor (SBR), (iii) the membrane bioreactor (MBR) and (iv) the aerated vertical-flow constructed wetland (AVFCW). This study results clearly demonstrated that the EU-certified small DWWT technologies are quite capable of treating high-strength wastewater and can provide high-quality treated water for safe reuse in rural communities of arid and semi-arid regions. During operation with high-strength wastewater with a mean inflow BOD_5, COD and TSS concentrations of 1532 ± 478, 2547 ± 830 and 546 ± 176 mg/L, a low mean BOD_5 (<10 mg/L), COD (<70 mg/L) and TSS (<15 mg/L) in the outflow of the four systems showed removal efficiency of BOD_5 (>99%), COD (>97%) and TSS (>97%), and met the maximum allowable limit value of water quality class A for reuse in agriculture according to Jordanian and Omani standard. The MBR showed almost a complete removal of *Escherichia coli* (*E. coli*) in a range of 6.1–6.9-log removal in the outflow during all three experimental phases and performed best for BOD_5, COD, TSS and pathogen removal when treating high-strength wastewater if properly maintained to prevent potential fouling and clogging of the membrane. Before the final permitting process, long-term monitoring under local temperature and climatic conditions as well as guidelines based on local needs (e.g., in Jordan, Oman, etc.) should be developed to guarantee a minimum level of performance standards of such small DWWT technologies and requirements for operation and maintenance (O&M).

Keywords: arid regions; rural communities; high-strength wastewater; decentralized technologies; pollutants removal; treated water quality; reuse; guidelines; extreme environmental scenarios

1. Introduction

Water scarcity is one of the main challenges to achieving sustainable development and economic growth, specifically in arid and semi-arid regions worldwide. Many rural and semi-urban areas in arid climates face severe stress on local water resources. Water scarcity is mainly due to rapid changes in climatic conditions, rapid population growth, influxes of refugees caused by war and regional political conflict, increased urbanization, water reallocation from rural to urban regions, rapid socio-economic development and higher standards of living [1–6].

Most of the countries in these water-scarce regions (in the Middle East, Gulf countries, North Africa, Southern Europe, etc.) are among the most highly water-stressed countries in the world and they are predicted to face serious shortages of water resources in the future [7–9]. Jordan is ranked among the poorest countries in the world in terms of water availability [10]. It has been forecasted that in the next 30 years, water-scarce countries such as Oman and the neighboring regions will run out of water [11]. A recent study demonstrates that 29% of the territory of the European Union (EU), except Italy, was affected by water scarcity conditions between the years 2000 and 2019 for at least one quarter of the year and faced the worst seasonal water scarcity in 2019 [12].

A critical factor for wastewater management in arid and semi-arid countries is the drastic difference in water consumption [13]. As of 2021, the daily water consumption in Europe ranges from 77 to 223 L per capita in selected European countries, where Malta ranked with the lowest daily water consumption of only 77 L per capita and Italy ranked first with the highest water consumption of 223 L per capita [14]. In contrast, the daily water consumption in remote rural areas of Jordan is fairly low (40–80 L per capita) and extremely low (25–50 L per capita) in informal settlements, which results in highly concentrated wastewater [15–17]. This is a common issue for wastewater treatment in countries with very limited water supply [13,18]. High water scarcity and hence low per capita consumption of water contributes to extremely high concentrations of organic matter and nutrients in raw wastewater (for example, BOD_5: 600–1500 mg/L, COD: 1000–2500 mg/L, TN: 80–150 mg/L, TP: 20–60 mg/L), which can be generated by public buildings and small-scale operations in rural communities such as schools, university campuses, office buildings, refugee and migrant camps, informal nomadic settlements, military bases, etc. [1,15,19–22].

In general, high-strength wastewater contains higher soluble organics than typical domestic wastewater [23,24], indicated by its elevated concentration ranges of BOD_5 and COD, fecal matter and nutrients (N and P). Such highly concentrated wastewater is a principal source of environmental pollution and represents vast technical and operational challenges to its treatment [25,26]. Insufficient treatment of such high-strength wastewater can severely reduce the dissolved oxygen (DO) levels in the water if released into the receiving aquatic bodies and may cause severe damage to the aquatic organisms; it can be threatening to the quality of the existing surface water bodies as well as groundwater resources [27,28]. Therefore, it must undergo one or more effective treatment processes before being safe enough to be disposed into any receiving aquatic bodies, such as lakes, rivers, or seas, or to be reused for other purposes (e.g., agricultural irrigation).

In remote rural or peri-urban areas with low income where construction of a centralized wastewater collection and treatment system is not economically feasible, decentralized systems are becoming a popular solution nowadays [29]. Small-scale wastewater treatment systems can be presented as cost-efficient decentralized solutions over centralized treatment systems due to low population densities and scattered settlements in rural areas [30]. High-strength wastewater with very high organic load may potentially impose operational problems in the treatment plants. Therefore, suitable decentralized wastewater treatment (DWWT) technologies need to be selected before successful implementation of treated wastewater reuse on a local scale [13]. Reuse of treated water has been approached as an alternative source of water in arid and semi-arid regions [6].

In principle, small DWWT technologies can vary in size and can include any available technology from simple passive anaerobic systems to technically highly complex solutions [31,32]. Conventional biological wastewater treatment systems include trickling filters (TFs), aerated biofilters (ABs), up-flow anaerobic sludge blankets (UASBs), rotating biological contactors (RBCs), submerged fixed bed reactors (SFBRs), anaerobic baffled reactors (ABRs), membrane bioreactors (MBRs), sequencing batch reactors (SBRs), moving bed biofilm reactors (MBBRs), nature-based solutions (NBSs) such as constructed wetland (CW), etc. [33–35]. These processes can effectively remove organic pollutants and nutrients from the wastewater. Challenges for implementing small DWWT technologies in arid and semi-arid regions are the abundance of marketable and internationally available technologies. Therefore, the decision makers face difficulties in assessing the sustainability of the technologies for a given local context.

In the EU, all smaller treatment systems (used for populations up to 50 PE) must be certified according to the European Standard DIN EN 12566-3 [36], which regulates a minimum standard of operation reliability and treatment performance. These state-of-the-art EU-certified smaller treatment plants have already proven their capability in overload and underload conditions in a test facility that has been authorized [37] and tested according to the EU testing range of the conventional wastewater parameter concentrations (BOD_5: 150–500 mg/L, COD: 300–1000 mg/L, TSS: 200–700 mg/L; NH_4-N: 22–80 mg/L and TP: 5–20 mg/L). Nevertheless, these plants are not designed and experimentally tested to treat wastewater beyond these EU testing ranges of the parameters.

High-strength wastewater with high organic and nutrient concentrations beyond the EU testing ranges (for example, BOD_5 > 500 mg/L, COD > 1000 mg/L, NH_4-N > 80 mg/L and TP > 20 mg/L) are not covered by the testing and certification procedures that are currently in place for small DWWT technologies in EU countries. Therefore, existing EU-certified smaller systems operating with high-strength wastewater feeding may not guarantee the fulfillment of the required effluent quality standards that are regulated by the countries in arid and semi-arid regions. Extremely high BOD_5, COD and nutrient concentrations in the influent wastewater may lead to a low treatment performance (i.e., poor effluent quality), as well as unacceptable technical and operational problems such as clogging, excessive sludge production, fouling, etc., and plants become biologically overloaded [1,38].

This study provides an assessment of four (4) small-scale DWWT technologies (4–8 PE) regarding their capability to treat high-strength wastewater effectively and their potential for adaptation to the needs of water-stressed regions. A few previous studies reported on the use of several technologies combined in series or coupled together to improve the treatment performance when treating heavily polluted domestic wastewater [39–41]. However, there is a dearth of critical and stand-alone investigation and reliable data on the treatment performance of small DWWT technologies treating high-strength wastewater and operating in parallel for a realistic comparison of their effluents. To our knowledge, no scientific studies have yet addressed and compared the treatment performance and the treated water quality of such smaller systems for agricultural reuse standards.

The aim of this study was to assess the treatment performance, effectiveness and applicability of small DWWT technologies treating high-strength wastewater and the potential reuse of the treated effluent in agriculture. For this, we investigated four (4) commercially available small DWWT technologies (4–8 PE) operating with various wastewater compositions (strengths). The specific objectives of this study were (i) to evaluate the influence of different wastewater compositions on the treatment performance of the selected DWWT technologies; (ii) to compare the effectiveness and applicability of the EU-certified DWWT technologies treating wastewater with extremely high organic and nutrient concentrations that were way beyond the EU testing ranges; and (iii) to compare the treated water quality collected from the effluents of different systems with the maximum allowable limits for agricultural reuse standards regulated in the countries from arid regions (for example, Jordan or Oman) and the EU. Knowledge gaps are highlighted, and suggestions for further research activities on small DWWT technologies are also provided.

2. Materials and Methods

2.1. Site Description

The site that was used to carry out the investigation is located at the Research and Demonstration Centre for decentralized wastewater treatment BDZ e.V. in Leipzig, Germany, where 13 unique small DWWT technologies (design capacity: 4–8 PE) simultaneously run with municipal wastewater under identical climate and operating conditions. The demonstration area (32 m × 17 m) of the BDZ consists of two rows of six (6) demonstration boxes (demo boxes) for small DWWT technologies. Apart from the MBR system (which is found outside of the demo-box area), each demo box has a dimension of 5 m × 7 m, and each of the technologies is carefully installed in those individual demo boxes (Figure 1b).

(**a**)

(**b**)

Figure 1. The research and demonstration site at BDZ e.V. in Leipzig, Germany: (**a**) a container-based mixing tank in connection with the dosing tank that contains modified wastewater of different strengths (existing municipal wastewater mixed with a solution of chemical additives); (**b**) the four selected small DWWT technologies used in this study.

Except for the NBS i.e., CW, the other demo boxes are covered, and thereby the technologies inside the demo boxes are protected against the weather effect. As backfill material, coarse gravel is used for easier installation and removal. To achieve a high degree of safety in the event of a leakage of the plants inside, a trough made of impermeable concrete is used as a foundation base. The MBR is installed in a container separately from the other 12 demo boxes at the site (Figure 1b).

All treatment systems are connected to the distribution systems for loading, sampling, as well as operation and maintenance from a closed central operating aisle. Wastewater is obtained from a nearby pumping station by using a pipe network and fed on a routine basis into a circulatory system ring line (wastewater pressure line). The ring line is able to supply all the wastewater treatment technologies with the same wastewater simultaneously. The supply via the ring line constantly ensures fresh wastewater for all the technologies installed at this site.

The outflow from the technologies as well as the excess water flow to a collection shaft and are transported back to the discharge shaft via a double pumping station. From there, it is combined with the wastewater from the local network in Leutzsch and goes to the centralized sewage treatment plant at Leipzig-Rosental for further treatment.

Each dosing system comprises a 30 L container, and the feeding takes place via two time-controlled pneumatic gate valves by means of a programmable logic controller (PLC). Feeding each system with wastewater to a freely selectable and country-specific daily flow pattern is possible by simulating different feeding scenarios (e.g., under- or overload conditions, electricity failure, etc.) in order to operate the units under realistic operating conditions.

A separate dosing unit (Figure 1a) can be used to alter the composition of the inflow wastewater. Therefore, a dosing station was built at the site for the preparation and simulation of different wastewater compositions [35]. It consists of a dosing tank with a capacity of 22 m^3, a second ring line to feed the technologies with the artificial wastewater, and a container-based mixing tank (800 L). The mixing tank comprises an agitator for rapid mixing of the supplied chemical additives. The existing municipal wastewater can be modified and the concentrations of certain wastewater parameters such as BOD_5, COD, TN, TP, etc. can be spiked with calculated dosing of chemical additives; complete artificial wastewater can also be prepared by using chemicals in supply water at the site. In order to avoid precipitation at the bottom of the dosing tank (22 m^3) or any changes of concentrations in the artificially prepared modified wastewater, an integrated agitator continuously operates for homogeneous mixing.

The operation of each DWWT technology with pure municipal wastewater from the local network (via ring line 1) or with artificially prepared wastewater from the dosing tank (using ring line 2) is freely selectable. A separate and fully automatic sample collection chamber with an individual cooling device is installed for each demo box. Representative 24 h mixed samples according to DIN EN 12566-3 [36] and random samples from the outlet and inlet of each DWWT technology can be collected.

2.2. Technology Description

A total of thirteen (13) small-scale DWWT technologies are operating at the site but only four (4) of them were selected for investigation in this study. Three (3) selected EU-certified commercial systems (with design capacity ranges from 4 to 6 PE) supplied by different companies were designated as (i) a moving bed biofilm reactor (MBBR), (ii) a sequencing batch reactor (SBR) and (iii) a membrane bioreactor (MBR). Another one, a nature-based solution (NBS) of 8 PE design capacity, was designated as an aerated vertical-flow constructed wetland (AVFCW) and built according to the German Standard DWA-A 262E [42]. A brief description and treatment process of the selected DWWT technologies are described in the following section.

2.2.1. Moving Bed Biofilm Reactor (MBBR)

One of the main characteristic features of the MBBR system is the movement of media or carrier materials in the fluids. The system incorporates the advantage of an attached growth process where microorganisms grow as a biofilm on the plastic carrier materials [43]. The movement of the support material within the biofilm reactor is achieved either by the agitation produced by aeration (aerobic process) or by mechanical stirrers [43,44]. The MBBR system used in this study essentially consists of three process stages: (1) primary treatment with integrated sludge storage, (2) biological treatment and (3) clarification. All these three process stages take place in an integrated container or tank made of concrete or plastic (single-tank plant). The integrated container contains the following three chambers for those three process stages: the primary treatment tank (1/2 of the container), the biofilm reactor for biological treatment (1/4 of the container) and for and the clarifier for secondary clarification (1/4 of the container).

Wastewater is fed to the first chamber for the pre-treatment (sedimentation stage), which also serves as a sludge storage chamber. The mechanically pre-treated wastewater is then fed to the bioreactor for biological treatment, which is based on the principle of the moving or suspended bed biofilm method. In this bioreactor chamber, a special and high-quality media or carrier material is placed in order to attach the biomass. On these carrier materials, the microorganisms grow by using the organic components of the wastewater as their food and convert it into ecologically harmless by-products. Compressed air is supplied to the biological treatment stage with time-controlled aerators. In order to prevent a wash-out of the carrier materials, this chamber is equipped with a slotted tube. The small treatment plant is equipped with a control unit, from which the compressor and the conveying device of the final settling tank are automatically controlled. To ensure optimum operation with minimum energy consumption, the compressor is operated intermittently (alternating operating and idle periods).

2.2.2. Sequencing Batch Reactor (SBR)

The SBR system used in this study functioned according to the conventional SBR process with different treatment stages or cycles [45]. The wastewater first passes through the preliminary treatment stage, which is connected to the biological stage through an opening in the partition wall. The mechanically treated water from the preliminary treatment stage flows through a submerged overflow baffler into the activation stage. The opening in the partition wall causes the water level in the complete plant to be set at the same level. Thus, the complete surface area of the plant is used as a buffer.

A float valve that is inserted in the activation stage directs the air either to the aeration facility at the bottom of the tank or to the airlift pump located higher up. During low water levels, the activation stage is stirred and aerated by the air input. Following an appropriate wastewater inflow, the valve switches on at a defined water level to the airlift pump. Following the switching off of the valve, the first surge affects the transportation as sludge return flow and is guided via an outlet opening into the buffer tank of the preliminary treatment stage. Then comes the settling phase, where the activated sludge sinks to the bottom and a clarified water zone is formed in the upper region. Following the ending of the settling, the airlift pump is in turn flushed with treated wastewater in several short internal surges. During the clarified water offtake, the outlet opening in the buffer tank is closed and the clarified water can discharge via the clarified water pipe.

The treated wastewater is pumped out until the lower switching point of the float valve is reached. When this is reached, the plant switches on again automatically with mechanical aeration, and a new cycle starts. The cyclic treatment process of the plant is controlled by water level. Depending on the mean water consumption, 1 to 3 cycles can take place per day. As the plant used in this study functions according to the volume of wastewater, even sudden fluctuations in the inflow are treated reliably. Moreover, as the preliminary treatment stage and the SBR chamber are connected, the wastewater has to be transported only once (to the outlet) and thus valuable energy is saved.

2.2.3. Membrane Bioreactor (MBR)

The MBR is a combined system of both biological and filtration processes and is potentially very efficient in treating a wide range of wastewater compositions [23]. The bioreactor acts as a biological treatment process, and the membrane acts as a filter within the filtration process. The MBR used in this study is a two-stage treatment process (max. daily wastewater volume: 0.6 m^3/d), which consists of a primary treatment tank and an aeration tank. In the primary treatment tank (which also serves as a wastewater storage tank), an aerated sieve separates biologically degradable coarse matter (e.g., feces and toilet paper) and non-dissolvable substances from the wastewater. A pump then conveys the separated wastewater into the aeration tank, where the organic matter in the wastewater is biologically degraded by microorganisms. An aeration system provides the oxygen needed for this process.

The treated wastewater then passes through micro-filtration membranes with a pore size of 0.4 μm for additional mechanical separation. The pores of the micro-filtration membranes are so small that they retain not only the suspended solids but also bacteria and other microorganisms in the reaction tank and remove them from the effluent. Therefore, the effluent discharged from the MBR system is an absolutely clear, odorless and safe liquid.

The dual-tank MBR system used in this study consists of two double-walled safety tanks (each tank footprint is 0.75 m × 1.15 m) with inlet height > 1.65 m, which can be easily installed in the residential building cellar, a garage or a garden shed, and can be adapted to most decentralized locations. In the standard application, the system is installed within a few hours and ready for immediate use. Due to automatic switching, the biological performance of the system is fully maintained, even during holiday periods of several weeks.

2.2.4. Aerated Vertical-Flow Constructed Wetland (AVFCW)

NBS, like CW, is an environmentally sustainable, socially accepted and cost-effective wastewater treatment technology which shows a strong potential for better secondary effluent treatment with low energy consumption and easy management [46–48]. Due to a higher oxygen transfer rate and small footprint, vertical-flow constructed wetlands have been widely applied in wastewater treatment [49,50]. Newly developed VFCWs integrated with active aeration augment the treatment capacity, which leads to a smaller footprint and hence reduces construction costs. It can be used in all types of climates and provide stable treatment performance all year round, and it is less vulnerable to clogging. When the system undergoes continuous aeration, an aerobic environment is created that helps to remove a high amount of organic matter as well as to achieve high nitrification [51].

For this study, an actively aerated vertical-flow constructed wetland (AVFCW) system (area: 8 m^2, design capacity: 8 PE) was built according to the German standard DWA-A 262E [42] at the site. The construction details were discussed by Rahman et al. [35]. In brief, the system is filled with 90 cm of coarse gravel (8–16 mm) as the main filter media, which is saturated, with a hydraulic retention time (HRT) of 3.5 days. The aeration system consists of irrigation lines on the wetland bottom [52]. Aeration is provided 24 h/d by a fishpond air pump (Aqua Medic, Mistral 4000; design capacity: 66 L/min) connected to the perforated irrigation lines that are evenly placed at the bottom of the wetland bed. The wastewater flow direction is vertical (from the top), and the homogeneous distribution of applied air from the bottom is expected to cause appropriate mixing and saturation with dissolved oxygen (DO) in the filter bed.

A three-chamber settling tank (volume: 2.9 m^3) made of concrete serves for the pre-treatment of wastewater, and a mechanical distribution unit (PE tank) containing a submersible pump enables periodic distribution to the wetland. Inflow distribution systems are installed evenly on top of the wetland bed (sub-surface flow) over the horizontal length, and the outflow is collected at the bottom of the tank in a perforated inversed T-shaped pipe connected to an elbow. The elbow discharges the effluent at a level around 5 cm below the gravel surface. The treated wastewater is then sent to a storage tank that is in close proximity to the wetland bed.

The wetland surface was evenly planted with *Phragmites australis* (at a density of 5 rhizomes/m^2). Before the experimental phases were started, the plants were allowed to grow and the biofilms to develop within the rhizosphere. During this acclimatization period of 8–10 weeks, the wetland was fed with a small amount of pre-treated wastewater from time to time and aerated continuously.

2.3. *Modified Wastewater Preparation*

Instead of preparing entirely artificial wastewater by using drinking water, the already existing municipal wastewater at the site was modified and the concentrations spiked by calculated dosing of several chemical additives. It was adjusted to target the concentration of certain parameters such as BOD_5, COD, TN, TP, TSS, etc. within the inflow wastewater in different experimental phases during this study.

Maisonnave et al. [53] showed the composition of artificial wastewater representing typical domestic wastewater (in mg/L): BOD_5, 400; COD, 880; SS, 350; TKN-N, 80; P, 10. The concentrations of the ingredients that were used (in mg/L) were as follows: glycine, 250; glycerol, 300; K_2HPO_4: 10; NH_4Cl, 60; Na_2HPO_4, 30. Based on the study by Maisonnave et al. [53], the composition of the ingredients was adapted and simulated with already existing municipal wastewater at the site. In general, the concentrations of the existing wastewater parameters were increased by adding organic carbon sources (glycine and glycerol) and nutrient sources (NH_4Cl, K_2HPO_4, Na_2HPO_4), and thereby modified wastewater of defined strength was prepared in this study.

After rapid and homogeneous mixing of the added chemicals in the mixing tank, the thick solution from the mixing tank was pumped to the dosing tank or wastewater storage tank (22 m^3) (Figure 1a). There, it was mixed with the already existing municipal wastewater of a specified volume, and the concentrations of the specific wastewater parameters (BOD_5, N, P, etc.) were increased. All the ingredients used in the preparation of the modified wastewater for spiking the concentrations were dissolved almost exclusively in water. To increase the concentrations of filterable substances or TSS, a pre-calculated amount of thick digested sludge from a nearby sludge storage tank was pumped directly to the dosing tank and added to the volume of existing wastewater.

After every filling and proper mixing of the freshly prepared modified wastewater, a sample was taken from the dosing tank using ring line 2 to assess the concentrations of typical wastewater parameters and thereby know the strength of the inflow wastewater that is fed to the selected technologies.

2.4. Operational Conditions

Based on low water consumption and hence low wastewater generation or discharge (volume) per capita in a day within the arid and semi-arid regions, wastewater compositions of three different strengths were simulated and prepared in three experimental phases (I, II and III) of this study. The same mean influent BOD_5, TN and TP load (in grams per capita and day) that are typical of the rural areas in arid regions were used and remained constant in the three experimental phases of this study. Depending on the water consumption on a daily basis, the concentrations of these parameters (BOD_5, TN and TP) vary, i.e., they either dilute when more volume of water is consumed and thereby more volume of wastewater is discharged or become strong when less volume of water is consumed and therefore less volume of wastewater is discharged. The wastewater discharge rates were incrementally reduced and spiked with chemical additives in order to increase the concentrations of these specific parameters in this study. Due to significant logistic issues, it was not possible to carry out this experiment in any rural areas of arid regions with extreme water scarcity (e.g., in Jordan), and therefore municipal wastewater at the site in Germany was used in a simulation to achieve the target concentrations of these specific parameters. In this study, the municipal wastewater discharge rate on a daily basis was assumed as 150 L per capita at the site [42]. Three mean BOD_5 concentrations were targeted as 300, 600 and 1200 mg/L in three experimental phases in Phase I, II and III, respectively. Similarly, three mean concentrations of 60, 120 and 240 mg/L TN and 10, 20 and 40 mg/L TP were targeted to be achieved within the inflow wastewater in experimental Phases I, II and III, respectively (Table 1).

Table 1. Mean influent BOD_5, TN, TP load and wastewater discharge volume used to simulate the targeted mean concentrations of BOD_5, TN and TP in three experimental phases (Phases I–III) in this study.

Experimental Phase	Wastewater Type	Wastewater Discharge [L/(PE × d)]	BOD_5 Load [g/(PE × d)] [c]	TN Load [g/(PE × d)] [c]	TP Load [g/(PE × d)] [c]	BOD_5 (mg/L)	TN (mg/L)	TP (mg/L)
Phase I	real [a]	150	45	9	1.5	300	60	10
Phase II	modified [b]	75	45	9	1.5	600	120	20
Phase III	modified [b]	37.5	45	9	1.5	1200	240	40

Notes: [a] existing municipal wastewater at the site; [b] existing wastewater mixed with a calculated amount of chemical additives; [c] adapted from Bertanza and Boiocchi [54].

For the EU certification process (including a practical test in a test field) of small DWWT technologies, the concentration range of the raw domestic wastewater parameters is already defined according to DIN EN 12566-3 [36]. The concentration ranges of the parameters for the EU practical testing and actual concentration ranges in the influent wastewater (strengths) in three experimental phases (Phases I–III) are shown in Table 2.

All four selected DWWT technologies were fed by the wastewater of the same influent strength as mentioned in each experimental phase and operated with the same adjusted daily flow pattern. For each cycle of feeding, the influent wastewater was fed to each system in intermittent mode with a fixed volume and time using an automatic time-controlled PLC. Table 3 shows the influent flow rates of the four selected DWWT technologies operated in three experimental phases in Phases I, II and III.

Table 2. The concentration ranges of the parameters for the EU testing and in the influent wastewater (strengths) used within the three experimental phases (Phases I–III).

Parameters	Unit	Concentration (Range)			
		EU Practical Test DIN EN 12566-3 [34]	Phase I [a] (Real Wastewater)	Phase II [b] (Modified Wastewater)	Phase III [c] (Modified Wastewater)
BOD_5	mg/L	150–500	250–400	500–800	1000–1900
COD	mg/L	300–1000	460–980	980–1550	1550–3400
TSS	mg/L	200–700	100–280	200–450	400–750
NH_4-N	mg/L	22–80	20–80	60–130	140–420
TP	mg/L	5–20	3–10	10–25	25–60

Notes: [a] number of samples, $n = 11$; [b] $n = 12$; [c] $n = 12$.

Table 3. Inflow feeding rates of the four selected systems operated during three experimental phases (Phases I–III) in this study.

System	Design Capacity (PE)	Influent Flow Rates (L/d)		
		Phase I (Real Wastewater)	Phase II (Modified Wastewater)	Phase III (Modified Wastewater)
MBBR	4	600	300	150
SBR	6	900	450	225
MBR	4	600	300	150
AVFCW	8	1200	600	300

To avoid freezing in the dosing tank (22 m^3), which operates under outside air temperature at the site, the whole investigation and operation of the selected technologies was paused for 3 weeks in winter when the outside air temperature was extremely low (below 0 °C). The duration of the whole investigation with three experimental phases was one year and the mean air temperature was measured in a wide range of 10–35 °C within the course of the study period at the site.

2.5. Sampling and Analysis

Throughout all three experimental phases (Phases I–III), the samples from the influent and the effluents of the four respective DWWT technologies were analyzed on a weekly routine basis for measuring the concentrations of the following parameters: BOD_5 (DIN 38409 H52, WTW OxiTOP®, manufactured by Xylem Analytics, Weilheim, Germany), COD (TNTplus™ 821/822, HR 20–1500 mg/L COD manufactured by HACH®, Düsseldorf, Germany), ammonium—nitrogen (NH_4-N) [55], nitrate–nitrogen (NO_3-N) [56], total nitrogen (TN) (TNTplus™ Vial Test TNT828, UHR 20-100 mg/L N, manufactured by HACH®, Düsseldorf, Germany), total phosphorous (TP) [57] and TSS (by using vacuum filtration). The spectrophotometer DR 2800 (Hach Lange, Duesseldorf, Germany) was used for reading the concentrations of COD and TN, according to the standard method for COD and TN tests specified by the manufacturer (HACH®, Düsseldorf, Germany). Dissolved oxygen (DO) (ConOx®, WTW Germany), water temperature (T °C) and pH electrode (SenTix® pH) using a handheld meter (Multi 350i®, WTW, Weilheim, Germany), redox potential (E_h) (electrode SenTix® ORP, WTW, Weilheim, Germany), electrical conductivity (EC) (conductivity meter Cond 330i, WTW, Weilheim, Germany) of the inflow and outflow samples were measured and *E. coli* ([58] were quantified simultaneously by using Colilert-18 Quanti-Tray™ method (IDEXX, Westbrook, ME, USA) to investigate treatment performances of the technologies and to assess the requirements for system maintenance.

Pollutant removal efficiency was calculated as concentration reduction or removal in percentage using the following Equation (1):

$$R = \frac{(c_{in} - c_{out})}{c_{in}} \times 100 \quad (1)$$

where R is the removal efficiency (%), c_{in} is the influent concentration [mg/L] and c_{out} is the effluent concentration [mg/L]. Based on available data from all 4 systems, we calculated and compared the pollutant removal efficiencies only with the concentrations. Random samples were taken from the systems on a weekly basis, and the removal efficiency was calculated as the weekly average in this study.

2.6. Data and Statistical Analysis

The concentrations of different parameters in the samples from the influent and effluent were reported in the format of mean ± SD in this study. Microsoft Excel 2013 software (product name and activation status: Microsoft Office Professional Plus 2013; version: Microsoft® Excel® 2013 (15.0.5571.1000) MSO (15.0.5571.1000) 32-bit) package was used to determine the mean and standard deviation of the sample concentrations. Statistical analysis was performed using a one-way analysis of variance (ANOVA) test with a 95% significance level of difference within Microsoft Excel 2013 to compare the treatment performance of the four selected DWWT technologies in terms of differences between mean concentrations at the influent and effluent, as well as mean removal efficiencies with various wastewater compositions or strengths. The statistical test results and differences were considered significant at p-values of less than 0.05 ($p < 0.05$).

3. Results and Discussion

3.1. Treatment Performance Assessment of the Four DWWT Technologies

Mean concentrations of different physico-chemical parameters and microbial quantifications at the inlet and outlet and thereby removal efficiencies of the four DWWT technologies during the whole experimental period are discussed in the following sections. Water quality parameters were statistically analyzed to determine any significant changes in the treatment performance of the four DWWT technologies when the influent wastewater strength was increased stepwise in the three experimental phases (Phases I–III).

Treated wastewater qualities from the outflow of four selected systems achieved in three experimental phases are also compared with the maximum allowable limit values for reuse in agriculture that are regulated in two arid climate countries Jordan and Oman. Both the Jordanian and Omani standards are also compared with the strictest EU standard for reclaimed water quality requirements in agricultural reuse as shown in Table 4.

Table 4. Allowable limit values and different classes of treated wastewater for reuse in Jordan according to Jordanian national standard JS 893/2006 adapted from Abdallat et al. [59]; maximum quality limits in Oman according to Oman Ministerial Decision 145/1993 [60]; and minimum reclaimed water quality requirements for agricultural irrigation in the EU [61] (n.p.: not provided).

Parameter	Unit	Allowable Limits for Reuse in Jordan			Maximum Wastewater Quality Limits in Oman		Minimum Reclaimed Water Quality Requirements According to the EU			
		Class A [a]	Class B [b]	Class C [c]	Class A [d]	Class B [e]	Class A [f]	Class B [g]	Class C [h]	Class D [i]
BOD_5	mg/L	30	200	300	15	20	10	25	25	25
COD	mg/L	100	500	500	150	200	n.p.	n.p.	n.p.	n.p.
TSS	mg/L	50	150	150	15	30	10	35	35	35
DO	mg/L	>2	n.p.	n.p.	n.p.	n.p.	n.p.	n.p.	n.p.	n.p.
pH	–	6–9	6–9	6–9	6–9	6–9	n.p.	n.p.	n.p.	n.p.
EC	µS/cm	n.p.	n.p.	n.p.	2000	2700	n.p.	n.p.	n.p.	n.p.
NH_4-N	mg/L	n.p.	n.p.	n.p.	5	10	n.p.	n.p.	n.p.	n.p.
NO_3-N	mg/L	30	45	45	50	50	n.p.	n.p.	n.p.	n.p.
TN	mg/L	45	70	70	n.p.	n.p.	n.p.	n.p.	n.p.	n.p.
TP	mg/L	n.p.	n.p.	n.p.	30	30	n.p.	n.p.	n.p.	n.p.
E. coli	MPN/100 mL (log_{10} *E.coli*)	100 (2.0)	1000 (3.0)	n.p.	200 (2.3)	1000 (3.0)	10 (1.0)	100 (2.0)	1000 (3.0)	10,000 (4.0)

[a] Cooked vegetables, parks and playgrounds within city limits (restricted irrigation); [b] fruit trees, sides of roads outside city limits and landscape (restricted irrigation); [c] field crops, industrial crops and forest trees; [d] vegetables likely to be eaten raw, fruit likely to be eaten raw and within two (2) weeks of any irrigation, public parks, hotel lawns and recreational areas with public access; [e] vegetables to be cooked or processed, fruit if no irrigation within two (2) weeks of cropping, fodder, cereal seed crops, pasture, no public access; [f] all food crops, including root crops consumed raw and food crops where the edible portion is in direct contact with reclaimed water (all irrigation methods allowed); [g] food crops consumed raw where the edible portion is produced above ground and is not in direct contact with reclaimed water and processed food crops and nonfood crops, including crops to feed dairy or meat-producing animals (all irrigation methods allowed); [h] food crops consumed raw where the edible portion is produced above ground and is not in direct contact with reclaimed water, processed food crops and nonfood crops including crops to feed dairy or meat-producing animals (drip irrigation or other irrigation methods); [i] industrial and energy crops and seeded crops (all irrigation methods allowed).

3.1.1. pH, E_h, DO, EC and T

Table 5 shows the mean inflow and outflow values of pH, E_h, T and DO concentrations in the four selected DWWT technologies that were observed during the three experimental phases of this study.

Table 5. Measured mean values of pH, E_h, EC, T and DO concentrations in the inflow and outflow of the four selected systems with different wastewater compositions in the three experimental phases (Phases I–III).

Experimental Phase	Sampling Point/System	pH (-)	E_h (mV)	DO (mg/L)	EC (μS/cm)	T (°C)	Number of Samples
I	Inflow	7.7 ± 0.4	−100 ± 53	0.4 ± 0.2	1240 ± 550	11.3 ± 2.6	8
	Outflow						
	MBBR	7.4 ± 0.2	295 ± 130	9.4 ± 1.3	1062 ± 438	9.9 ± 2.9	8
	SBR	7.2 ± 0.3	299 ± 116	10.7 ± 0.8	1090 ± 442	9.9 ± 2.7	8
	MBR	7.0 ± 0.3	294 ± 130	5.8 ± 2.3	1044 ± 440	13.8 ± 2.0	8
	AVFCW	6.9 ± 0.5	286 ± 103	10.5 ± 1.5	1116 ± 455	8.2 ± 3.5	8
II	Inflow	7.0 ± 0.5	−224 ± 61	1.1 ± 1.0	1815 ± 281	19.6 ± 3.1	12
	Outflow						
	MBBR	6.8 ± 0.4	180 ± 25	8.2 ± 1.4	1377 ± 268	17.1 ± 4.3	12
	SBR	6.6 ± 0.7	186 ± 22	7.8 ± 1.8	1300 ± 184	17.7 ± 3.7	12
	MBR	7.8 ± 0.3	98 ± 92	3.9 ± 2.2	1478 ± 182	19.9 ± 2.4	12
	AVFCW	7.2 ± 0.4	180 ± 32	7.7 ± 2.1	1276 ± 292	13.2 ± 4.8	12
III	Inflow	7.2 ± 0.5	−227 ± 49	0.2 ± 0.1	3474 ± 858	20.3 ± 4.1	12
	Outflow						
	MBBR	6.9 ± 0.8	209 ± 22	7.3 ± 3.1	1915 ± 427	16.1 ± 3.7	12
	SBR	5.8 ± 0.5	222 ± 29	9.1 ± 1.1	2170 ± 523	15.5 ± 3.4	12
	MBR	5.9 ± 1.0	243 ± 33	8.0 ± 1.4	2263 ± 831	18.4 ± 3.3	12
	AVFCW	6.2 ± 0.5	230 ± 46	8.6 ± 2.8	2599 ± 833	11.6 ± 3.8	12

The statistical analysis showed significant ($p < 0.05$) differences in pH, E_h, EC and T values and DO concentrations between the three influent wastewater compositions used in the three experimental phases from Phase I to Phase III in this study.

The pH of the inflow wastewater in all three experimental phases was in the range of 5.3–8.4 with a mean value of 7.7 ± 0.4, 7.0 ± 0.5 and 7.2 ± 0.5 in Phases I, II and II, respectively. In general, the mean pH value was significantly reduced ($p < 0.05$) in the outflow of all four systems as compared to the inflow in all three experimental phases. During operation with high-strength wastewater in Phase III with a mean pH value of 7.2 ± 0.5 in the inflow, a significant reduction in mean pH ($p < 0.05$) in the range of 5.8–6.9 was observed in the outflow of all four selected systems. As compared to Jordanian and Omani standards of pH in a range of 6.0–9.0 for all classes of treated water, only the SBR and MBR systems showed a mean pH of <6.0 in the outlet and did not meet the standard when operated with high-strength wastewater in Phase III. Prior to the reuse of such treated water with a slightly low pH, some pH adjustment may potentially be needed.

Low redox potential in a range of −340 to −31 mV was recorded in the inflow samples from the dosing tank during all three experimental phases in this study. A general trend of increasing the mean redox value in the range from 98 to 299 mV in the outflow of the four systems suggested that aerobic conditions prevailed within all the systems (Table 5). The lowest mean value of −227 ± 49 mV was recorded in the inflow samples of high-strength modified wastewater in Phase III, as compared to the other two phases (Phases I and II). During the operation with high-strength wastewater in Phase III with a mean redox potential of −227 ± 51 mV in the inflow, significantly higher ($p < 0.05$) mean redox values were recorded in the range of 209–243 mV in the outflow of the four systems. The observed increase in redox potential at the outlet of all the systems was perhaps due to an overall decrease in organic load [58]. A comparatively low mean redox potential of 98 ± 92 mV was observed in the outflow of the MBR system in Phase II.

As expected, very low mean DO concentrations in the range of 0.2–1.1 mg/L were analyzed in the inflow samples of the three experimental phases. During operation with different wastewater compositions in Phases I–III, mean DO concentrations increased substantially in a range of 7.3–10.7 mg/L in the outlet of the MBBR, SBR and AVFCW systems. High DO concentration at the outlet clearly showed an abundance of oxygen within these systems and potentially favored redox conditions necessary for the oxidation of the wastewater pollutants. Only the MBR showed a comparatively low DO of 5.8 ± 2.3 mg/L and 3.9 ± 2.2 mg/L in the outflow in Phase I and Phase II, respectively. After the maintenance of the MBR system, it showed relatively higher DO with a mean concentration of 8.0 ± 1.5 mg/L in the outflow in Phase III, as compared to the previous two experimental phases (Phases I and II). However, mean DO concentrations at the outlet of all four systems met the allowable limit concentration of treated water quality class A for agricultural reuse recommended by the Jordanian standard, which is >2.0 mg/L (Table 4).

Mean EC in the inflow wastewater was significantly increased ($p < 0.05$) stepwise and recorded as 1240 ± 550, 1815 ± 281 and 3474 ± 858 μS/cm in experimental Phases I, II and III, respectively (Table 5). A general trend of decreasing the mean value of EC within the outflow of the four systems was observed. During the operation with high-strength wastewater in Phase III, mean EC showed significant changes ($p < 0.05$) in the inflow and outflow of all systems. EC is a common indicator of salinity and only the MBBR system with a mean EC of 1915 ± 427 μS/cm in the outflow met class A water quality (<2000 μS/cm) according to the Omani standard when treating high-strength wastewater in this study. The effluent from an MBBR system treating laundry wastewater also exhibited high EC in the range of 1800–2000 μS/cm [62]. Mean ECs recorded in the outlet of the other three systems were in compliance with the class B water quality limit (<2700 μS/cm). To protect the soil from high salinity, direct reuse of the treated effluents from the four systems with a comparatively high EC may not be possible, and therefore innovative irrigation methods (e.g., sub-surface irrigation) can be suggested.

The mean temperature of the inflow wastewater was recorded as 11.3 ± 2.6, 19.6 ± 3.1 and 20.3 ± 4.1 °C during the experimental Phases I, II and III, respectively (Table 5). In general, the mean temperature in the outflow of the MBBR, SBR and AVFCW systems decreased slightly in comparison with the mean inflow temperature. Comparatively low water temperatures in the range of 8.2–13.2 °C were observed in the outflow of the AVFCW system during the three phases (Phases I–III). The mean water temperature increased slightly in the outflow of the MBR system in Phases I and II but decreased in Phase III as compared to the mean inflow water temperatures.

3.1.2. Removal of BOD_5 and COD

Treatment performance with respect to pollutant removal efficiency and the concentration profile of BOD_5 and COD from the inflow and outflow of the four selected DWWT technologies during the three experimental phases are shown in Table 6 and Figure 2.

The results of a very low mean BOD_5 in the outflow of all four systems showed highly efficient removal efficiency (>99%) in all three experimental phases in this study (Table 6). Significant reduction ($p < 0.05$) of BOD_5 concentrations was observed at the outlet of all systems as compared to the respective inflow in three phases. In general, BOD_5 concentration dynamics in the outflow of four systems showed no significant changes ($p < 0.05$) with different wastewater strengths as inflow (Figure 2). No potential impact of increased wastewater strength was also observed on BOD_5 removal performance. Only a fluctuating tendency in the outflow BOD_5 concentrations from the systems was observed (Figure 2). During operation with high-strength wastewater in experimental Phase III with a mean BOD_5 concentration of 1532 ± 478 mg/L in the inflow, the mean BOD_5 concentration of <10 mg/L in the outflow of the four systems accounted for a highly efficient BOD_5 removal (>99%) and successfully met the maximum allowable limit value of class A water quality requirements suggested by the Jordanian standard for restricted irrigation (<30 mg/L), the Omani standard (<15 mg/L) and the EU regulations (<10 mg/L) for reuse in agricultural irrigation purposes.

As can be seen in Figure 2, the outflow COD concentrations from the systems slightly raised and fluctuated with increasing COD concentrations in the inflow from Phase I to Phase III. However, the overall mean COD concentrations in the inflow were significantly decreased ($p < 0.05$) in the outflow from all four systems, with an average removal efficiency in the range of 93–98% in all three experimental phases. During Phase III with high-strength wastewater as inflow with a mean COD concentration of 2547 ± 830 mg/L, the mean COD concentrations in the outflow from all systems resulted in a highly efficient mean COD removal of >97% and were much lower than the maximum allowable COD limit value for class A water quality according to Jordanian standards (<100 mg/L) as well as Omani standards (<150 mg/L). No COD standard of reclaimed water quality for agricultural irrigation purposes is suggested by the EU. High-strength wastewater is rich in organic matter, which is a very good source of carbon and nutrients (N and P) for the microbes to produce energy and synthesize chemical products [63], and hence highly efficient COD removal was observed in this study.

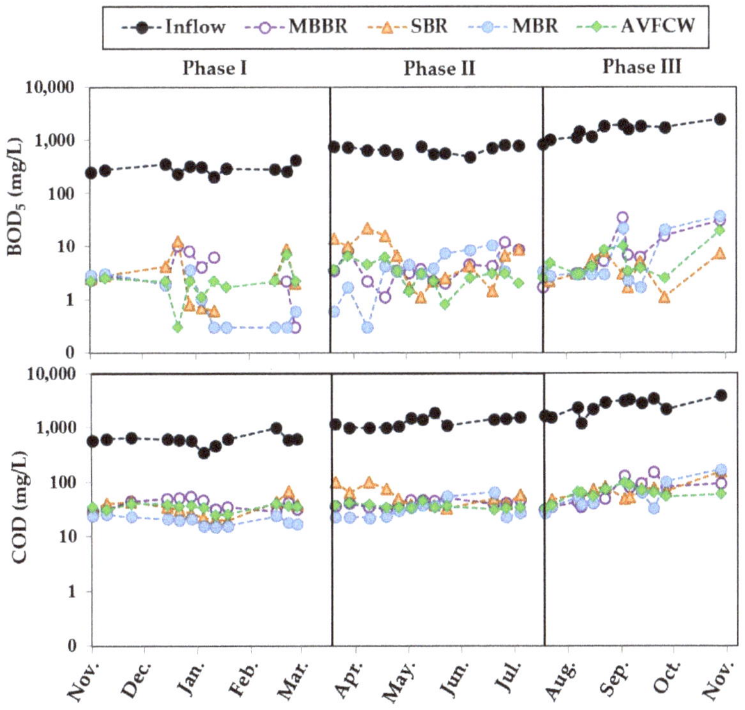

Figure 2. The dynamics of BOD_5 and COD concentrations that were observed in the inflow and outflow of the four DWWT technologies in the three experimental phases (Phases I–III) in this study.

Table 6. Treatment performance with different wastewater compositions observed by analyzing the mean concentrations of the conventional wastewater parameters at the inflow and outflow of the four selected DWWT technologies in the three experimental phases (Phases I–III).

Experimental Phase	Sampling Point/ System	BOD_5 (mg/L)	COD (mg/L)	TSS (mg/L)	NH_4-N (mg/L)	NO_3-N (mg/L)	TN (mg/L)	TP (mg/L)	log10 *E. coli* (MPN/100 mL)	Number of Samples
I	Inflow	292 ± 60	608 ± 152	207 ± 64	44.2 ± 19	0.5 ± 0.1	60 ± 19	7.9 ± 3	7.1 ± 0.3	11
	Outflow									
	MBBR	4 ± 3 (99 [a])	41 ± 9 (93 [a])	7 ± 3 (97 [a])	2.8 ± 2 (94 [a])	21 ± 4	28 ± 7 (54 [a])	6.2 ± 1 (22 [a])	4.8 ± 0.5 (2.3 [b])	10
	SBR	4 ± 3 (99)	35 ± 14 (94)	6 ± 5 (97)	0.6 ± 0.5 (99)	33 ± 9	38 ± 9 (37)	6.6 ± 2 (17)	3.9 ± 0.4 (3.2)	10
	MBR	1 ± 0.8 (99)	20 ± 4 (97)	1.2 ± 1 (99)	0.8 ± 0.7 (98)	27 ± 10	29 ± 12 (52)	5.4 ± 2 (32)	0.2 ± 0.1 (6.9)	10
	AVFCW	2 ± 1.7 (99)	34 ± 6 (94)	3 ± 1 (99)	3.9 ± 3.8 (91)	40 ± 9	44 ± 9 (27)	4.2 ± 1 (48)	4.7 ± 0.6 (2.4)	11
II	Inflow	664 ± 107	1288 ± 277	310 ± 85	88.4 ± 27	0.5 ± 0.1	123 ± 23	18.8 ± 5	6.9 ± 0.4	12
	Outflow									
	MBBR	5 ± 3 (99 [a])	41 ± 6 (97 [a])	4 ± 1 (99 [a])	3.8 ± 3 (96 [a])	41 ± 7	47 ± 12 (62 [a])	13 ± 3 (29 [a])	3.6 ± 0.6 (3.3 [b])	12
	SBR	7 ± 6 (99)	57 ± 23 (96)	13 ± 9 (96)	1.1 ± 0.7 (99)	43 ± 20	51 ± 23 (59)	14 ± 3 (27)	4.1 ± 1.0 (2.8)	12
	MBR	4 ± 3 (99)	33 ± 14 (97)	2.1 ± 2 (99)	49 ± 38 (45)	3.8 ± 3	46 ± 32 (63)	8 ± 6 (58)	0.1 ± 0.1 (6.8)	12
	AVFCW	3 ± 2 (99)	36 ± 4 (97)	3 ± 1 (99)	1.0 ± 0.5 (99)	31 ± 14	33 ± 13 (73)	11 ± 2 (44)	4.2 ± 0.8 (2.7)	12
III	Inflow	1532 ± 478	2547 ± 830	546 ± 176	267 ± 110	0.5 ± 0.1	367 ± 133	45.3 ± 12	6.2 ± 0.6	12
	Outflow									
	MBBR	10 ± 9 (99 [a])	70 ± 39 (97 [a])	13 ± 9 (98 [a])	33 ± 30 (87 [a])	46 ± 13	89 ± 35 (76 [a])	20 ± 3 (56 [a])	3.8 ± 0.6 (2.4 [b])	12
	SBR	4 ± 2 (99)	70 ± 30 (97)	15 ± 7 (97)	53 ± 38 (80)	83 ± 34	190 ± 71 (48)	24 ± 5 (48)	2.8 ± 0.8 (3.4)	11
	MBR	9 ± 7 (99)	62 ± 46 (98)	4.5 ± 2 (99)	60 ± 36 (78)	66 ± 31	143 ± 59 (61)	37 ± 15 (18)	0.1 ± 0.1 (6.1)	12
	AVFCW	6 ± 5 (99)	64 ± 18 (97)	8 ± 3 (98)	46 ± 33 (83)	101 ± 33	166 ± 75 (55)	14 ± 5 (69)	3.6 ± 0.5 (2.6)	11

[a] mean percentage removal efficiency in all the following parentheses; [b] log-removal in all the following parentheses.

The BOD_5 and COD from high-strength wastewater can be degraded by both aerobic and anaerobic bacteria that can be potentially grown within the four DWWT technologies, and aerobic degradation with a high DO concentration is also fast and important [49]. Limited oxygen supply restricts COD degradation for wastewater treatment [64], and in this study, the mean DO concentrations in the range of 7.3–9.1 mg/L in the outlet of the systems were observed in experimental Phase III operating with high-strength wastewater (Table 5). These relatively high DO concentrations inside the four selected systems were the potential reason for highly efficient BOD_5 and COD removal. Zhao et al. [65] showed that with the increasing influent wastewater strength, the limitation of oxygen became more obvious and resulted in decreasing COD removal efficiencies in vertical flow constructed wetlands (VFCWs) without aeration. However, in this study, the air supplied via aeration facilities within the systems provided the oxygen needed to degrade the organic matter with increasing strength of inflow wastewater, and therefore BOD_5 and COD degradation were never restricted due to a limited supply of oxygen.

3.1.3. TSS Removal

The TSS concentration profile from the inflow and outflow of the four systems with various wastewater compositions in three experimental phases is shown in Figure 3.

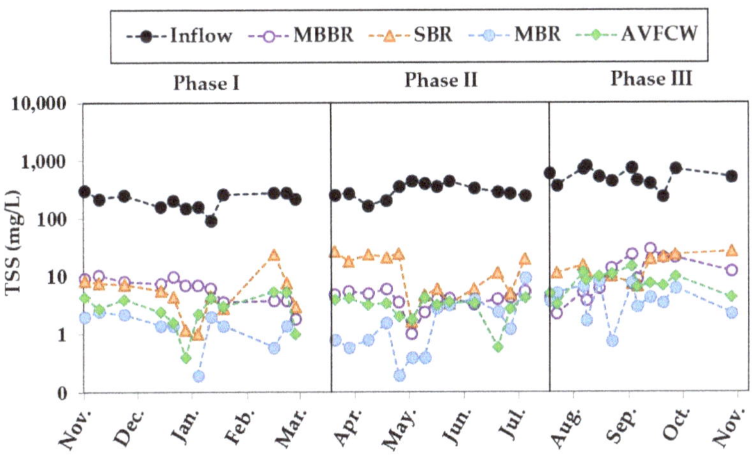

Figure 3. The dynamics of TSS concentration within the inflow and outflow of the four DWWT technologies observed in the three experimental phases (Phases I–III).

The dynamics of outflow TSS concentrations from all systems showed a fluctuating trend in all experimental phases as compared to the inflow. The SBR system showed a slight rise of TSS concentration in the outflow when the system was fed with more concentrated wastewater in Phases II and III. However, the mean inflow TSS concentrations of 207 $\pm$ 64, 310 $\pm$ 85 and 546 $\pm$ 176 mg/L in three experimental Phases I, II and III were reduced to mean outflow TSS concentrations in the range of 1.2–15 mg/L, which accounted for highly efficient mean TSS removal of >96% in all four systems in this study (Table 6). Therefore, all four systems proved to be highly effective at removing suspended solids from high-strength wastewater with a very high TSS concentration and met the treated water quality class A for TSS according to Jordanian (<50 mg/L) and Omani standards (<15 mg/L). However, a few measurements with TSS > 20 mg/L in the outflow of the MBBR and SBR systems exceeded class A water quality standard (<15 mg/L) and met class B standard (<30 mg/L) according to Omani regulations (Figure 3). The MBR and AVFCW systems showed highly efficient TSS removal and the mean outflow TSS concentrations in all three experimental phases even met the strictest EU standard for class A water quality (<10 mg/L) recommended for reuse in agriculture (Tables 4 and 6). A similar trend in the effluent TSS concentrations

(<15 mg/L) for a pilot-scale recirculating vertical flow constructed wetland was reported by Sklarz et al. [66].

3.1.4. NH_4-N, NO_3-N and TN Removal

The treatment performance of NH_4-N, NO_3-N and TP and their dynamics in the inflow and outflow of the four DWWT technologies during the three experimental phases are shown in Table 6 and Figure 4, respectively.

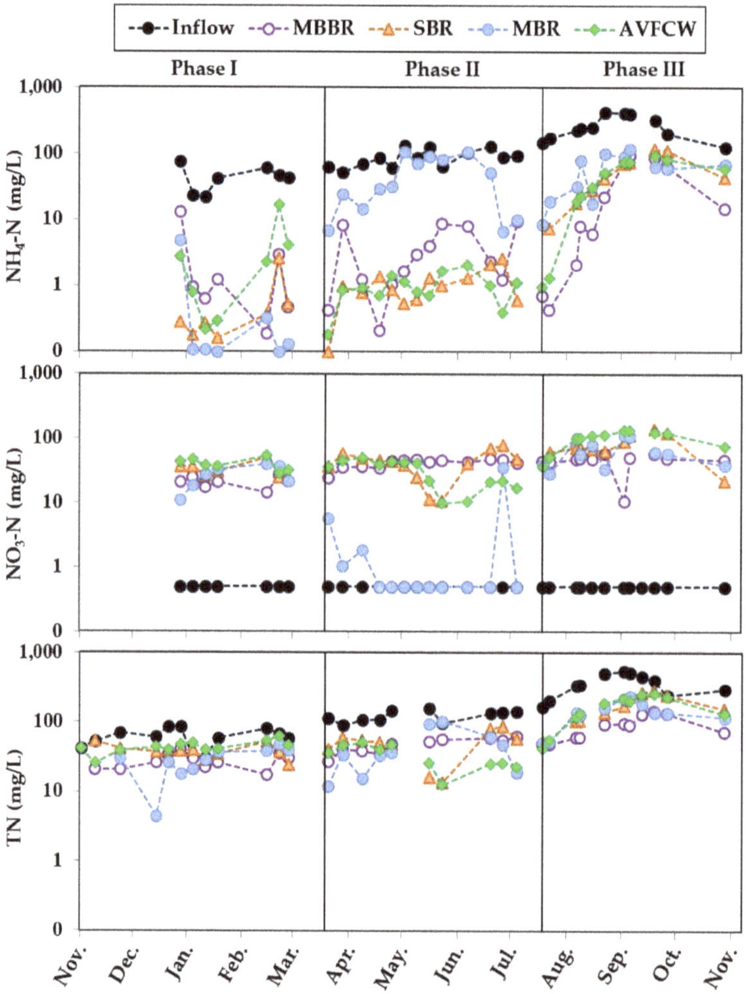

Figure 4. The dynamics of NH_4-N, NO_3-N and TN concentrations in the inflow and outflow of the four selected DWWT technologies during the three experimental phases (Phases I–III).

The four systems were almost completely nitrified by converting ammonia–nitrogen (NH_4-N) to nitrate–nitrogen (NO_3-N) in experimental Phase I and also in Phase II, except for the MBR system, which required sufficient maintenance. In general, the results demonstrated enhanced nitrification and therefore highly efficient removal of NH_4-N in the range of 91–99% in experimental Phases I and II in this study. Mean concentrations of NH_4-N within the outflow also achieved class A water quality standard for NH_4-N (<5 mg/L) according to Omani guidelines for reuse. The only exception was the MBR system in Phase II, where a high NH_4-N mean concentration of 49 ± 38 mg/L in the outflow demonstrated a mean NH_4-N removal efficiency of 45% only. Low DO with a mean concentration of 3.9 ± 2.2 mg/L in the outlet of the MBR system during this phase was potentially associated with a low nitrification rate [67], and therefore only a low NO_3-N mean concentration of 3.8 ± 3 mg/L was observed (Table 6). This was probably due to the insufficient maintenance of the MBR system and clogging of the membrane during this phase. After changing the membrane and some necessary maintenance work needed for the MBR system during this phase (Phase II), the outflow NH_4-N concentration decreased remarkably and showed better NH_4-N removal again (Figure 4).

When operating with high-strength wastewater in experimental Phase III, relatively high NH_4-N mean concentrations in the outflow suggested that there was room for optimizing NH_4-N removal performance within the systems. In experimental Phase III with a very high NH_4-N concentration in the inflow, the dynamics of NH_4-N concentration showed a sharp rise in the outflow of all four systems (Figure 4). A mean inflow NH_4-N concentration of 267 ± 110 mg/L was attributed to a mean outflow concentration in the range of 33–60 mg/L, which resulted in a NH_4-N removal efficiency with a great extent and showed in the range of 78–87% within the four systems. (Table 6). However, a slightly lower mean NH_4-N removal (78–87%) and at the same time, a highly efficient mean COD removal of >97% in all four systems in experimental Phase III was consistent with the results achieved by Li et al. [68]. Previous results indicated that artificial aeration increased the DO concentrations in the AVFCW treating heavily polluted water, which significantly favored the removal of organic matter and NH_4-N [51].

The main removal process of NH_4-N is nitrification, where NH_4-N is converted into NO_3-N. This process needs high DO concentration [48] and afterward, this NO_3-N is transformed into N_2O or N_2 via a denitrification process in order to be permanently removed from the system [69]. The four selected DWWT technologies showed higher mean DO concentration in the outflow (7.3–9.1 mg/L) during operation with high-strength wastewater in this investigation. A relatively higher NO_3-N with a mean concentration in the range of 46–101 mg/L was observed in the four systems in the experimental phase, and this NO_3-N was shown as the predominant form of nitrogen in the effluent in experimental Phase III with high-strength wastewater (Table 6). Only the MBBR system outflow mean NO_3-N concentration (46 ± 13 mg/L) achieved the treated water quality of Omani class A standard (<50 mg/L) for reuse in agriculture. The reason for a comparatively low mean NO_3-N concentration in the effluent might be due to the fact that the MBBR with a thick biofilm potentially allows for some anoxic environment that promotes denitrification. The mean outflow concentrations of the other three systems did not comply with the allowable limit suggested by the Omani standard. Nitrification usually becomes the limiting step for nitrogen removal in conventional VFCWs due to insufficient DO [70]. However, in this study, the AVFCW system with active aeration provided sufficient DO (8.6 ± 2.8 mg/L), and this was also in agreement with other artificially aerated constructed wetlands with saturated DO levels in the range of 8–11 mg/L [71]. Therefore, it is well documented that all four technologies in this study were sufficiently aerated within the systems.

Mean inflow TN concentrations of 60 ± 19, 123 ± 23 and 367 ± 133 mg/L were reduced in the outflow, which produced a mean TN removal efficiency in the range of 27–76% within the four systems in experimental Phases I, II and III, respectively (Table 6). With the increase of inflow wastewater strength in experimental Phases II and III, relatively high mean removal of TN within the range of 48–76% was detected in all four systems. TN removal of only 55% was also observed using an aerated CW system in another study [71]. A relatively high TN in the outflow dynamics (Figure 4) and mean concentrations in the range of 89–190 mg/L were observed in experimental Phase III with high-strength wastewater with a mean inflow TN concentration of 374 ± 138 mg/L (Table 6). This was potentially due to the overestimation of TN (target TN was 240 mg/L, Table 1) and thereby overdose of the chemicals for increasing TN concentration in the inflow within this phase.

A relatively high NO_3-N mean concentration (46–101 mg/L) in the outflow of all four systems suggested high nitrification and prevailed aerobic conditions that potentially inhibited denitrification, and therefore a relatively low mean TN removal efficiency was observed in high-strength wastewater as inflow in experimental Phase III. Therefore, an effective TN removal did not occur, presumably due to a lack of available carbon source and anoxic conditions (e.g., lack of denitrification) within the four systems. Together with the knowledge that the presence of oxygen inhibits both the synthesis and activity of denitrification enzymes [72], this study's results recommend the denitrification process within the artificially aerated systems as the minor nitrogen removal pathway. The removal of TN up to 96% was recorded in a laboratory-scale aerated constructed wetland through configurational differences where only the front end of the wetland bed was aerated, thus providing anoxic zones in the other end [70]. Thus, optimum NO_3-N or TN removal could be achieved by limiting the aeration to a fraction inside the systems. Intermittent aeration can potentially contribute to the DO fluctuations and can form both aerobic and anoxic regions inside the systems to promote higher TN removal. Nevertheless, limited TN removal in vertical flow wetlands (only 20–30%) is commonly acknowledged in the literature as well [73]. However, an unintentional overdose of nitrogen compounds in the modified wastewater inflow also played a role in a relatively low TN removal by the systems when operating with high-strength wastewater, and none of the outflow mean TN concentrations were in compliance with the Jordanian class A (<45 mg/L) or even class B standard (<70 mg/L) in experimental Phase III within this study (Tables 4 and 6).

3.1.5. TP Removal

During the experimental phases from I to III with a mean inflow TP concentration of 7.9 ± 3, 18.8 ± 5 and 45.3 ± 12 mg/L, the mean TP removal efficiency in the range of 17–69% in the outflow of the four systems was observed (Table 6). The dynamics of inflow and outflow TP concentrations in the three experimental phases with different wastewater compositions are also shown in Figure 5.

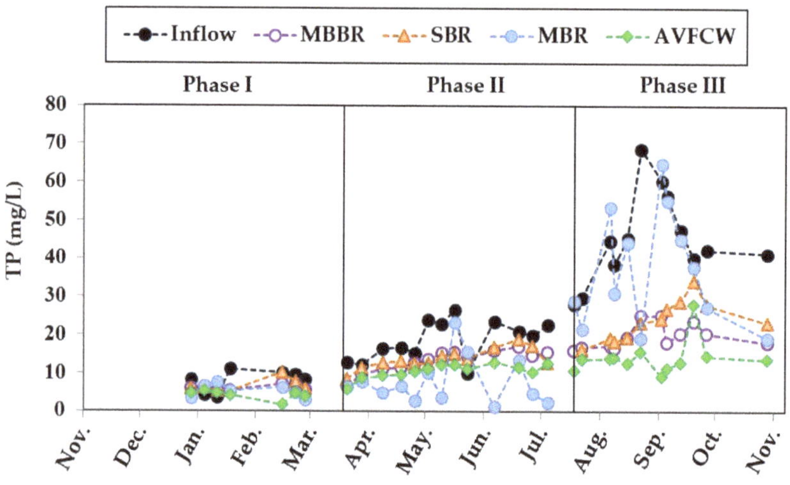

Figure 5. TP concentration profile observed in the inflow and outflow of the four DWWT technologies during the three experimental phases (Phases I–III).

With the increase of inflow wastewater strength, specifically operating with high-strength wastewater in experimental Phase III, relatively high removal of TP within the range of 48–69% was observed in the systems, except for the MBR with a mean TP removal efficiency of 18% only (Table 6). The rapid increase of TP concentration in the inflow also presumably resulted in a remarkable increasing trend of higher TP concentrations or fluctuations in the outflow dynamics of all four systems (Figure 5). Therefore, increasing the organic loading may potentially have led to improved phosphorus removal in this study. Except for the MBR system, the mean concentrations of TP in the effluent of the MBBR, SBR and AVFCW systems were significantly below the Omani standard of class A, which is 30 mg/L for reuse in agriculture (Table 4).

3.1.6. Removal of *E. coli*

The profiles of *E. coli* counted within the inflow and outflow samples in all three experimental phases are shown in Figure 6.

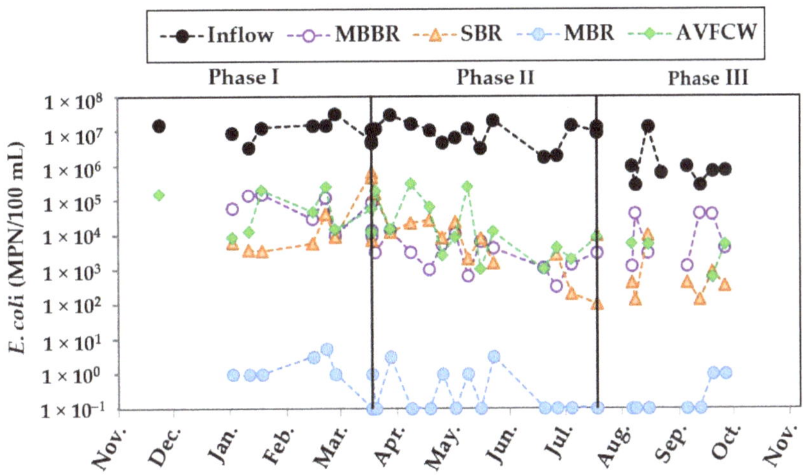

Figure 6. *E. coli* concentration in the wastewater inflow and outflow of the four technologies during the three experimental phases (Phases I–III) in this study.

During operation in experimental Phases I, II and III with the mean inflow *E. coli* of 7.1 ± 0.3, 6.9 ± 0.4 and 6.2 ± 0.6 log10 (MPN/100 mL), respectively, decreased down to a mean outflow *E. coli* in a range of 0.1–4.8 log10 (MPN/100 mL). This accounted for a mean removal of *E. coli* in a range of 2.3–6.9-log removal from the four systems in this study (Table 6). During the experimental Phases II and III with a stepwise increase of wastewater strength, the *E. coli* count demonstrated a slight downward trend in the outflow dynamics of all four systems (Figure 6). As expected, the MBR system showed almost a complete removal of *E. coli* in a range of 6.1–6.9-log removal in the outflow during all three experimental phases and performed the best for pathogen removal as compared to the other three systems in this study. The other three systems (MBBR, SBR and AVFCW) achieved *E. coli* removal in the range of 2.8–4.8-log removal.

Apart from the MBR (<1 MPN/100 mL), in experimental Phase III with high-strength wastewater, the *E. coli* counts from the other three systems were not in accordance with the agricultural reuse standard of class A water quality as suggested by the Jordanian, Omani and European Union regulations. A mean outflow from the MBBR and AVFCW systems with 3.8 ± 0.6 and 3.6 ± 0.5 log10 *E. coli*, respectively, met the water quality standard of class D (<10,000 MPN/100 mL) for agricultural reuse suggested by the EU regulations (Tables 4 and 6). The SBR system with 2.8 ± 0.8 log10 *E. coli* met the treated water quality class C (<1000 MPN/100 mL) for agricultural reuse according to the Jordanian, Omani and EU standards.

E. coli removal in the AVFCW was somehow limited, with an average reduction of 2.6-log removal. This is partly similar to *E. coli* removal reported by Headley et al. [74], with saturated vertical flow wetlands with aeration as 2.1-log removal. For higher pathogen removal, more optimization and final polishing would be necessary for the MBBR, SBR and AVFCW systems. A slow-sand filtration as a post-treatment step, particularly for the AVFCW system, or the inclusion of an ultraviolet (UV) lamp at the outlet would have been very effective for pathogen removal, depending on the required treated water quality class for reuse. The AVFCW can also be considered in combination with a horizontal flow CW for better *E. coli* and nitrogen removal.

It is important to highlight that, in this study, the *E. coli* counts in the modified high-strength wastewaters were reduced instead of getting higher in more concentrated wastewaters (Figure 6). Since the wastewater discharge rates were incrementally reduced (halved) in each phase, and the reduced volume was simulated by adding chemicals to increase the concentrations of specific parameters, the number of *E. coli* was reduced in more concentrated wastewater. In reality, it was difficult to artificially increase the number of *E. coli* in the influent of modified high-strength wastewater, and this is potentially an important limitation of this study that should not be overlooked.

3.2. Comparing the Effectiveness of the Selected DWWT Technologies

The reuse of treated wastewater in agricultural irrigation should ensure compliance with higher water quality classes or categories (class A, B, C or D) according to local or international standards. To achieve this, effective treatment technologies need to be selected, operated and practiced on various decentralized applications; reuse protocols or guidelines should also be standardized [75].

In general, the four selected small DWWT technologies used in this study performed efficiently when treating high-strength wastewater with extremely high organic and nutrient concentrations, which were significantly beyond the EU practical testing ranges. Results also showed very high NH_4-N removal efficiencies. However, the technologies did not remove NO_3-N, TN, TP and *E.coli* to a great extent when treating high-strength wastewater.

To compare the DWWT technologies, one important advantage of the MBR system is that it has the ability to be amazingly effective in removing pollutants and can be an ideal option for decentralized wastewater treatment and safe reuse applications [76]. The MBR systems provide excellent effluent quality with reduced footprint and lesser sludge generation as compared to other conventional DWWT technologies [77]. Despite that, their major disadvantage is associated with membrane fouling [23], which potentially increases maintenance and operating costs or membrane regeneration costs. The MBR system used in this study also demonstrated very good performance in treating high-strength wastewater and produced high-quality treated water with efficient removal of BOD_5, COD and TSS, and nearly complete removal of *E. coli*. However, the membrane fouling led to a complete clogging of the membrane, which was finally solved by replacing the clogged membrane with a new one. It clearly indicates that regular maintenance work needs to be performed to maintain a continuously high-performing MBR system. Otherwise, the membrane may lead to potential clogging due to high organic load, thus reducing treatment performance and requiring expensive membrane replacement for further water quality improvement. Consistent maintenance of the SBR and MBBR systems is also necessary for proper operation.

The AVFCW used in this investigation demonstrated a highly promising NBS with a very efficient treatment performance when treating high-strength wastewater. The presence of plants (*Phragmites*) and their intricate root structure may play a role in a higher mean DO within such intensified constructed wetland systems. The plants showed some stress when operating with high-strength wastewater in experimental Phase III, and their growth decreased gradually under extreme weather, for example, during hot summer days and during winter. However, the system proved to be very robust, requires simple operation and maintenance, and can be ideal for decentralized applications in small communities in rural areas. Further research should aim to optimize the system in order to increase TN and *E. coli* removal and to determine how aerated systems respond under intermittent aeration.

Treated water quality is an important factor for various reuse purposes (e.g., agricultural irrigation), and therefore assessing the local regulations and allowable limit values for the safe reuse of treated wastewater is very important. Such regulations and guidelines on small DWWT technologies would potentially recognize the importance of decentralized wastewater treatment for arid/semi-arid regions (e.g., Jordan or Oman) and would definitely be helpful in promoting decentralized wastewater treatment and reuse on a local scale [13]. A good compilation of a variety of existing national standards or guidelines for treated wastewater reuse can be found in [78].

This study's results propose that small DWWT technologies can be implemented for wastewater treatment and agricultural reuse purposes in the remote areas of arid/semi-arid regions, which can provide an additional water resource for countries with high water scarcity such as Jordan, Oman, etc. However, long-term operation under local climatic conditions and further optimization of these DWWT technologies should be mandatory for more sustainability [31] and to meet local regulatory limit values for safe agricultural reuse of the treated effluent.

4. Conclusions and Recommendations

This study has clearly shown that the EU-certified small DWWT technologies can efficiently treat high-strength wastewater with concentrations beyond the EU-testing ranges (i.e., BOD_5 > 500 mg/L; COD > 1000 mg/L; TSS > 700 mg/L) and can be promising wastewater treatment alternatives for small communities in the rural areas of arid/semi-arid regions. During the three experimental phases with different wastewater compositions, all four selected systems showed highly efficient removal of BOD_5, COD, TSS and NH_4-N, but showed a rather limited performance in the case of NO_3-N, TN, TP and *E. coli* (except MBR) removal. However, comparatively more NO_3-N and TN were removed in the MBBR system, while more *E. coli* were removed in the MBR system. The final outflow mean BOD_5, COD and TSS concentrations were in compliance with class A standards as outlined in the Jordanian and Omani national standards for agricultural reuse. Maximum allowable limit values for Class A reuse standard were not achieved for other recommended parameters by the systems but met other quality classes or categories (e.g., class B, C and D). In general, the selected DWWT technologies can efficiently remove organic pollutants from high-strength wastewater and reduce the substantial risk of pathogens; and the treated effluent can be reused for both restricted and unrestricted irrigation purposes (depending on the water quality requirements). The stepwise increase of inflow wastewater strengths showed neither any remarkable impact on the overall treatment performance of the DWWT technologies nor a limiting factor for stable operation if sufficiently maintained. However, long-term experimental operation and further optimization as well as potential post-treatment steps should be necessary. Further research should aim to simultaneously meet the highest quality of all reuse standard parameters, including *E. coli*, which could potentially be achieved with a subsequent treatment step such as a sand filter or the use of an ultraviolet (UV) disinfection unit.

Treating wastewater using small DWWT technologies in combination with the safe reuse of the treated effluents for agricultural irrigational purposes can potentially create noticeable environmental and economic benefits in arid and semi-arid regions. These DWWT technologies are not only designed to operate at small scale for a smaller community or household basis to reduce the effects of wastewater disposal on the nearby environment and to protect public health but also increase the potential for reuse the treated wastewater in agriculture and food production, depending on the local community, reuse regulations, appropriate system selection and technical feasibility. Therefore, this study fosters a broader perspective on the treatment performance of small DWWT technologies and helps to overcome the difficulties that the decision makers and administrators face in assessing, permitting and implementing such systems in a given local context.

A localized certification guideline for the EU-certified DWWT technologies based on local needs (e.g., in Jordan and Oman) and socio-economic and climatic conditions should be mandatory. Guidelines to ensure minimum quality and performance standards of DWWT technologies for wastewater treatment and reuse, requirements for O&M of such systems, requirements for local training and administrative protocol should be developed before the final permitting process. Applying a methodology for the up-scaling of such small DWWT technologies to ensure conformity between the pilot test and full-scale technology is also recommended.

Overall, this study demonstrates the possibilities of using small DWWT technologies to transform high-strength wastewater from a major environmental and health hazard into a potentially clean and attractive resource.

Author Contributions: Conceptualization, M.v.A. and R.A.M.; methodology, M.v.A.; software, K.Z.R.; formal analysis, K.Z.R.; investigation, S.A.S. and K.B.; resources, K.B.; data curation, S.A.S. and K.Z.R.; writing—original draft preparation, K.Z.R.; writing—review and editing, J.F. and M.A.R.; visualization, S.A.S. and K.Z.R.; supervision, R.A.M., M.v.A., M.A.R. and J.F.; project administration, R.A.M. and M.A.R.; funding acquisition, R.A.M., M.v.A. and J.F. All authors have read and agreed to the published version of the manuscript.

Funding: This research work was partly funded by the German Federal Ministry of Education & Research (BMBF) within the framework of the SMART-MOVE joint research project (grant identifier: 02WM1355B) and the Ministry of Higher Education, Research and Innovation (MoHERI) of the Sultanate of Oman through the pilot project "IATI-WW" (Contract No.: TRC/RCP/15/002), as well as co-financed by means of taxation based on the budget adopted by the representatives of the State Parliament (Landtag) of Saxony (funding no.: 100702517) through the RegioWasser project.

Data Availability Statement: The original contributions presented in the study are included in the article, further inquiries can be directed to the corresponding author.

Acknowledgments: The authors are very thankful and glad to acknowledge Peter Mosig from the Helmholtz Center for Environmental Research (UFZ) for his important contributions and valuable support in the design, construction, operation and maintenance, and sample collection, and Grit Weichert for her outstanding assistance in chemical analysis in the laboratory. The authors would also like to thank the BDZ e.V. administration for their infrastructure support and the involved German companies for providing the technical support needed for their respective wastewater treatment technologies. The authors also acknowledge Ralf Zimmer, Dirk Baum, the laboratory staff from the Municipal Waterworks Leipzig GmbH (Leipzig Waterworks) and Elmar Dorgeloh from the Testing Institute for Wastewater Technology (PIA GmbH) in Aachen, Germany, for their immense support and cooperation. Finally, the authors would like to thank and express their appreciation to Lucie Moeller for her critical review and editing of the manuscript.

Conflicts of Interest: The authors declare no conflicts of interest.

References

1. Abdulla, F.A.; Alfarra, A.; Qdais, H.A.; Sonneveld, B. Evaluation of Wastewater Treatment Plants in Jordan and Suitability for Reuse. *Acad. J. Environ. Sci.* **2016**, *4*, 111–117.
2. Garcia, X.; Pargament, D. Reusing wastewater to cope with water scarcity: Economic, social and environmental considerations for decision-making. *Resour. Conserv. Recycl.* **2015**, *101*, 154–166. [CrossRef]
3. Garrick, D.; Stefano, L.D.; Yu, W.; Jorgensen, I.; O'Donnel, E.; Turley, L.; Aguilar-Barajas, I.; Dai, X.; de Souza Leão, R.; Punjabi, B. Rural water for thirsty cities: A systematic review of water reallocation from rural to urban regions. *Environ. Res. Lett.* **2019**, *14*, 043003. [CrossRef]
4. Sowers, J.; Vengosh, A.; Weinthal, E. Climate change, water resources, and the politics of adaptation in the Middle East and North Africa. *Clim. Change* **2011**, *104*, 599–627. [CrossRef]
5. Liang, L.; Wang, Z.; Li, J. The effect of urbanization on environmental pollution in rapidly developing urban agglomerations. *J. Clean. Prod.* **2019**, *237*, 117649. [CrossRef]
6. Akhoundi, A.; Nazif, S. Sustainability assessment of wastewater reuse alternatives using the evidential reasoning approach. *J. Clean. Prod.* **2018**, *195*, 1350–1376. [CrossRef]
7. Aleisa, E.; Al-Zubari, W. Wastewater reuse in the countries of the Gulf Cooperation Council (GCC): The lost opportunity. *Env. Monitor. Assess.* **2017**, *189*, 553. [CrossRef] [PubMed]
8. Baawain, M.; Sana, A.; Al-Yahyai, R.; Amoatey, P. Sustainable reuse options of treated effluents: A case study from Muscat, Oman. *J. Environ. Eng. Sci.* **2019**, *14*, 195–202. [CrossRef]

9. Berbel, J.; Expósito, A.; Gutiérrez-Martín, C. Effects of the Irrigation Modernization in Spain 2002–2015. *Water Resour. Manag.* **2019**, *33*, 1835–1849. [CrossRef]
10. Matouq, M.; El Hasan, T.; Al Bilbisi, H.; Abdelhadi, M.; Hindiyeh, M.; Eslaiman, S.; Duheisat, S. The climate change implication on Jordan: A case study using GIS and artificial neural networks for weather casting. *J. Sci. Taibah Univ.* **2013**, *7*, 44–55. [CrossRef]
11. Odhiambo, G.O. Water scarcity in the Arabian Peninsula and socio-economic implications. *Appl. Water Sci.* **2017**, *7*, 2479–2492. [CrossRef]
12. European Environment Agency (EEA): Water Scarcity Conditions in Europe (Water Exploitation Index Plus), 13 January 2023. Available online: https://www.eea.europa.eu/en/analysis/indicators/use-of-freshwater-resources-in-europe-1 (accessed on 27 May 2024).
13. Nivala, J.; Abdallat, G.; Aubron, T.; Al-Zreiqat, I.; Abbassi, B.; Wu, G.M.; van Afferden, M.; Müller, R.A. Vertical flow constructed wetlands for decentralized wastewater treatment in Jordan: Optimization of total nitrogen removal. *Sci. Total Environ.* **2019**, *671*, 495–504. [CrossRef]
14. Salas, E.B. Statista. Average Residential Drinking Water Daily Consumption Per Capita in Selected European Countries in 2021. In *Drinking Water Daily Consumption Per Capita in Europe 2021, by Country*; Erick Burgueño Salas, Statista GmbH: Berlin, Germany, 2023. Available online: https://www.statista.com/statistics/1393712/drinking-water-consumption-per-capita-daily-europe/ (accessed on 26 June 2024).
15. Uleimat, A. Wastewater Production, Treatment, and Use in Jordan. Ministry of Water & Irrigation Water Authority of Jordan. In Proceedings of the Second Regional Workshop 'Safe Use of Wastewater in Agriculture', New Delhi, India, 16–18 May 2012. Available online: https://www.ais.unwater.org/ais/pluginfile.php/356/mod_page/content/111/Jordan_ULIMAT_ICID_AHMA_2012-1.pdf (accessed on 10 May 2024).
16. Salameh, E.; Shteiwi, M.; Al Raggad, M. Chapter 3: Patterns of water use. In *Water Resources of Jordan: Political, Social and Economic Implications of Scarce Water Resources*; Springer International Publishing AG: Cham, Switzerland, 2018; Volume 1, pp. 61–66.
17. MWI. *National Water Strategy 2016–2025*; Ministry of Water and Irrigation: Amman, Jordan, 2016. Available online: https://faolex.fao.org/docs/pdf/jor156264E.pdf (accessed on 15 May 2024).
18. Metcalf and Eddy Inc. *Wastewater Engineering: Treatment, Disposal, and Reuse*, 4th ed.; McGraw-Hill: New York, NY, USA, 2003.
19. Ammary, B.Y. Wastewater reuse in Jordan: Present status and future plans. *Desalination* **2007**, *211*, 164–176. [CrossRef]
20. Baawain, M.S.; Al-Omairi, A.; Choudri, B.S. Characterization of domestic wastewater treatment in Oman from three different regions and current implications of treated effluents. *Environ. Monit. Assess.* **2014**, *186*, 2701–2716. [CrossRef]
21. van Afferden, M.; Cardona, J.A.; Rahman, K.Z.; Daoud, R.; Headley, T.; Kilani, Z.; Subah, A.; Mueller, R.A. A step towards Decentralized Wastewater Management in the Lower Jordan Rift Valley. *Water Sci. Technol. Water Supply* **2010**, *61*, 3117–3128. [CrossRef]
22. Al-Mulqui, H.; Bataineh, F.; Malkawi, S. Wastewater Reuse—The Hashemite Kingdom of Jordan. Water Demand Management Forum on Wastewater Reuse, 26–27 March 2002, Rabat, Morocco. In *Managing Water Demand: Policies, Practices and Lessons from the Middle East and North Africa*; Baroudy, E., Lahlou, A.A., Attia, B., Eds.; IWA Publishing: London, UK, 2005. Available online: https://idrc-crdi.ca/sites/default/files/openebooks/187-6/index.html (accessed on 4 June 2024).
23. Mutamim, N.S.A.; Noor, Z.Z.; Abu Hassan, M.A.; Olsson, G. Application of membrane bioreactor technology in treating high strength industrial wastewater: A performance review. *Desalination* **2012**, *305*, 1–11. [CrossRef]
24. Liew, Y.X.; Chan, Y.J.; Manickam, S.; Chong, M.F.; Chong, S.; Tiong, T.J.; Lim, J.W.; Pan, G.T. Enzymatic pretreatment to enhance anaerobic bioconversion of high strength wastewater to biogas: A review. *Sci. Total Environ.* **2020**, *713*, 136373. [CrossRef]
25. Musa, M.A.; Idrus, S.; Man, H.C.; Daud, N.N.N. Performance Comparison of Conventional and Modified Upflow Anaerobic Sludge Blanket (UASB) Reactors Treating High-Strength Cattle Slaughterhouse Wastewater. *Water* **2019**, *11*, 806. [CrossRef]
26. Bustillo-Lecompte, C.F.; Mehrvar, M.; Quiñones-Bolaños, E. Cost-Effectiveness Analysis of TOC Removal from Slaughter-house Wastewater Using Combined Anaerobic-Aerobic and UV/H_2O_2 Processes. *J. Environ. Manag.* **2014**, *134*, 145–152. [CrossRef]
27. Ozturk, D.; Yilmaz, A.E. Treatment of Slaughter house Wastewater with the Electrochemical Oxidation Process: Role of Operating Parameters on Treatment Efficiency and Energy Consumption. *J. Water Process Eng.* **2019**, *31*, 100834. [CrossRef]
28. Scherer, L.; Pfster, S. Global Biodiversity loss by freshwater consumption and eutrophication from swiss food consumption. *Environ. Sci. Technol.* **2016**, *50*, 7019–7028. [CrossRef]
29. Capodaglio, A.G. Integrated, Decentralized Wastewater Management for Resource Recovery in Rural and Peri-Urban Areas. *Resources* **2017**, *6*, 22. [CrossRef]
30. Khurelbaatar, G.; Al Marzuqi, B.; van Afferden, M.; Müller, R.A.; Friesen, J. Data reduced method for cost comparison of wastewater management scenarios—Case study for two settlements in Jordan and Oman. *Front. Environ. Sci.* **2021**, *9*, 626634. [CrossRef]
31. Capodaglio, A.G.; Callegari, A.; Cecconet, D.; Molognoni, D. Sustainability of decentralized wastewater treatment technologies. *Water Pract. Technol.* **2017**, *12*, 463–477. [CrossRef]
32. Reynaud, N.; Buckley, C. Field-data on parameters relevant for design, operation and monitoring of communal decentralized wastewater treatment systems (DEWATS). *Water Pract. Technol.* **2015**, *10*, 787–798. [CrossRef]
33. Pabbati, R.; Jhansi, V.; Reddy, K.V. Conventional Wastewater Treatment Processes. In *Advances in the Domain of Environmental Biotechnology*; Springer: Singapore, 2021; pp. 455–479, ISSN 2662-1681. [CrossRef]

34. Thomas, B.D.; Uludag-Demirer, S.; Frost, H.; Liu, Y.; Dusenbury, J.S.; Liao, W. Decentralized high-strength wastewater treatment using a compact aerobic baffled bioreactor. *J. Environ. Manag.* **2022**, *305*, 114281. [CrossRef]
35. Rahman, K.Z.; Al Saadi, S.; Al Rawahi, M.; Knappe, J.; van Afferden, M.; Moeller, L.; Bernhard, K.; Müller, R.A. A multi-functional nature-based solution (NBS) for greywater treatment and reuse at the same plot. *Ecol. Eng.* **2023**, *191*, 106952. [CrossRef]
36. *DIN EN 12556-3*; Small Wastewater Treatment Systems for up to 50 PT—Part 3: Packaged and/or Site Assembled Domestic Wastewater Treatment Plants. European Committee for Standardization: Brussels, Belgium, 2005. (In English)
37. PIA GmbH. Testing Institute for Wastewater Technology GmbH at RWTH Aachen University e.V., CE tests according to EN 12566. Available online: https://www.pia-gmbh.com/en/services/wastewater-treatment-plants#part-6-prefabricated-treatment-units-for-septic-tank-effluent (accessed on 27 May 2024).
38. Hirooka, K.; Asano, R.; Yokoyama, A.; Okazaki, M.; Sakamoto, A.; Nakai, Y. Reduction in excess sludge production in a dairy wastewater treatment plant via nozzle-cavitation treatment: Case study of an on-farm wastewater treatment plant. *Bioresour. Technol.* **2009**, *100*, 3161–3166. [CrossRef]
39. Nguyen, A.V.; Pham, N.T.; Nguyen, T.H.; Morel, A.; Tonderski, K. Improved septic tank with constructed wetland, a promising decentralized wastewater treatment alternative in Vietnam. In Proceedings of the Paper XI-RCS-07-30 NOWRA 16th Annual Technical Education Conference & Exposition, Baltimore, MD, USA, 10–14 March 2007.
40. Villegas-Gómez, J.D.; Jhonniers-Guerrero, E.; Castaño-Rojas, J.M.; Paredes-Cuervo, D. Septic Tank (ST)-Up Flow Anaerobic Filter (UAF)-Subsurface Flow Constructed Wetland (SSF-CW) systems aimed at wastewater treatment in small localities in Colombia. *Rev. Téc. Fac. Ing. Univ. Zulia* **2006**, *29*, 269–281.
41. de Anda, J.; López-López, A.; Villegas-García, E.; Valdivia-Aviña, K. High-Strength Domestic Wastewater Treatment and Reuse with Onsite Passive Methods. *Water* **2018**, *10*, 99. [CrossRef]
42. *Standard DWA-A 262E*; Principles for Dimensioning, Construction and Operation of Wastewater Treatment Plants with Planted and Unplanted Filters for Treatment of Domestic and Municipal Wastewater. The German Association for Water, Wastewater and Waste (DWA): Hennef, Germany, 2017; ISBN 978-3-88721-641-2.
43. Ødegaard, H. Innovations in wastewater treatment: The moving bed bioflm process. *Water Sci. Technol.* **2006**, *53*, 17–33. [CrossRef] [PubMed]
44. Leyva-Díaz, J.C.; Martín-Pascual, J.; Poyatos, J.M. Moving bed bioflm reactor to treat wastewater. *Int. J. Environ. Sci. Technol.* **2017**, *14*, 881–910. [CrossRef]
45. Dutta, A.; Sudipta Sarkar, S. Sequencing Batch Reactor for Wastewater Treatment: Recent Advances. *Curr. Pollut. Rep.* **2015**, *1*, 177–190. [CrossRef]
46. Vymazal, J. Constructed wetlands for wastewater treatment: Five decades of experience. *Environ. Sci. Technol.* **2010**, *45*, 61–69. [CrossRef] [PubMed]
47. Wu, S.; Carvalho, P.N.; Muller, J.A.; Manoj, V.R.; Dong, R. Sanitation in constructed wetlands: A review on the removal of human pathogens and fecal indicators. *Sci. Total Environ.* **2016**, *541*, 8–22. [CrossRef] [PubMed]
48. Zhu, L.; Takala, J.; Hiltunen, E.; Li, Z.; Kristianto, Y. Comparison of vertical-flow constructed wetlands with and without supplementary aeration treating decentralized domestic wastewater. *Environ. Technol.* **2013**, *34*, 53–60. [CrossRef] [PubMed]
49. Saeed, T.; Sun, G. A review on nitrogen and organics removal mechanisms in subsurface flow constructed wetlands: Dependency on environmental parameters, operating conditions and supporting media. *J. Environ. Manag.* **2012**, *112*, 429–448. [CrossRef]
50. Zhang, D.; Gersberg, R.M.; Keat, T.S. Constructed wetlands in China. *Ecol. Eng.* **2009**, *35*, 1367–1378. [CrossRef]
51. Dong, H.; Qiang, Z.; Li, T.; Jin, H.; Chen, W. Effect of artificial aeration on the performance of vertical-flow constructed wetland treating heavily polluted river water. *J. Environ. Sci.* **2012**, *24*, 596–601. [CrossRef]
52. Nivala, J.; Wallace, S.; Headley, T.; Kassa, K.; Brix, H.; van Afferden, M.; Müller, R.A. Oxygen transfer and consumption in subsurface flow treatment wetlands. *Ecol. Eng.* **2013**, *61(Part B)*, 544–554. [CrossRef]
53. Maisonnave, V.; Lepimpec, P.; Cauchi, A.; Vignoles, C.; Deleris, S. Eaux usées artificielles: Une voie nouvelle pour les essais. *Tech. Sci. Méthodes* **2011**, *6*, 18–30. (In French) [CrossRef]
54. Bertanza, G.; Boiocchi, R. Interpreting per capita loads of organic matter and nutrients in municipal wastewater: A study on 168 Italian agglomerations. *Sci. Total Environ.* **2022**, *819*, 153236. [CrossRef]
55. *EN ISO 11732*; Water Quality—Determination of Ammonium Nitrogen—Method by Flow Analysis (CFA and FIA) and Spectrometric Detection. ISO: London, UK, 2005.
56. *EN ISO 10304-1*; Water Quality—Determination of Dissolved Anions by Liquid Chromatography of Ions—Part 1: Determination of Bromide, Chloride, Fluoride, Nitrate, Nitrite, Phosphate and Sulfate. ISO: London, UK, 2007.
57. *EN ISO 15681-1*; Water Quality—Determination of Orthophosphate and Total Phosphorus Contents by Flow Analysis (FIA and CFA)—Part 1: Method by Flow Injection Analysis (FIA). ISO: London, UK, 2003.
58. *ISO 9308-2*; Water Quality—Enumeration of Escherichia coli and Coliform Bacteria—Part 2: Most Probable Number Method. ISO: London, UK, 2012.
59. Abdallat, G.A.; Salameh, E.; Shteiwi, M.; Bardaweel, S. Pharmaceuticals as Emerging Pollutants in the Reclaimed Wastewater Used in Irrigation and Their Effects on Plants, Soils, and Groundwater. *Water* **2022**, *14*, 1560. [CrossRef]

60. Ministry of Regional Municipalities and Environment (MRME). Sultanate of Oman. Regulations for Wastewater Reuse and Discharge. Ministerial Decision 145/1993, 13 June 1993. Available online: https://www.pdo.co.om/hseforcontractors/Environment/Documents/Oman%20Laws/Misterial%20Decision%20-%20Guidelines/Wastewater%20Reuse%20and%20Discharge.pdf (accessed on 14 May 2024).
61. European Union. Regulation (EU) 2020/741 of the European Parliament and of the council 25 May 2020 on Minimum Requirements for Water Reuse. *Off. J. Eur. Union* **2020**, *OJ L 177*, 32–55. Available online: https://eur-lex.europa.eu/legal-content/EN/TXT/PDF/?uri=CELEX:32020R0741&from=EN (accessed on 14 May 2024).
62. Mozia, S.; Janus, M.; Bering, S.; Tarnowski, K.; Mazur, J.; Szymański, K.; Morawski, A.W. Hybrid System Coupling Moving Bed Bioreactor with UV/O_3 Oxidation and Membrane Separation Units for Treatment of Industrial Laundry Wastewater. *Materials* **2020**, *13*, 2648. [CrossRef]
63. Bustamante, M.; Liao, W. A self-sustaining high-strength wastewater treatment system using solar-bio-hybrid power generation. *Bioresour. Technol.* **2017**, *234*, 415–423. [CrossRef]
64. Bezbaruah, A.N.; Zhang, T.C. Quantification of oxygen release by bulrush (*Scirpus validus*) roots in a constructed treatment wetland. *Biotechnol. Bioeng.* **2005**, *89*, 308–318. [CrossRef]
65. Zhao, C.; Wang, G.; Xu, X.; Yang, Y.; Yang, F. Long-term operation of oxygen-limiting membrane bioreactor (MBR) for the development of simultaneous partial nitrification, anammox and denitrification (SNAD) process. *Environ. Technol.* **2018**, *39*, 2193–2202. [CrossRef]
66. Sklarz, M.Y.; Gross, A.; Yakirevich, A.; Soares, M.I.M. A Recirculating Vertical Flow Constructed Wetland for the Treatment of Domestic Wastewater. *Desalination* **2009**, *246*, 617–624. [CrossRef]
67. Dotro, G.; Jefferson, B.; Jones, M.; Vale, P.; Cartmell, E.; Stephenson, T. A review of the impact and potential of intermittent aeration on continuous flow nitrifying activated sludge. *Environ. Technol.* **2011**, *32*, 1685–1697. [CrossRef]
68. Li, J.; Meng, J.; Li, J.; Wang, C.; Deng, K.; Sun, K.; Buelna, G. The effect and biological mechanism of COD/TN ratio on nitrogen removal in a novel upflow microaerobic sludge reactor treating manure-free piggery wastewater. *Bioresour. Technol.* **2016**, *209*, 360–368. [CrossRef]
69. Wu, H.; Zhang, J.; Ngo, H.H.; Guo, W.; Hu, Z.; Liang, S.; Fan, J.; Liu, H. A review on the sustainability of constructed wetlands for wastewater treatment: Design and operation. *Bioresour. Technol.* **2015**, *175*, 594–601. [CrossRef]
70. Maltais-Landry, G.; Maranger, R.; Brissona, J. Effect of artificial aeration and macrophyte species on nitrogen cycling and gas flux in constructed wetlands. *Ecol. Eng.* **2009**, *35*, 221–229. [CrossRef]
71. Butterworth, E.; Dotro, G.; Jones, M.; Richards, A.; Onunkwo, P.; Narroway, Y.; Jefferson, B. Effect of artificial aeration on tertiary nitrification in a full-scale subsurface horizontal flow constructed wetland. *Ecol. Eng.* **2013**, *54*, 236–244. [CrossRef]
72. Colliver, B.B.; Stephenson, T. Production of nitrogen oxide and dinitrogen oxide by autotrophic nitrifiers. *Biotechnol. Adv.* **2000**, *18*, 219–232. [CrossRef]
73. Dotro, G.; Langergraber, G.; Molle, P.; Nivala, J.; Puigagut, J.; Stein, O.; von Sperling, M. Biological Wastewater Treatment. In *Treatment Wetlands*; IWA Publishing: London, UK, 2017; Volume 7.
74. Headley, T.; Nivala, J.; Kassa, K.; Olsson, L.; Wallace, S.; Brix, H.; van Afferden, M.; Müller, R. *Escherichia coli* removal and internal dynamics in subsurface flow ecotechnologies: Effects of design and plants. *Ecol. Eng.* **2013**, *61*, 564–574. [CrossRef]
75. Al Hamedi, F.H.; Kandhan, K.; Liu, Y.; Ren, M.; Jaleel, A.; Alyafei, M.A.M. Wastewater Irrigation: A Promising Way for Future Sustainable Agriculture and Food Security in the United Arab Emirates. *Water* **2023**, *15*, 2284. [CrossRef]
76. Bérubé, P. Chapter 9 Membrane Bioreactors: Theory and Applications to Wastewater Reuse. *Sustain. Sci. Eng.* **2010**, *2*, 255–292.
77. Hamza, R.A.; Iorhemen, O.T.; Tay, J.H. Advances in biological systems for the treatment of high-strength wastewater. *J. Water Process. Eng.* **2016**, *10*, 128–142. [CrossRef]
78. Shoushtarian, F.; Negahban-Azar, M. Worldwide Regulations and Guidelines for Agricultural Water Reuse: A Critical Review. *Water* **2020**, *12*, 971. [CrossRef]

environments

MDPI

Article

Plant-Wide Models for Optimizing the Operation and Maintenance of BTEX-Contaminated Wastewater Treatment and Reuse

Dániel Bencsik [1,2], Tanush Wadhawan [1], Ferenc Házi [1] and Tamás Karches [3,*]

[1] Dynamita, SARL, 2015 Route d'Aiglun, 06910 Sigale, France; daniel@dynamita.com (D.B.); tanush@dynamita.com (T.W.)
[2] Doctoral School of Military Engineering, Ludovika University of Public Service, Hungária krt. 9-11, H-1101 Budapest, Hungary
[3] Department of Water Supply and Sewerage, Ludovika University of Public Service, 12-14 Bajcsy-Zsilinszky u., H-6500 Baja, Hungary
* Correspondence: karches.tamas@uni-nke.hu

Abstract: Benzene, toluene, ethylbenzene and xylenes, collectively known as BTEX compounds, are significant emerging contaminants in municipal wastewater. Stricter effluent quality regulations necessitate their removal, especially with concerns about organic micropollutant concentrations. Water scarcity further underscores the need for wastewater treatment to ensure safe agricultural or drinking water supplies. Although biological treatment partially reduces BTEX levels through processes like biodegradation and sorption, additional purification using physico-chemical methods is crucial for substantial reduction. This paper aims to outline plant-wide simulation methods for treating BTEX-contaminated sewage and facilitating reuse, adhering to IWA Good Modelling Practice Guidelines. The model, built upon the MiniSumo process model, incorporates equations detailing BTEX metabolism and removal kinetics, informed by an extensive literature review. Using a variant of the Benchmark Simulation Model with granular activated carbon for water reuse, the study examines strategies for improving effluent quality and minimizing operational costs. These strategies include adjusting the sludge retention time and airflow to enhance BTEX degradation and stripping, respectively, and comparing maintenance approaches for the GAC tower.

Keywords: activated carbon; BTEX; micropollutant removal; plant-wide modelling; wastewater

1. Introduction

The consumption of raw materials, as well as emissions, tends to get concentrated within specific areas as human communities evolve and grow. Due to the continuous human impact on the environment, special attention must be paid to the operation of public works that are crucial for a sustainable modern civilization. Certain substances that may be hazardous to human health must not be allowed to reach consumers or to adversely affect environmental conditions. The pollutants building up throughout municipal sewer systems appear at water resource recovery facilities.

The practice of wastewater use in agriculture has become increasingly important due to the growing demand for water resources and the need for sustainable and efficient agricultural practices. The main reason is that it could help reduce the pressure on limited freshwater sources and thus mitigate the impacts of climate change by conserving the water quantity and reusing the nutrients from the treated wastewater. Low-income countries pose substantial risks by the direct use of untreated wastewater or the indirect use of polluted water courses [1]. Scheierling et al. (2011) gave guidelines for developing a strategic plan on treated wastewater and highlighted the necessity of a paradigm shift for decision makers in wastewater treatment [2]. Identifying the risk of being exposed to pathogens and heavy metals and the salinization of soil can be easily seen as drawbacks. Additionally, contaminants of emerging concern were given special consideration, taking into account the known or perceived environmental and health risks associated with these contaminants [3]. Pedrero et al. (2010) found that with proper salts management, treated municipal wastewater appears to be a viable water resource for irrigating citrus trees, and no other concerns are related to treated municipal wastewater [4], whereas Jaramillo and Restrepo (2017) stated that the absence of a quantitative assessment of the microbiological risk, specifically related to the concentration of helminths, is the essential component that is necessary for the appropriate adoption of agricultural reuse [5]. However, the utilization of untreated wastewater is prevalent in particular areas due to the unavailability of an appropriate alternative water supply [6].

Among the volatile organic compounds (VOCs) that are released into the environment from both natural and manufactured sources, benzene, toluene, ethylbenzene and xylene isomers (BTEX) are considered to be among the most significant [7]. Due to their volatile nature, attention should be given to the health of the personnel working at and people living in the vicinity of treatment plants. To measure the concentration of BTEX compounds in the atmosphere surrounding wastewater treatment plants, air quality has been monitored in numerous studies. In an effort to assess the extent of exposure of wastewater treatment plant (WWTP) workers to BTEX compounds, Dehghani et al. (2022) conducted a study. The study revealed that the workers were indeed exposed to BTEX compounds, particularly benzene, in certain cases exceeding the permissible reference values. Thus, it is imperative to recognize WWTPs as a source of BTEX emissions, and workers employed in these facilities may be at risk of exposure to these volatile compounds [8]. Numerous studies have investigated the potential adverse impacts of BTEX compounds on human health, including their effects on organs such as the nervous system, liver, heart and kidneys [9].

During biological wastewater treatment, BTEX pollutants are removed through various pathways such as biodegradation, sorption and volatilization. However, these pathways only offer partial reduction in comparison to the initial influent concentrations. For a substantial reduction in BTEX pollutants, further purification of the biologically treated water is necessary, primarily through physico-chemical methods. Advanced Oxidation Processes (AOPs) use highly reactive chemical species, such as hydroxyl radicals, to degrade or transform contaminants into less harmful compounds. AOPs have been shown to be effective in removing BTEX compounds from wastewater due to their ability to break down organic compounds into smaller, less toxic molecules [10,11].

Biofilm-based treatment processes have been employed in many cases; the microbial community—that attaches to a solid surface—is established and grown on a support material. During the process, the BTEX compounds diffuse into the biofilm, where they are metabolized by the microorganisms. Mello et al. (2019) achieved 99% removal efficiency by applying attached biomass-supported activated carbon media [12]. The same reduction rate was achieved in horizontal-flow anaerobic immobilized biomass reactors where a denitrifying consortium was used as an inoculum to form a biofilm [13,14]. Although advanced techniques such as the use of membranes [15], microalgae [16] or carbon nanotubes [17] have been developed for BTEX removal, the study did not aim to provide a comprehensive review of the state-of-the-art techniques in this area; the focus was more on the kinetics of BTEX biodegradation.

Fundamentally, the presence of dissolved oxygen (usually introduced through mechanical aeration) and chemically bound oxygen (in anoxic conditions) or the complete absence of oxygen (in anaerobic conditions) define the environmental conditions that support the growth and survival of biomass. In each environment, the BTEX removal kinetics were investigated; high biodegradation could be expected in aerobic conditions [18], while the buildup of toxic metabolites led to the incomplete degradation of both xylene and benzene, whether they were present as individual components or as a mixture. The addition of another pollutant always resulted in competition and inhibition, with xylene being the most inhibitory [19], whilst the external carbon source could cause enrichment on BTEX degradation by bacterial cultures [20]. Anaerobic conditions could be developed in packed bed reactors or deeper layers of the biofilm, which could also show significant biodegradation [21].

Lee et al. (1998) made significant steps towards modelling a highly hydrophobic compound in wastewater, and the authors explained that the biodegradation of a compound that has been adsorbed to activated sludge directly enhances its biodegradation and diminishes its discharge with the waste activated sludge. Additionally, the evaporation of compounds from the primary and secondary systems' surfaces is significant for compounds that possess moderate to high volatilities [22]. Pomiès et al. (2012) gathered 18 mathematical approaches to modelling micropollutant removal from wastewater, and they concluded that most authors had held the view that only dissolved micropollutants are biodegradable, although no experimental evidence has been found to support this assertion. Simplified approaches based on first-order kinetics cannot take into account advanced metabolisms and the connection between the various stages of treatment (such as between mainstream and sludge line unit operations) or respond to process dynamics during operation. Incorporating particulate phase data into models would be extremely valuable in carrying out a comprehensive mass balance calculation, particularly for hydrophobic substances, which would include their fate in sludge. Furthermore, the recent incorporation of co-metabolism into models has revealed an interesting correlation between the biodegradation of macropollutants and micropollutants. These modelling tools are beneficial in conducting research to comprehend the mechanistic behavior of micropollutant removal, but they are not yet ready to be used in process design [23].

The objective of this paper is to develop numerical methodology for simulating the treatment and reuse of BTEX-contaminated sewage throughout a facility using dynamic mathematical models that can adapt to diurnal and seasonal variations and changes in operational approaches. The novelty of the study lies in its application of a modeling approach to a full-plant configuration, which investigates the potential for operational cost minimization. This is achieved by adjusting the sludge retention time or air flow to enhance BTEX degradation and stripping, respectively.

Modelling wastewater treatment processes is complex and involves several submodels, such as a hydrodynamic model, the characterization of wastewater components [24], a phase separation model [25], a gas dissolution model, a process control model and a biokinetic model. Biokinetic modelling is a widely used approach to describing the biological processes that occur in wastewater treatment systems. Activated Sludge Models (ASMs) are a type of biokinetic model that simulate the behavior of microorganisms in the activated sludge process, one of the most common wastewater treatment technologies [26]. Plant-wide models, on the other hand, provide simulation capabilities of more advanced—such as granular and biofilm-based—treatment processes, feature more biological processes such as anaerobic digestion, enable pH, precipitation and gas transfer calculations and allow for more complex plant configurations that may include mechanical pretreatment, sidestream treatment or even effluent wastewater recycling. They can further provide oversight of different aspects of facilities such as energy management and operational costs, and additionally, they take into account equipment operational limitations regarding industrial safety [27].

The International Water Association (IWA) published guidelines for good modelling practice in the field of wastewater treatment. These guidelines provide a framework for developing and applying mathematical models to treatment systems, aiming to ensure that models are used in a rigorous and consistent manner. The guidelines cover topics such as model development, verification and validation, sensitivity analysis and uncertainty analysis [28].

The aim of this study is to develop and present a methodology for robust calculations regarding the fate of BTEX compounds in biological wastewater treatment, which—through the use of plant-wide modelling—can be coupled with the concepts of physico-chemical treatment for effluent polishing and water reuse. The following chapters describe model development efforts that consisted of setting up a biokinetic model matrix and the introduction of gas transfer equations, all to be integrated into (and used alongside) existing concepts for plant-wide simulation.

2. Materials and Methods

2.1. Mathematical Model Development

The newly introduced sub-models were implemented according to a review of the scientific literature, equipment manufacturer specifications and experience gained throughout industrial modelling projects. Following a review of the related literature, it was decided that the most significant BTEX removal mechanisms—biochemical reactions and gas–liquid equilibrium—can be most efficiently coded in the Gujer matrix format [29], containing the stoichiometric coefficients and process rates, from which ordinary differential equations can be expanded and solved using computer algorithms. In this study, the rate expressions for BTEX modelling are added to the existing matrix of the commercially available MiniSumo model (see Tables A1–A13) They can be linked to any other similar biokinetic model; however, we note that the presented nomenclature regarding the state variables corresponds to the logic of MiniSumo. For the purpose of this study, this model was chosen as the basis for the sake of simulation speed and transparency, since it is a simple yet well calibrated plant-wide model that focuses on the most important and general aspects of water resource recovery—one-step nitrification and denitrification, anaerobic digestion, the oxygen requirement and sludge production [30].

2.1.1. Biodegradation

The biological reactions are interpreted using an ASM-type concept based mainly on the Monod-kinetics of saturation and inhibition, which, combined with a hydraulic sub-model (considering the amount of liquid entering individual compartments of unit processes), form volume-integrated scalar transport equations that define the mass balance throughout the plant configuration with simplified hydrodynamics. The component balance in Equation (1) describes the concentration change in a liquid phase component i (e.g., benzene) in time within a completely stirred tank unit with the liquid volume of V_r. This ordinary differential equation (ODE) is generalized, and variables marked with the lower index must be expanded for each state variable in the model to be integrated over time by a solver algorithm. The liquid phase states generalized as L_i may further be categorized based on the particle size, S_i designating soluble components, C_i applied for colloids and X_i applied for particulates as follows:

$$\frac{dL_i}{dt} = \frac{F_{L_{i,in}} - F_{L_{i,out}} + rateF_{L_i}}{V_r}, \tag{1}$$

The mass flow or load to a tank is calculated knowing the influent concentration of the state variable—$L_{i,in}$—entering the tank, by Equation (2).

$$F_{L_{i,in}} = Q_{in} \cdot L_{i,in}, \tag{2}$$

In a similar manner, the component mass flow leaving the tank is interpreted by Equation (3).

$$F_{L_{i,out}} = Q_{out} \cdot L_i, \tag{3}$$

The kinetic reaction rate of the state variable—interpreted in the dimension of concentration change—expressed as rate $_i$ in Equation (4), is derived from the kinetic matrix by multiplying a stoichiometric coefficient ($v_{j,i}$, regarding the interaction of the given process j with the state variable i) by process rates (r_j for the process j), summing them up for every case where a given process affects the state variable (based on reading the matrix) [29].

$$rate_i = \sum_{j=1}^{m} v_{j,i} \cdot r_j, \tag{4}$$

Equation (5) determines $rateF_i$, the mass rate of the process variable, which was seen earlier (regarding the liquid phase) as contributing to the component balance from Equation (1).

$$rateF_i = V_r \cdot rate_i, \tag{5}$$

The results of the research conducted by Kasi et al. (2012) were used for the concept of BTEX degradation kinetics [20], to be implemented in Gujer matrix format for this study. Appendix A contains the final format of the developed matrix (Tables A1–A10). It is assumed that one common group of heterotrophic biomass (present in all ASM-type models) metabolizes the BTEX contaminants, either oxidizing them aerobically using oxygen or removing them in an anoxic environment with only nitrite or nitrate available as the electron acceptor; otherwise, they are degraded through fermentation when neither of the previously mentioned electron acceptors are available. The reaction rate for aerobic pathways is higher than those of non-aerated conditions.

In this study, the process model MiniSumo was modified to include the presented BTEX-related biological reactions and gas exchange processes. MiniSumo is a full-plant model developed for simulating oxygen uptake rates and sludge production. It considers the aerobic, anoxic and anaerobic removal of soluble and particulate biodegradable organics, nitrification, denitrification, methanogenesis and phosphorus removal through chemical addition [30]. Regarding the model extension, BTEX-related ordinary heterotrophic organism (OHO) growth processes have to be introduced into a model that already contains the state variables oxygen, nitrite and nitrate (either separately or combined), as well as volatile fatty acids. The latter variable, for instance, is not originally featured in MiniSumo; therefore, it was necessary to include VFA production and uptake (along with the related missing biochemical processes, such as the fermentation of the readily biodegradable substrate) from the matrix of the Sumo1 process model [31] in our example.

Eight new state variables are included in the model matrix to represent both dissolved and gaseous states of the BTEX compounds in $g_{COD}\ m^{-3}$ concentration units. These include dissolved benzene (S_{BENE}), dissolved toluene (S_{TENE}), dissolved ethylbenzene (S_{EBENE}), dissolved xylene (S_{XENE}), gaseous benzene (G_{BENE}), gaseous toluene (G_{TENE}), gaseous ethylbenzene (G_{EBENE}) and gaseous xylene (G_{XENE}). All the dissolved BTEX compounds undergo biodegradation under anaerobic, anoxic and aerobic conditions carried out by Ordinary Heterotrophic Organisms (X_{OHO}) biomass. All the biological reactions are parallel reactions that completely mineralize the BTEX compounds to carbon dioxide and biomass. A complete list of rates and the model structure are disclosed as supplementary information in Appendix A (Tables A1–A10).

The temperature-dependent growth rates of the added metabolic processes can be calculated using Arrhenius-type functions, with this example study applying the MiniSumo implementation for heterotrophs. The kinetic rates and stochiometric yields associated with the biological reactions are described in Appendix B (Table A11), relying on the experimental data of Kasi et al. (2012) [20].

2.1.2. Gas–Liquid Transfer

It is important to discuss the fate of BTEX components emitted from the wastewater into the atmosphere. The equilibrium between the vapor and liquid phases for these volatile compounds is simulated using Fick's two-film theory, analogous to existing gas transfer processes in MiniSumo, serving as an example of model development for absorption and desorption. The impact of equilibrium processes on state variables is described within the process model's Gujer matrix. We note that, within the gas transfer equations, standard conditions regarding gases are interpreted as 101,325 Pa for pressure (p_{NTP}) and 293.15 K for temperature ($T_{NTP,K}$).

The internal gas phase components are simulated in units of mass per liquid volume. This makes mass and component balancing substantially simpler (compared to using the gas volume as the basis). Equation (6) describes the component balance for gas i.

$$\frac{dG_i}{dt} = \frac{F_{G_{i,air,inp}} - F_{G_{i,air,outp}} + rateF_{G_i}}{V_r}, \tag{6}$$

$F_{Gi,air,inp}$ denotes the mass flow of gas i from the air supply and is calculated based on the composition of the input gas and the air flow based on the ideal gas law, as shown by Equation (7). Appendix C elaborates on the stoichiometry behind the equivalent molar mass $MM_{EQ,i}$ of each gaseous BTEX component derived from the theoretical oxygen demand.

$$F_{G_{i,air,inp}} = \frac{G_{i,air,inp}/100 \cdot p_{NTP} \cdot Q_{air,NTP} \cdot MM_{EQ,i}}{R \cdot T_{NTP,K}}, \tag{7}$$

For gas phase mass balancing, the output gas mass flow $F_{Gi,air,outp}$ is calculated by Equation (8), where the concentrations expressed per liquid volume must be converted per volume of gas.

$$F_{G_{i,air,outp}} = G_i \cdot \frac{V_r}{V_{gas,NTP}} \cdot Q_{gas,outp,NTP}, \tag{8}$$

The volume of the gas phase in the reactors is estimated based on the gas hold-up fraction ε_{gas}. The gas hold-up is an input parameter, for which this paper uses a value of 0.01 $m^3{}_{gas}\ m^{-3}$ in case of aerated conditions and 0.001 $m^3{}_{gas}\ m^{-3}$ regarding non-aerated conditions [32]. The gas volumes at field conditions and standard conditions (V_{gas} and $V_{gas,NTP}$, respectively) are calculated by Equations (9) and (10).

$$V_{gas} = \frac{V_r}{\frac{1}{\varepsilon_{gas}} - 1}, \tag{9}$$

$$V_{gas,NTP} = \frac{V_{gas} \cdot p_{gas} \cdot T_{NTP,K}}{p_{NTP} \cdot T_{air,K}}, \tag{10}$$

The gas phase pressure p_{gas} is determined by Equation (11) in accordance with EPA guidelines on dissolved oxygen saturation calculations [33].

$$p_{gas} = \left(p_{air} + h_{sat,eff} \cdot \rho_{H_2O} \cdot g - p_{v,T}\right) \cdot \frac{p_{NTP}}{p_{NTP} - p_{v,T}}, \tag{11}$$

The vapor pressure $p_{v,T}$ is calculated by Equation (12) based on the water temperature and Antoine-coefficients, where a multiplication factor is used for unit conversion from mmHg to Pa.

$$p_{v,T} = 10^{(A_{H_2O} - B_{H_2O}/(T + C_{H_2O}))} \cdot 133.322, \tag{12}$$

The air pressure p_{air} is calculated by the barometric Equation (13), considering the elevation of the facility above sea level h_{sea} (the molar mass of air MM_{air} is converted from the unit g mol^{-1} to kg mol^{-1}). The parameter L_{air} for the temperature lapse rate applies a value of 0.0065 K m^{-1} [34].

$$p_{air} = p_{NTP} \cdot \left(1 - \frac{L_{air} \cdot h_{sea}}{T_{air,K}}\right)^{\frac{g \cdot MM_{air}/1000}{R \cdot L_{air}}}, \tag{13}$$

The calculation of the effective saturation depth depends on whether the tank is aerated, in which case $h_{sat,eff}$ is calculated by Equation (14), and in case there is no aeration, the diffuser submergence h_{diff} is replaced by the total side water depth (h_r). The saturation depth fraction $f_{h,sat,eff}$ may be assigned according to various approaches; the mid-depth concept provides reasonable accuracy in diffused aeration [35].

$$h_{sat,eff} = h_{diff} \cdot f_{h,sat,eff}, \tag{14}$$

For the diffuser depth calculation, the diffuser mounting height is subtracted from the liquid height, according to Equation (15).

$$h_{diff} = h_r - h_{diff,floor} \tag{15}$$

Unlike liquid phase hydraulic balancing, where the effluent flow is assumed to be equal to the influent, gas phase volumetric balancing needs to take into account the input air flow, as well as the gas transfer activity in the volumetric unit. Equation (16) determines the outgoing gas volumetric flow, $Q_{gas,outp,NTP}$, which, physically, cannot be negative; this shall be reflected in the model code (e.g., using a maximum function).

$$Q_{gas,outp,NTP} = Q_{air,NTP} + Q_{gas,transfer,NTP}, \tag{16}$$

The volumetric flow of the gas transfer $Q_{gas,transfer,NTP}$ is derived from the summation of the individual gas component transfer molar flows, calculated by Equation (17).

$$Q_{gas,transfer,NTP} = \sum_i \frac{rateF_{G_i}}{MM_{EQ,i}} \cdot \frac{R \cdot T_{NTP,K}}{p_{NTP}}, \tag{17}$$

Two gas transfer routes are modelled: between the atmosphere and water surface and between the bulk-of-the liquid and bubbles. According to Fick's first law, the process rate involving the transfer of gas i between the atmosphere and water surface is described by Equation (18), with the driving force of either dissolution or stripping is determined by the difference in the saturation concentration and the liquid phase concentration. The transfer rate of exchange at the gas bubble interface is fundamentally analogous, described by Equation (19). These equations are listed for processes involving the individual gas phase BTEX states in the Gujer matrix, located within Appendix A.

$$r_j = k_L a_{i,bub} \cdot (S_{i,bub,sat} - L_i), \tag{18}$$

$$r_j = k_L a_{i,sur} \cdot (S_{i,sur,sat} - L_i), \tag{19}$$

Regarding gas transfer through bubbles, the volumetric liquid-side mass transfer coefficients $k_L a_{i,bub,st,cw}$—in clean water for a standard 20 °C temperature—and $k_L a_{i,bub}$—in wastewater at the field temperature—are calculated by Equations (20) and (21).

$$k_L a_{i,bub,st,cw} = k_{L,i,bub,st,cw} \cdot a_{bub}, \tag{20}$$

$$k_L a_{i,bub} = k_L a_{i,bub,st,cw} \cdot \alpha \cdot \theta^{(T-20)}, \tag{21}$$

The interfacial transfer area a_{bub} is calculated by Equation (22) on a geometrical basis. Based on the literature review, in this study, the bubble Sauter mean diameter d_{bub} is input as 0.003 m for aerated conditions and as 0.01 m under non-aerated conditions [32,36,37].

$$a_{bub} = \frac{6 \cdot \frac{V_{gas}}{V_r}}{d_{bub}}, \tag{22}$$

It shall be noted that in cases of fully covered tanks, atmospheric gas transfer is not modelled; the specific surface area a_{bub} is extended with the specific contact area between the water surface and the headspace, as shown by Equation (23).

$$a_{bub} = \frac{6 \cdot \frac{V_{gas}}{V_r}}{d_{bub}} + \frac{A_r \cdot f_{wave}}{V_r} \tag{23}$$

On the basis of the Higbie penetration theory, values of the liquid-side mass transfer coefficient $k_{L,i,bub,st,cw}$ are calculated using the respective diffusivities of volatile and soluble gas components by Equation (24) [38], relying on $k_{L,O2,bub,st,cw}$ as a model parameter with the input value of 0.54 m h^{-1} [39]. BTEX-specific diffusion coefficients were averaged from values found in the scientific literature [40–42] and are disclosed in Appendix C (Table A12) The parameter for the fraction in the liquid side $f_{kL,i}$ equals 1 for all gases in this study, as they diffuse slowly through the liquid film, with the exception of 0.05 in the case of ammonia, due to its very high solubility [43].

$$k_{L,i,bub,st,cw} = f_{k_{L,i}} \cdot \sqrt{\frac{D_{i,25}}{D_{O_2,25}}} \cdot k_{L,O_2,bub,st,cw} \tag{24}$$

Throughout aerated compartments, the α correction factor is modelled dynamically from process variables representative of wastewater characteristics and loads, relying on a method published for increased accuracy in oxygen transfer and air flow requirement predictions [44]. The θ compensation factor for simulating Arrhenius-type temperature sensitivity in mass transfer is taken as 1.024, following standard procedures in wastewater treatment design [33].

For aerated conditions, regarding oxygen, $k_La_{O2,bub}$ is calculated by Equation (25), incorporating a further correction factor for diffuser fouling, an input parameter of aerated unit processes [45], which, in practice, is best adjusted knowing the diffuser age and the time of the last cleaning procedure [46]. For non-aerated conditions (and regarding coarse bubbles), the correction factor for fouling is not interpreted (equal to one).

$$k_La_{O_2,bub} = k_La_{O_2,bub,st,cw} \cdot \alpha \cdot F \cdot \theta^{(T-20)} \tag{25}$$

Regarding oxygen, under aerated conditions, $k_La_{O2,bub,st,cw}$ directly relates to the clean water performance of diffusers; thus, it is derived—according to Equation (26)—from the standard oxygen transfer rate involving air bubbles, $SOTR_{bub}$.

$$k_La_{O2,bub,st,cw} = \frac{SOTR_{bub}}{S_{O_2,bub,sat,st,cw} \cdot V_r}, \tag{26}$$

$SOTR_{bub}$ is derived from the specific standard oxygen transfer efficiency SSOTE, which describes the diffused aerator characteristics in clean water, as Equation (27) shows.

$$SOTR_{bub} = SSOTE/100 \cdot h_{diff} \cdot F_{GO_2,air,inp}, \tag{27}$$

The prediction of SSOTE is implemented according to a novel concept proposed in the form of Equation (28), as a function of the diffuser-specific air flux $Q_{air,NTP,sp}$ [47].

$$SSOTE = \frac{\left(\left(SSOTE_0 - SSOTE_{asym}\right) \cdot e^{-exp_{SSOTE} \cdot Q_{air,NTP,sp}} + SSOTE_{asym}\right) \cdot corr_{d,diff}}{corr_{h,diff}}, \tag{28}$$

This proposed concept simulates the phenomena concerning fine bubble aerators—where SSOTE is known to drop significantly with an increased air flow rate [48]—and regarding coarse bubble diffusers—which show an increasing trend of SSOTE with an increasing air flow [49]. The specific air flow per diffuser, $Q_{air,NTP,sp}$ is calculated by Equation (29).

$$Q_{air,NTP,sp} = \frac{Q_{air,NTP}}{n_{diff}}, \tag{29}$$

SSOTE is corrected to account for the diffuser submergence and density. In the case of coarse bubble aeration, the correction for the diffuser density has no practical meaning, and thus, it is neglected; however, in fine bubble aeration, increasing the diffuser floor coverage will raise the SSOTE [50]. The correction for diffuser depth is implemented to reflect the phenomenon that the absolute transfer efficiency (SOTE) does not increase linearly with depth, especially in tanks deeper than 8 m [51].

The correction term for diffuser density, $corr_{d,diff}$, is calculated by Equation (30).

$$corr_{d,diff} = \left(\frac{d_{diff}}{div_{d,diff}}\right)^{pow_{d,diff}}, \tag{30}$$

The correction term for diffuser submergence, $corr_{h,diff}$, is calculated by Equation (31).

$$corr_{h,diff} = coeff_{lead,h,diff} \cdot h_{diff}{}^{pow_{h,diff}} + coeff_{lin,h,diff} \cdot h_{diff} + 1, \tag{31}$$

The diffuser floor density d_{diff} is the ratio of the total diffuser area to the tank surface, as Equation (32) shows.

$$d_{diff} = \frac{n_{diff} \cdot A_{diff,sp}}{A_r}, \tag{32}$$

The method for clean water aeration efficiency estimation relies on an exponential function with asymptotes, instead of setting strict Boolean-type threshold levels for SSOTE, so that the air flux may safely be extrapolated outside of the model calibration range. This avoids potential numerical issues when utilized in a dynamic simulator, with the further advantage of not carrying over conceptual errors in SOTE, when combining it with factors describing the process water performance. The model parameters were fine-tuned based on data collected from numerous diffuser tests for various equipment types [33,52]. Appendix C contains the parameter set used for this paper, characterizing ceramic discs.

At the atmospheric interface, calculating the volumetric liquid-side mass transfer coefficients involves the specific mass transfer coefficients (k_L) estimated for the liquid surface. The clean water $k_La_{i,sur,st,cw}$ at standard conditions and the process water $k_La_{i,sur}$ for field conditions are calculated by Equations (33) and (34), respectively.

$$k_La_{i,sur,st,cw} = k_{L,i,sur,st,cw} \cdot a_{sur}, \tag{33}$$

$$k_La_{i,sur} = k_La_{i,sur,st,cw} \cdot \alpha \cdot \theta^{(T-20)}, \tag{34}$$

The liquid-side mass transfer coefficient regarding the water surface $k_{L,i,sur}$ is determined by Equation (35) using the same principle as for the gas bubble interface, as previously shown by Equation (24).

$$k_{L,i,sur} = f_{k_{L,i}} \cdot \sqrt{\frac{D_{i,25}}{D_{O_2,25}}} \cdot k_{L,O_2,sur}, \tag{35}$$

The specific surface area of the liquid a_{sur} is calculated by Equation (36), incorporating the multiplication factors f_{cover} for reactor coverage and f_{wave} for turbulence (waviness). The model inputs of 0.54 m h^{-1} for $k_{L,O2,sur}$ and 1.9 for f_{wave} in this study are adjusted based on the typical measured values of $k_La_{O2,sur}$ found in the relevant literature [53].

$$a_{sur} = \frac{A_r \cdot (1 - f_{cover}) \cdot f_{wave}}{V_r}, \tag{36}$$

The water surface area A_r is defined by basin geometry, according to Equation (37).

$$A_r = V_r / h_r, \tag{37}$$

According to Fick's law, as illustrated previously by Equations (18) and (19), the saturation concentration of a component determines whether the gas–liquid transfer process is driven towards dissolution or stripping. They are calculated by Henry's law based on the partial pressure of a component, with temperature dependency correction implemented using the van't Hoff equation [54]. The saturation concentrations of gases attributed to the liquid–gas bubble interface $S_{i,bub,sat,st,cw}$, standardized at 20 °C for clean water (thus requiring temperature conversion from 25 °C, used as the basis of Henry's law), are modelled based on Equation (38). The variable for process water at field conditions, $S_{i,bub,sat,st}$, is expressed by Equation (39). The values of the Henry's law model parameters regarding BTEX contaminants were averaged in the process of the literature review [55–57] and are listed in Appendix C (Table A12).

$$S_{i,bub,sat,st,cw} = p_{partial,i,bub,st} \cdot Henry_{i,25} \cdot e^{Henry_{i,dt} \cdot ((1/T_{NTP,K}) - (1/T_{SATP,K}))} \cdot MM_{EQ,i}, \tag{38}$$

$$S_{i,bub,sat} = \beta \cdot p_{partial,i,bub} \cdot Henry_{i,25} \cdot e^{Henry_{i,dt} \cdot ((1/T_K) - (1/T_{SATP,K}))} \cdot MM_{EQ,i}, \tag{39}$$

The β correction factor for impurities in process water uses an input parameter value of 0.95, which is typical of municipal wastewater [33].

With regard to the bubble–water interface, partial pressures for standard conditions, $p_{partial,I,bub,st}$, and for process conditions, $p_{partial,i,bub}$, are calculated by Equations (40) and (41), respectively.

$$p_{partial,i,bub,st} = \frac{G_i / MM_{EQ,i}}{n_{gas,bub}} \cdot p_{st,h,sat,eff}, \tag{40}$$

$$p_{partial,i,bub} = \frac{G_i / MM_{EQ,i}}{n_{gas,bub}} \cdot p_{gas} \tag{41}$$

Equation (42) calculates the molar quantity of gas bubbles per unit liquid volume, $n_{gas,bub}$.

$$n_{gas,bub} = \frac{V_{gas,NTP} \cdot p_{NTP}}{V_r \cdot R \cdot T_{NTP,K}}, \tag{42}$$

As explained by Equation (43), the pressure at standard conditions, $p_{st,h,sat,eff}$, is also compensated for the effective saturation depth, as diffuser testing in clean water involves the design submergence [58].

$$p_{st,h,sat,eff} = \left(p_{NTP} + h_{sat,eff} \cdot \rho_{H_2O} \cdot g - p_{v,T}\right) \cdot \frac{p_{NTP}}{p_{NTP} - p_{v,T}} \tag{43}$$

The off-gas composition in molar or volume percent units (interchangeable according to Avogadro's law) may be calculated using Equation (44), expressing $G_{i,percent}$, derived from the individual gas component molar concentrations and their sum.

$$G_{i,percent} = \frac{G_i / MM_{EQ,i}}{\sum_i G_i / MM_{EQ,i}} \cdot 100, \tag{44}$$

The saturation concentrations for the atmosphere–liquid interface, $S_{i,sur,sat,st,cw}$ in the case of clean water and $S_{i,sur,sat}$ regarding process water, are calculated by Equations (45) and (46).

$$S_{i,sur,sat,st,cw} = p_{partial,i,sur,st} \cdot Henry_{i,25} \cdot e^{Henry_{i,dt} \cdot ((1/T_{NTP,K}) - (1/T_{SATP,K}))} \cdot MM_{EQ,i}, \tag{45}$$

$$S_{i,sur,sat} = \beta \cdot p_{partial,i,sur} \cdot Henry_{i,25} \cdot e^{Henry_{i,dt} \cdot ((1/T_K) - (1/T_{SATP,K}))} \cdot MM_{EQ,i}, \tag{46}$$

For the atmospheric saturation concentration calculations, the composition of the atmosphere around the water surface is required. The compositional figures $G_{i,atm}$ are defined as model constants, measured in volumetric percentages. Equations (47) and (48) quantify the partial pressure of gases in the atmosphere, $p_{partial,i,sur,st}$ for standardized conditions and $p_{partial,i,sur}$ for field conditions.

$$p_{partial,i,sur,st} = G_{i,atm}/100 \cdot p_{NTP}, \tag{47}$$

$$p_{partial,i,sur} = G_{i,atm}/100 \cdot p_{air}, \tag{48}$$

To sum up, the gas phase composition and the amount of gas dissolving or stripping are dependent on individual gas state variables that impact one another, according to the laws of mass balance. In accordance, the sufficient injection of air or nitrogen gas production will result in BTEX components stripping out of the liquid.

2.1.3. Adsorption on Granular Activated Carbon

Describing the simulated mechanism of BTEX contaminants binding to activated carbon granules is essential. The developed methodology in this study features a granular activated carbon sub-model with an engineer-oriented concept; the GAC media consumption is estimated according to the bed volume and TOC adsorption capacity specified.

With the goal of shortcut steady-state runs, the GAC unit operation process model may be set up using a continuous process flow approach, purely based on algebraic equations, and focusing on the plain instantaneous mass-balances of states, with the effluent quality estimated for operational phases with loading. For more robust dynamic modelling aimed at evaluating maintenance or control strategies, a cyclical process flow model setup is implemented. The latter features asymmetric breakthrough curves for quantifying the effluent quality based on the accumulated pollutant load and requires an ODE solver for the integrated mass balance calculation.

Water treatment using GAC adsorption is applicable to dissolved organic components. The adsorbable states in the extended MiniSumo biokinetic model presented in this chapter include VFA, the liquid phase BTEX variables, soluble unbiodegradable and readily biodegradable substrates as well as soluble biodegradable organic N. The unit process implementation does not include a reactive volume, with the assumption being that no biodegradation of adsorbed COD occurs. This approach was selected as appropriately sufficient for the main target of this study, which applies GAC as a tertiary treatment step, after readily biodegradable and slowly biodegradable substrates have mostly been removed biologically and mechanically, respectively.

It is assumed that, per unit mass of AC, a specified amount of soluble organic pollutants—expressed in unit mass of C—can be adsorbed, until the medium reaches its breakthrough capacity. Thus, in order to quantify the number of adsorbed components, their respective mass concentrations have to be converted to carbon equivalent units. This involves stoichiometric conversions from unit g_{COD} into unit g_C regarding the carbonaceous state variables and from unit g_N into g_C in case of organic nitrogen. The COD of organic substrates are interpreted using the concept of theoretical oxygen demand. Appendix D (Table A13) lists the ratios of COD (and N) to TOC. The continuous process flow model instantaneously calculates the carbon equivalent for all adsorbed components on the GAC bed volume based on the influent concentrations, using Equation (49). The removal efficiency $Rem_{GAC,i}$ was assigned to be 99% regarding BTEX components, 92% for a soluble unbiodegradable substrate and 90% in the case of VFA, as well as soluble biodegradable carbon and nitrogen [59,60].

$$EQ_{C,ad,total} = \sum_{i} L_{i,in} \cdot Rem_{GAC,i}/i_{C,i}, \tag{49}$$

As shown by Equation (50), from the summed carbon equivalent variable $EQ_{C,ad,total}$ and the influent flow, the algebraic model can calculate the average replacement cycle frequency, knowing the set bed volume V_{ac} and the TOC breakthrough capacity BTC.

$$N_{repl} = \frac{Q_{in} \cdot EQ_{C,ad,total}}{BTC \cdot V_{ac}}, \tag{50}$$

Using Equation (51), BTC is converted from the breakpoint-related TOC adsorption capacity BTC_m in the mass fraction unit and the apparent density of the media ρ_{ac}, both of which are design settings of the GAC model.

$$BTC = BTC_m \cdot \rho_{ac}, \tag{51}$$

The required mass flow of the replaced GAC media can be quantified by Equation (52), according to the mass of activated carbon filled per cycle $M_{ac,cycle}$.

$$F_{ac} = N_{repl} \cdot M_{ac,cycle}, \tag{52}$$

The mass of GAC filled into the towers within each cycle corresponds to the bed volume, as Equation (53) explains.

$$M_{ac,cycle} = V_{ac} \cdot \rho_{ac}, \tag{53}$$

The replacement frequency can also be fixed based on the lifespan of the AC media. The cyclical process flow model schedules replacement events either based on the lifespan or relying on breakthrough-based or volumetric load-based maintenance logic. In the case of the breakthrough-based approach, either the amount of the total adsorbed carbon per bed volume $EQ_{C,ad,total}$ is monitored and compared to the BTC or the effluent TOC is monitored and compared to a TOC concentration threshold to trigger the bed replacement procedure. Regarding the volumetric load-based logic, the influent loading expressed in the number of bed volumes is registered and initiates the replacement process when it reaches the target volume ratio of the load per bed.

In the dynamic GAC model, $EQ_{C,ad,total}$ is expressed directly as a sum of the accumulated C equivalent mass of pollutants per bed volume, following Equation (54).

$$EQ_{C,ad,total} = \sum_{i} L_{i,ad} / i_{C,i}, \tag{54}$$

$L_{i,ad}$,, the adsorbed mass per bed volume of state variable i, is determined by differential equations that have to be solved by a numerical integrator. The derivative of $L_{i,ad}$ is expressed by Equation (55), describing GAC operational loading phases. The authors note that the same equation can also be used to account for potential impurities from backwash streams.

$$\frac{dL_{i,ad}}{dt} = \frac{F_{L_{i,in}} \cdot Rem_{GAC,i,actual}}{V_{ac}}, \tag{55}$$

During bed replacement phases, the adsorbed state variable masses per bed volume are nullified, so the change in $L_{i,ad}$ is described simply using a zero-order rate expression, as seen in Equation (56). The accumulated mass per bed volume prior to the replacement $L_{i,ad,pre}$ shall be registered at the start of the replacement event, which is linearly decreased towards 0 throughout the duration of the bed replacement t_{repl}.

$$\frac{dL_{i,ad}}{dt} = \frac{-L_{i,ad,pre}}{t_{repl}}, \tag{56}$$

The actual removal efficiency of organic constituents, $Rem_{GAC,i,actual}$, depends on how saturated the granular medium is; based on $EQ_{C,ad,total}$, the accumulated adsorbed components, a logistic saturation function expresses the drop compared to the defined initial removal ratio $Rem_{GAC,i}$, as Equation (57) shows.

$$Rem_{GAC,i,actual} = Rem_{GAC,i} \cdot \left(1 - \frac{1}{1 + e^{(C_{mid} - EQ_{C,ad,total}) \cdot sl_{break}}}\right), \quad (57)$$

Applying the logistic formula would, in itself, generate a symmetric S-shaped curve; therefore, as a function of $EQ_{C,ad,total}$, the midpoint concentration C_{mid} is corrected for the asymmetry involving breakthrough curves, as pointed out by Equation (58).

$$C_{mid} = C_{mid,symm} \cdot \left(-magn_{mid,asymm} \cdot \frac{\frac{EQ_{C,ad,total}}{BTC}^{pow_{mid,asymm}}}{\frac{EQ_{C,ad,total}}{BTC}^{pow_{mid,asymm}} + 1} + \frac{magn_{mid,asymm}}{2} + 1\right), \quad (58)$$

$C_{mid,symm}$, the midpoint concentration of the breakthrough curve without asymmetry correction, is derived from the BTC, the breakpoint fraction f_{break} and the breakthrough curve slope sl_{break}, using Equation (59). The drop in removal efficiency at the breakthrough point adsorption capacity is defined by f_{break}. The slope and the two model parameters of the asymmetry correction term $pow_{mid,asymm}$ and $magn_{mid,asymm}$ were adjusted based on measured data involving breakthrough curves found in the relevant literature, and they can be re-estimated depending on the type of GAC media applied for the treatment [61,62]. The values of the parameters are disclosed in Appendix D (Table A13).

$$C_{mid,symm} = \ln \frac{1/f_{break} - 1}{sl_{break}} + BTC, \quad (59)$$

The actual removal efficiency calculated using the asymmetric sigmoid expression also serves as the basis of estimating effluent state variable concentrations, as shown by Equation (60). The model with continuous process flow implementation does not consider the saturated status of the bed, so it uses the ideal removal efficiency ratio input instead. It is important to note that this equation only applies to soluble organic states; other dissolved compounds (such as ammonia and orthophosphate) do not have removal efficiencies specified and will pass through the GAC unit without treatment.

$$L_{i,out} = \frac{F_{L_{i,in}} \cdot (1 - Rem_{GAC,i,actual})}{Q_{out}}, \quad (60)$$

Although in practice, TSS loading to the GAC process should be minimal, activated carbon beds can also act as filtration media. The developed model includes an instantaneous mass balance separator approach for the mechanical separation of particulate state variables. This was implemented following the principle applied for the sand filter unit process in the simulation software used for demonstrations in this study, Sumo22 [30]. Regarding the cyclical process flow model setup, the deposition of suspended matter on the bed volume is considered in the mass balance, with derivatives calculated analogously to Equation (55). Backwash events can also be scheduled by specifying the loading phase duration. Similar to the concept of wastage during replacement events quantified by Equation (56), the decrease in the captured particulate mass per volume due to backwashing is determined using simplified zero-order expressions, relating to the backwash duration.

2.2. Model Configuration

For the demonstration of BTEX modelling applications, a variant of the BSM1 test configuration [63] was set up in the Sumo22 simulation software and used to evaluate potential operational cost savings through controlling the sludge retention time or by air stripping and through comparing strategies of GAC tower maintenance. The chosen technological sequence is a simulation benchmark that defines a relatively simple standard water resource recovery facility with influent loads provided, aimed at evaluating system performance based on test procedures, according to a given set of evaluation criteria. The BSM1 plant comprises a five-compartment activated sludge reactor cascade, including two anoxic zones and three aerobic zones. This setup combines nitrification with pre-denitrification, a widely adopted arrangement for achieving biological nitrogen removal. Downstream of the reactor cascade, the configuration features a secondary clarifier for phase separation [63]. Table 1 summarizes the main characteristics of the benchmarking facility, regarding average dry weather conditions.

Table 1. Main input parameters of the BSM1 test configuration.

Parameter	Value	Unit
	Influent properties	
Flow	18,446	$m^3\ d^{-1}$
COD	360	$g_{COD}\ m^{-3}$
Filtered COD	144	$g_{COD}\ m^{-3}$
TOC	114	$g_C\ m^{-3}$
TKN	47	$g_N\ m^{-3}$
NH_4-N	30	$g_N\ m^{-3}$
	Tank dimensions	
Anoxic 1 zone volume	1000	m^3
Anoxic 2 zone volume	1000	m^3
Aerobic 1 zone volume	1333	m^3
Aerobic 2 zone volume	1333	m^3
Aerobic 3 zone volume	1333	m^3
Clarifier surface area	1500	m^2
Clarifier depth	4	m

In addition to the BSM1 layout, Sumo22's empirical blower and pump models were set up to assess energy and associated costs regarding aeration, return activated sludge pumping, nitrified mixed liquor recycle pumping as well as backwash pumping. Solving the rate expression and component balance equations of variables in wastewater treatment technologies requires a simulation environment based on biokinetics. From an operational strategic point of view, process modelling can reveal cost-down opportunities by experimenting with control strategies and pointing out necessary interventions for complying with environmental regulations. Sumo22 is a multipurpose, open-source environment that has been developed specifically for dynamic environmental modelling, with particular emphasis on municipal and industrial wastewater treatment plants. Sumo22 can simulate a wide variety of BNR (biological nutrient removal) plant configurations [30].

According to the model layout prepared for the study, portrayed by Figure 1, as a fictive retrofit of the municipal plant with effluent purification, the scope of the standard BSM1 configuration was extended with a granular activated carbon process and a microfiltration step prior to GAC for the sake of mechanical protection against the remaining solids overflowing from the secondary clarifiers. In the granular activated carbon unit process model, the replacement frequency of the exhausted bed is determined based on the TOC adsorption capacity of the media and the load of soluble organics removed. It can be run in either a quasi steady-state mode, where the AC demand is quantified from an instantaneous carbon mass balance, or in dynamic mode, where it is calculated cyclically, based on the effluent quality predicted using an empirical asymmetric breakthrough curve. The total bed volume of the towers was sized as 50 m^3; with the effluent quality and quantity associated with the BSM1 use case, this ought to require bed replacement roughly every month, assuming that the bed density is 450 kg m^{-3} and that the soluble organic carbon adsorption capacity of the bed is 0.2 $g_C\ g_{AC}^{-1}$, estimated based on operational expertise. The MF and GAC units are backwashed in half-hour and daily cycles, respectively, for industrial safety reasons against solids depositing on the filter and the granular media. We note here that bed replacement can be calculated automatically in Sumo22 from the mass flow of the BTEX compounds removed, simulating how—in a full-scale treatment plant—the operators would remove exhausted granules and reinstall fresh granules periodically, whenever the breakpoint of the media is reached. After sedimentation, an additional re-aeration zone with the hydraulic retention time of 0.3 h—with coarse bubble aeration—is used to maintain sufficient dissolved oxygen levels required for the effluent quality. The influent quality regarding BTEX concentrations was characterized according to typical levels in municipal wastewater, containing 303 μg L^{-1} of benzene, 290 μg L^{-1} of toluene, 249 μg L^{-1} of ethylbenzene and 933 μg L^{-1} of xylene [64].

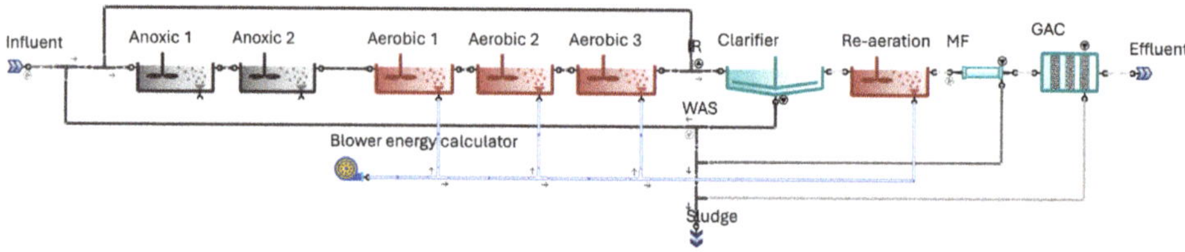

Figure 1. Modified BSM1 test configuration for modelling examples of BTEX removal. Arrows represent the direction of the fluid flow.

Using the scenario analysis functionality of Sumo22, the targeted sludge retention time and the air flow to the re-aeration tanks were selected as the input parameters to be changed in between the simulation runs in order to efficiently evaluate the changes in the blower energy demand, the yearly requirement of GAC and the amount of BTEX chemicals removed. SRT control is handled according to the BSM1 scenario guidelines; Sumo22 automatically adjusts the wasted sludge flow rate, varying the activated sludge solids concentration to meet the SRT target.

3. Results and Discussion

The simulation and results interpretation involved preparing the model output setup for exporting variables linked to pollutant loads, composition profiles, effluent quality, biodegradation, stripping and adsorption rates as well as calculations on resources, regarding aeration requirement and activated carbon consumption. Sumo22's combined global and local solvers were used to handle the integrated calculations necessary for plant-wide modelling. Based on a steady-state simulation featuring a BSM1 baseline scenario, Figure 2 displays a profile of the summed BTEX component uptake and stripping rates throughout the biological treatment, expressed in the mass flow of COD per reactor cell volume.

The BTEX biodegradation and stripping rate profiles represent the plug-flow characteristics of the treatment plant; degradation is most significant at the first anoxic cell because, here, the BTEX compounds serve as substrates for the biomass in the highest concentration. There is a decline in biological uptake in the second anoxic zone as substrate availability becomes more limited, but as the wastewater enters the first aerated compartment, the degradation rate is boosted due to the faster aerobic reactions. The elimination of BTEX compounds through gas exchange is mainly visible at aerobic zones due to the contribution of air stripping driving out greater mass flows of volatile contaminants; stripping from anoxic compartments is mostly limited to the interface of the liquid surface and the atmosphere. Throughout the aerated cells, the profiles of removal rates attributed to both biodegradation and stripping correspond well with the decreasing concentration gradient of BTEX pollutants during activated sludge treatment.

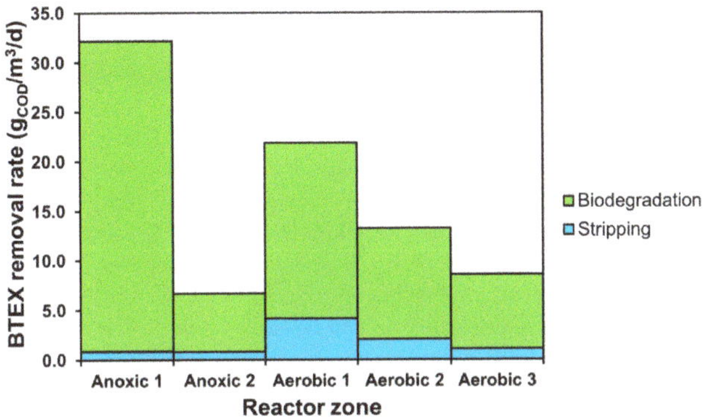

Figure 2. Combined BTEX component removal rate profile.

In this section on demonstrated studies, the analysis is primarily centered around examining the model concept in detail based on SRT, followed by an evaluation of the effect of air stripping and concluding with a presentation of GAC operational strategy modelling.

3.1. SRT-Based Scenarios

In the first modelling experiment, the test configuration ran at a steady state in dry weather conditions; using Sumo22's combined global and local solvers, a series of runs were performed with varied SRT to obtain the blower power consumption and the required mass flow of the GAC bed to be replaced. From these variables and the raw influent characteristics regarding BSM1 specifications, the blower energy and activated carbon demand specific to the influent BOD were plotted, as shown in Figure 3, to provide more generic results irrespective of the benchmarking plant capacity.

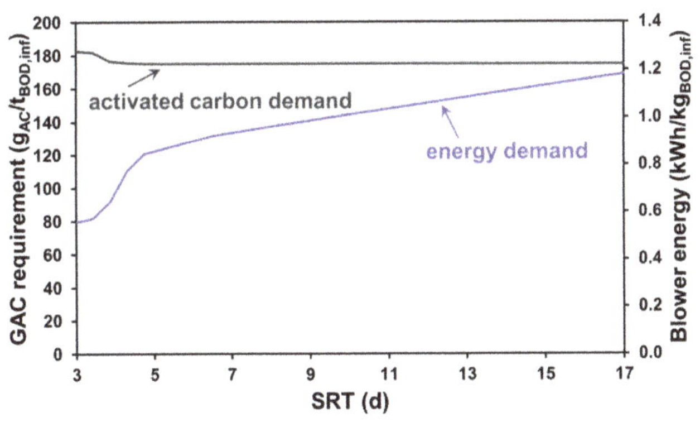

Figure 3. Specific AC demand and blower energy demand for SRT-based scenarios.

The results show that the reduction in demand for the GAC replacement is most significant within the range of 3 to 6 days of SRT. However, this range would not be typical for a water reclamation plant, as even at warm/moderate water temperatures, longer retention times are needed for proper nutrient removal. It typically takes between 10 and 15 days for a thorough nitrification–denitrification process. If the SRT falls below this range, the process remains incomplete. Conversely, if the SRT exceeds this range, the plant's energy efficiency is compromised [65]. After 6 days, the further decrease in the activated carbon demand is not significant, given that, proportionally, a much higher performance is demanded from the blowers to meet the dissolved oxygen concentration target at higher biomass concentrations. The reason behind this is that most of the organic load treated by the towers is of the soluble unbiodegradable organic fraction, the decomposition of which cannot be boosted by raising the sludge residence time.

Apart from the operating cost, another important aspect of micropollutant removal is the effluent quality. The BTEX concentrations in the treated water were also plotted against the same range of SRT, as displayed in Figure 4. The difference in BTEX removal is most significant between the SRT range of 3 to 6 days, as a greater load of contaminants, which were not exposed to the biomass for a sufficient amount of time to be treated, bleed through the GAC bed. Consequently, it is important to note that the activated sludge system must always be operated responsibly, not allowing the SRT to drop below the necessary level for nutrient removal, as the activated carbon treatment step might not adsorb the surplus in the load sufficiently to meet regulations, depending on the use case (e.g., reuse for agriculture, drinking water supply). Benstoem et al. [61] noted a decrease in micropollutant removal efficiency due to the improper functioning of a secondary clarifier, resulting in an increase in the average TSS concentration from 6 mg/L to a peak of 30 mg/L [66].

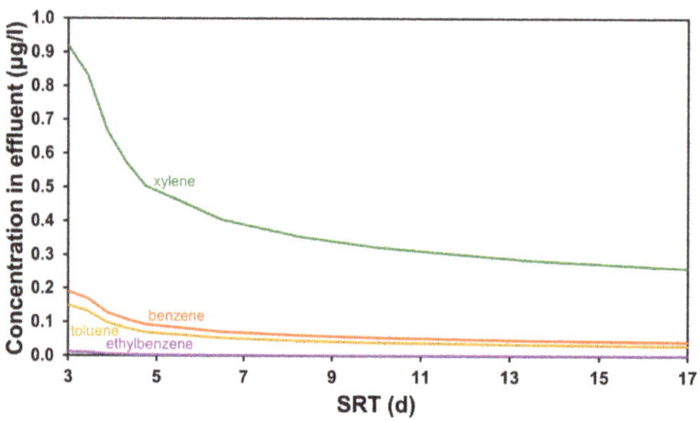

Figure 4. Effluent quality concerning BTEX compounds for SRT-based scenarios.

For interpreting the results of the modelling scenarios in more detail, the contributors to the BTEX metabolism throughout the treatment plant are shown in Figure 5, based on comparing the total mass flow of the compounds biodegraded, stripped and adsorbed to the influent BTEX load. The mass flows are interpreted in equivalent units of chemical oxygen demand. When discussing the COD percent removal of BTEX pollutants, according to the modelling study, biodegradation is predicted to be the most significant process—by raising the plant-wide SRT, a steep increase can be achieved in the extent of anoxic and anaerobic biodegradation. The increased biodegradation is in accordance with decreased stripping and the reduction in the adsorbed mass flows on the activated carbon beds since the remaining micropollutant load, to be removed by gas transfer or effluent polishing, drops.

However, as deduced from the results in Figure 4, most of the load fed to GAC treatment is in the form of soluble unbiodegradable compounds; therefore, there is only limited cost down capability in minimizing the bed replacement frequency through intensifying the activated sludge process. It can be concluded that, the more complete the nutrient removal is, the better the effluent quality is, including lower BTEX concentrations. However, the increased demand for aeration is more significant than the amount of activated carbon saved; thus, an actual SRT target is practically determined by how stringent the water quality requirements are. However, ultra-short SRT processes (<4 days) aiming for carbon removal from wastewater demonstrate high efficiency in organic matter separation and in enhancing energy recovery rates in wastewater treatment plants [67].

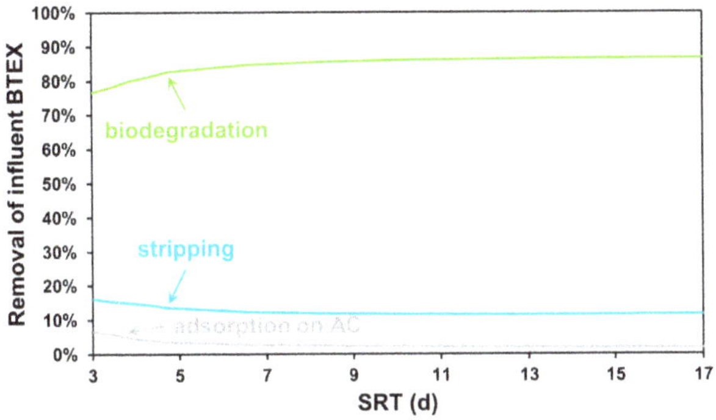

Figure 5. Contribution of BTEX removal processes for SRT-based scenarios.

3.2. The Effect of Aeration Intensity

In another demonstrated case study of the developed numerical methods, the scenario analysis was conducted with varied aeration intensities of the facility's post-aeration tanks for the different steady-state runs. The decision to modify air flow to this re-aeration process—rather than changing DO setpoints in the activated sludge trains—came down to the fact that oxygen transfer will be more effective in secondary clarifier effluent water than in mixed liquor due to the absence of biomass oxygen uptake and higher α factors. The volume to be aerated is also much smaller this way, with sufficient HRT for the reaction rate of stripping. Figure 6 presents these scenarios' results on BOD-specific aeration energy and activated carbon demand as a function of superficial gas velocity—the air flow expressed per re-aeration tank surface area.

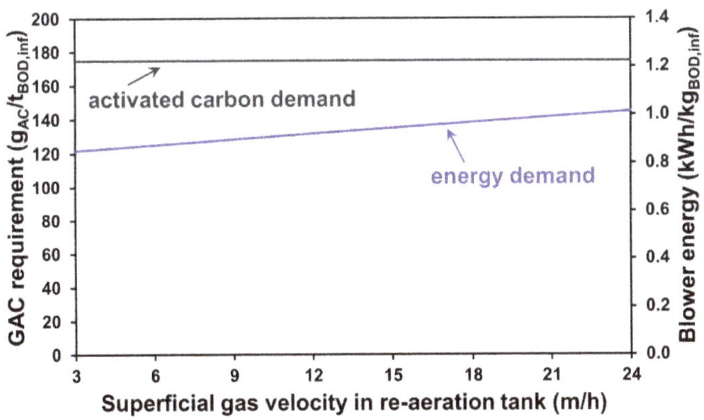

Figure 6. Specific AC demand and blower energy demand for air flux-based scenarios.

By raising the air supply to the post-aeration tanks, it was revealed that despite the less significant energy requirement increase compared to the trend seen through raising the SRT, it only provides negligible cost-saving opportunities in activated carbon usage.

The overall percent removal of BTEX chemicals is already quite high (99%), even with the lowest superficial gas velocity, and given that the majority of the organic load to be polished by GAC is not volatile and cannot be removed through air stripping, there is not much room for improvement by removing some of the remaining BTEX components. It must be noted that the water samples investigated in this paper are in the context of municipal wastewater treatment plants, with the inlet concentration of BTEX compounds being marginal compared to that of other pollutants that the technology focuses on removing. The proposal of the enhanced stripping of volatile pollutants from much more concentrated industrial wastewater, especially in the petroleum industry [68], to reduce the frequency of GAC bed replacement could make much more sense.

The impact of stripping enhancement on effluent BTEX concentrations is presented in Figure 7. It may be a viable method for stabilizing the effluent water quality regarding micropollutants through a relatively fast intervention. Regarding environmental security, emergencies in wastewater plants—such as operational difficulties with biological treatment—may cause a sudden load increase in BTEX contaminants fed to GAC columns, in which case online monitoring could alert operators to make a swift change in the air supply to the tanks prior to the adsorption process, preventing a violation of the tertiary treatment effluent water quality limits.

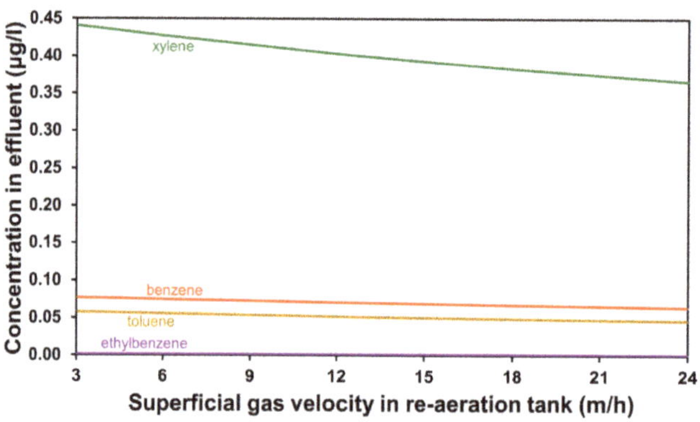

Figure 7. Effluent quality concerning BTEX compounds for air flux-based scenarios.

Figure 8 shows the removal percentage of the influent COD-based BTEX load through individual treatment mechanisms, as a function of the varied air flux. There is effectively no change in the contribution by biodegradation, since the re-aeration zone is located after the bioreactors, following the separation of biomass from the liquid phase by secondary clarification. Enhanced air stripping promotes the removal of the remaining BTEX constituents following biological treatment, whose concentrations are much smaller than those of the other remaining organic contaminants, therefore not being able to provide further minimization of the activated carbon requirement but potentially being useful for stabilizing water quality as a safe alternative to comply with effluent regulations. The fractional contribution of the adsorption process is inversely proportional to this trend. Figures 5 and 8 emphasize that the developed plant-wide modelling tools also provide insight into variables that cannot be directly measured, which, in conclusion, can give engineers and operators a better process understanding. This is a useful feature of modelling tools that are calibrated and virtually linked to actual facilities in real time (e.g., through digital twin implementations) [69].

The authors stress that this paper focuses on the general implementation of a methodology for plant-wide process simulation involving the fate of BTEX pollutants in biological wastewater treatment, along with options of physico-chemical effluent polishing, and the experimental cases presented are for demonstrating the feasibility of the model applications in municipal wastewater treatment. Therefore, these results shall not be generalized to other sources of municipal wastewater and certainly not to industrial treatment. The verification of robustness and accuracy shall be performed as a future effort in modelling studies, not on a benchmarking tool but against measured data from existing facilities.

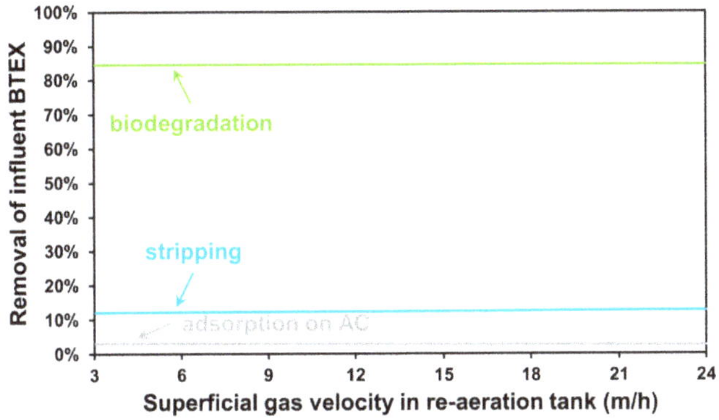

Figure 8. Contribution of BTEX removal processes for air flux-based scenarios.

3.3. GAC Operational Strategies

A third modelling case featured dynamic runs to compare two different GAC operational strategies, as illustrated by Figure 9. Scheduling bed replacement on a fixed time interval basis may be beneficial for sustaining stable effluent quality; however, this is not a cost-effective option, as some useful carbon adsorption capacity is lost. The use of effluent monitoring (e.g., applying online TOC probes) to synchronize bed replacement with the breakthrough capacity of the bed, i.e., when a certain effluent concentration limit is reached, may enable a more than 15% operational cost reduction involving the GAC process.

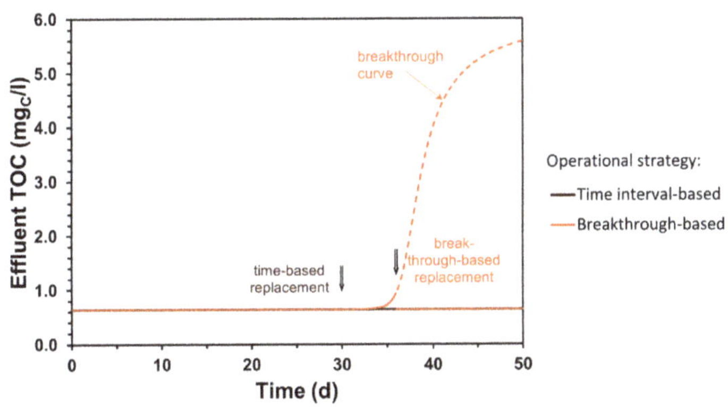

Figure 9. Illustration of GAC tower operational strategies.

Zietzschmann et al. (2016) [70] conducted small-scale tests with GAC to assess the adsorption of organic micropollutants (OMP) from sewage effluent. They observed that as long as influent concentrations remained below specific thresholds, the relative breakthrough behavior remained unaffected. Consequently, the capacity for OMP removal was directly correlated with the influent OMP concentration. However, the authors noted that using the specific throughput of dissolved organic carbon alone was insufficient to generate superimposed breakthrough curves [70]. In practice, when operating a full-scale facility, complete breakthrough curves shall never be able to be registered, as stable effluent quality must be maintained. We emphasize that the purpose of this research is not to generate accurate breakthrough curves but to provide a robust simulation of the breakpoint based on the adsorption capacity specification, thus aiding in decision-making about scheduling GAC bed replacement and the evaluation of operational and control strategies.

4. Conclusions

In this paper, the proposed kinetic concepts have been shown to provide a reasonable framework for simulating the fate of BTEX compounds in water resource recovery facilities. They can be merged with existing biokinetic models to be run in mathematical simulation software environments, as potential methods of effluent quality estimation and operational optimization in the areas of cost and resources.

The model-based benchmarking tests of this study demonstrated the applicability of the presented concepts for plant-wide modelling by investigating the impacts of operational parameters in the activated sludge process on the removal efficiency of BTEX chemicals, alongside the demand for granular activated carbon usage in effluent polishing. The first studied case revealed that the reduction in bed replacement frequency is most significant within the range of 3 to 6 days of SRT. Above 6 days, the further decrease in the AC requirement is less significant, and proportionally, a higher blower power is required to meet the DO setpoint at higher MLSS concentrations. Since BTEX constituents are only present at low concentrations in municipal settings, potential cost reductions in GAC bed replacement—through optimizing the biological treatment—may not be significant due to the higher fraction of inert and non-volatile material fed to GAC treatment, which cannot be removed through biodegradation and stripping, respectively. However, according to another studied example case, enhanced aeration for increased stripping prior to the GAC process may be a useful approach to preventing the deterioration of effluent quality regarding BTEX compounds in case of operational emergencies. The demonstration of dynamic simulation to compare operational strategies showed that synchronizing bed replacement with the activated carbon breakthrough capacity—based on the effluent quality—may result in more than 15% operational cost savings regarding the GAC process.

Goals for future studies include linking the model with further processes in water reuse. Efforts shall also be focused on model developments for off-gas treatment, considering the health risks—especially of benzene—in the atmosphere. Although the kinetic model parameters were set up based on values found in the scientific literature, future studies should aim for thorough model validation based on measured data from various facilities in municipal sewage treatment. As an outlook on degradation modelling with regard to granular activated carbon, future development efforts are dedicated to enhancing the GAC adsorption methodology with biomass activity included, enabling the simulation of BAC applications and lowered loading to an AC bed. The focus of potential further research could also be placed on applications in industrial wastewater treatment with a high BTEX content.

Author Contributions: Conceptualization, D.B. and T.K.; methodology, D.B., T.W. and F.H.; software, D.B., T.W. and F.H.; data curation, D.B., T.W. and F.H.; writing—original draft preparation, D.B. and T.K.; writing—review and editing, D.B. and T.K.; visualization, D.B.; supervision, T.K. All authors have read and agreed to the published version of the manuscript.

Funding: Project no. TKP2021-NVA-18 has been implemented with the support provided by the Ministry of Culture and Innovation of Hungary from the National Research, Development and Innovation Fund, financed under the TKP2021-NVA funding scheme.

Data Availability Statement: The data are contained within the article.

Conflicts of Interest: The authors declare no conflicts of interest.

Nomenclature

α	alpha (wastewater/clean water) correction factor for mass transfer coefficient
a_{bub}	specific contact area between the gas bubble surface and liquid phase [$m^2\ m^{-3}$]
a_{sur}	specific contact area between the surface gas and liquid phase [$m^2\ m^{-3}$]
$A_{diff,sp}$	area per diffuser [m^2]

A_r	liquid surface [m^2]
β	beta (wastewater/clean water) correction factor for the saturation concentration
BTC	TOC breakthrough capacity (in concentration unit) [$g_C\ m^{-3}$]
BTC_m	TOC adsorption capacity (at breakpoint, in mass fraction unit) [$g_C\ g_{AC}^{-1}$]
C_{mid}	midpoint concentration of breakthrough curve, with asymmetry correction [$g_C\ m^{-3}$]
$C_{mid,symm}$	midpoint concentration of curve, without asymmetry correction [$g_C\ m^{-3}$]
$coeff_{lead,h,diff}$	leading coefficient in a diffuser submergence correction term [m^{-1}]
$coeff_{lin,h,diff}$	linear coefficient in a diffuser submergence correction term [m^{-1}]
d_{bub}	bubble Sauter mean diameter [m]
d_{diff}	diffuser density [$m^2\ m^{-2}$]
$D_{i,25}$	diffusion coefficient of gas state variable i in water [$m^2\ d^{-1}$]
$div_{d,diff}$	divisor value in a diffuser density correction term [$m^2\ m^{-2}$]
ε	gas hold-up [$m^3{}_{gas}\ m^{-3}$]
$EQ_{C,ad,total}$	carbon equivalent for all adsorbed components on a GAC bed [g C m^{-3}]
exp_{SSOTE}	exponent in SSOTE correlation [$d\ m^{-3}{}_{gas}$]
F	diffuser fouling factor
F_{ac}	replaced activated carbon mass flow [$g\ d^{-1}$]
f_{cover}	covered fraction of the reactor surface
F_{Gi}	mass flow of gas phase state variable i [$g\ d^{-1}$]
$f_{h,sat,eff}$	effective saturation depth fraction
$f_{kL,i}$	fraction in the liquid side for the mass transfer of gas state variable i
F_{Li}	mass flow of liquid phase state variable i [$g\ d^{-1}$]
f_{wave}	waviness factor
G_i	concentration of gas phase state variable i in off-gas, per liquid volume [$g\ m^{-3}$]
$G_{i,air,inp}$	concentration of gas phase state variable i in the air input [$\%V\ V^{-1}$]
$G_{i,atm}$	concentration of gas phase state variable i in the atmosphere [$\%V\ V^{-1}$]
$G_{i,percent}$	concentration of gas phase state variable i in off-gas, percentage [$\%V\ V^{-1}$]
h_{diff}	diffuser submergence [m]
$h_{diff,floor}$	diffuser height from floor [m]
$Henry_{i,dt}$	temperature dependency factor for Henry coefficient of gas i [K]
$Henry_{i,SATP}$	Henry coefficient of gas i, standard (SATP) temperature (25 °C) [$mol\ m^{-3}\ Pa^{-1}$]
h_r	reactor depth [m]
$h_{sat,eff}$	effective saturation depth [m]
h_{sea}	elevation above sea level [m]
$i_{C,i}$	equivalent mass of soluble organic state variable i per unit mass of carbon [$g\ g_C^{-1}$]
$k_{L,i,bub,st,cw}$	liquid-side mass transfer coefficient for gas bubbles, standard conditions [$m\ d^{-1}$]
$k_{L,i,sur,st,cw}$	liquid-side mass transfer coefficient for liquid surface, standard conditions [$m\ d^{-1}$]
$k_La_{i,bub}$	volumetric mass transfer coefficient for gas bubbles, field conditions [d^{-1}]
$k_La_{i,bub,st,cw}$	volumetric mass transfer coefficient for gas bubbles, standard conditions [d^{-1}]
$k_La_{i,sur}$	volumetric mass transfer coefficient for liquid surface, field conditions [d^{-1}]
$k_La_{i,sur,st,cw}$	volumetric mass transfer coefficient for liquid surface, standard conditions [d^{-1}]
L_{air}	temperature lapse rate for air pressure calculation [$K\ m^{-1}$]
L_i	concentration of liquid phase state variable i [$g\ m^{-3}$]
$L_{i,ad}$	adsorbed soluble organic state variable i mass per bed volume [$g\ m^{-3}$]
$M_{ac,cycle}$	mass of activated carbon filled per cycle [g]
$magn_{mid,asymm}$	magnitude of the breakthrough curve midpoint asymmetry correction term
MM_{air}	molar mass of air [$g\ mol^{-1}$]
$MM_{EQ,i}$	equivalent molar mass of gas phase state variable i [$g\ mol^{-1}$]
n_{diff}	number of diffusers
$n_{gas,bub}$	molar quantity of gas bubbles per unit liquid volume [$mol\ m^{-3}$]
N_{repl}	activated carbon bed replacement cycle frequency [d^{-1}]
p_{air}	air pressure at field elevation [Pa]
p_{gas}	gas phase pressure [Pa]
p_{NTP}	pressure at standard (NTP) conditions (101,325 Pa) [Pa]
$pow_{d,diff}$	power value in a diffuser density correction term
$pow_{h,diff}$	power value in a diffuser submergence correction term
$pow_{mid,asymm}$	power of the breakthrough curve midpoint asymmetry correction term
$p_{partial,i,bub}$	partial pressure of gas state variable i in the gas phase [Pa]

$p_{partial,i,bub,st}$	partial pressure of gas state variable i in the gas phase, standard conditions [Pa]
$p_{partial,i,sur}$	partial pressure of gas state variable i in the atmosphere [Pa]
$p_{partial,i,sur,st}$	partial pressure of gas state variable i in the atmosphere, standard conditions [Pa]
$p_{st,h,sat,eff}$	pressure at standard conditions and effective saturation depth [Pa]
$p_{v,T}$	saturated vapor pressure of water at temperature T [Pa]
θ	Arrhenius temperature correction factor for the mass transfer coefficient
Q	volumetric flow of wastewater [$m^3\ d^{-1}$]
$Q_{air,NTP}$	air flow at standard (NTP) conditions [$m^3{}_{gas}\ d^{-1}$]
$Q_{air,NTP,sp}$	air flow per diffuser at standard (NTP) conditions [$m^3{}_{gas}\ d^{-1}$]
$Q_{gas,transfer,NTP}$	gas transfer flow at standard (NTP) conditions [$m^3{}_{gas}\ d^{-1}$]
$Q_{gas,outp,NTP}$	off-gas flow at standard (NTP) conditions [$m^3{}_{gas}\ d^{-1}$]
ρ_{ac}	apparent density of granular activated carbon [$g_{AC}\ m^{-3}$]
$rateF_i$	mass rate of state variable i [$g\ d^{-1}$]
$rate_i$	reaction rate for the state variable [$g\ m^{-3}\ d^{-1}$]
$Rem_{GAC,i}$	removal ratio of soluble organic state variable i by granular activated carbon
r_j	process rate regarding process j (from Gujer matrix) [$g\ m^{-3}\ d^{-1}$]
$S_{i,bub,sat}$	saturation concentration at the gas bubble interface [$g\ m^{-3}$]
$S_{i,bub,sat,st,cw}$	saturation concentration at the gas bubble interface, standard conditions [$g\ m^{-3}$]
$S_{i,sur,sat}$	saturation concentration at the atmospheric interface [$g\ m^{-3}$]
$S_{i,sur,sat,st,cw}$	saturation concentration at the atmospheric interface, standard conditions [$g\ m^{-3}$]
sl_{break}	slope of the breakthrough curve [$m^3\ g_C{}^{-1}$]
S_{O2}	dissolved oxygen concentration [$g_{O2}\ m^{-3}$]
$SOTR_{bub}$	standard oxygen transfer rate from bubbles [$g\ d^{-1}$]
SSOTE	specific standard oxygen transfer efficiency [$\%\ m^{-1}$]
$SSOTE_0$	intercept in SSOTE correlation [$\%\ m^{-1}$]
$SSOTE_{asym}$	asymptote in SSOTE correlation [$\%\ m^{-1}$]
T	liquid temperature [°C]
$T_{air,K}$	field air temperature [K]
T_K	liquid temperature in an SI unit [K]
$T_{NTP,K}$	temperature at standard (NTP) conditions (20 °C) [K]
t_{repl}	duration of activated carbon bed replacement [d]
$T_{SATP,K}$	temperature at standard (SATP) conditions (25 °C) [K]
V_{ac}	activated carbon bed volume [m^3]
V_{gas}	gas phase volume [$m^3{}_{gas}$]
$V_{gas,NTP}$	gas phase volume at standard (NTP) conditions [$m^3{}_{gas}$]
$v_{j,i}$	stoichiometric coefficient of state variable i in process j
V_r	reactive volume [m^3]

Appendix A. Gujer Matrix Development

The following tables show the MiniSumo biokinetic model structure, alongside BTEX-related model adjustments and the affected cells of the Gujer matrix based on reference [29]. All process rates are interpreted in $g_{COD}\ m^{-3}\ d^{-1}$ units, with the exception of the elimination of surfactants, which is interpreted in d^{-1}.

Table A1. Gujer Matrix—part 1.

Symbol	Process Name
1	OHO growth on VFAs, O_2
2	OHO growth on VFAs, NO_x
3	OHO growth on benzene, O_2
4	OHO growth on benzene, NO_x
5	OHO growth on toluene, O_2
6	OHO growth on toluene, NO_x
7	OHO growth on ethylbenzene, O_2
8	OHO growth on ethylbenzene, NO_x
9	OHO growth on xylene, O_2
10	OHO growth on xylene, NO_x

Table A1. *Cont.*

Symbol	Process Name
11	OHO growth on S_B, O_2
12	OHO growth on S_B, NO_x
13	S_B fermentation with high VFA (OHO growth, anaerobic)
14	S_B fermentation with low VFA (OHO growth, anaerobic)
15	Benzene fermentation with low VFA (OHO growth, anaerobic)
16	Toluene fermentation with low VFA (OHO growth, anaerobic)
17	Ethylbenzene fermentation with low VFA (OHO growth, anaerobic)
18	Xylene fermentation with low VFA (OHO growth, anaerobic)
19	OHO decay
20	NITO growth
21	NITO decay
22	AMETO growth
23	AMETO decay
24	HMETO growth
25	HMETO decay
26	X_B hydrolysis
27	X_B anaerobic hydrolysis (fermentation)
28	$S_{N,B}$ ammonification
29	NO_x assimilative reduction
30	FeP precipitation
31	FeP redissolution
32	AlP precipitation
33	AlP redissolution
34	Elimination of surfactants
35	Methane gas transfer—bubbles
36	Hydrogen gas transfer—bubbles
37	Oxygen gas transfer—bubbles
38	Nitrogen gas transfer—bubbles
39	Benzene gas transfer—bubbles
40	Toluene gas transfer—bubbles
41	Ethylbenzene gas transfer—bubbles
42	Xylene gas transfer—bubbles
43	Methane gas transfer—surface
44	Hydrogen gas transfer—surface
45	Oxygen gas transfer—surface
46	Nitrogen gas transfer—surface
47	Benzene gas transfer—surface
48	Toluene gas transfer—surface
49	Ethylbenzene gas transfer—surface
50	Xylene gas transfer—surface

Table A2. Gujer Matrix—part 2.

	S_{BENE}	S_{TENE}	S_{EBENE}	S_{XENE}	S_B	X_B	S_U	X_U	X_E	X_{OHO}
1										1
2										1
3	$-1/Y_{OHO,BTEX,ox}$									1
4	$-1/Y_{OHO,BTEX,anox}$									1
5		$-1/Y_{OHO,BTEX,ox}$								1
6		$-1/Y_{OHO,BTEX,anox}$								1
7			$-1/Y_{OHO,BTEX,ox}$							1
8			$-1/Y_{OHO,BTEX,anox}$							1
9				$-1/Y_{OHO,BTEX,ox}$						1
10				$-1/Y_{OHO,BTEX,anox}$						1
11					$-1/Y_{OHO,SB,ox}$					1
12					$-1/Y_{OHO,SB,anox}$					1
13					$-1/Y_{OHO,SB,ana}$					1
14					$-1/Y_{OHO,SB,ana}$					1
15	$-1/Y_{OHO,BTEX,ana}$									1
16		$-1/Y_{OHO,BTEX,ana}$								1
17			$-1/Y_{OHO,BTEX,ana}$							1
18				$-1/Y_{OHO,BTEX,ana}$						1
19						$1 - f_E$			f_E	-1
21						$1 - f_E$			f_E	
23						$1 - f_E$			f_E	
25						$1 - f_E$			f_E	
26					1	-1				
27					$1 - f_{H2}$	-1				
29										$-EEQ_{NO3} \times X_{OHO}/X_{BIO,kin}$
39	1									
40		1								
41			1							
42				1						
47	1									
48		1								
49			1							
50				1						

Table A3. Gujer Matrix—part 3.

	S_{VFA}
1	$-1/Y_{OHO,VFA,ox}$
2	$-1/Y_{OHO,VFA,anox}$
13	$(1 - Y_{OHO,SB,ana} - Y_{OHO,H2,ana,high})/Y_{OHO,SB,ana}$
14	$(1 - Y_{OHO,SB,ana} - Y_{OHO,H2,ana,low})/Y_{OHO,SB,ana}$
15	$(1 - Y_{OHO,SB,ana} - Y_{OHO,H2,ana,low})/Y_{OHO,SB,ana}$
16	$(1 - Y_{OHO,SB,ana} - Y_{OHO,H2,ana,low})/Y_{OHO,SB,ana}$
17	$(1 - Y_{OHO,SB,ana} - Y_{OHO,H2,ana,low})/Y_{OHO,SB,ana}$
18	$(1 - Y_{OHO,SB,ana} - Y_{OHO,H2,ana,low})/Y_{OHO,SB,ana}$
22	$-1/Y_{AMETO}$

Table A4. Gujer Matrix—part 4.

	X_{NITO}	X_{AMETO}	X_{HMETO}
20	1		
21	−1		
22		1	
23		−1	
24			1
25			−1
29	$-EEQ_{NO3} \times X_{NITO}/X_{BIO,kin}$	$-EEQ_{NO3} \times X_{AMETO}/X_{BIO,kin}$	$-EEQ_{NO3} \times X_{HMETO}/X_{BIO,kin}$

Table A5. Gujer Matrix—part 5.

	S_{NHx}	S_{NOx}	S_{N2}
1	$-i_{N,BIO}$		
2	$-i_{N,BIO}$	$-(1 - Y_{OHO,VFA,anox})/(EEQ_{N2,NO3} \times Y_{OHO,VFA,anox})$	$(1 - Y_{OHO,VFA,anox})/(EEQ_{N2,NO3} \times Y_{OHO,VFA,anox})$
3	$-i_{N,BIO}$		
4	$-i_{N,BIO}$	$-(1 - Y_{OHO,BTEX,anox})/(EEQ_{N2,NO3} \times Y_{OHO,BTEX,anox})$	$(1 - Y_{OHO,BTEX,anox})/(EEQ_{N2,NO3} \times Y_{OHO,BTEX,anox})$
5	$-i_{N,BIO}$		
6	$-i_{N,BIO}$	$-(1 - Y_{OHO,BTEX,anox})/(EEQ_{N2,NO3} \times Y_{OHO,BTEX,anox})$	$(1 - Y_{OHO,BTEX,anox})/(EEQ_{N2,NO3} \times Y_{OHO,BTEX,anox})$
7	$-i_{N,BIO}$		
8	$-i_{N,BIO}$	$-(1 - Y_{OHO,BTEX,anox})/(EEQ_{N2,NO3} \times Y_{OHO,BTEX,anox})$	$(1 - Y_{OHO,BTEX,anox})/(EEQ_{N2,NO3} \times Y_{OHO,BTEX,anox})$
9	$-i_{N,BIO}$		
10	$-i_{N,BIO}$	$-(1 - Y_{OHO,BTEX,anox})/(EEQ_{N2,NO3} \times Y_{OHO,BTEX,anox})$	$(1 - Y_{OHO,BTEX,anox})/(EEQ_{N2,NO3} \times Y_{OHO,BTEX,anox})$
11	$-i_{N,BIO}$		
12	$-i_{N,BIO}$	$-(1 - Y_{OHO,SB,anox})/(EEQ_{N2,NO3} \times Y_{OHO,SB,anox})$	$(1 - Y_{OHO,SB,anox})/(EEQ_{N2,NO3} \times Y_{OHO,SB,anox})$
13	$-i_{N,BIO}$		
14	$-i_{N,BIO}$		
15	$-i_{N,BIO}$		
16	$-i_{N,BIO}$		
17	$-i_{N,BIO}$		
18	$-i_{N,BIO}$		
19	$-f_E \times (i_{N,XE} - i_{N,BIO})$		
20	$-1/Y_{NITO} - i_{N,BIO}$	$1/Y_{NITO}$	
21	$-f_E \times (i_{N,XE} - i_{N,BIO})$		
22	$-i_{N,BIO}$		
23	$-f_E \times (i_{N,XE} - i_{N,BIO})$		
24	$-i_{N,BIO}$		
25	$-f_E \times (i_{N,XE} - i_{N,BIO})$		
28	1		
29	$1 + EEQ_{NO3} \times i_{N,BIO}$	−1	
38			1
46			1

Table A6. Gujer Matrix—part 6.

	$S_{N,B}$	$X_{N,B}$	S_{PO4}	$X_{P,B}$	S_{O2}	S_{CH4}	S_{H2}
1			$-i_{P,BIO}$		$-(1-Y_{OHO,VFA,ox})/Y_{OHO,VFA,ox}$		
2			$-i_{P,BIO}$				
3			$-i_{P,BIO}$		$-(1-Y_{OHO,BTEX,ox})/Y_{OHO,BTEX,ox}$		
4			$-i_{P,BIO}$				
5			$-i_{P,BIO}$		$-(1-Y_{OHO,BTEX,ox})/Y_{OHO,BTEX,ox}$		
6			$-i_{P,BIO}$				
7			$-i_{P,BIO}$		$-(1-Y_{OHO,BTEX,ox})/Y_{OHO,BTEX,ox}$		
8			$-i_{P,BIO}$				
9			$-i_{P,BIO}$		$-(1-Y_{OHO,BTEX,ox})/Y_{OHO,BTEX,ox}$		
10			$-i_{P,BIO}$				
11			$-i_{P,BIO}$		$-(1-Y_{OHO,SB,ox})/Y_{OHO,SB,ox}$		
12			$-i_{P,BIO}$				
13			$-i_{P,BIO}$				$Y_{OHO,H2,ana,high}/Y_{OHO,SB,ana}$
14			$-i_{P,BIO}$				$Y_{OHO,H2,ana,low}/Y_{OHO,SB,ana}$
15			$-i_{P,BIO}$				$Y_{OHO,H2,ana,low}/Y_{OHO,BTEX,ana}$
16			$-i_{P,BIO}$				$Y_{OHO,H2,ana,low}/Y_{OHO,BTEX,ana}$
17			$-i_{P,BIO}$				$Y_{OHO,H2,ana,low}/Y_{OHO,BTEX,ana}$
18			$-i_{P,BIO}$				$Y_{OHO,H2,ana,low}/Y_{OHO,BTEX,ana}$
19		$(1-f_E)\times i_{N,BIO}$		$(1-f_E)\times i_{P,BIO}$			
20			$-i_{P,BIO}$		$-(EEQ_{NO3}-Y_{NITO})/Y_{NITO}$		
21		$(1-f_E)\times i_{N,BIO}$		$(1-f_E)\times i_{P,BIO}$			
22			$-i_{P,BIO}$			$(1-Y_{AMETO})/Y_{AMETO}$	
23		$(1-f_E)\times i_{N,BIO}$		$(1-f_E)\times i_{P,BIO}$			
24			$-i_{P,BIO}$			$(1-Y_{HMETO})/Y_{HMETO}$	$-1/Y_{HMETO}$
25		$(1-f_E)\times i_{N,BIO}$		$(1-f_E)\times i_{P,BIO}$			
26	$X_{N,B}/X_B$	$-X_{N,B}/X_B$	$X_{P,B}/X_B$	$-X_{P,B}/X_B$			
27	$X_{N,B}/X_B$	$-X_{N,B}/X_B$	$X_{P,B}/X_B$	$-X_{P,B}/X_B$			f_{H2}
28	-1						
29			$EEQ_{NO3}\times i_{P,BIO}$				
30			$-f_{P,Fe}$				
31			$f_{P,Fe}$				
32			$-f_{P,Al}$				
33			$f_{P,Al}$				
35						1	
36							1
37					1		
43						1	
44							1
45					1		

Table A7. Gujer Matrix—part 7.

	S_{ALK}	X_{FeOH}	X_{FeP}	X_{AlOH}	X_{AlP}
1	$(-i_{N,BIO} \times CH_{NHx} - i_{P,BIO} \times CH_{PO4})$				
2	$(-(1 - Y_{OHO,VFA,anox})/(EEQ_{N2,NO3} \times Y_{OHO,VFA,anox}) \times CH_{NO3} - i_{N,BIO} \times CH_{NHx} - i_{P,BIO} \times CH_{PO4})$				
3	$(-i_{N,BIO} \times CH_{NHx} - i_{P,BIO} \times CH_{PO4})$				
4	$(-(1 - Y_{OHO,BTEX,anox})/(EEQ_{N2,NO3} \times Y_{OHO,BTEX,anox}) \times CH_{NO3} - i_{N,BIO} \times CH_{NHx} - i_{P,BIO} \times CH_{PO4})$				
5	$(-i_{N,BIO} \times CH_{NHx} - i_{P,BIO} \times CH_{PO4})$				
6	$(-(1 - Y_{OHO,BTEX,anox})/(EEQ_{N2,NO3} \times Y_{OHO,BTEX,anox}) \times CH_{NO3} - i_{N,BIO} \times CH_{NHx} - i_{P,BIO} \times CH_{PO4})$				
7	$(-i_{N,BIO} \times CH_{NHx} - i_{P,BIO} \times CH_{PO4})$				
8	$(-(1 - Y_{OHO,BTEX,anox})/(EEQ_{N2,NO3} \times Y_{OHO,BTEX,anox}) \times CH_{NO3} - i_{N,BIO} \times CH_{NHx} - i_{P,BIO} \times CH_{PO4})$				
9	$(-i_{N,BIO} \times CH_{NHx} - i_{P,BIO} \times CH_{PO4})$				
10	$(-(1 - Y_{OHO,BTEX,anox})/(EEQ_{N2,NO3} \times Y_{OHO,BTEX,anox}) \times CH_{NO3} - i_{N,BIO} \times CH_{NHx} - i_{P,BIO} \times CH_{PO4})$				
11	$(-i_{N,BIO} \times CH_{NHx} - i_{P,BIO} \times CH_{PO4})$				
12	$(-(1 - Y_{OHO,SB,anox})/(EEQ_{N2,NO3} \times Y_{OHO,SB,anox}) \times CH_{NO3} - i_{N,BIO} \times CH_{NHx} - i_{P,BIO} \times CH_{PO4})$				
13	$(-i_{N,BIO} \times CH_{NHx} - i_{P,BIO} \times CH_{PO4})$				
14	$(-i_{N,BIO} \times CH_{NHx} - i_{P,BIO} \times CH_{PO4})$				
15	$(-i_{N,BIO} \times CH_{NHx} - i_{P,BIO} \times CH_{PO4})$				
16	$(-i_{N,BIO} \times CH_{NHx} - i_{P,BIO} \times CH_{PO4})$				
17	$(-i_{N,BIO} \times CH_{NHx} - i_{P,BIO} \times CH_{PO4})$				
18	$(-i_{N,BIO} \times CH_{NHx} - i_{P,BIO} \times CH_{PO4})$				
19	$-f_E \times (i_{N,XE} - i_{N,BIO}) \times CH_{NHx}$				
20	$((-1/Y_{NITO} - i_{N,BIO}) \times CH_{NHx} + 1/Y_{NITO} \times CH_{NO3} - i_{P,BIO} \times CH_{PO4})$				
21	$-f_E \times (i_{N,XE} - i_{N,BIO}) \times CH_{NHx}$				
22	$(-i_{N,BIO} \times CH_{NHx} - i_{P,BIO} \times CH_{PO4})$				
23	$-f_E \times (i_{N,XE} - i_{N,BIO}) \times CH_{NHx}$				
24	$(-i_{N,BIO} \times CH_{NHx} - i_{P,BIO} \times CH_{PO4})$				
25	$-f_E \times (i_{N,XE} - i_{N,BIO}) \times CH_{NHx}$				
26	$X_{P,B}/X_B \times CH_{PO4}$				
27	$X_{P,B}/X_B \times CH_{PO4}$				
28	CH_{NHx}				
29	$((1 + EEQ_{NO3} \times i_{N,BIO}) \times CH_{NHx} - CH_{NO3} + EEQ_{NO3} \times i_{P,BIO} \times CH_{PO4})$				
30	$-f_{P,Fe} \times CH_{PO4}$	−1	1		
31	$f_{P,Fe} \times CH_{PO4}$	1	−1		
32	$-f_{P,Al} \times CH_{PO4}$			−1	1
33	$f_{P,Al} \times CH_{PO4}$			1	−1

Table A8. Gujer Matrix—part 8.

	S_{ALPHA}	G_{CH4}	G_{H2}	G_{O2}	G_{N2}	G_{BENE}	G_{TENE}	G_{EBENE}	G_{XENE}
34	1								
35		−1							
36			−1						
37				−1					
38					−1				
39						−1			
40							−1		
41								−1	
42									−1

Table A9. Gujer Matrix—part 9.

	Rate
1	$\mu_{OHO,T} \times X_{OHO} \times \mathrm{Msat}_{SVFA,KVFA} \times \mathrm{Msat}_{SO2,KO2,OHO} \times \mathrm{Msat}_{SNHx,KNHx,BIO} \times \mathrm{Msat}_{SPO4,KPO4,BIO} \times \mathrm{Msat}_{SALK,KALK}$
2	$\mu_{OHO,T} \times X_{OHO} \times \eta_{OHO,anox} \times \mathrm{Msat}_{SVFA,KVFA} \times \mathrm{Msat}_{SNOx,KNOx,OHO} \times \mathrm{Minh}_{SO2,KO2,OHO} \times \mathrm{Msat}_{SNHx,KNHx,BIO} \times \mathrm{Msat}_{SPO4,KPO4,BIO} \times \mathrm{Msat}_{SALK,KALK}$
3	$\mu_{OHO,BENE,T} \times X_{OHO} \times \mathrm{Msat}_{SBENE,KBENE} \times \mathrm{Msat}_{SO2,KO2,OHO} \times \mathrm{Msat}_{SNHx,KNHx,BIO} \times \mathrm{Msat}_{SPO4,KPO4,BIO} \times \mathrm{Msat}_{SALK,KALK}$
4	$\mu_{OHO,BENE,T} \times X_{OHO} \times \eta_{OHO,anox} \times \mathrm{Msat}_{SBENE,KBENE} \times \mathrm{Msat}_{SNOx,KNOx,OHO} \times \mathrm{Minh}_{SO2,KO2,OHO} \times \mathrm{Msat}_{SNHx,KNHx,BIO} \times \mathrm{Msat}_{SPO4,KPO4,BIO} \times \mathrm{Msat}_{SALK,KALK}$
5	$\mu_{OHO,TENE,T} \times X_{OHO} \times \mathrm{Msat}_{STENE,KTENE} \times \mathrm{Msat}_{SO2,KO2,OHO} \times \mathrm{Msat}_{SNHx,KNHx,BIO} \times \mathrm{Msat}_{SPO4,KPO4,BIO} \times \mathrm{Msat}_{SALK,KALK}$
6	$\mu_{OHO,TENE,T} \times X_{OHO} \times \eta_{OHO,anox} \times \mathrm{Msat}_{STENE,KTENE} \times \mathrm{Msat}_{SNOx,KNOx,OHO} \times \mathrm{Minh}_{SO2,KO2,OHO} \times \mathrm{Msat}_{SNHx,KNHx,BIO} \times \mathrm{Msat}_{SPO4,KPO4,BIO} \times \mathrm{Msat}_{SALK,KALK}$
7	$\mu_{OHO,EBENE,T} \times X_{OHO} \times \mathrm{Msat}_{SEBENE,KEBENE} \times \mathrm{Msat}_{SO2,KO2,OHO} \times \mathrm{Msat}_{SNHx,KNHx,BIO} \times \mathrm{Msat}_{SPO4,KPO4,BIO} \times \mathrm{Msat}_{SALK,KALK}$
8	$\mu_{OHO,EBENE,T} \times X_{OHO} \times \eta_{OHO,anox} \times \mathrm{Msat}_{SEBENE,KEBENE} \times \mathrm{Msat}_{SNOx,KNOx,OHO} \times \mathrm{Minh}_{SO2,KO2,OHO} \times \mathrm{Msat}_{SNHx,KNHx,BIO} \times \mathrm{Msat}_{SPO4,KPO4,BIO} \times \mathrm{Msat}_{SALK,KALK}$
9	$\mu_{OHO,XENE,T} \times X_{OHO} \times \mathrm{Msat}_{SXENE,KXENE} \times \mathrm{Msat}_{SO2,KO2,OHO} \times \mathrm{Msat}_{SNHx,KNHx,BIO} \times \mathrm{Msat}_{SPO4,KPO4,BIO} \times \mathrm{Msat}_{SALK,KALK}$
10	$\mu_{OHO,XENE,T} \times X_{OHO} \times \eta_{OHO,anox} \times \mathrm{Msat}_{SXENE,KXENE} \times \mathrm{Msat}_{SNOx,KNOx,OHO} \times \mathrm{Minh}_{SO2,KO2,OHO} \times \mathrm{Msat}_{SNHx,KNHx,BIO} \times \mathrm{Msat}_{SPO4,KPO4,BIO} \times \mathrm{Msat}_{SALK,KALK}$
11	$\mu_{OHO,T} \times \mathrm{Msat}_{SB,KSB} \times \mathrm{Minh}_{SVFA,KVFA} \times X_{OHO} \times \mathrm{Msat}_{SO2,KO2,OHO} \times \mathrm{Msat}_{SNHx,KNHx,BIO} \times \mathrm{Msat}_{SPO4,KPO4,BIO} \times \mathrm{Msat}_{SALK,KALK}$
12	$\mu_{OHO,T} \times \eta_{OHO,anox} \times \mathrm{Msat}_{SB,KSB} \times \mathrm{Minh}_{SVFA,KVFA} \times X_{OHO} \times \mathrm{Msat}_{SNOx,KNOx,OHO} \times \mathrm{Minh}_{SO2,KO2,OHO} \times \mathrm{Msat}_{SNHx,KNHx,BIO} \times \mathrm{Msat}_{SPO4,KPO4,BIO}$
13	$\mu_{FERM,OHO,T} \times X_{OHO} \times \mathrm{Logsat}_{SVFA,KVFA,FERM} \times \mathrm{Msat}_{SB,KSB,ana} \times \mathrm{Minh}_{SO2,KO2,OHO} \times \mathrm{Minh}_{SNOx,KNOx,OHO} \times \mathrm{Msat}_{SNHx,KNHx,BIO} \times \mathrm{Msat}_{SPO4,KPO4,BIO} \times \mathrm{Msat}_{SALK,KALK}$
14	$\mu_{FERM,OHO,T} \times X_{OHO} \times \mathrm{Loginh}_{SVFA,KVFA,FERM} \times \mathrm{Msat}_{SB,KSB,ana} \times \mathrm{Minh}_{SO2,KO2,OHO} \times \mathrm{Minh}_{SNOx,KNOx,OHO} \times \mathrm{Msat}_{SNHx,KNHx,BIO} \times \mathrm{Msat}_{SPO4,KPO4,BIO} \times \mathrm{Msat}_{SALK,KALK}$
15	$\mu_{FERM,OHO,BENE,T} \times X_{OHO} \times \mathrm{Loginh}_{SVFA,KVFA,FERM} \times \mathrm{Msat}_{SBENE,KBENE,ana} \times \mathrm{Minh}_{SO2,KO2,OHO} \times \mathrm{Minh}_{SNOx,KNOx,OHO} \times \mathrm{Msat}_{SNHx,KNHx,BIO} \times \mathrm{Msat}_{SPO4,KPO4,BIO} \times \mathrm{Msat}_{SALK,KALK}$
16	$\mu_{FERM,OHO,TENE,T} \times X_{OHO} \times \mathrm{Loginh}_{SVFA,KVFA,FERM} \times \mathrm{Msat}_{STENE,KTENE,ana} \times \mathrm{Minh}_{SO2,KO2,OHO} \times \mathrm{Minh}_{SNOx,KNOx,OHO} \times \mathrm{Msat}_{SNHx,KNHx,BIO} \times \mathrm{Msat}_{SPO4,KPO4,BIO} \times \mathrm{Msat}_{SALK,KALK}$
17	$\mu_{FERM,OHO,EBENE,T} \times X_{OHO} \times \mathrm{Loginh}_{SVFA,KVFA,FERM} \times \mathrm{Msat}_{SEBENE,KEBENE,ana} \times \mathrm{Minh}_{SO2,KO2,OHO} \times \mathrm{Minh}_{SNOx,KNOx,OHO} \times \mathrm{Msat}_{SNHx,KNHx,BIO} \times \mathrm{Msat}_{SPO4,KPO4,BIO} \times \mathrm{Msat}_{SALK,KALK}$
18	$\mu_{FERM,OHO,XENE,T} \times X_{OHO} \times \mathrm{Loginh}_{SVFA,KVFA,FERM} \times \mathrm{Msat}_{SXENE,KXENE,ana} \times \mathrm{Minh}_{SO2,KO2,OHO} \times \mathrm{Minh}_{SNOx,KNOx,OHO} \times \mathrm{Msat}_{SNHx,KNHx,BIO} \times \mathrm{Msat}_{SPO4,KPO4,BIO} \times \mathrm{Msat}_{SALK,KALK}$
19	$b_{OHO,T} \times X_{OHO} \times (\mathrm{Msat}_{SO2,KO2,OHO} + \eta_{b,anox} \times \mathrm{Msat}_{SNOx,KNOx,OHO} \times \mathrm{Minh}_{SO2,KO2,OHO} + \eta_{b,ana} \times \mathrm{Minh}_{SNOx,KNOx,OHO} \times \mathrm{Minh}_{SO2,KO2,OHO})$
20	$\mu_{NITO,T} \times \mathrm{Msat}_{SNHx,KNHx,NITO} \times X_{NITO} \times \mathrm{Msat}_{SO2,KO2,NITO} \times \mathrm{Msat}_{SPO4,KPO4,BIO} \times \mathrm{Msat}_{SALK,KALK}$
21	$b_{NITO,T} \times X_{NITO} \times (\mathrm{Msat}_{SO2,KO2,NITO} + \eta_{b,anox} \times \mathrm{Msat}_{SNOx,KNOx,NITO} \times \mathrm{Minh}_{SO2,KO2,NITO} + \eta_{b,ana} \times \mathrm{Minh}_{SNOx,KNOx,NITO} \times \mathrm{Minh}_{SO2,KO2,NITO} + m_{tox,ana})$
22	$\mu_{AMETO,T} \times \mathrm{Hsat}_{SVFA,AMETO} \times X_{AMETO} \times \mathrm{Minh}_{SO2,KiO2,AMETO} \times \mathrm{Minh}_{SNOx,KNOx,AMETO} \times \mathrm{Msat}_{SNHx,KNHx,BIO} \times \mathrm{Msat}_{SPO4,KPO4,BIO} \times \mathrm{Msat}_{SALK,KALK}$
23	$b_{AMETO,T} \times X_{AMETO} \times (\mathrm{Msat}_{SO2,KiO2,AMETO} + \eta_{b,anox} \times \mathrm{Msat}_{SNOx,KNOx,AMETO} \times \mathrm{Minh}_{SO2,KiO2,AMETO} + \eta_{b,ana} \times \mathrm{Minh}_{SNOx,KNOx,AMETO} \times \mathrm{Minh}_{SO2,KiO2,AMETO})$

Table A9. *Cont.*

	Rate
24	$\mu_{HMETO,T} \times Msat_{SH2,KH2,HMETO} \times X_{HMETO} \times Minh_{SO2,KiO2,HMETO} \times Minh_{SNOx,KNOx,HMETO} \times Msat_{SNHx,KNHx,BIO} \times Msat_{SPO4,KPO4,BIO} \times Msat_{SALK,KALK}$
25	$b_{HMETO,T} \times X_{HMETO} \times (Msat_{SO2,KiO2,HMETO} + \eta_{b,anox} \times Msat_{SNOx,KNOx,HMETO} \times Minh_{SO2,KiO2,HMETO} + \eta_{b,ana} \times Minh_{SNOx,KNOx,HMETO} \times Minh_{SO2,KiO2,HMETO})$
26	$q_{HYD,T} \times X_{BIO,kin} \times MRsat_{XB,XBIO,kin,KHYD} \times (Msat_{SO2,KO2,OHO} + \eta_{b,anox} \times Msat_{SNOx,KNOx,OHO} \times Minh_{SO2,KO2,OHO}) \times Msat_{SALK,KALK}$
27	$q_{HYD,T} \times X_{BIO,kin} \times MRsat_{XB,XBIO,kin,KHYD} \times \eta_{b,ana} \times Minh_{SNOx,KNOx,OHO} \times Minh_{SO2,KO2,OHO} \times Msat_{SALK,KALK}$
28	$q_{AMMON,T} \times S_{N,B} \times X_{BIO,kin}$
29	$q_{ASSIM,T} \times Msat_{SNOx,KNOx,ASSIM} \times Minh_{SNHx,KiNHx,ASSIM} \times X_{BIO,kin}$
30	$q_{FeOH,PREC,Me} \times S_{PO4} \times X_{FeOH}$
31	$q_{FeOH,DISSOL,Me} \times X_{FeP} \times Msat_{SALK,KALK}$
32	$q_{AlOH,PREC,Me} \times S_{PO4} \times X_{AlOH}$
33	$q_{AlOH,DISSOL,Me} \times X_{AlP} \times Msat_{SALK,KALK}$
34	$q_{ALPHA,O2} \times X_{VSS} \times damp_{ALPHA} \times (S_{ALPHA,sat} - S_{ALPHA})$
35	$k_La_{GCH4,bub} \times (S_{GCH4,bub,sat} - S_{CH4})$
36	$k_La_{GH2,bub} \times (S_{GH2,bub,sat} - S_{H2})$
37	$k_La_{GO2,bub} \times (S_{GO2,bub,sat} - S_{O2})$
38	$k_La_{GN2,bub} \times (S_{GN2,bub,sat} - S_{N2})$
39	$k_La_{GBENE,bub} \times (S_{GBENE,bub,sat} - S_{BENE})$
40	$k_La_{GTENE,bub} \times (S_{GTENE,bub,sat} - S_{TENE})$
41	$k_La_{GEBENE,bub} \times (S_{GEBENE,bub,sat} - S_{EBENE})$
42	$k_La_{GXENE,bub} \times (S_{GXENE,bub,sat} - S_{XENE})$
43	$k_La_{GCH4,sur} \times (S_{GCH4,sur,sat} - S_{CH4})$
44	$k_La_{GH2,sur} \times (S_{GH2,sur,sat} - S_{H2})$
45	$k_La_{GO2,sur} \times (S_{GO2,sur,sat} - S_{O2})$
46	$k_La_{GN2,sur} \times (S_{GN2,sur,sat} - S_{N2})$
47	$k_La_{GBENE,sur} \times (S_{GBENE,sur,sat} - S_{BENE})$
48	$k_La_{GTENE,sur} \times (S_{GTENE,sur,sat} - S_{TENE})$
49	$k_La_{GEBENE,sur} \times (S_{GEBENE,sur,sat} - S_{EBENE})$
50	$k_La_{GXENE,sur} \times (S_{GXENE,sur,sat} - S_{XENE})$

where the following functions were used:

Table A10. Nomenclature of the functions used in Gujer Matrix.

Symbol	Name	Expression
Msat(var; k)	Monod saturation	var/(k + var)
Minh(var; k)	Monod inhibition	k/(k + var)
MRsat(s;x;k)	Monod ratio saturation	(s/x)/(s/x + k)
Logsat(var; halfval; slope)	Logistic saturation	1/(1 + Exp((halfval − var) × slope))
Loginh(var; halfval; slope)	Logistic inhibition	1/(1 + Exp((var − halfval) × slope))
Hsat(var; halfval; halfinh)	Haldane equation	var/(halfval + var + (var^2/halfinh))

Appendix B. BTEX Kinetic and Stoichiometric Model Parameters

The following tables only show the BTEX-related model parameters introduced in this study. For a complete list of parameters, please refer to the MiniSumo process model.

Table A11. BTEX kinetic and Stoichiometric Model Parameters.

Ordinary Heterotrophic Organism Kinetics (OHO)			
Symbol	**Name**	**Value**	**Unit**
$\mu_{OHO,BENE}$	Maximum specific growth rate of OHOs on benzene	0.006	d^{-1}
$\mu_{OHO,TENE}$	Maximum specific growth rate of OHOs on toluene	0.014	d^{-1}
$\mu_{OHO,EBENE}$	Maximum specific growth rate of OHOs on ethylbenzene	0.014	d^{-1}
$\mu_{OHO,XENE}$	Maximum specific growth rate of OHOs on xylene	0.010	d^{-1}
$\mu_{FERM,OHO,BENE}$	Fermentation growth rate of OHOs on benzene	0.0030	d^{-1}
$\mu_{FERM,OHO,TENE}$	Fermentation growth rate of OHOs on toluene	0.0042	d^{-1}
$\mu_{FERM,OHO,EBENE}$	Fermentation growth rate of OHOs on ethylbenzene	0.0035	d^{-1}
$\mu_{FERM,OHO,XENE}$	Fermentation growth rate of OHOs on xylene	0.0050	d^{-1}
K_{BENE}	Half-saturation of benzene for OHOs	6.8	$g_{COD}\ m^{-3}$
K_{TENE}	Half-saturation of toluene for OHOs	14.8	$g_{COD}\ m^{-3}$
K_{EBENE}	Half-saturation of ethylbenzene for OHOs	3.8	$g_{COD}\ m^{-3}$
K_{XENE}	Half-saturation of xylene for OHOs	17.6	$g_{COD}\ m^{-3}$
$K_{BENE,ana}$	Half-saturation of benzene in fermentation by OHOs	238	$g_{COD}\ m^{-3}$
$K_{TENE,ana}$	Half-saturation of toluene in fermentation by OHOs	310	$g_{COD}\ m^{-3}$
$K_{EBENE,ana}$	Half-saturation of ethylbenzene in fermentation by OHOs	67	$g_{COD}\ m^{-3}$
$K_{XENE,ana}$	Half-saturation of xylene in fermentation by OHOs	615	$g_{COD}\ m^{-3}$
	Stoichiometric yields		
Symbol	**Name**	**Value**	**Unit**
$Y_{OHO,BTEX,ox}$	Yield of OHOs on BTEX under aerobic conditions	0.55	$g\ X_{OHO}\ g\ S_{BTEX}^{-1}$
$Y_{OHO,BTEX,anox}$	Yield of OHOs on BTEX under anoxic conditions	0.35	$g\ X_{OHO}\ g\ S_{BTEX}^{-1}$
$Y_{OHO,BTEX,ana}$	Yield of OHOs on BTEX under anaerobic conditions	0.10	$g\ X_{OHO}\ g\ S_{BTEX}^{-1}$

Appendix C. Gas Transfer, Aeration and BTEX Model Parameters

Table A12. Gas Transfer, Aeration and BTEX Model Parameters.

Henry Coefficients			
Symbol	**Name**	**Value**	**Unit**
$Henry_{BENE,25}$	Henry coefficient for benzene at 25 °C	1.70×10^{-3}	$mol\ m^{-3}\ Pa^{-1}$
$Henry_{BENE,dt}$	Henry's law temperature dependency factor of benzene	4150	K
$Henry_{TENE,25}$	Henry coefficient for toluene at 25 °C	1.50×10^{-3}	$mol\ m^{-3}\ Pa^{-1}$
$Henry_{TENE,dt}$	Henry's law temperature dependency factor of toluene	4150	K
$Henry_{EBENE,25}$	Henry coefficient for ethylbenzene at 25 °C	1.27×10^{-3}	$mol\ m^{-3}\ Pa^{-1}$
$Henry_{EBENE,dt}$	Henry's law temperature dependency factor of ethylbenzene	5100	K
$Henry_{XENE,25}$	Henry coefficient for xylene at 25 °C	1.56×10^{-3}	$mol\ m^{-3}\ Pa^{-1}$
$Henry_{XENE,dt}$	Henry's law temperature dependency factor of xylene	4083	K
	Diffusion coefficients		
Symbol	**Name**	**Value**	**Unit**
$D_{BENE,25}$	Diffusion coefficient of benzene in water at 25 °C	9.13×10^{-5}	$m^2\ d^{-1}$
$D_{TENE,25}$	Diffusion coefficient of toluene in water at 25 °C	7.89×10^{-5}	$m^2\ d^{-1}$
$D_{EBENE,25}$	Diffusion coefficient of ethylbenzene in water at 25 °C	7.27×10^{-5}	$m^2\ d^{-1}$
$D_{XENE,25}$	Diffusion coefficient of xylene in water at 25 °C	7.08×10^{-5}	$m^2\ d^{-1}$
	Oxygen transfer efficiency correlation parameters		
Symbol	**Name**	**Value**	**Unit**
$SSOTE_0$	Intercept in SSOTE correlation	7.77	$\%\ m^{-1}$
exp_{SSOTE}	Exponent (absolute value) in SSOTE correlation	0.01041	$d\ m^{-3}{}_{gas}$
$SSOTE_{asym}$	Asymptote in SSOTE correlation	5.75	$\%\ m^{-1}$

Table A12. *Cont.*

	Henry Coefficients		
$div_{d,diff}$	Divisor value in a diffuser density correction term	0.1173	$m^2\ m^{-2}$
$pow_{d,diff}$	Power value in a diffuser density correction term	0.1329	
$coeff_{lead,h,diff}$	Leading coefficient in a diffuser submergence correction term	0.011	m^{-1}
$pow_{h,diff}$	Power value in a diffuser submergence correction term	1.6031	
$coeff_{lin,h,diff}$	Linear coefficient in a diffuser submergence correction term	−0.0229	m^{-1}
	Specific molecular masses		
Symbol	**Name**	**Value**	**Unit**
$MM_{EQ,GBENE}$	Equivalent molar mass of benzene	239.97	$g_{COD}\ mol^{-1}$
$MM_{EQ,GTENE}$	Equivalent molar mass of toluene	287.96	$g_{COD}\ mol^{-1}$
$MM_{EQ,GEBENE}$	Equivalent molar mass of ethylbenzene	335.95	$g_{COD}\ mol^{-1}$
$MM_{EQ,GXENE}$	Equivalent molar mass of xylene	335.95	$g_{COD}\ mol^{-1}$

Appendix D. GAC Model Parameters

Table A13. GAC Model Parameters.

	State Variable Equivalent Mass Ratios to Carbon		
Symbol	**Name**	**Value**	**Unit**
$i_{C,VFA}$	COD-to-carbon-mass ratio of VFA	5.33	$g_{COD}\ g_C^{-1}$
$i_{C,BENE}$	COD-to-carbon-mass ratio of benzene	19.98	$g_{COD}\ g_C^{-1}$
$i_{C,TENE}$	COD-to-carbon-mass ratio of toluene	23.98	$g_{COD}\ g_C^{-1}$
$i_{C,EBENE}$	COD-to-carbon-mass ratio of ethylbenzene	27.97	$g_{COD}\ g_C^{-1}$
$i_{C,XENE}$	COD-to-carbon-mass ratio of xylene	27.97	$g_{COD}\ g_C^{-1}$
$i_{C,SB}$	COD-to-carbon-mass ratio of readily biodegradable substrate	3.20	$g_{COD}\ g_C^{-1}$
$i_{C,SU}$	COD-to-carbon-mass ratio of soluble unbiodegradable organics	2.80	$g_{COD}\ g_C^{-1}$
$i_{C,SN,B}$	Nitrogen-to-carbon-mass ratio of soluble biodegradable organic N	1.17	$g_N\ g_C^{-1}$
	Breakthrough curve parameters		
Symbol	**Name**	**Value**	**Unit**
f_{break}	Breakpoint fraction	0.05	
sl_{break}	Breakthrough curve slope	0.00015	$m^3\ g^{-1}$
$pow_{mid,asymm}$	Power of the midpoint asymmetry correction term	20.00	
$magn_{mid,asymm}$	Magnitude of the midpoint asymmetry correction term	0.50	

References

1. Scheierling, S.M.; Bartone, C.; Mara, D.D.; Drechsel, P. *Improving Wastewater Use in Agriculture: An Emerging Priority*; World Bank Policy Research Working Paper Series 5412; The World Bank: Washington, DC, USA, 2010. [CrossRef]
2. Scheierling, S.M.; Bartone, C.R.; Mara, D.D.; Drechsel, P. Towards an agenda for improving wastewater use in agriculture. *Water Int.* **2011**, *36*, 420–440. [CrossRef]
3. Ofori, S.; Puškáčová, A.; Růžičková, I.; Wanner, J. Treated wastewater reuse for irrigation: Pros and cons. *Sci. Total Environ.* **2021**, *760*, 144026. [CrossRef] [PubMed]
4. Pedrero, F.; Kalavrouziotis, I.; Alarcón, J.J.; Koukoulakis, P.; Asano, T. Use of treated municipal wastewater in irrigated agriculture—Review of some practices in Spain and Greece. *Agric. Water Manag.* **2010**, *97*, 1233–1241. [CrossRef]
5. Jaramillo, M.F.; Restrepo, I. Wastewater Reuse in Agriculture: A Review about Its Limitations and Benefits. *Sustainability* **2017**, *9*, 1734. [CrossRef]
6. Khalil, S.; Kakar, M.K. Agricultural use of untreated urban wastewater in Pakistan. *Asian J. Agric. Rural Dev.* **2011**, *1*, 21–26.
7. Sheikholeslami, Z.; Kebria, D.Y.; Qaderi, F. Nanoparticle for degradation of BTEX in produced water; an experimental procedure. *J. Mol. Liq.* **2018**, *264*, 476–482. [CrossRef]
8. Dehghani, M.; Abbasi, A.; Taherzadeh, Z.; Dehghani, S. Exposure assessment of wastewater treatment plant employees to BTEX: A biological monitoring approach. *Sci. Rep.* **2022**, *12*, 21433. [CrossRef]
9. Davidson, C.J.; Hannigan, J.H.; Bowen, S.E. Effects of inhaled combined Benzene, Toluene, Ethylbenzene, and Xylenes (BTEX): Toward an environmental exposure model. *Environ. Toxicol. Pharmacol.* **2021**, *81*, 103518. [CrossRef]
10. Dhivakar, V.; Rajan, T. BTEX compounds removal from wastewater by using UV&UV/H_2O_2 process. *Int. J. Recent Eng. Sci.* **2018**, *5*, 22–25. [CrossRef]

11. Elmobarak, W.F.; Hameed, B.H.; Almomani, F.; Abdullah, A.Z. A Review on the Treatment of Petroleum Refinery Wastewater Using Advanced Oxidation Processes. *Catalysts* **2021**, *11*, 782. [CrossRef]
12. Mello, J.M.M.; Brandao, H.L.; Valerio, A.; de Souza, A.A.U.; de Oliveira, D.; da Silva, A. Biodegradation of BTEX compounds from petrochemical wastewater: Kinetic and toxicity. *J. Water Process Eng.* **2019**, *32*, 100914. [CrossRef]
13. Gusmão, V.R.; Martins, T.H.; Chinalia, F.A.; Sakamoto, I.K.; HenriqueThiemann, O.; Varesche, M.B.A. BTEX and ethanol removal in horizontal-flow anaerobic immobilized biomass reactor, under denitrifying condition. *Process Biochem.* **2006**, *41*, 1391–1400. [CrossRef]
14. Kuyukina, M.S.; Krivoruchko, A.V.; Ivshina, I.B. Advanced Bioreactor Treatments of Hydrocarbon-Containing Wastewater. *Appl. Sci.* **2020**, *10*, 831. [CrossRef]
15. Trusek-Holownia, A.; Noworyta, A. Advanced treatment of wastewater with BTEX. *Desalination Water Treat.* **2012**, *50*, 440–445. [CrossRef]
16. Takáčová, A.; Smolinská, M.; Semerád, M.; Matúš, P. Degradation of BTEX by microalgae *Parachlorella kessleri*. *Pet. Coal* **2015**, *57*, 2.
17. Anjum, H.; Johari, K.; Gnanasundaram, N.; Ganesapillai, M.; Arunagiri, A.; Regupathi, I.; Thanabalan, M. A review on adsorptive removal of oil pollutants (BTEX) from wastewater using carbon nanotubes. *J. Mol. Liq.* **2019**, *277*, 1005–1025. [CrossRef]
18. El-Naas, M.H.; Acio, J.A.; El Telib, A.E. Aerobic biodegradation of BTEX: Progresses and prospects. *J. Environ. Chem. Eng.* **2014**, *2*, 1104–1122. [CrossRef]
19. Carvajal, A.; Akmirza, I.; Navia, D.; Pérez, R.; Muñoz, R.; Lebrero, R. Anoxic denitrification of BTEX: Biodegradation kinetics and pollutant interactions. *J. Environ. Manag.* **2018**, *214*, 125–136. [CrossRef] [PubMed]
20. Kasi, M.; Wadhawan, T.; McEvoy, J.; Padmanabhan, G.; Khan, E. Effect of Carbon Source during Enrichment on BTEX Degradation by Anaerobic Mixed Bacterial Cultures. *Biodegradation* **2013**, *24*, 279–293. [CrossRef] [PubMed]
21. de Nardi, I.R.; Zaiat, M.; Foresti, E. Kinetics of BTEX degradation in a packed-bed anaerobic reactor. *Biodegradation* **2007**, *18*, 83. [CrossRef]
22. Lee, K.C.; Rittmann, B.E.; Shi, J.; McAvoy, D. Advanced steady-state model for the fate of hydrophobic and volatile compounds in activated sludge. *Water Environ. Res.* **1998**, *70*, 1118–1131. [CrossRef]
23. Pomiès, M.; Wisniewski, C.; Choubert, J.M.; Coquery, M. Modelling of micropollutant removal in biological wastewater treatments: A review. *Sci. Total Environ.* **2012**, *443*, 733–748. [CrossRef]
24. Orhon, D.; Çokgör, E.U. COD fractionation in wastewater characterization—The state of the art. *J. Chem. Technol. Biotechnol.* **1997**, *68*, 283–293. [CrossRef]
25. Patry, G.G.; Takács, I. Settling of flocculent suspensions in secondary clarifiers. *Water Res.* **1992**, *26*, 473–479. [CrossRef]
26. Henze, M.; Gujer, W.; Mino, T.; Matsuo, T.; Wentzel, M.C.; Marais, G.V.R.; Van Loosdrecht, M.C. Activated sludge model no. 2d, ASM2d. *Water Sci. Technol.* **1999**, *39*, 165–182. [CrossRef]
27. Arnell, M.; Ahlström, M.; Wärff, C.; Saagi, R.; Jeppsson, U. Plant-Wide Modelling and Analysis of WWTP Temperature Dynamics for Sustainable Heat Recovery from Wastewater. *Water Sci. Technol.* **2021**, *84*, 1023–1036. [CrossRef]
28. Rieger, L.; Gillot, S.; Langergraber, G.; Ohtsuki, T.; Shaw, A.; Takács, I. *Guidelines for Using Activated Sludge Models*; IWA Publishing: London, UK, 2012; ISBN 978-1-84339-174-6.
29. Henze, M.; Grady, L., Jr.; Gujer, W.; Marais, G.; Matsuo, T. *Activated Sludge Model No 1*; IAWPRC Publishing: London, UK, 1987.
30. Dynamita. *Sumo22 User Manual*; Dynamita SARL: Sigale, France, 2022.
31. Gazsó, Z.; Házi, F.; Kenyeres, I.; Váci, L. Full-Scale Wastewater Treatment Plant Simulation for Real-Time Optimization. *Water Pract. Technol.* **2017**, *12*, 848–856. [CrossRef]
32. Herrmann-Heber, R.; Reinecke, S.F.; Hampel, U. Dynamic Aeration for Improved Oxygen Mass Transfer in the Wastewater Treatment Process. *Chem. Eng. J.* **2019**, *386*, 122068. [CrossRef]
33. U.S. Environmental Protection Agency. *Design Manual: Fine Pore Aeration Systems*; EPA/625/1-89/023; U.S. Environmental Protection Agency: Cincinnati, OH, USA, 1989.
34. Sincero, A.P.; Sincero, G.A. *Physical-Chemical Treatment of Water and Wastewater*, 1st ed.; CRC Press: Boca Raton, FL, USA, 2002; ISBN 978-1-58716-124-7.
35. Stenstrom, M.K.; Leu, S.-Y.; Jiang, P. Theory to Practice: Oxygen Transfer and the New ASCE Standard. *Proc. Water Environ. Fed.* **2006**, *7*, 4838–4852. [CrossRef]
36. DelSontro, T.; McGinnis, D.; Wehrli, B.; Ostrovsky, I. Size Does Matter: Importance of Large Bubbles and Small-Scale Hot Spots for Methane Transport. *Environ. Sci. Technol.* **2015**, *49*, 1268–1276. [CrossRef]
37. Jensen, M.B.; Kofoed, M.V.W.; Fischer, K.; Voigt, N.V.; Agneessens, L.M.; Batstone, D.J.; Ottosen, L.D.M. Venturi-Type Injection System as a Potential H2 Mass Transfer Technology for Full-Scale in Situ Biomethanation. *Appl. Energy* **2018**, *222*, 840–846. [CrossRef]
38. Roberts, P.V.; Munz, C.; Dändliker, P. Modeling Volatile Organic Solute Removal by Surface and Bubble Aeration. *J. Water Pollut. Control Fed.* **1984**, *56*, 157–163.
39. Khalil, A.; Rosso, D.; DeGroot, C.T. Effects of Flow Velocity and Bubble Size Distribution on Oxygen Mass Transfer in Bubble Column Reactors—A Critical Evaluation of the Computational Fluid Dynamics-Population Balance Model. *Water Environ. Res.* **2021**, *93*, 2274–2297. [CrossRef] [PubMed]
40. Hayduk, W.; Laudie, H. Prediction of Diffusion Coefficients for Nonelectrolytes in Dilute Aqueous Solutions. *AIChE J.* **1974**, *20*, 611–615. [CrossRef]

41. Yaws, C.L. *Handbook of Transport Property Data: Viscosity, Thermal Conductivity, and Diffusion Coefficients of Liquids and Gases*; Library of Physico-Chemical Property Data; Gulf Publishing Company: Houston, TX, USA, 1995; ISBN 978-0-88415-392-4.
42. New Jersey Department of Environmental Protection. *Chemical Properties for Calculation of Impact to Ground Water Soil Remediation Standards*; New Jersey Department of Environmental Protection: Trenton, NJ, USA, 2008. [CrossRef]
43. Batstone, D.; Flores-Alsina, X. *Generalised Physicochemical Model (PCM) for Wastewater Processes*; IWA Publishing: London, UK, 2022; ISBN 978-1-78040-982-5.
44. Bencsik, D.; Takács, I.; Rosso, D. Dynamic Alpha Factors: Prediction in Time and Evolution along Reactors. *Water Res.* **2022**, *216*, 118339. [CrossRef]
45. U.S. Environmental Protection Agency. *Summary Report: Fine Pore (Fine Bubble) Aeration Systems*; EPA/625/8-85/010; U.S. Environmental Protection Agency: Cincinnati, OH, USA, 1989.
46. Jiang, L.-M.; Chen, L.; Zhou, Z.; Sun, D.; Li, Y.; Zhang, M.; Liu, Y.; Du, S.; Chen, G.; Yao, J. Fouling Characterization and Aeration Performance Recovery of Fine-Pore Diffusers Operated for 10 Years in a Full-Scale Wastewater Treatment Plant. *Bioresour. Technol.* **2020**, *307*, 123197. [CrossRef]
47. Bencsik, D.; Wadhawan, T.; Takács, I.; Bott, C.; Rosso, D. Improved Aeration Modelling Using Innovative Concepts for Prediction of Key Factors in Oxygen Transfer. In Proceedings of the Innovations in Process Engineering 2021, Virtual, 9–10 and 15–16 June 2022; Water Environment Federation: Alexandria, VA, USA, 2021.
48. Morgan, P.F.; Bewtra, J.K. Air diffuser efficiencies. *J. Water Pollut. Control Fed.* **1960**, *32*, 1047–1059.
49. Eckenfelder, W.W. Factors affecting the aeration efficiency of sewage and industrial wastes. *Sew. Ind. Wastes* **1959**, *31*, 60–70.
50. Wagner, M.R.; Pöpel, H.J. Oxygen transfer and aeration efficiency—Influence of diffuser submergence, diffuser density, and blower type. *Water Sci. Technol.* **1998**, *38*, 1–6. [CrossRef]
51. Pöpel, H.J.; Wagner, M.R. Modelling of Oxygen Transfer in Deep Diffused-Aeration Tanks and Comparison with Full-Scale Plant Data. *Water Sci. Technol.* **1994**, *30*, 71–80. [CrossRef]
52. Rosso, D.; Iranpour, R.; Stenstrom, M.K. Fifteen Years of Offgas Transfer Efficiency Measurements on Fine-Pore Aerators: Key Role of Sludge Age and Normalized Air Flux. *Water Environ. Res.* **2005**, *77*, 266–273. [CrossRef] [PubMed]
53. Plósz, B.G.; Jobbágy, A.; Grady, C.P.L. Factors Influencing Deterioration of Denitrification by Oxygen Entering an Anoxic Reactor through the Surface. *Water Res.* **2003**, *37*, 853–863. [CrossRef] [PubMed]
54. Sander, R. Compilation of Henry's Law Constants (Version 4.0) for Water as Solvent. *Atmos. Chem. Phys.* **2015**, *15*, 4399–4981. [CrossRef]
55. Mackay, D.; Shiu, W.Y. A Critical Review of Henry's Law Constants for Chemicals of Environmental Interest. *J. Phys. Chem. Ref. Data* **1981**, *10*, 1175–1199. [CrossRef]
56. Staudinger, J.; Roberts, P.V. A Critical Review of Henry's Law Constants for Environmental Applications. *Crit. Rev. Environ. Sci. Technol.* **1996**, *26*, 205–297. [CrossRef]
57. Staudinger, J.; Roberts, P.V. A Critical Compilation of Henry's Law Constant Temperature Dependence Relations for Organic Compounds in Dilute Aqueous Solutions. *Chemosphere* **2001**, *44*, 561–576. [CrossRef] [PubMed]
58. Water Environment Federation. *Design of Municipal Wastewater Treatment Plants MOP 8*, 5th ed.; McGraw-Hill Education: Alexandria, VA, USA, 2009; ISBN 978-0-07-166358-8.
59. Cecen, F.; Aktas, Ö. *Activated Carbon for Water and Wastewater Treatment: Integration of Adsorption and Biological Treatment*; WILEY-VCH: Weinheim, Germany, 2011; ISBN 978-3-527-63945-8.
60. Benedek, P.; Major, V.; Takács, I. Mathematical Model Suggested for a Carbon-Activated Sludge System. *Water Res.* **1985**, *19*, 407–413. [CrossRef]
61. Sun, Y.-X.; Yang, Z.; Ye, T.; Shi, N.; Tian, Y. Evaluation of the Treatment of Reverse Osmosis Concentrates from Municipal Wastewater Reclamation by Coagulation and Granular Activated Carbon Adsorption. *Environ. Sci. Pollut. Res.* **2016**, *23*, 13543–13553. [CrossRef] [PubMed]
62. Vahala, R.; Rintala, J.; Järvinen, A. Removal of TOC by Two-Step GAC Filtration in Cold Humic Waters. In Proceedings of the NOM Workshop, Poitiers, France, 18 September 1996.
63. Alex, J.; Benedetti, L.; Copp, J.; Gernaey, K.; Jeppsson, U.; Nopens, I.; Pons, M.; Rieger, L.; Rosen, C.; Steyer, J.-P. *Benchmark Simulation Model No. 1 (BSM1)*; Report by the IWA Taskgroup on Benchmarking of Control Strategies for WWTPs; Lund University: Lund, Sweden, 2008.
64. Mrowiec, B. Effect of BTX on Biological Treatment of Sewage. *Environ. Prot. Eng.* **2009**, *35*, 197–206.
65. Smith, R.C.; Elger, S.O.; Mleziva, S. Implementation of solids retention time (SRT) control in wastewater treatment. *Xylem Anal.* **2015**, *20*, 1–6.
66. Benstoem, F.; Nahrstedt, A.; Boehler, M.; Knopp, G.; Montag, D.; Siegrist, H.; Pinnekamp, J. Performance of granular activated carbon to remove micropollutants from municipal wastewater—A meta-analysis of pilot-and large-scale studies. *Chemosphere* **2017**, *185*, 105–118. [CrossRef] [PubMed]
67. Zha, Y.; Wang, Y.; Liu, S.; Liu, S.; Yang, Y.; Jiang, H.; Zhang, Y.; Qi, L.; Wang, H. Adsorption characteristics of organics in the effluent of ultra-short SRT wastewater treatment by single-walled, multi-walled, and graphitized multi-walled carbon nanotubes. *Sci. Rep.* **2018**, *8*, 17245. [CrossRef] [PubMed]

68. Costa, A.S.; Romão, L.P.C.; Araújo, B.R.; Lucas, S.C.O.; Maciel, S.T.A.; Wisniewski Jr, A.; Alexandre, M.D.R. Environmental strategies to remove volatile aromatic fractions (BTEX) from petroleum industry wastewater using biomass. *Bioresour. Technol.* **2012**, *105*, 31–39. [CrossRef] [PubMed]
69. Torfs, E.; Nicolaï, N.; Daneshgar, S.; Copp, J.B.; Haimi, H.; Ikumi, D.; Johnson, B.; Plosz, B.B.; Snowling, S.; Townley, L.R.; et al. The Transition of WRRF Models to Digital Twin Applications. *Water Sci. Technol.* **2022**, *85*, 2840–2853. [CrossRef]
70. Zietzschmann, F.; Stützer, C.; Jekel, M. Granular activated carbon adsorption of organic micro-pollutants in drinking water and treated wastewater–aligning breakthrough curves and capacities. *Water Res.* **2016**, *92*, 180–187. [CrossRef]

environments

MDPI

Article

The Effect of Foam-Recycle on Ammonium Removal by Aerobic Denitrification Using *Alcaligenes faecalis* No. 4

Kwanyong Lee [1] and Hung-Soo Joo [2,*]

1 Department of Bio and Environmental Health, Dongnam Health University, Suwon 16328, Republic of Korea; kwanyonglee@dongnam.ac.kr

2 Department of Environmental Engineering, Anyang University, Anyang 14028, Republic of Korea

* Correspondence: hjoo@anyang.ac.kr or hungsoo.joo@gmail.com; Tel.: +82-31-463-1291

Abstract: Aerobic denitrifier *Alcaligenes faecalis* No. 4 removes ammonium-nitrogen to nitrogen gas via denitrification in a single aerobic condition. In our previous studies, factors such as ammonium removal rate, denitrification ratio, and cell growth were tested in various conditions. The removal pathway from ammonium to nitrogen gas still needs to be determined in detail. To clarify this pathway of *Alcaligenes faecalis* No. 4, we in this study investigated the effects of several factors on ammonium removal, such as foam-recycle, initial pH, initial ammonium concentration, and airflow rate. Denitrification ratio was improved by up to 23% through foam-recycle. The improvement of the denitrification ratio was resulted by the higher enzyme activity of hydroxylamine oxidoreductase (HAO) in the produced foam, which was about 28 times higher than that in the culture broth (i.e., without foam-recycle). The stripped ammonia was significantly high (above pH 9). The initial ammonium concentration and airflow rate also influenced the denitrification ratio.

Keywords: ammonium removal; heterotrophic nitrification; aerobic denitrification; foam-recycle; hydroxylamine oxidoreductase

1. Introduction

Traditionally, biochemical or biological ammonium removal from wastewater is commonly carried out using autotrophic nitrification by nitrifiers under aerobic conditions and heterotrophic denitrification by denitrifying bacteria under anaerobic or anoxic conditions [1]. The disadvantages of this approach are mainly caused by autotrophic nitrifiers. Since those are slow growers and are vulnerable to organic matter, two separate reactors or two-chambers (aerobic and anaerobic) are needed to convert NH_4^+-N to N_2 [2–6]. As an alternative to this traditional approach, microorganisms that have faster growth rates than autotrophic nitrifiers and the ability to convert ammonium to N_2 by simultaneously heterotrophic nitrification and aerobic denitrification have attracted significant attention [7–13]. These organisms assimilate a part of nitrogen and organic carbon to their cell mass, and their assimilation ratio commonly increases at high ratios of C/N, resulting in decrease in denitrification ratio [14,15]. In our previous studies, the bacterium *A. faecalis* No. 4 (No. 4) has such heterotrophic nitrification and aerobic denitrification abilities, as well as rapid growth rates (doubles every 2.4 h) [16]. No. 4 denitrified 40~50%, assimilated 50%, and nitrified less than 10% of removed ammonium [16]. The ammonium removal rate of No. 4 was estimated at 20~30 mg-N/L·h, which was also significantly higher than that of other bacteria with similar abilities [16,17].

Operational conditions such as initial pH, initial ammonium concentration, airflow rate, etc. may influence the performance of ammonium removal processes. Furthermore, ammonia stripping is generally affected by pH conditions in the aerated cultivation, since pH controls the concentration of free ammonia. An increase in pH increases free ammonia [18,19], which leads to increase in ammonia stripping via aeration. In aerobic operation conditions, excessive oxygen or initial ammonium and carbon concentrations may also affect heterotrophic nitrification and aerobic denitrification [20–27]. Our previous studies revealed that No. 4 characteristically produced significant foam over the aerated culture medium [16]. The viable cell number of No. 4 in the foam was relatively higher than that in the culture broth, implying that foam-recycle may be a potential or key strategy for maintaining high cell mass in the reactor. Therefore, foam-recycle would consequently promote nitrogen removal efficiency and this foam-recycle will be a key operation condition in the aerobic cultivation of No. 4.

There are several enzymes which plays important role in biochemical nitrogen cycles. In those enzymes, the hydroxylamine oxidoreductase (HAO) and nitrite reductase (NIR) are involved in the denitrification process. Traditionally in this process, HAO oxidizes hydroxylamine to nitrite then NIR denitrifies nitrite under anaerobic condition. A few past studies have demonstrated that hydroxylamine was removed through HAO generated by heterotrophic nitrifier-aerobic denitrifier species [28–32]. HAO produced by *A. faecalis* strain TUD was reported to influence the denitrification directly to N_2 from ammonium under aerobic conditions [33–35]. *A. faecalis* No. 4 was shown to oxidize ammonium to hydroxylamine, nitrite and nitrate, simultaneously to denitrify ammonium to nitrogen and nitrous dioxide under aerobic conditions [16]. However, ammonium removal pathway based on enzymes mediated by No. 4 was not clarified, especially the activities of HAO and NIR are suspected to be in play in those processes.

In this study, removal characteristics of high-strength ammonium wastewater using *A. faecalis* No. 4 was studied under several operating conditions, such as foam-recycle, initial pH, airflow rate, and initial ammonium concentration. Ammonium removal rate and denitrification ratio were compared to evaluate the effects of those operation conditions. To further elucidate the activity of heterotrophic nitrification and aerobic denitrification by No. 4, the activities of enzymes HAO and NIR were also investigated.

2. Materials and Methods

2.1. Microorganism Used

A. faecalis No. 4 isolated from sewage sludge was stocked in 25% glycerol solution at −80 °C. The stock solution was inoculated into the culture medium, and cultivated for approximately 24 h. The resulting culture was then inoculated into a fresh medium and cultivated for 16 h before being used as the seed culture.

2.2. Medium

The basic medium formulating the synthetic wastewater was prepared by dissolving 14 g [K_2HPO_4], 6 g [KH_2PO_4], 51 g [trisodium citrate dihydrate], 6 g [$(NH_4)_2SO_4$], 0.2 g [$MgSO_4 \cdot 7H_2O$] and 2 mL [trace mineral solution], in 1 L distilled water. The C/N ratio was adjusted to around 10 which was optimum condition for No. 4 growth as determined in a previous study [16]. Trace mineral solution contained 57.1 g [EDTA·2Na], 3.9 g [$ZnSO_4 \cdot 7H_2O$], 7 g [$CaCl_2 \cdot 2H_2O$], 5.1 g [$MnCl_2 \cdot 4H_2O$], 5.0 g [$FeSO_4 \cdot 7H_2O$], 1.1 g [$(NH_4)_6Mo_7O_{24} \cdot 4H_2O$], 1.6 g [$CuSO_4 \cdot 5H_2O$], and 1.6 g [$CoCl_2 \cdot 6H_2O$ at pH 6.0], per volumetric liter. To determine the viable cell number of No. 4, L agar medium containing 1% peptone, 0.5% NaCl, 0.5% yeast extract, and 1.5% agar (pH 7.0) was used.

2.3. Aerated Batch Cultivation

The aerated batch reactor system shown in Figure 1 was used throughout this study [16]. A total of 360 mL of membrane filtered medium was poured into a 500 mL conical flask and an aliquot of 5% seed culture of No. 4 was inoculated and stirred with a magnetic stirrer at 200-cycles/min. Air was supplied from a compressor at the flow rate of 0.5 L/min. Samples of 2 mL each were taken every 8 or 12 h from the reactor for chemical analysis, measurement of pH and for microbial count. Ammonia stripped from the reactor by the aeration air was captured in a 35 mM H_2SO_4 solution. In previous studies, we had observed foam generation and accumulation in an aerated batch cultivation system, which turned out to be extremely rich in No. 4 biomass. We hypothesized that foam recycle could enhance this process, thus the effect of foam-recycle compared with no-foam-recycle was also investigated in this study. Foam was allowed to accumulate in a foam holder, and subsequently was continuously recycled back into the reactor using a peristaltic pump. Different conditions of initial pH, airflow rate, and ammonium concentrations were also tested to determine the effects of varying these factors on the denitrification of No. 4 under the condition of foam-recycle. The initial pH-conditions of the medium were sequentially set at 6, 7 and 8 using 2 N NaOH and 2 N HCl, while inlet ammonium concentration was adjusted at 1037~1122 ppm, and airflow rate was fixed at 0.5 L/min. To examine the effects of initial ammonium concentration and airflow rate, ammonium concentrations were sequentially set to 250, 1000, and 3000 ppm $NH_4{}^+$-N by adjusting amounts of sodium citrate and ammonium sulfate, and effect of airflow rates was test at 0.2, 0.5, and 0.8 L/min. Both the effects of ammonium concentrations and airflow rate were tested at a fixed pH condition of 7.0.

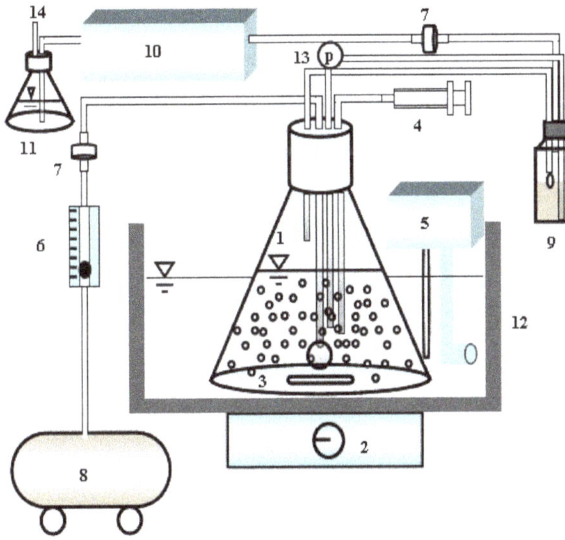

Figure 1. A schematic diagram of an aerated batch reactor used in this study. (1: reactor (500 mL conical flask), 2: stirrer, 3: magnetic bar, 4: sampling syringe, 5: temperature controller, 6: flow meter, 7: air filter, 8: air compressor, 9: foam holder, 10: NOx analyzer, 11: ammonia trap bottle, 12: water bath, 13: peristaltic pump (p), 14: gas effluent).

2.4. Shaking-Flask Cultivation

To verify metabolic properties of hydroxylamine and nitrite those related with HAO and NIR, hydroxylamine and sodium nitrite were used instead of ammonium sulfate. NH_2OH-N and $NO_2{}^-$-N concentrations were adjusted to about 120~130 ppm and 120~170 ppm, respectively. The amount of citrate as a carbon source was adjusted to provide a C/N ratio of 10. A 150 mL of 0.2 μm membrane-filtered medium was set in a 500 mL shaking flask and 2% (0.3 mL) separated foam or 5% (0.75 mL) No. 4 culture broth were inoculated to start the cultivation with similar cell numbers. During incubation at 30 °C and shaking at 120 spm (strokes per minute), 2 mL of samples were taken at 1, 3, 6, and 16 h after seeding for chemical analyses.

2.5. Activities of Hydroxylamine Oxidoreductase (HAO) and Nitrite Reductase (NIR)

To determine the ability of HAO on the reduction of hydroxylamine in the foam and in the culture without foam-recycle, HAO activity was determined by the reduction of $K_3Fe(C/N)_6$ at 400 nm [31]. The molar absorption coefficient (ε) of $K_3Fe(C/N)_6$ was

1/mM·cm. The reaction mixture contained 5 mM [Tris/HCl buffer] at pH 8.0, 100 μM [$K_3Fe(C/N)_6$], 1 mM [NH_2OH], and 400 μM [EDTA]. The optical density at 660nm (OD_{660}) of the No. 4 cells of the separated foam were adjusted to about 0.9 by dilution with 5 mM [Tris/HCl buffer] and centrifuged at 10,000 × rotation per minute (rpm) for 10 min. The separated cells were suspended in Tris/HCl buffer and sonicated for 2 min, while the supernatants were used for HAO analysis. The reaction of the supernatant with $K_3Fe(C/N)_6$ was carried out for 2 min. The viable cell number was used for calculation of enzyme activity per colony forming unit (cfu).

Nitrite reductase (NIR) activity was determined using previously reported methods [34,35]. Briefly, test tubes were loaded with 0.675 mL of 100 mM sodium phosphate buffer (pH 7.0), 0.4 mL of 10 mM sodium nitrite, 0.5 mL of 3 mM methyl viologen, and 0.125 mL of separated foam or the culture. The reaction was started by adding 0.3 mL of 100 mM sodium hydrosulfite dissolved in 50 mM sodium bicarbonate. Control tests were conducted without foam or culture broth. After the test tubes were sealed with a cap, pre-incubation of the reaction mixture was performed at 37 °C for 5 min under anaerobic conditions and the reaction mixture was shaken until complete decolorization. A 0.05 mL of the reaction mixture was mixed with 2.95 mL sterilized water, and then 1 mL of 1% sulfanilamide and 1 mL of 0.02% *N*-1-naphthylenediamine were added. After allowing the reaction mixture to stand at room temperature for 20 min, nitrite detection was performed by UV-VIS spectrophotometer at 540 nm (UV-2400PC, Shimadzu, Kyoto, Japan).

2.6. Analytical Methods

Culture samples were centrifuged at 10,000 rpm, and filtered through 0.2 μm membrane filters. The filtrate was used for the analyses of NH_4^+-N and nitrification products (NH_2OH, NO_2^-, NO_3^-), nitrous oxide (NO), nitrogen dioxide (NO_2), and citrate. Concentrations of NH_4^+-N was analyzed by indophenol method [36–38], NH_2OH was measured according to the method by Fear and Burrell using spectrophotometer [39]. The NO_2^-, NO_3^- and citrate were measured by ion chromatography (HIC-6A, Shimadzu, column: Shim-pack IC-A1, Shimadzu, at flow rate of 1.5 mL min^{-1}, buffer solution containing 2.5 mmol phtalate and 2.4 mmol trisaminomethane, and at oven temperature of 40 °C). Optical density to check growth rate and No. 4 biomass at 660 nm (OD_{660nm}) was measured by a spectrophotometer (UV-1200, Shimadzu). The HAO and NIR activities were measured by UV-VIS recording spectrophotometer (UV-2400PC, Shimadzu). Viable cell number was counted on L agar plate and expressed as colony forming units (cfu). Intracellular nitrogen content (mg-N/L) was calculated from viable cell number and nitrogen content (%) obtained from elemental analysis of dry cell mass. Dry cell was prepared by centrifugation of culture broth, then washing the cells with distilled water, and finally drying the washed cell at 105 °C for 24 h. Denitrification ratio was calculated by subtracting the amount of total nitrifying products (NH_2OH-N, NO_2^--N, NO_3^--N), amount of nitrogen in dry cell mass (intracellular nitrogen), and stripped ammonia from removed ammonium. The NO and NO_2 were analyzed by a NOx analyzer (AP1 200A, Riken Keiki Co., Ltd., Tokyo, Japan).

3. Results and Discussion

3.1. Effect of Foam-Recycle

In our previous studies, ammonium was oxidized to hydroxylamine, nitrite, and nitrate under aerobic condition by A. faecalis No. 4, providing these intermediate substrates for subsequent denitrification [16,17,40]. Hydroxylamine was utilized without citrate consumption [16]. Less than 10% of consumed hydroxylamine was converted to nitrite implying that most of hydroxylamine was converted to denitrified products (such as N_2 and N_2O, more than 80% N_2 production) [16]. When No. 4 was cultivated under aerobic condition, significant foaming was observed. This foam was collected in a separate foam holder (Figure 1) and continuously recycled back into the reactor. About 10~20 mL of foam was obtained in the foam holder based on the 360 mL of culture medium. The OD_{660nm} of the separated foam was significantly higher than that in the normal culture broth of No. 4.

The results of the aerated batch experiments performed with and without foam-recycle are shown in Figures 2 and 3, respectively. In both cases, the consumption of carbon (75~83% citrate was consumed), pH increased and growth of No. 4 (2~3 × 10^{10} cfu) were similar. However, the production of hydroxylamine and nitrite were significantly different. In the culture without foam-recycle, No. 4 produced 82 ppm NO_2^--N in 110 h, while a maximum of 7 ppm hydroxylamine was observed at 24 h point, after that it decreased (Figure 3B). However, in the culture with foam-recycle, No. 4 produced lower NO_2^--N (13 ppm) in 93 h but extremely low production of hydroxylamine was observed (Figure 2B). If hydroxylamine produced was rapidly converted to denitrified products, almost no residual hydroxylamine was detected in the culture broth. Table 1 shows nitrogen mass balance. Ammonium removal rate was 22 mg-N/L·h and denitrification ratio was 39% in the culture without foam-recycle. A similar or slightly higher ammonium removal rate of 24 mg-N/L·h was observed in the culture with foam-recycle. However, denitrification ratio was considerably higher (48%); approximately 9% (23% on 39% basis in the culture) without foam-recycle higher than the culture without foam-recycle.

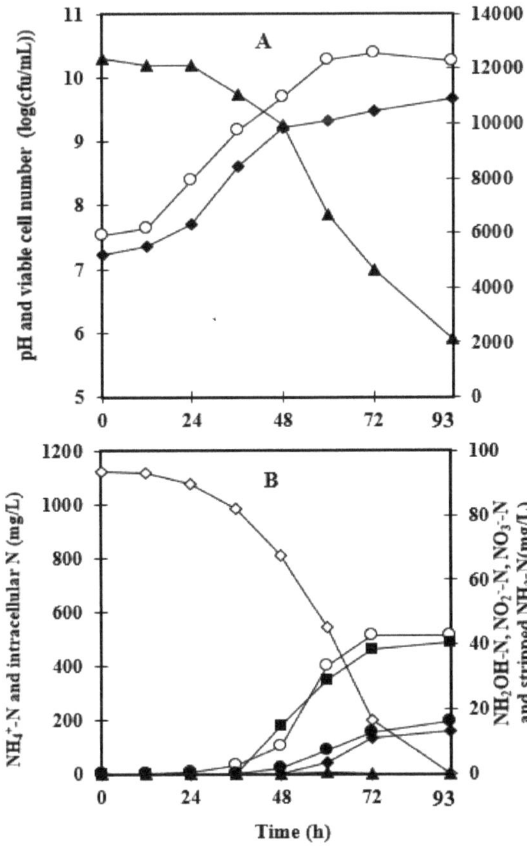

Figure 2. Changes of nitrogen species and substrates, the pH in the culture medium, citrate concentration, and viable cell numbers in the aerated batch experiments with foam recycling (mean values of two replications). (**A**): pH (◆), viable cell number (○) and citrate-C (▲), (**B**): NH_4^+-N (◇), intracellular nitrogen (○), NH_2OH-N (▲), NO_2^--N (◆), stripped NH_3-N (■), and NO_3^--N (•).

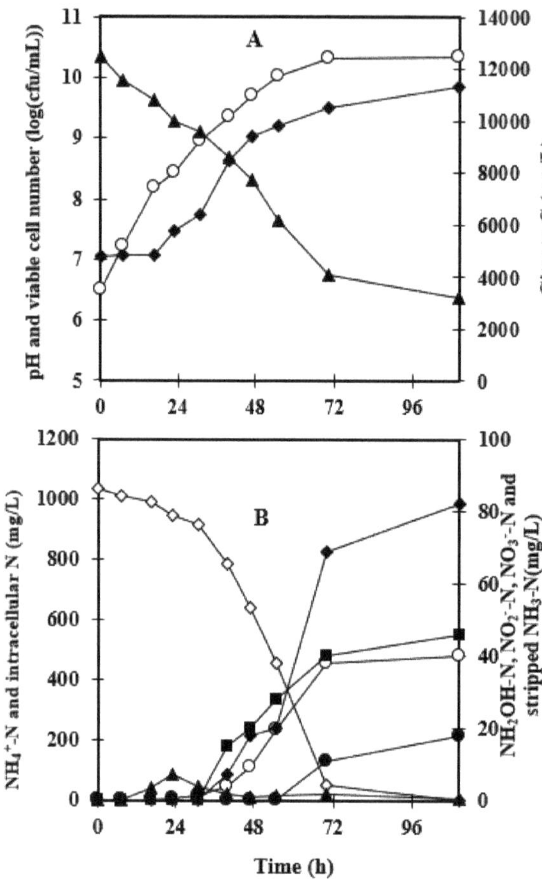

Figure 3. Changes of nitrogen species and substrates, the pH in the culture medium, citrate concentration, and viable cell numbers in the aerated batch experiments without foam recycling (mean values of two replications). (**A**): pH (◆), viable cell number (○) and citrate-C (▲), (**B**): NH_4^+-N (◇), intracellular nitrogen (○), NH_2OH-N (▲), NO_2^--N (◆), stripped NH_3-N (■) and NO_3^--N (•).

Table 1. Nitrogen balance in aerated batch experiments of *A. faecalis* No. 4 with and without foam recycling (Initial pH: 7, the time elasped: 110 h).

	Without Foam Recycling	Foam Recycling
Initial NH_4^+-N (mg/L)	1037 ± 22	1122 ± 5.5
Removed N (mg/L)	1037 ± 22	1122 ± 5.5
Intracellular N (mg/L)	481 ± 9.8	509 ± 13
Nitrification products (mg-N/L)	101 ± 8.7	30 ± 4.3
Stripped N (mg/L)	46 ± 3.6	41 ± 4.3
Denitrification ratio (%) *	39 ± 1.1	48 ± 0.4
Maximum ammonium removal rate (mg-N/L·h)	26 ± 0.9	25 ± 1.2

*: Denitrification ratio is percentage of the removed ammonium based on initial ammonium concentration.

3.2. Consumption of Other Nitrogen Sources

Figures 4 and 5 present the results of No. 4's ability to metabolize other nitrogen sources in the culture with addition of either hydroxylamine or nitrite in the basic medium instead of ammonium source. Figures 4 and 5 show the results with 2% foam-recycle and with 5% No. 4 culture broth (without foam-recycle), respectively. According to Figure 4A, hydroxylamine decreased from the beginning of incubation, and about 9 ppm nitrite was produced in 6 h after consumption of 110 ppm hydroxylamine, but no nitrate was detected. Hydroxylamine decreased at the rate of 1.6×10^{-10} mg-N/cfu × h ((74.54 mg-NH_2OH-N/L) ÷ (7.8×10^7 cfu/mL) ÷ 6 h). Nitrite produced from hydroxylamine decreased slightly at 8 to 16 h, and no production in nitrate was observed. While in Figure 4B, when 116 ppm of NO_2^--N was added to basic medium, only a decreased of 8 ppm of NO_2^--N was observed in 16 h and neither NO_3^- nor hydroxylamine was detected. In the experiment using the culture broth (without foam-recycle) of Figure 5, hydroxylamine decline rate (7.1×10^{-11} mg-N/cfu × h = ((71.46 mg-NH_2OH-N/L) ÷ (6.3×10^7 cfu/mL) ÷ 16 h)) was significantly slower than in the former experiment (Figure 4A). And the NO_2^--N consumption in this experiment (Figures 4B and 5B) was similar to that in both experiments and no consumptions of carbon and nitrogen sources were observed. These results suggest that *A. faecalis* No. 4 either with foam-recycle or without foam-recycle converts most of hydroxylamine to denitrified products and a small portion of hydroxylamine to nitrite. This fact can be explained that denitrification activities between the cultures with foam-recycle and without foam-recycle were different in Table 1, indicating 48% and 39%, respectively.

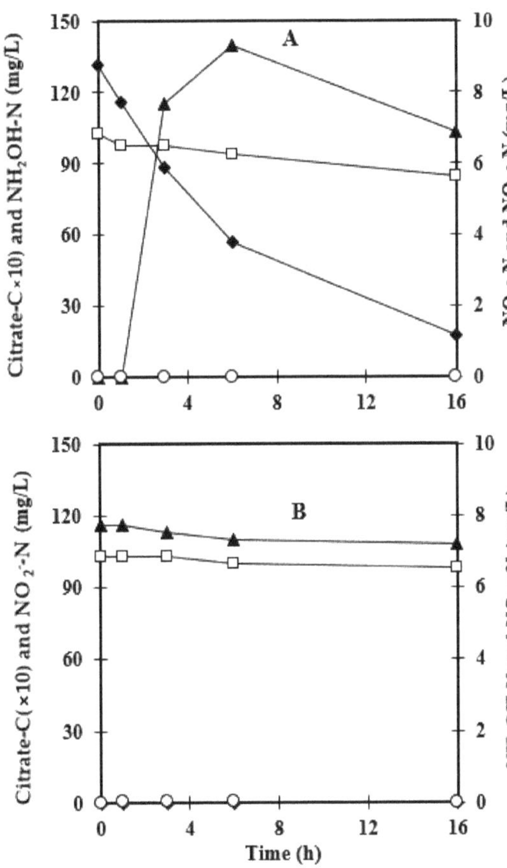

Figure 4. Removal of nitrogen spieces (with hydroxylamine (**A**) and nitrite (**B**) instead of ammonium) and carbon consumption in batch reactors with produced foam. Hydroxylamine (◆), nitrite (▲), nitrate (○), and citrate (□).

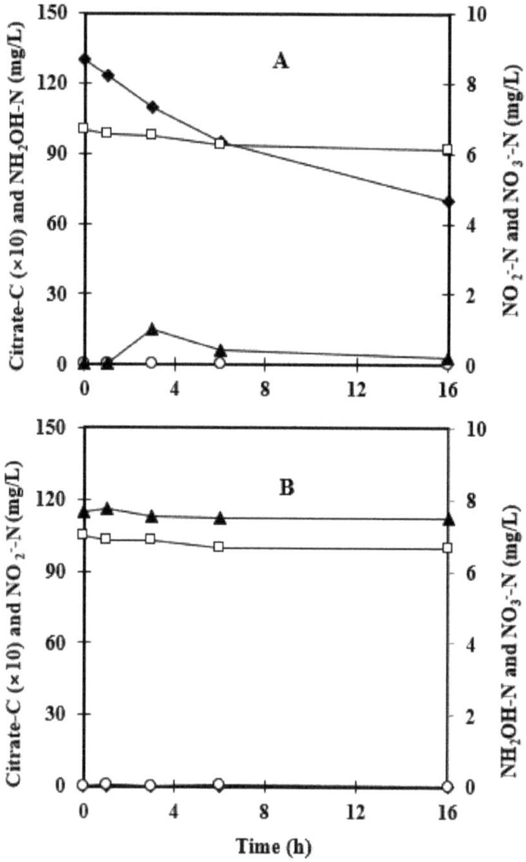

Figure 5. Removal of nitrogen spieces (with hydroxylamine (**A**) and nitrite (**B**) instead of ammonium) and carbon consumption in batch reactors with No. 4 culture. Hydroxylamine (◆), nitrite (▲), nitrate (○), and citrate (□).

3.3. Enzyme Activity

Table 2 shows hydroxylamine oxidoreductase (HAO) activities between the cultures with foam-recycle and without foam-recycle. Measurements of hydroxylamine oxidoreductase (HAO) activity confirmed the existence of HAO activity in both cultures with foam-recycle and without foam-recycle. The HAO activity in the foam was significantly higher than that in the culture without foam-recycle. The activity of HAO per 1 mL of culture broth with foam-recycle was about 28 times higher than that in the culture without foam-recycle and the activity per cfu in the foam was about five times than that in the culture without foam-recycle.

Table 2. Hydroxylamine oxidoreductase activities (HAO) in separated foam and No. 4 culture media.

	Foam	Culture of No. 4
Cell number (cfu/mL)	$2.1 \pm 2.1 \times 10^{10}$	$4.0 \pm 1.2 \times 10^{9}$
Activity (U/mL) *	34.3 ± 3.2	1.25 ± 0.1
Specific Activity (U/cfu)	1.5×10^{-9}	3.3×10^{-10}

* One unit (U) is the amount of the enzyme, which reduce 1 μM-$K_3Fe(C/N)_6$/min.

Measurements of hydroxylamine oxidoreductase (HAO) and nitrite oxidoreductase (NIR) activities to evaluate their involvement in the denitrification process were adopted from previous studies [14,33,35,41–45] The activity of HAO by No. 4 was 10~100 times higher than the previously reported activity levels [33,41,45]. The reason behind this higher activity is not clear yet, but the cells attached to foam significantly enhanced the denitrification ratio with higher HAO activity. This is primarily due to the fact No. 4 (aerobic bacteria) more functionally exhibited their aerobic denitrification ability on the cell-air interface of foam surface.

Although *A. faecalis* is known to denitrify hydroxylamine [33,35,45], denitrification from nitrite has not been reported before in past studies. Figure 6 shows nitrite reductase (NIR) activity measured under anaerobic condition using methyl viologen [36,46]. Activity of NIR was observed both in foam and foam-free culture (No. 4 culture broth). The NIR activity in the culture without foam-recycle was about 25% of that in the culture with foam-recycle. However, NIR activity (nmol-nitrite/μL-foam · min) was significantly lower, compared to HAO activities (Table 2) and NIR activity produced by other microorganisms under anaerobic conditions [47]. The likelihood that NIR activity is associated with denitrification under aerobic conditions is negligible, and therefore the dominant denitrification pathway of No. 4 is most likely via hydroxylamine under HAO.

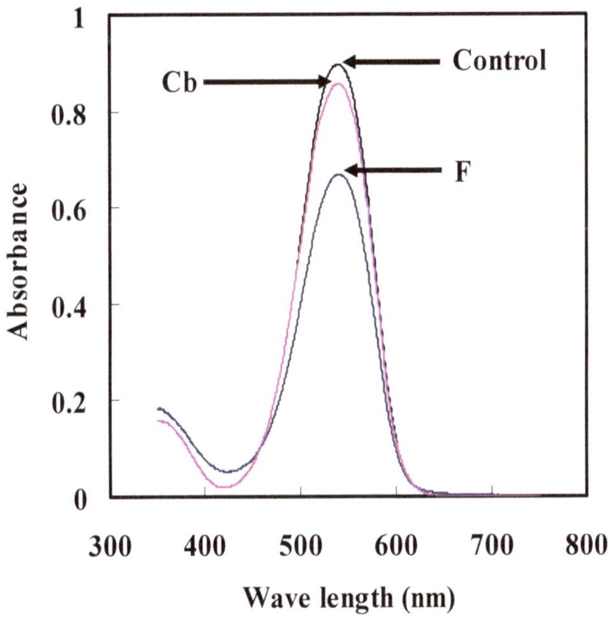

Figure 6. Spectra of nitrite reducing activity (NIR activity). Control (no addition of enzyme), F (separated foam from aerated batch cultivation by No. 4), Cb (No. 4 culture broth).

3.4. Effect of Initial Ammonium Load and Airflow Rate

The effect of initial ammonium concentration on ammonium removal was investigated at 250, 1000, and 3000 ppm under different airflow rates. The nitrogen mass balances for these experiments are shown in Table 3. Enhanced denitrification ratio and ammonium removal rate at the higher ammonium concentrations (1000 and 3000 ppm) were observed at higher airflow rates (0.5 and 0.8 L/min) compared with that at lower airflow rate of 0.2 L/min.

Table 3. Nitrogen balance in different airflow rate and initial ammonium concentrations.

	Airflow Rate (L/min)	Initial Ammonium Level (mg-N/L)		
		3000	**1000**	**250**
Initial N (mg/L)	0.2	2965	1167	217
Removed N (mg/L) (%)		990 (33)	617 (53)	217 (100)
Intracellular N (mg/L) (%) *		378 (38)	318 (52)	110 (51)
Nitrification N (mg/L) (%) *		19 (2)	14 (2)	13 (6)
Stripped N (mg/L) (%) *		289 (29)	40 (6)	2.5 (1)
Denitrification ratio (%) *		31	40	42
Maximum ammonium removal rate (mg-N/L·h)		13	11	11
Initial N (mg/L)	0.5	2975	1190	218
Removed N (mg/L) (%)		2697 (91)	1063 (89)	218 (100)
Intracellular N (mg/L) (%) *		546 (20)	397 (37)	124 (57)
Nitrification N (mg/L (%) *		39 (2)	29 (3)	21 (10)
Stripped N (mg/L) (%) *		333 (12)	59 (6)	5 (2)
Denitrification ratio (%) *		66	54	31
Maximum ammonium removal rate (mg-N/L·h)		28	19	13
Initial N (mg/L)	0.8	2967	1177	225
Removed N (mg/L) (%)		2967 (100)	1104 (94)	225 (100)
Intracellular N (mg/L) (%) *		668 (23)	404 (37)	128 (57)
Nitrification N (mg/L) (%) *		56 (2)	38 (3)	24 (11)
Stripped N (mg/L) (%) *		354 (12)	72 (6)	4 (2)
Denitrification ratio (%) *		64	54	31
Maximum ammonium removal rate (mg-N/L·h)		39	21	15

* Percentages based on removed ammonium. The cultivation times in 3000 ppm, 1000 ppm, and 250 ppm of initial ammonium concentrations (mg-N/L) were 137 h, 80 h and 34 h, respectively.

Denitrification ratio at 3000 ppm initial ammonium level was more than 60% and the ammonium removal rate was the highest. These results indicated that the airflow rate is critical for denitrification ability of No. 4 especially for high-strength ammonium wastewater. Higher ammonium removal rate was achieved at elevated ammonium concentration levels and air supply. Ammonia stripping was also significant at higher ammonium concentrations (i.e., ammonia stripping increased with increase in ammonium concentration). The highest ammonia (170 ppm) which was about 15% of initial ammonium concentration was stripped out at pH 8, while only 79 and 41 ppm of NH_3-N were stripped out at pH 6 and 7, respectively. Ammonium concentration had a higher influence on ammonia stripping than airflow rate as is evident based on data presented in Table 3. Approximately 10~30% ammonia was stripped when initial of NH_4^+-N was 3000 ppm at the three airflow rates, suggesting that nitrogen loss via air stripping should be a concern during treatment of high-strength ammonium wastewater.

3.5. Effect of pH

The nitrogen mass balance and ammonium removal rate after the 96 h of aerated batch process at different initial pH-conditions are shown in Table 4. Ammonium removal rates at pH 7 and 6 were 25 mg-N/L·h, and 20 mg-N/L·h, respectively, which were higher than that at pH 8. The denitrification ratio at pH 6 was higher than that at either pH 7. However, similar denitrification ratios at pH 6 and 7 suggested that the initial pH in aerobic cultivation with No. 4 should be maintained in the range of 6 to 7. Ammonia stripping was significantly dependent on pH-conditions (Figure 7). Extremely low ammonia stripping occurred below pH of 8.5, but this situation changed rapidly in higher pH connditions and ammonia stripping became significant above pH 9. This phenomenon has been well documented in past research and this was atributed to the increase (or release) in the concentration of free ammonia with increase in the pH [11,48]. The reported pH values of high-strength ammonium wastewaters range from 7.5 to 8.5 [49–53]. Given that 7.5 to 8.5 is natural pH range of high-strength wastewater, it is quite apparent that pH control at around 7 would be the optimal compromise for No. 4 to exhibit the maximum capability of denitrification and also to minimize ammonia stripping.

Table 4. Nitrogen balance at different initial pH-conditions.

	pH 6	pH 7	pH 8
Initial NH_4^+-N (mg/L)	1186	1122	1169
Removed N (mg/L)	1165	1122	1159
Intracellular N (mg/L)	424	509	521
Nitrification N (mg/L)	25	30	21
Stripped N (mg/L)	79	41	170
Denitrification ratio (%) *	55	48	39
Maximum ammonium removal rate (mg-N/L)	20	25	15

* Denitrification ratio is percentage against removed ammonium and there were no changes at pH 5.

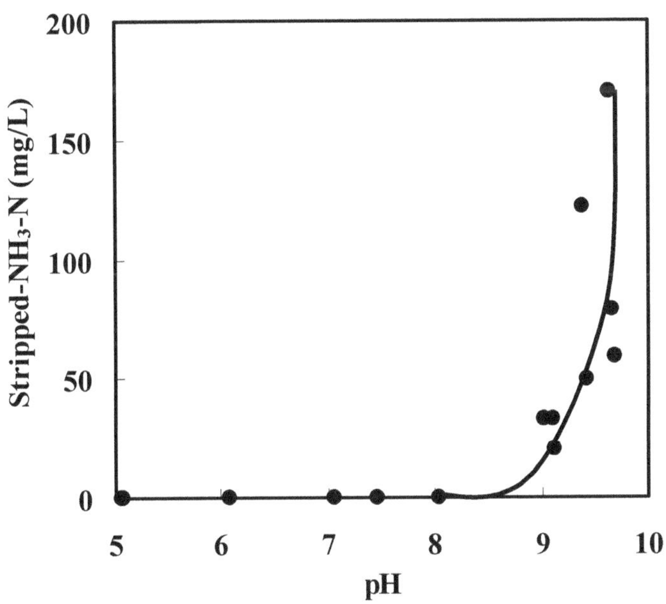

Figure 7. Ammonia stripping with pH change during the aerated batch experiments (simulated based on data presented in Table 4).

The total nitrogen mass balance in the ammonium removal and the removal pathway by No. 4 is shown in Figure 8. This figure was prepared based on the data presented in Figure 2 and some results from previous research [16]. No. 4 removed 51% of ammonium by means of heterotrophic-nitrification and 45% by assimilation into cells. At least 48% of removed ammonium (51%) was converted to denitrified products by HAO activity which was significantly high in the produced foam. Except for a part of assimilation, most of the ammonium (87%) denitrified via hydroxyamine.

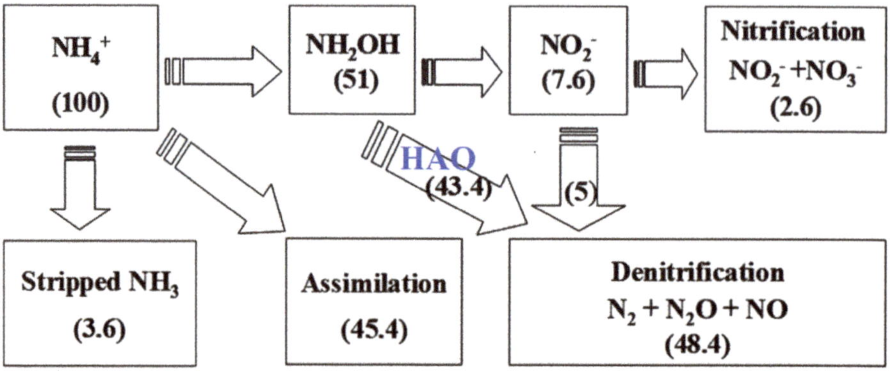

Figure 8. Nitrogen balance and removal pathway in the heterotrophic nitrification-aerobic denitrification by *A. faecalis* No. 4.

4. Conclusions

Factors that affect ammonium removal rate and denitrification ratio of No. 4 include the following foam-recycle, initial pH, airflow rate, and initial ammonium concentration were investigated in this study. Ammonium was first oxidized to hydroxylamine, nitrite, and nitrate under aerobic condition by No. 4, providing these intermediate substrates for subsequent denitrification. Hydroxylamine was utilized without citrate consumption. Less than 10% of consumed hydroxylamine was converted to nitrite, implying that most of hydroxylamine was converted to denitrified products (such as N_2 and N_2O).

Measurements of hydroxylamine oxidoreducatse (HAO) activity confirmed the existence of HAO activity both in the foam and the culture broth. In autotrophic nitrifiers, HAO was observed to convert hydroxylamine to NO_2^-. However, in *A. faecalis* which conducts heterotrophic nitrification and aerobic denitrification, HAO is believed to work also as an enzyme to convert hydroxylamine to denitrified product via HNO. In No. 4 biochemical processes, HAO converted more than 90% of hydroxylamine to denitrified products and 10% to nitrite. The activity of HAO in the foam was significantly higher than that in the culture broth.

A higher ammonium removal rate was achieved at elevated ammonium levels and air supply. Ammonia stripping, however, was significant at higher ammonium concentration (i.e., ammonia stripping increased with increase in ammonium concentration).

No. 4 removed 51% of ammonium by means of heterotrophic-nitrification and 45% by assimilation into cells. At least 48% of removed ammonium was converted to denitrified products. Evidently, under controlled conditions, No. 4 can be used in the treatment of wastewaters with high concentrations of ammonium, such as a piggery and dairy wastewater. Future research to advance the current work should be conducted with these kinds of wastewaters.

Author Contributions: Conceptualization, methodology, formal analysis, investigation, writing review and editing, H.-S.J.; data curation and writing-original draft preparation, K.L. All authors have read and agreed to the published version of the manuscript.

Funding: This work was supported by the research project of "Cooperative Research Program for Agriculture Science & Technology Development (Project No. PJ017075)" Rural Development Administration of Korea and the support of Particulate Matter Management Specialized Graduate Program through the Korea Environmental Industry & Technology Institute (KEITI) funded by the Ministry of Environment (MOE).

Data Availability Statement: Data from this study are available from the corresponding author upon reasonable request.

Conflicts of Interest: The authors declare no conflict of interest.

References

1. McCatry, P.L. What is the Best Biological Process for Nitrogen Removal: When and Why? *Environ. Sci. Technol.* **2018**, *52*, 3835–3841. [CrossRef]
2. Hooper, A.B.; Vannelli, T.; Bergmann, D.J.; Arociero, D.M. Enzymology of the oxidation of ammonia to nitrite by bacteria. *Antonie Van Leeuwenhoek* **1997**, *71*, 59–67. [CrossRef] [PubMed]
3. Van Loosdrecht, M.C.M.; Jetten, M.S.M. Microbiological conversion in nitrogen removal. *Water Sci. Technol.* **1998**, *38*, 1–7. [CrossRef]
4. Itokawa, H.; Hanaki, K.; Matsuo, T. Nitrous oxide production in high-loading biological nitrogen removal process under low COD/N ratio condition. *Water Res.* **2001**, *35*, 657–664. [CrossRef]
5. Wrage, N.; Velthof, G.L.; van Beusichem, M.L.; Oenema, O. Role of nitrifer denitrification in the production of nitrous oxide. *Soil Biol. Biochem.* **2001**, *33*, 1723–1732. [CrossRef]
6. Carrera, J.; Baeza, J.A.; Vicent, T.; Lafuente, J. Biological nitrogen removal of high-strength ammonium industrial wastewater with two-sludge system. *Water Res.* **2003**, *37*, 4211–4221. [CrossRef]
7. Jetten, M.S.M.; Logemann, S.; Muyzer, G.; Robertson, L.A.; de Vries, S.; van Loosdrecht, M.C.M.; Kuenen, J.G. Novel principles in the microbial conversion of nitrogen compounds. *Antonie Van Leeuwenhoek* **1997**, *71*, 75–93. [CrossRef]
8. Gupta, A.B.; Gupta, S.K. Simultaneous carbon and nitrogen removal in mixed culture aerobic RBC biofilm. *Water Res.* **1999**, *33*, 556–561. [CrossRef]
9. Gupta, A.B.; Gupta, S.K. Simultaneous carbon and nitrogen removal from high strength domestic wastewater in an aerobic RBC biofilm. *Water Res.* **2001**, *35*, 1714–1722. [CrossRef]
10. Schmidt, I.; Sliekers, O.; Schmid, M.; Bock, E.; Fuerst, J.; Kuenen, J.G.; Jetten, M.S.M.; Strous, M. New concepts of microbial treatment processes for the nitrogen removal in wastewater. *FEMS Microbiol. Rev.* **2003**, *772*, 481–492. [CrossRef]
11. Khin, T.; Annachhatre, A.P. Novel microbial nitrogen removal processes. *Biotechnol. Adv.* **2004**, *22*, 519–532. [CrossRef] [PubMed]
12. Chen, L.; Chen, L.; Pan, D.; Lin, H.; Ren, Y.; Zhang, J.; Zhou, B.; Lin, J.; Lin, J. Heterotrophic nitrification and related functional gene expression characteristics of *Alcaligenes faecalis* SDU20 with the potential use in swine wastewater treatment. *Bioprocess Biosyst. Eng.* **2021**, *44*, 2035–2050. [CrossRef] [PubMed]
13. Chen, J.; Xu, J.; Zhang, S.; Liu, F.; Peng, J.; Peng, Y.; Wu, J. Nitrogen removal characteristics of a novel heterotrophic nitrification and aerobic denitrification bacteria, *Alcaligenes faecalis* WT14. *J. Environ. Manag.* **2021**, *282*, 111961. [CrossRef] [PubMed]
14. Jetten, M.S.M.; de Bruijn, P.; Kuenen, G. Hydroxylamine metabolism in Pseudomonas PB16: Involvement of a novel hydroxylamine oxidoreductase. *Antonie Van Leeuwenhoek* **1997**, *71*, 69–74. [CrossRef]
15. Wang, Y.; Singh, R.P.; Geng, C.; Fu, D. Carbon to nitrogen ratio influence on the performance of bioretention for wastewater treatment. *Environ. Sci. Pollut. Res.* **2020**, *27*, 17652–17660. [CrossRef]
16. Joo, H.S.; Hirai, M.; Shoda, M. Characteristics of ammonium removal by hetetrophic nitrification-aerobic denitrification by *Alcaligenes faecalis* No. 4. *J. Biosci. Bioeng.* **2005**, *100*, 184–191. [CrossRef]
17. Joo, H.S.; Hirai, M.; Shoda, M. Piggery wastewater treatment using *Alcaligenes faecalis* strain No. 4 with heterotrophic nitrification and aerobic denitrification. *Water Res.* **2006**, *40*, 3029–3036. [CrossRef]
18. Yang, S.F.; Tay, J.H.; Liu, Y. Inhibition of free ammonia to the formation of aerobic granules. *Biochem. Eng. J.* **2004**, *17*, 41–48. [CrossRef]
19. Liu, Y.; Ngo, H.H.; Guo, W.; Peng, L.; Wnag, D.; Ni, B. The roles of free ammonia (FA) in biological wastewater treatment processes: A review. *Environ. Int.* **2019**, *123*, 10–19. [CrossRef]
20. Liu, Y.; Capdeville, B. Response pattern of nitrifying biofilm reactor to shock loading. *Biotechnol. Lett.* **1994**, *16*, 655–660. [CrossRef]
21. Kim, N.J.; Sugano, Y.; Hirai, M.; Shoda, M. Removal of a high load of ammonia gas by a marine bacterium, *Vibrio alginolyticus*. *J. Biosci. Bioeng.* **2000**, *90*, 410–415. [CrossRef] [PubMed]
22. Uemoto, H.; Ando, A.; Saiki, H. Effect of oxygen concentration on nitrogen removal by *Nitrosomonas europaea* and *Paracoccus denitrificans* immobilized within tubular polymeric gel. *J. Biosci. Bioeng.* **2000**, *6*, 654–660. [CrossRef]
23. Su, J.J.; Liu, B.Y.; Liu, C.Y. Comparison of aerobic denitrification under high oxygen atmosphere by *Thiosphaera pantotropha* ATCC 35512 and *Pseudomonas stutzeri* SU2 newly isolated from the activated sludge of a piggery wastewater treatment system. *J. Appl. Microbiol.* **2001**, *90*, 457–462. [CrossRef] [PubMed]
24. Shan, H.; Obbard, J.P. Ammonia removal from prawn aquaculture water using immobilized nitrifying bacteria. *Appl. Microbiol. Biotechnol.* **2001**, *57*, 791–798. [CrossRef] [PubMed]
25. Matsuzaka, E.; Nomura, N.; Nakajima-Kambe, T.; Okada, N.; Nakahara, T. A simple screening procedure for heterotrophic nitrifying bacteria with oxygen-tolerant denitrification activity. *J. Biosci. Bioeng.* **2003**, *95*, 409–411. [CrossRef]
26. Obaja, D.; Mace, S.; Sans, C.C.; Alvarez, J.M. Nitrification, denitrification and biological phosphorus removal in piggery wastewater using a sequencing batch reactor. *Bioresour. Technol.* **2003**, *87*, 103–111. [CrossRef]
27. Du, X.; Wang, J.; Jegatheesan, V.; Shi, G. Dissolved Oxygen Control in Activated Sludge Process Using a Neural Network-Based Adaptive PID Algorithm. *Appl. Sci.* **2018**, *8*, 261. [CrossRef]
28. Padhi, S.K.; Tripathy, S.; Mohanty, S.; Maiti, N.K. Aerobic and heterotrophic nitrogen removal by Enterobacter cloacae CF-S27 with efficient utilization of hydroxylamine. *Bioresour. Technol.* **2017**, *232*, 285–296. [CrossRef]
29. Yang, Y.; Lin, E.; Huang, S. Heterotrophic nitrogen removal in Bacillus sp. K5: Involvement of a novel hydroxylamine oxidase. *Water Sci. Technol.* **2017**, *76*, 3461–3467. [CrossRef]

30. Liu, Y.; Ni, G.-M.; Wu, M.-R.; Li, S.-S.; Miao, L.-L.; Liu, Z.-P. Photobacterium sp. NNA4, an efficient hydroxylamine-transforming heterotrophic nitrifier/aerobic denitrifier. *J. Biosci. Bioeng.* **2019**, *128*, 64–71. [CrossRef]
31. Huang, F.; Pan, L.; He, Z.; Zhang, M.; Zhang, M. Identification, interactions, nitrogen removal pathways and performances of culturable heterotrophic nitrification-aerobic denitrification bacteria from mariculture water by using cell culture and metagenomics. *Sci. Total Environ.* **2020**, *732*, 13268. [CrossRef] [PubMed]
32. Zhang, M.; Li, A.; Yao, Q.; Wu, Q.; Zhu, H. Nitrogen removal characteristics of a versatile heterotrophic nitrifying-aerobic denitrifying bacterium, *Pseudomonas bauzanensis* DN13-1, isolated from deep-sea sediment. *Bioresour. Technol.* **2020**, *305*, 122626. [CrossRef] [PubMed]
33. Otte, S.; Schalk, J.; Kuenen, J.G.; Jetten, M.S.M. Hydroxylamine oxidation and subsequent nitrous oxide production by the heterotrophic ammonia oxidizer *Alcaligenes faecalis*. *Appl. Microbiol. Biotechnol.* **1999**, *51*, 255–261. [CrossRef]
34. Zhao, B.; An, Q.; Liang, H.Y.; Guo, J.S. N_2O and N_2 production during heterotrophic nitrification by *Alcaligenes faecalis* strain NR. *Bioresour. Technol.* **2012**, *116*, 379–385. [CrossRef] [PubMed]
35. Wang, Y.; Chen, H.; Liu, Y.-X.; Ren, R.-P.; Lv, Y.-K. Effect of temperature, salinity, heavy metals, ammonium concentration, pH and dissolved oxygen on ammonium removal by an aerobic nitrifier. *RCS Adv.* **2015**, *5*, 79988–79996. [CrossRef]
36. Suruga, K.; Murakami, K.; Taniyama, Y.; Hama, T.; Chida, H.; Satoh, T.; Yamada, S.; Hakamata, W.; Kawachi, R.; Isogai, Y.; et al. A novel microperoxidase activity: Methyl viologen-linked nitrite reducing activity of microperoxidase. *Biochem. Biophys. Res. Commun.* **2004**, *315*, 815–822. [CrossRef]
37. Da Silva, S.; Cosnier, S.; Almeida, M.G.; Moura, J.J.G. An efficient poly(pyrrole-viologen)-nitrite reductase biosensor for the mediated detection of nitrite. *Electrochem. Commun.* **2004**, *6*, 404–408. [CrossRef]
38. *The Indophenol Blue Absorptiometric Method (JIS) k-0102, Japanese Industrial Standard*; Japanese Standard Association: Tokyo, Japan, 1991; Volume 42, pp. 134–145. Available online: https://www.jsa.or.jp/en/ (accessed on 23 July 2023).
39. Frear, D.S.; Burrell, R.C. Spectrophotometric methods for determining hydroxylamine reductase activity in higher plants. *Anal. Chem.* **1955**, *27*, 1664–1665. [CrossRef]
40. Wako, R. *Removal Characteristics of Ammonium by Heterotrophic Nitrifier Alcaligenes faecalis No. 4*; Tokyo Institute of Technology: Yokohama, Japan, 2002.
41. Hooper, A.B.; Balny, C. Reaction of oxygen with hydroxylamine oxidoreductase of Nitrosomonas fast kinetics. *FEBS Lett.* **1982**, *144*, 229–303. [CrossRef]
42. Sasaki, S.; Karube, I.; Hirota, N.; Arikawa, Y.; Nishiyama, M.; Kukimoto, M.; Horinouchi, S.; Beppu, T. Application of nitrite reductase from *Alcaligenes faecalis* S-6 for nitrite measurement. *Biosens. Bioelectron.* **1998**, *13*, 1–5. [CrossRef]
43. Song, B.; Ward, B.B. Nitrite Reductase genes in halobenzoate degrading denitrifying bacteria. *FEMS Microbiol. Ecol.* **2003**, *43*, 349–357. [CrossRef] [PubMed]
44. Chen, H.; Zhao, X.; Cheng, Y.; Jiang, M.; Li, X.; Xue, G. Iron Robustly Stimulates Simultaneous Nitrification and Denitrification Under Aerobic Conditions. *Environ. Sci. Technol.* **2018**, *52*, 1404–1412. [CrossRef] [PubMed]
45. Joo, H.S.; Lee, K.; Shoda, M.; Mwangi, N.P. Application of *Alcaligenes faecalis* No. 4 for treatment of high-strength ammonium wastewater. *Environ. Eng. Manag. J.* **2020**, *19*, 589–598. Available online: http://www.eemj.icpm.tuiasi.ro/ (accessed on 21 June 2023).
46. Bueno, E.; Sit, B.; Waldor, M.K.; Cava, F. Anaerobic nitrate reduction divergently governs population expansion of the enteropathogen *Vibrio cholerae*. *Nat. Microbiol.* **2018**, *3*, 1346–1353. [CrossRef] [PubMed]
47. Polcyn, W.; Lucinski, R. Aerobic and anaerobic nitrate and nitrite reduction in free-living cells of *Bradyrhizobium* sp. (Lupinus). *FEMS Microbiol. Lett.* **2003**, *226*, 331–337. [CrossRef] [PubMed]
48. Mohammed-Nour, A.; Al-Sewailem, M.; El-Naggar, A.H. The Influence of Alkalization and Temperature on Ammonia Recovery from Cow Manure and the Chemical Properties of the Effluents. *Sustainability* **2019**, *1*, 2441. [CrossRef]
49. Beline, F.; Martinez, J.; Marol, C.; Guiraud, G. Application of the ^{15}N technique to determine the contributions of nitrification and denitrification to the flux of nitrous oxide from aerated pig slurry. *Water Res.* **2001**, *35*, 2774–2778. [CrossRef]
50. Pollice, A.; Tandoi, V.; Lestingi, C. Influence of aeration and sludge retention time on ammonium oxidation to nitrite and nitrate. *Water Res.* **2002**, *36*, 2541–2546. [CrossRef]
51. Uygur, A.; Kargi, F. Biological nutrient removal from pre-treated landfill leachate in a sequencing batch reactor. *J. Environ. Manag.* **2004**, *71*, 9–14. [CrossRef]
52. Obaja, D.; Mace, S.; Alvarez, J.M. Biological nutrient removal by a sequencing batch reactor (SBR) using an internal organic carbon source in digested piggery wastewater. *Bioresour. Technol.* **2005**, *96*, 7–14. [CrossRef]
53. Neerackal, G.M.; Ndegwa, P.M.; Joo, H.S.J.; Harrison, H. Manure-pH management for ammonia emission from dairy barns and liquid manure storage. *Appl. Eng. Agric.* **2017**, *33*, 235–242. [CrossRef]

environments

MDPI

Article

Statistical Characterization of Full-Scale Thermophilic Biological Systems to Inform Process Optimization

Maria Cristina Collivignarelli [1,2], Stefano Bellazzi [1,*], Francesca Maria Caccamo [1], Marco Sordi [3], Barbara Crotti [3], Alessandro Abbà [4] and Marco Baldi [5,*]

1 Department of Civil Engineering and Architecture, University of Pavia, Via Ferrata 3, 27100 Pavia, Italy; mcristina.collivignarelli@unipv.it (M.C.C.); francescamaria.caccamo01@universitadipavia.it (F.M.C.)
2 Interdepartmental Centre for Water Research, University of Pavia, 27100 Pavia, Italy
3 ASMia S.R.L., Mortara, Via Tiziano Vecellio 540, 27036 Mortara, Italy; m.sordi@asmortara.eu (M.S.); b.crotti@asmortara.eu (B.C.)
4 Department of Civil, Environmental, Architectural Engineering and Mathematics, University of Brescia, Via Branze 43, 25123 Brescia, Italy; alessandro.abba@unibs.it
5 Department of Chemistry, University of Pavia, Via Puccini 66, 27011 Belgioioso, Italy
* Correspondence: stefano.bellazzi01@universitadipavia.it (S.B.); marco.carlo.baldi@gmail.com (M.B.); Tel.: +39-0382-985314 (S.B.)

Abstract: This paper focuses on using a novel approach to assess the statistical variability of management data from an aerobic thermophilic biological plant (AWTP) utilizing a fluidized bed biological reactor. A proper statistical characterization of full-scale thermophilic biological systems, in fact, may inform process optimization in the light of a future automation of treatment plants. We present a case study that spans the period from 2018 to 2023 and encompasses various high-strength aqueous waste (AW) in continuous mode. Key aspects of the proposed analytical approach include: (i) utilizing advanced descriptive statistics, such as violin graphs, to depict the variability of monitored parameters over five years; (ii) conducting correlation analyses (Spearman and Pearson correlation matrices) specifically focusing on nitrogenous forms within the AW; (iii) applying multivariate statistical analysis to assess the correlation between pollutants released and the plant's energy and oxygen consumption; and (iv) reconstructing parameter trends by considering periodic and random components, thus enhancing the understanding of the system's behavior over time. The findings presented in this paper offer valuable insights into the performance and optimization of AWTPs, potentially leading to a proper planning of the loads and consequent feeding of the plants. If properly enacted, our approach may provide a significant contribution to the field of aqueous waste management.

Keywords: environmental engineering; WWTPs; thermophilic; aqueous waste

1. Introduction

The treatment of aqueous waste (AW) represents a critical issue due to the high variability of their physical-chemical characteristics and the possible presence of emerging conservative micro-pollutants, such as perfluoroalkyl substances [1], absorbable organic halogens (AOX) and clorides. Conventional biological activated sludge (CAS) processes are not able to achieve a sufficient performance to remove some types of conservative pollutants; in some cases the biota of the plant may even need to be purged since it is no longer active [2,3].

In today's landscape, numerous additional treatments are available to increase the acceptance of conventional biological processes, reducing the effects of acute or chronic toxicity due to the presence of highly refractory organic compounds. The treatments proposed are usually of a physico-chemical nature and therefore often involve the use of various chemical reagents [4,5].

Despite their widespread use, physical-chemical treatments present several drawbacks, such as high operating costs linked to chemical reagents, the need for highly specialized personnel in their management, and finally the high carbon dioxide emission factors for production processes and for the reactions of the reagents themselves [6,7]. For these reasons, biological processes are preferable given their lower environmental impact. They have these advantages with respect to physical-chemical treatments: easy of management, lower costs, and lower CO_2 emission factors [8].

Among biological treatments, the scientific community agrees that thermophilic ones can have high removal yields even in the case of recalcitrant pollutants [9,10]. Considering to the high performance and bio-resistance to some recalcitrant pollutants, the scientific community has identified thermophilic biomasses as those suitable for the treatment of aqueous waste [11]. The main issues with thermophilic treatment are poor biomass settling and the potential absence or inhibition of nitrification. These challenges can be addressed by using membranes for better solid separation and implementing post-treatment processes to encourage nitrification under more favorable conditions. Despite these drawbacks, thermophilic treatment offers advantages such as faster sludge digestion and pathogen destruction, requiring careful management for effective waste treatment [12].

Mesophilic bacteria are used in the conventional wastewater treatment plants (WWTPs); their working range is between 15 and 30 °C. Thermophilic biota are usually exploited for the treatment of AW; they provide optimal working conditions at temperatures between 50–60 °C [13]. For example, in previous works, a thermophilic biological process combined with a membrane ultrafiltration system (TMR—Thermophilic Membrane Reactor) has been used for both the treatment of aqueous wastes and the minimization of biological sludge produced by the WWTPs [14,15]. The great elasticity of this treatment makes it suitable to be applied to different substrates in terms of both chemical and rheological characteristics [16]. This technology has demonstrated resistance to some conservative parameters and micropollutants and it has been applied in two full-scale plants [17]. As a matter of fact, TMR combines the robustness of physico-chemical treatments with the advantages of biological ones, thus allowing a production of equivalent CO_2 lower than standard physico-chemical treatments [18].

So far, the statistical-based characterization of aerobic thermophilic biological plant (AWTP) has received relatively poor attention in the literature. Aguado et al. [19] applied a multivariate statistical approach to derive alerts conditions in the management of the plant. The same strategy has been presented in Liu et al. [20], after variable reduction based on the principal component analysis (PCA). Wang et al. [21] reported a statistical study based on the dynamic simulation to achieve model predictive control. O'Brien and Teather [22] developed a predictive model successfully applied to estimate the pollutant concentrations in a biological system (activated sludge).

In this paper an approach to evaluate the statistical variability of management data of an AWTP at full-scale is proposed. The main novelty stands in the increase of the number of environmental variables that we consider characterizing a newly developed plant, based on a thermophilic fluidized bed biological reactor. The approach is designed to inform process optimization in the light of a future automation of treatment plants, by potentially leading to a proper planning of the loads and consequent feeding of the plants.

The main points of interest of this analysis are:

- The application of advanced descriptive statistics (violin graphs) to describe the variability of each parameter monitored over the five years; this allowed reporting univariate probability distributions of the values observed based on the density profile kernel estimation;
- Correlation analysis (Spearman and Pearson correlation matrices) of the nitrogenous forms;
- The application of multivariate statistical analysis to evaluate the correlation between the different pollutants released and the energy and oxygen consumption of the plant to provide a useful tool to optimize TMR;
- The reconstruction of the trends of the parameters studied taking into account periodic and random components [23].

- In this work, we present a case-study that analyzes the data coming from monitoring a full-scale aerobic thermophilic biological plant in the period 2018 (last trimester)—2023 (first trimester); along this period the plant treated different high-strength AW in continuous mode.

2. Materials and Methods

2.1. TMR Configuration

The monitored TMR plant was usually fed with a wide range of AW such as landfill leachate, wastewater with high contents of organic content, salts, solvents, pharmaceutical compounds and with very different pH values (strong acid or alkaline wastewater). The water line is composed by both physico-chemical and biological treatments (Figure 1). This article focuses only on the thermophilic system composed of a thermophilic biological reactor (TBR) followed by ultrafiltration (UF) membranes placed upstream of a traditional CAS for municipal sewerage treatment.

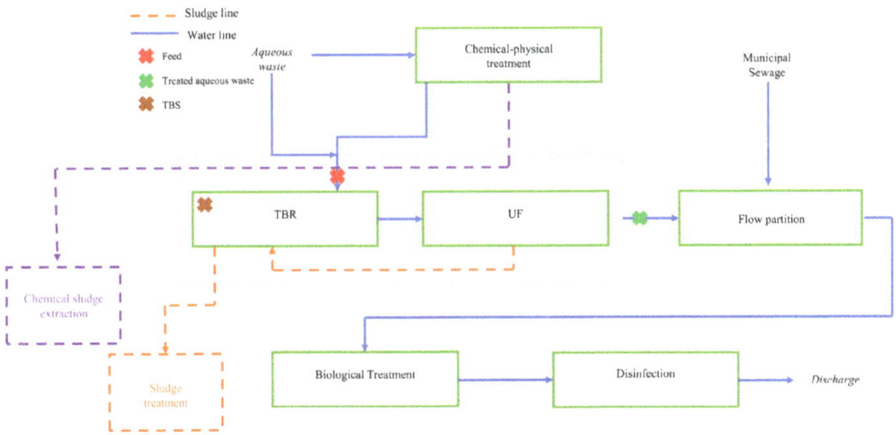

Figure 1. Scheme of the TMR and sampling points (X). TBR: Thermophilic biological reactor; UF: Ultrafiltration.

As seen from the plant scheme, AW have two treatment options based on their characteristics. Waste is pre-treated in batches in two physico-chemical compartments to safeguard the thermophilic biomass (T = 48 °C). The TBR has a volume of 1400 m^3 and is characterized by an average inlet flow rate Qin = 170 m^3d^{-1}, with a hydraulic retention time HRT of almost 8.0 d, as reported by Collivignarelli et al. the hydraulic retention time was studied in an experimental work using a pilot plant [9]. To improve the performance of the sludge and obtain a more compact biomass, a dissolved oxygen concentration of 1.5–2 mg L^{-1} is guaranteed in the reactor by injecting pure oxygen. Inside the biological tank, different areas have different concentrations of oxygen; this management strategy allows for both the aerobic removal of organic substance in the zone of high oxygen concentration (2 mgO_2 L^{-1}) and denitrification in the anoxic zone (<1 mgO_2 L^{-1}). As mentioned previously, the TBR is coupled to an ultrafiltration system due to the poor settleability of the thermophilic biomass probably caused by its carbohydrate content [24]. The platform is therefore equipped with two parallel UF units. The UF unit is composed of 6 vessels with 99 ceramic membranes (cut-off 300 kDa) with an operating pressure between 3 and 5 bar. As shown in Figure 1, the UF retentate is recirculated into the biological reservoir thus ensuring low excess sludge extraction.

2.2. Monitoring Plan

Monitoring activity focused on TMR from 2018 to 2023, sampling the inlet substrate and ultrafiltration permeate (Figure 1) once a week for five years. Analyzes of COD, nitrogen forms (TN–total nitrogen, N-NO_x–sum of nitrite nitrogen and nitrate nitrogen, N-NH_4^+–ammonia, and N_{org}–organic nitrogen), total phosphorus (TP), non-ionic surfactants (TAS) and anionic surfactants (MBAS) were measured according to standard methods [25]. The flow rates ingoing and outgoing the plant, the consumption of pure oxygen necessary for the degradation of the organic matter (O_2), and the energy consumption (kWh) were weekly monitored.

2.3. Data Processing

It was decided to evaluate the statistical variability through violin graphs for all of the input parameters of the plant. Violin graphs are an evolution of box plots, reporting the density profile of the observed values for a univariate distribution in the form of a kernel density plot. The kernel density plot [26] of the data is reported symmetrically by both sides of the distribution and this explains the typical shape of the graphs, which gives them their name [27].

To determine the removal efficiency of COD, nitrogen and total phosphorus, mass balances were performed weekly based on fed and extracted loads.

To obtain reliable modeling, the data were normalized by the maximum value of the series to obtain a series of homogeneous values between 0 and 1 [17]. Pearson and Spearman rank correlations were used to calculate the binary correlations between the nitrogenous forms monitored within the plant.

To evaluate energy and oxygen consumptions during the aerobic process, all input parameters were studied using a multiple linear regression analysis.

A stepwise regression method has been performed with the MATLAB function stepwisefit [28]; this method selects the variables that are most strongly correlated jointly with the reagent (pure oxygen input) and energy consumption of the system. Given the *i*-th parameter xi, it is described with the model:

$$x_i = \alpha_i + \sum_{j \neq i} \beta_{ij} x_j \quad (1)$$

If m parameters are measured, the stepwise procedure allows selecting the $m_i \leq m - 1$ parameters that are more statistically significant to predict each x_i.

To evaluate the presence of a random component and a periodic component in the parameters loaded to the aqueous waste treatment platform an algorithm called STL ("Seasonal and Trend decomposition using Loess") has been applied [29]. STL is a technique used for time series analysis and forecasting [30]. The objective was therefore to decompose a time series into easy-to-interpret components such as trends, seasonality, and noise. We will look for monthly seasonal components based on the variability of the influential loads and the trend in the five years of interest.

3. Results and Discussion

3.1. Monitoring of the Biological Thermophilic System-Violin Plot

To evaluate the distribution of the parameters entering the TBR industrial reactor, it was decided to create violin plot that show a boxplot divided for each parameter and the Kernel distribution reported symmetrically within it.

As can be seen in Figure 2, the association between the two numerical calculations shown (boxplot and Kernel distribution) effectively schematizes the great variability of the 8 pollutants in going to TMR. As well depicted in the trend of the distributions, the variability in terms of input parameters increased from 2018 to 2023 for polluting parameters such as N-NO_3^-, TAS and N_{org}. However, by observing the graph relating to TP it is possible to deduce the loading mode of the reactor with the presence of some peaks uniformly distributed over the various years as shown by the relative violin graphs; in the last quarter analyzed this aspect is less evident. However, with respect to the measurement of COD, an effort can be seen by the plant manager to homogenize the organic load entering the biological reactor.

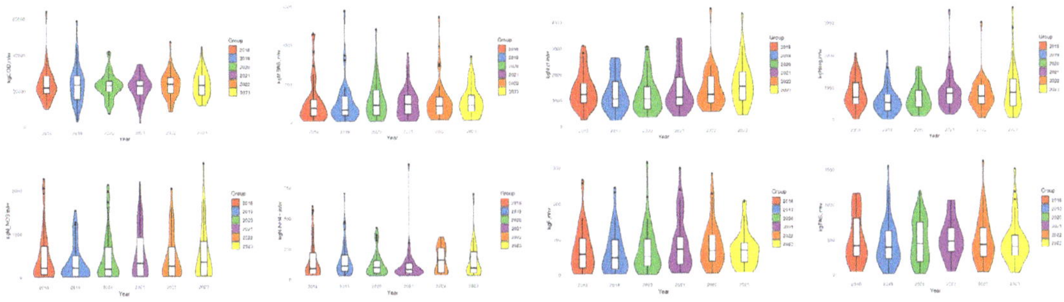

Figure 2. Violin Plot of the loads [kg/week] fed to the TMR relating to the following pollutants: COD, Ntot, N-NO_3^-, N-NH_4^+, N_{org}, TP, TAS, MBAS.

In Figure 3 it is possible to appreciate through the violin plots relating to the outputs from TMR the treatment effectiveness of this process which combines thermophilic biodegradation with an ultrafiltration treatment [31]. By observing the value in kg/week of each parameter output from the TMR it is possible to note the high removal yields shown subsequently in Table 1. The only parameter which has an output load value higher than the input is ammonia due of the strong ammonification process of organic nitrogen in thermophilic conditions [17]. The trend of the distribution of outgoing loads divided into the 6 years shown is instead useful for showing a second strong point of this technology [32]. The input loads presented a large variability at the input as demonstrated by the distributions in Figure 2, the outputs instead present a very low variability value, thus underlining on the one hand the great elasticity and resilience of the thermophilic system and on the other the very good applicability of this system placed in series with a CAS which requires homogeneity of the polluting parameters at input [33].

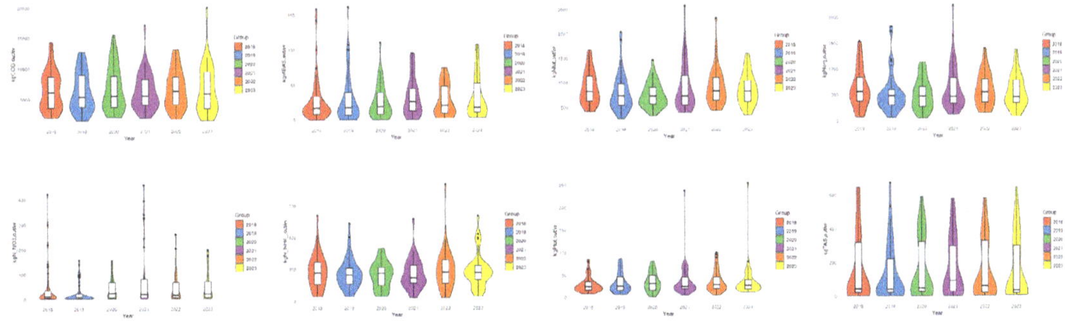

Figure 3. Violin Plot of the loads in output to the TBR relating to the parameters COD, Ntot, N-NO_x, N-NH_4^+, Norg, Ptot, TAS, MBAS [kg/week].

Table 1 shows the pollutant loads introduced and the performance of the thermophilic biological process. This analyzed biomass can treat high COD loads [13], the performance slightly dropped after 2019, reaching an average value over the 5 years of 71.8%. Nitrites and nitrates represent a particularly important parameter for the plant which aims to treat aqueous waste [34], the performance value obtained increased after an initial acclimatization phase reaching a value of 81.3%. The removal yield relating to ammonia is negative. This is due to the fact that, as already known by the scientific community, under thermophilic conditions a strong process of ammonification of organic nitrogen occurs. As already described by observing the graphs of the distributions at the entrance to the biological reactor, the TP presents a high variability value, despite this an average value of 55.3% was obtained over the five years, the accumulation of phosphorus in the sludge in inorganic form occurs mainly thanks to the formation of hydroxyapatite in pure oxygen aeration conditions [35]. The capacity of TMR therefore allows it to accumulate phosphorus within the sludge in the solid phase, facilitating possible recovery in agriculture depending on their content of persistent pollutants such as heavy-metals.

The effect on the removal of TAS and MBAS was also evaluated. The analyzed process allowed obtaining removal yields of 77.4% and 84.0%, respectively, demonstrating the good biodegradability of these types of surfactants in thermophilic conditions. Yields related to MBAS are generally higher due to their structure being more easily attacked by microorganisms [36]. The removal can be attributed to thermophilic biological degradation in the TAR reactor and not to filtration on the subsequent UF membrane, typical molecular weights of surfactants ranged from 0.2–0.4 kDa, depending on their molecular structure, and that the molecular weight limit of the UF membrane used in this study was 30 kDa, most of the surfactants (TAS and MBAS) could pass the pores of the UF membranes [37].

Table 1. Fed pollutants, loads and performance and other operational parameters of thermophilic system; *n* stands for the amount of data. The values are report plus or minus the standard deviation. The parameters read as follows: Chemical Oxygen Demand (COD), Nitric and Nitrous Nitrogen (N-NOX), Ammoniacal Nitrogen (N-NH_4^+), Organic Nitrogen (Norg), Total Nitrogen (TN), Total Phosphorus (TP), Non-ionic surfactants (TAS), Anionic surfactants (MBAS).

Parameter	2018 [n = 20]	2019 [n = 52]	2020 [n = 48]	2021 [n = 52]	2022 [n = 52]	2023 [n = 13]	2018–2023 [n = 237]
COD IN [kg week^{-1}]	20,842 ± 3332	19,174 ± 1756	20,038 ± 1671	25,735 ± 2508	26,502 ± 2531	25,484 ± 2971	22,928 ± 1061
μ COD [%]	78.8 ± 4.3	80.9 ± 2.3	73.4 ± 2.9	66.2 ± 3.0	63.1 ± 3.1	74.1 ± 1.9	71.8
N-NO_X IN [kg week^{-1}]	138.8 ± 72.8	235.0 ± 64.3	277.9 ± 141.7	849.6 ± 157.6	575.4 ± 187.7	789.7 ± 330.8	469 ± 73
μ N-NO_X [%]	73.2 ± 10.3	79.4 ± 6.9	76.6 ± 9.5	89.1 ± 3.4	81.0 ± 8.6	79.3 ± 7.0	81.3
N-NH_4^+ IN [kg week^{-1}]	208.3 ± 53.7	141.9 ± 34.6	111.7 ± 36.8	123.6 ± 47.8	132.9 ± 24.7	260.7 ± 63.7	140 ± 17.1
μ N-NH_4^+	−51.3 ± 8.3	−58.5 ± 7.3	−66.3 ± 6.7	−51.5 ± 9.3	−52.5 ± 8.1	−16.7 ± 3.2	−55.8
Norg IN [kg week^{-1}]	688.5 ± 190.9	564.5 ± 88.4	749.6 ± 160.1	894.8 ± 137.0	989.3 ± 182.0	989.3 ± 182.0	776 ± 69
μ Norg [%]	43.2 ± 13.6	54.0 ± 5.9	39.8 ± 8.6	30.9 ± 6.7	32.8 ± 5.7	29.0 ± 4.4	40.5
TN IN [kg week^{-1}]	1020.1 ± 210.7	950.8 ± 127.5	1166.3 ± 222.9	1834.6 ± 207.2	1662.8 ± 205.0	1624.5 ± 250.4	1376 ± 96
μ Ntot [%]	24.7 ± 10.7	34.4.6 ± 6.2	42.7 ± 8.2	41.2 ± 6.0	39.1 ± 5.1	23.0 ± 9.2	37.5
TP IN [kg week^{-1}]	62.1 ± 27.9	103.2 ± 20.9	65.7 ± 18.2	72.9 ± 14.2	76.9 ± 15.2	17.5 ± 14.9	75 ± 8
μ Ptot [%]	61.9 ± 12.1	65.2 ± 6.7	58.8 ± 8.1	49.9 ± 8.2	45.9 ± 6.8	22.9 ± 9.2	55.3
TAS IN [kg w^{-1}]	246.3 ± 66.9	475.7 ± 81.3	402.9 ± 60.5	605.6 ± 82.7	639.5 ± 100.5	101.1 ± 40.3	486 ± 41
μ TAS [%]	85.2 ± 5.2	91.2 ± 3.2	84.6 ± 6.1	69.4 ± 5.9	54.9 ± 6.4	40.8 ± 3.9	77.4
MBAS IN [kg week^{-1}]	115.0 ± 63.3	252.0 ± 75.4	311.9 ± 67.9	325.0 ± 74.8	279.3 ± 52.1	52.8 ± 24.8	267 ± 32
μ MBAS [%]	82.1 ± 9.3	91.2 ± 3.0	89.2 ± 3.9	76.6 ± 6.0	80.6 ± 4.4	70.9 ± 9.8	84.0

3.2. Data Analysis

The Pearson moment correlation coefficient and the Spearman rank correlation coefficient were calculated to evaluate a possible correlation between the monitored nitrogenous forms in thermophilic biological process input (Table 2).

This analysis allows us to have an indication of the composition of the total nitrogen loaded and the possible relationship between the various forms of nitrogen. The analysis of Pearson's correlation index highlighted a positive correlation between Norg introduced into the plant and TN (0.57). The correlation between the value of nitric and nitrous nitrogen and the total nitrogen entering the plant was 0.66 while the correlation between ammoniacal nitrogen and the monitored total was only 0.2. From the Person index it can therefore be deduced that the most significant amount of nitrogen in terms of total nitrogen was that attributed to nitric and nitrous nitrogen. It is also possible to note some negative correlations between the monitored nitrogenous forms (NH4-NO_x) and organic nitrogen, suggesting a treatment upstream of the biological process and a possible interference in the measurement method [38].

Table 2. Pearson correlation matrix.

	Norg IN	TN IN	N-NH_4^+ IN	N-NO_x IN
Norg IN	1.00	\	\	\
TN IN	0.57	1.00	\	\
N-NH_4^+ IN	−0.06	0.20	1.00	\
N-NO_x IN	−0.17	0.66	0.10	1.00

The Spearman correlation matrix shows very similar results (Table 3). A strong positive correlation can be observed between nitrates input into the system and total nitrogen input. This correlation can be explained by considering that nitrates constitute the largest share of the forms of nitrogen treated by the thermophilic biological system.

Table 3. Spearman correlation matrix.

	Norg IN	TN IN	N-NH_4^+ IN	N-NO_x IN
Norg IN	1.00	\	\	\
TN IN	0.45	1.00	\	\
N-NH_4^+ IN	−0.12	0.25	1.00	\
N-NO_x IN	−0.19	0.69	0.24	1.00

From the analysis of the two correlation matrices, a positive correlation emerges between all of the non-organic nitrogenous forms entering the biological reactor with the total nitrogen value. This analysis suggests the possibility of estimating the composition of the nitrogenous forms entering the plant based on the measurement of total nitrogen, usually simpler than that relating to nitrates.

According to Correlation Matrixes, Pearson (P) and Spearman (S) is interesting since if you have S > P, it means you have a monotonic but non-linear correlation [39]. The results could help in the future in the correct modeling of the observed parameters, providing information on the type of correlation between the data, whether linear or monotonic.

3.3. Multivariate Linear Regressions

Table 4 shows the matrix of the coefficients of the best multiple linear regressions able to describe the possible relationship between the parameters of the substrates entering the biological reactor and the chemical, pure oxygen, and energy consumption. The regressions are obtained by means of a stepwise search, as implemented in the MATLAB Statistics and Machine Learning toolbox [40].

As shown in Table 4, it was found that the value of oxygen needed by the system is linearly correlated with the input of organic nitrogen (B = 0.1382), i.e., around 7 kg of organic nitrogen correspond to 1 kg of oxygen consumption. As reported by Table 1, the analysis shows that the oxidation process of organic nitrogen corresponds to the highest rate of oxygen consumption, thus leading to an increase of the ammoniacal nitrogen. This regression has a low Pval value (0.0175) and a root mean square deviation (Rmse) value of 0.1226, i.e., well below the maximum value. This demonstrates a good applicability of the model as well as for forecasting purposes. The same approach was used to study the value of electricity consumption. The parameters selected by the model were total nitrogen and anionic surfactant. In this case the Pval value was very low (around 10^{-7}) showing excellent statistical significance of the model, with a Rmse value of 0.1042.

Table 4. Coefficients of multivariate linear regressions. N corresponds to the parameter selected by the model, being 1 equal to yes and 0 to no; Bare the linear regression coefficients; Pvals are the P-values, i.e., the probability of obtaining a result equal to or more extreme than the observed one, assuming that the null hypothesis is true (Pvals are computed using the F-statistic); finally Root Mean Square Error (Rmse) evaluates the performance of a forecasting model by comparing the prediction to the observed or actual data.

	COD, in [kg/week]	**N_{org}, in [kg/week]**	**TN, in [kg/week]**	**N−NH4, in [kg/week]**	**N−NO3, in [kg/week]**	**TP, in [kg/week]**	**TAS, in [kg/week]**	**MBAS, in [kg/week]**	**KgO_2 in [kgO_2/week]**
N	0	1	0	0	0	0	0	0	
B	\	0.1382	\	\	\	\	\	\	
Pval	0.0175								
Rmse	0.1226								
	COD, in [kg/week]	**N−Org, in [kg/week]**	**N−tot, in [kg/week]**	**N−NH4, in [kg/week]**	**N−NO3, in [kg/week]**	**P, in [kg/week]**	**TAS, in [kg/week]**	**MBAS, in [kg/week]**	**kWh in [kWh/week]**
N	0	0	1	0	0	0	0	1	
B	\	\	0.1297	\	\	\	\	0.2643	
Pval	71.44×10^{-7}								
Rmse	0.1042								

This type of analysis allows the AW utilities manager to be able to forecast the electrical and oxygen consumption values of the system based on the input parameters.

The proposal involves utilizing future linear regression analysis to establish a connection between oxygen consumption and removal efficiencies. This aims to set a minimum operational threshold for plant performance or adherence to discharge limits. Let us note that flow rate is an important determinant of the energy consumption of all MBR systems. In our experimental setting, however, the variability of the flow rate between output data was quite low, and therefore it was not included in the regression. In case of a different experimental setting, one may consider modeling as a first regression the relationships between energy consumption and flow rate, and then perform a new regression study, as we did, on the residuals between the first regression and the data. The suggestion to enlarge the number of regressors extends to exploring also other variables such as temperature, pH, or operational factors influencing oxygen consumption and pollutant removal. The suggested approach of correlating oxygen consumption with removal efficiencies offers a comprehensive perspective on plant performance, facilitating the establishment of minimum operational objectives.

3.4. Time Series Decomposition

Time series decomposition is a data analysis technique used to extract trend and seasonal components of a time series. Its main purpose is thus to highlight and separate long-term effects from periodical ones. This strategy is often used to break down, analyze, filter, and interpret complex data in a meaningful way for research, forecasting, or decision-making purposes. In detail, we have performed the following analysis:

1. Decomposition of time series: we have assumed an additive decomposition model, which assumes that a time series Y can be written as follows: Y = T + S + R, where T is a trend component, S is a seasonal component and R is a random component. For finding the components we have exploited the algorithm called STL ("Seasonal and Trend decomposition using Loess") [29] a robust approach able to deal with handle any type of seasonality, and capable of handling outliers and non-linear trends. In

our case, we looked for yearly seasonality (i.e., we wanted to highlight if there were regularities in the loadings of the same month over different years).

2. Data visualization: STL decomposition was used to visualize data by plotting long term trends and monthly seasonality.

The results helped to identify regularities that can be effectively exploited by the plant managers to plan the input loads to the biological reactor, forecasting and therefore minimizing fluctuations of the input loads [41].

To explain the practical impact of this analysis, we can concentrate our attention to the panel related to N-NO_3 in Figure 4. We can see that the original time series is quite noisy. However, the trend analysis clearly shows an increase starting on July 2020, with a peak in January 2022 and a following decrease in the last year. Moreover, the cyclic component highlights a low value in June (as it happens also for other pollutants) and a higher values in February and November. The combination of these aspects can be used to optimize the plant configuration at large, from the plants settings to the personnel shifts.

Looking at the results shown in Figure 4, it is possible to note that COD, TN, N-NO_3, N_{org} and TAS show a long-term increasing trend. Moreover, there is high long-term variability for N-NH_4^+, as well as for MBAS. The analysis of the seasonal components highlights that February and July are high loading months in terms of all input quantities, while August has low loadings for several inputs: COD, TN, MBAS, and N_{org}.

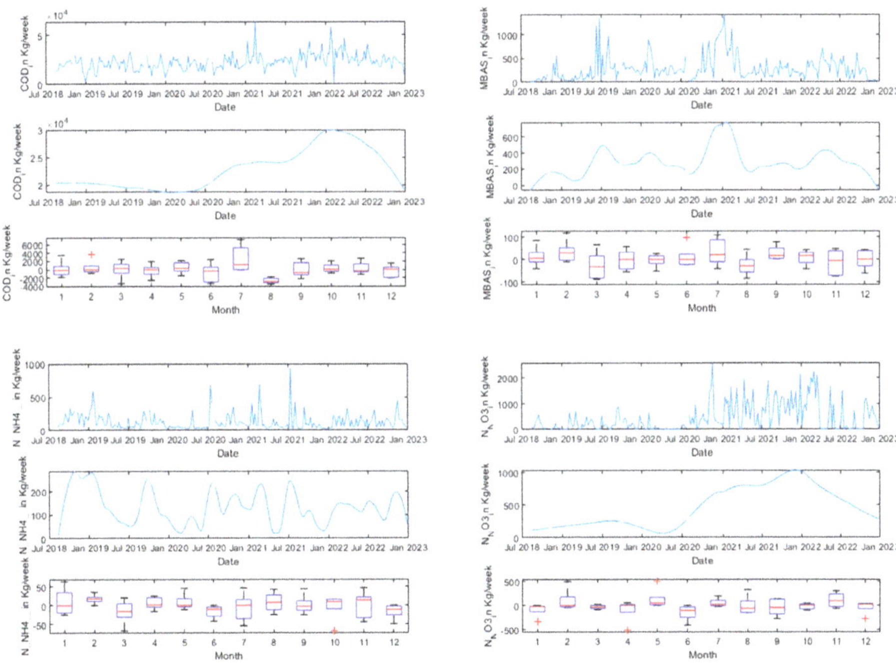

Figure 4. *Cont.*

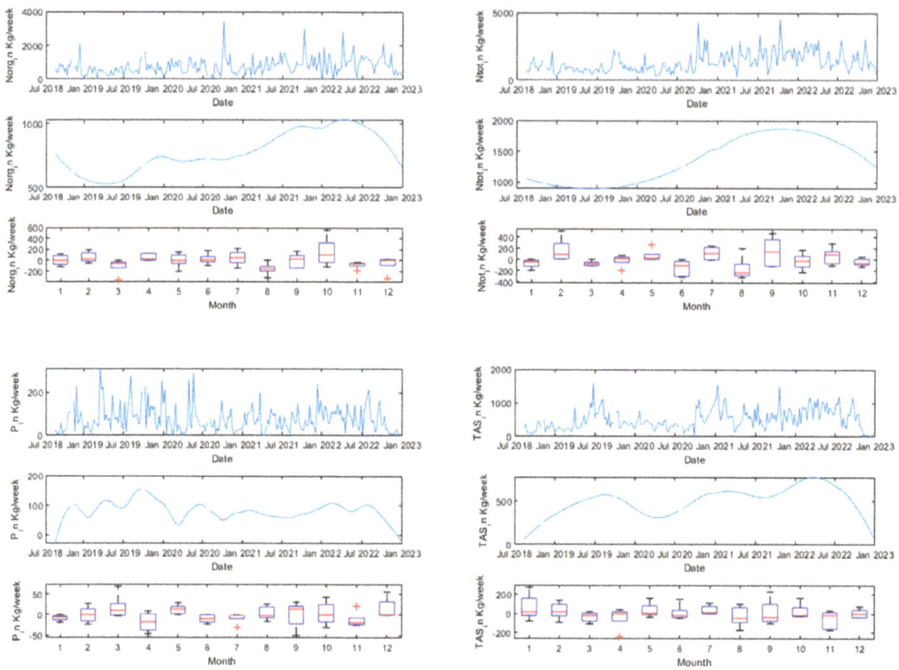

Figure 4. Time series decomposition of the loads fed to the TBR relating to the parameters COD, Ntot, Ptot, MBAS, N−NH_4^+, Norg, N−NO3, TAS [kg/week]. Each subplot shows the original data (**upper plot**), the long-term trend (**middle plot**) and monthly boxplot over the years, where the blue line shows the boxplots and the red lines the average value (**lower plot**).

4. Future Outlooks and Conclusions

A full-scale TMR has a great elasticity in the examined AWTP throughout the five years of the research. An approach to evaluate the statistical variability of management data was investigated to derive alerts conditions in the management of the plant. Advanced descriptive statistics (violin graphs) highlight the treatment effectiveness of this process which combines thermophilic biodegradation with an ultrafiltration treatment; the loading mode of the reactor with the presence of some peaks uniformly distributed over the various years. Correlation matrixes (Pearson (P) and Spearman (S)) help in the future in the correct modeling of the observed parameters, providing information on the type of correlation between the data, whether linear or monotonic. Multiple linear regressions allow the AW utilities manager to be able to forecast the electrical and oxygen consumption values of the system based on the input parameters. Time series decomposition helped to identify regularities that can be effectively exploited by the plant managers to plan the input loads to the biological reactor, forecasting and therefore minimizing fluctuations of the input loads. Clearly, the approach presented in the paper needs to be further applied to other systems, to assess its potentials in different contexts. New analysis may provide insights about the methodological choices and suggest different statistical methods to highlight the main plant characteristics.

For future work, we will explore the different ways for effectively optimizing the treatment plant by looking both at the "loading" and "feeding" aspects. First, an important action is to try to reduce the variability of the loads in terms of physico-chemical properties and volumetric quantity. This may be possible by planning ex-ante the load based on the analysis of the historical loads of the plants. Second, a proper calibration of the feeding of pollutants may minimize oxygen and energy consumption during the oxidation process. Again, a proper statistical analysis of the monitoring data is essential to achieve the optimization goals.

Author Contributions: Conceptualization: M.C.C. and S.B.; methodology: M.C.C., F.M.C. and S.B.; validation: M.S., B.C. and M.B.; formal analysis: S.B.; investigation: A.A., S.B. and M.B.; resources: M.C.C. and S.B.; data curation: S.B. and F.M.C.; writing—original draft preparation: S.B., A.A. and B.C.; writing—review and editing: S.B. and F.M.C.; visualization: S.B., M.B., F.M.C. and A.A.; supervision: M.C.C. and S.B.; funding acquisition: M.C.C. All authors have read and agreed to the published version of the manuscript.

Funding: The authors thank Department of Civil Engineering and Architecture of Pavia for providing financial support to the reviews activities.

Data Availability Statement: Data are contained within the article.

Acknowledgments: We thank the Itelyum group and AsMortara's ownership of the plant for granting us the availability of the data. Finally, we appreciate the work of all volunteers who contributed to sample processing and data collection.

Conflicts of Interest: The authors declare no conflicts of interest. The funders had no role in the design of the study; in the collection, analyses, or interpretation of data; in the writing of the manuscript; or in the decision to publish the results.

References

1. Ata, R.; Merdan, G.F.; Töre, G.Y. Activated Sludge Process for Refractory Pollutants Removal. In *Removal of Refractory Pollutants from Wastewater Treatment Plants*; CRC Press: Boca Raton, FL, USA, 2021.
2. Buttiglieri, G.; Knepper, T.P. Removal of Emerging Contaminants in Wastewater Treatment: Conventional Activated Sludge Treatment. In *Emerging Contaminants from Industrial and Municipal Waste: Removal Technologies*; Springer: Berlin/Heidelberg, Germany, 2008; pp. 1–35.
3. Gurung, K.; Ncibi, M.C.; Fontmorin, J.M. Incorporating Submerged MBR in Conventional Activated Sludge Process for Municipal Wastewater Treatment: A Feasibility and Performance Assessment. *J. Membr. Sci. Technol.* **2016**, *6*, 1000158. [CrossRef]
4. Camargo, F.P.; Sérgio Tonello, P.; dos Santos, A.C.A.; Duarte, I.C.S. Removal of Toxic Metals from Sewage Sludge through Chemical, Physical, and Biological Treatments—A Review. *Water Air Soil Pollut.* **2016**, *227*, 433. [CrossRef]
5. Robinson, S.; Abdullah, S.Z.; Bérubé, P.; Le-Clech, P. Ageing of Membranes for Water Treatment: Linking Changes to Performance. *J. Membr. Sci.* **2016**, *503*, 177–187. [CrossRef]
6. Lackner, K.S.; Wendt, C.H.; Butt, D.P.; Joyce, E.L.; Sharp, D.H. Carbon Dioxide Disposal in Carbonate Minerals. *Energy* **1995**, *20*, 1153–1170. [CrossRef]
7. Nguyen, T.K.L.; Ngo, H.H.; Guo, W.; Chang, S.W.; Nguyen, D.D.; Nghiem, L.D.; Liu, Y.; Ni, B.; Hai, F.I. Insight into Greenhouse Gases Emissions from the Two Popular Treatment Technologies in Municipal Wastewater Treatment Processes. *Sci. Total Environ.* **2019**, *671*, 1302–1313. [CrossRef]
8. Hagos, K.; Zong, J.; Li, D.; Liu, C.; Lu, X. Anaerobic Co-Digestion Process for Biogas Production: Progress, Challenges and Perspectives. *Renew. Sustain. Energy Rev.* **2017**, *76*, 1485–1496. [CrossRef]
9. Collivignarelli, M.C.; Abbà, A.; Bertanza, G. Treatment of High Strength Pharmaceutical Wastewaters in a Thermophilic Aerobic Membrane Reactor (TAMR). *Water Res.* **2014**, *63*, 190–198. [CrossRef]
10. Simstich, B.; Beimfohr, C.; Horn, H. Lab Scale Experiments Using a Submerged MBR under Thermophilic Aerobic Conditions for the Treatment of Paper Mill Deinking Wastewater. *Bioresour. Technol.* **2012**, *122*, 11–16. [CrossRef]
11. Yang, S.-J.; Kataeva, I.; Hamilton-Brehm, S.D.; Engle, N.L.; Tschaplinski, T.J.; Doeppke, C.; Davis, M.; Westpheling, J.; Adams, M.W.W. Efficient Degradation of Lignocellulosic Plant Biomass, without Pretreatment, by the Thermophilic Anaerobe "*Anaerocellum thermophilum*" DSM 6725. *Appl. Environ. Microbiol.* **2009**, *75*, 4762–4769. [CrossRef]
12. Collivignarelli, M.C.; Abbà, A.; Frattarola, A.; Manenti, S.; Todeschini, S.; Bertanza, G.; Pedrazzani, R. Treatment of Aqueous Wastes by Means of Thermophilic Aerobic Membrane Reactor (TAMR) and Nanofiltration (NF): Process Auditing of a Full-Scale Plant. *Environ. Monit. Assess.* **2019**, *191*, 708. [CrossRef]
13. LaPara, T.M.; Nakatsu, C.H.; Pantea, L.; Alleman, J.E. Phylogenetic Analysis of Bacterial Communities in Mesophilic and Thermophilic Bioreactors Treating Pharmaceutical Wastewater. *Appl. Environ. Microbiol.* **2000**, *66*, 3951–3959. [CrossRef] [PubMed]
14. Collivignarelli, M.C.; Abbà, A.; Bertanza, G.; Setti, M.; Barbieri, G.; Frattarola, A. Integrating Novel (Thermophilic Aerobic Membrane Reactor-TAMR) and Conventional (Conventional Activated Sludge-CAS) Biological Processes for the Treatment of High Strength Aqueous Wastes. *Bioresour. Technol.* **2018**, *255*, 213–219. [CrossRef] [PubMed]
15. Collivignarelli, M.C.; Abbà, A.; Carnevale Miino, M.; Caccamo, F.M.; Argiolas, S.; Bellazzi, S.; Baldi, M.; Bertanza, G. Strong Minimization of Biological Sludge Production and Enhancement of Phosphorus Bioavailability with a Thermophilic Biological Fluidized Bed Reactor. *Process Saf. Environ. Prot.* **2021**, *155*, 262–276. [CrossRef]

16. Collivignarelli, M.C.; Todeschini, S.; Bellazzi, S.; Carnevale Miino, M.; Caccamo, F.M.; Calatroni, S.; Baldi, M.; Manenti, S. Understanding the Influence of Diverse Non-Volatile Media on Rheological Properties of Thermophilic Biological Sludge and Evaluation of Its Thixotropic Behaviour. *Appl. Sci.* **2022**, *12*, 5198. [CrossRef]
17. Collivignarelli, M.C.; Pedrazzani, R.; Bellazzi, S.; Carnevale Miino, M.; Caccamo, F.M.; Baldi, M.; Abbà, A.; Bertanza, G. Numerical Analysis of a Full-Scale Thermophilic Biological System and Investigation of Nitrate and Ammonia Fates. *Appl. Sci.* **2022**, *12*, 6952. [CrossRef]
18. Boldrin, A.; Neidel, T.L.; Damgaard, A.; Bhander, G.S.; Møller, J.; Christensen, T.H. Modelling of Environmental Impacts from Biological Treatment of Organic Municipal Waste in EASEWASTE. *Waste Manag.* **2011**, *31*, 619–630. [CrossRef]
19. Aguado, D.; Rosen, C. Multivariate Statistical Monitoring of Continuous Wastewater Treatment Plants. *Eng. Appl. Artif. Intell.* **2008**, *21*, 1080–1091. [CrossRef]
20. Liu, Y.; Pan, Y.; Sun, Z.; Huang, D. Statistical Monitoring of Wastewater Treatment Plants Using Variational Bayesian PCA. *Ind. Eng. Chem. Res.* **2014**, *53*, 3272–3282. [CrossRef]
21. Wang, X.; Ratnaweera, H.; Holm, J.A.; Olsbu, V. Statistical Monitoring and Dynamic Simulation of a Wastewater Treatment Plant: A Combined Approach to Achieve Model Predictive Control. *J. Environ. Manag.* **2017**, *193*, 1–7. [CrossRef]
22. O'Brien, G.J.; Teather, E.W. A Dynamic Model for Predicting Effluent Concentrations of Organic Priority Pollutants from an Industrial Wastewater Treatment Plant. *Water Environ. Res.* **1995**, *67*, 935–942. [CrossRef]
23. Arismendy, L.; Cárdenas, C.; Gómez, D.; Maturana, A.; Mejía, R.; Quintero M., C.G. Intelligent System for the Predictive Analysis of an Industrial Wastewater Treatment Process. *Sustainability* **2020**, *12*, 6348. [CrossRef]
24. Piterina, A.V.; Bartlett, J.; Pembroke, J.T. Morphological Characterisation of ATAD Thermophilic Sludge; Sludge Ecology and Settleability. *Water Res.* **2011**, *45*, 3427–3438. [CrossRef]
25. Rice, E.W.; Bridgewater, L.; American Public Health Association (Eds.) *Standard Methods for the Examination of Water and Wastewater*; American Public Health Association: Washington, DC, USA, 2012.
26. Aitchison, J.; Lauder, I.J. Kernel Density Estimation for Compositional Data. *Appl. Stat.* **1985**, *34*, 129. [CrossRef]
27. Azami, H.; Sarrafzadeh, M.H.; Mehrnia, M.R. Influence of Sludge Rheological Properties on the Membrane Fouling in Submerged Membrane Bioreactor. *Desalin. Water Treat.* **2011**, *34*, 117–122. [CrossRef]
28. Siwek, K.; Osowski, S. Data Mining Methods for Prediction of Air Pollution. *Int. J. Appl. Math. Comput. Sci.* **2016**, *26*, 467–478. [CrossRef]
29. Cleveland, R.B.; Cleveland, W.S.; McRae, J.E.; Terpenning, I. STL: A Seasonal-Trend Decomposition. *J. Off. Stat.* **1990**, *6*, 3–73.
30. Golyandina, N.; Zhigljavsky, A. *Singular Spectrum Analysis for Time Series*; SpringerBriefs in Statistics; Springer: Berlin/Heidelberg, Germany, 2013; ISBN 978-3-642-34912-6.
31. Lopetegui, J.; Sancho, L. Aerated Thermophilic Biological Treatment with Membrane Ultrafiltration: Alternative to Conventional Technologies Treating Paper Mill Effluents. *Water Supply* **2003**, *3*, 245–252. [CrossRef]
32. Pianosi, F.; Wagener, T. Distribution-Based Sensitivity Analysis from a Generic Input-Output Sample. *Environ. Model. Softw.* **2018**, *108*, 197–207. [CrossRef]
33. Fatone, F.; Di Fabio, S.; Bolzonella, D.; Cecchi, F. Fate of Aromatic Hydrocarbons in Italian Municipal Wastewater Systems: An Overview of Wastewater Treatment Using Conventional Activated-Sludge Processes (CASP) and Membrane Bioreactors (MBRs). *Water Res.* **2011**, *45*, 93–104. [CrossRef]
34. Siciliano, A.; Limonti, C.; Curcio, G.M.; Molinari, R. Advances in Struvite Precipitation Technologies for Nutrients Removal and Recovery from Aqueous Waste and Wastewater. *Sustainability* **2020**, *12*, 7538. [CrossRef]
35. Mañas, A.; Biscans, B.; Spérandio, M. Biologically Induced Phosphorus Precipitation in Aerobic Granular Sludge Process. *Water Res.* **2011**, *45*, 3776–3786. [CrossRef] [PubMed]
36. Zhu, F.-J.; Ma, W.-L.; Xu, T.-F.; Ding, Y.; Zhao, X.; Li, W.-L.; Liu, L.-Y.; Song, W.-W.; Li, Y.-F.; Zhang, Z.-F. Removal Characteristic of Surfactants in Typical Industrial and Domestic Wastewater Treatment Plants in Northeast China. *Ecotoxicol. Environ. Saf.* **2018**, *153*, 84–90. [CrossRef]
37. Chang, I.-S.; Chung, C.-M.; Han, S.-H. Treatment of Oily Wastewater by Ultrafiltration and Ozone. *Desalination* **2001**, *133*, 225–232. [CrossRef]
38. Arogo, J.; Westerman, P.W.; Heber, A.J.; Robarge, W.P.; Classen, J.J. Ammonia in Animal Production—A Review. In Proceedings of the 2001 Annual Meeting of the American Society of Association Executives, Philadelphia, PA, USA, 4–7 August 2001.
39. de Winter, J.C.F.; Gosling, S.D.; Potter, J. Comparing the Pearson and Spearman Correlation Coefficients across Distributions and Sample Sizes: A Tutorial Using Simulations and Empirical Data. *Psychol. Methods* **2016**, *21*, 273–290. [CrossRef] [PubMed]
40. Draper, N.R.; Smith, H. *Applied Regression Analysis*; John Wiley and Sons: Hoboken, NJ, USA, 1998.
41. Cheng, T.; Dairi, A.; Harrou, F.; Sun, Y.; Leiknes, T. Monitoring Influent Conditions of Wastewater Treatment Plants by Nonlinear Data-Based Techniques. *IEEE Access* **2019**, *7*, 108827–108837. [CrossRef]

environments

MDPI

Article

Heterogeneous Activation of Persulfate by Nickel Oxide/Strontium Carbonate Composite for Sulfamethoxazole Degradation in Water

Despoina Jessica Skempi [1], Konstantinos Kouvelis [1], Athanasia Petala [2], Georgios Bampos [1,*] and Zacharias Frontistis [3]

1 Department of Chemical Engineering, University of Patras, GR-26504 Patras, Greece; kozeta007@hotmail.com (D.J.S.); kos_kouv@chemeng.upatras.gr (K.K.)
2 Department of Environment, Ionian University, GR-29100 Zakynthos, Greece; apetala@ionio.gr
3 Department of Chemical Engineering, University of Western Macedonia, GR-50132 Kozani, Greece; zfronistis@uowm.gr
* Correspondence: geoba@chemeng.upatras.gr

Abstract: The development of efficient heterogeneous persulfate activators is one of the main research topics in the wastewater treatment area. The present work deals with the heterogeneous activation of sodium persulfate (SPS) using nickel oxide/strontium carbonate ($NiO/SrCO_3$) for the degradation of sulfamethoxazole (SMX), a representative compound from the group of antibiotics. Results showed that $NiO/SrCO_3$ exhibited high performance towards the activation of SPS, leading to SMX elimination in brief time spans. The impact of SPS (25–100 mg/L), $NiO/SrCO_3$ (50–250 mg/L), and SMX (0.25–3.00 mg/L) concentration, and initial pH on the decomposition of SMX was further examined. Experiments were also conducted in real matrices such as secondary effluent and bottled water, revealing the existence of retarding phenomena compared to ultrapure water. This behavior was further investigated with the addition of bicarbonates, chlorides, or humic acid in ultrapure water. It was found that organic matter significantly hampered SMX removal. The role of the main radicals (hydroxyl and sulfate radicals) was determined using appropriate radical traps (methanol and tert-butanol). These quenching experiments combined with the conducted electrochemical measurements revealed that both a radical and a non-radical mechanism contribute to the decomposition of SMX.

Keywords: persulfate-based advanced oxidation processes (sr-aops); water matrix; catalytic materials; pharmaceuticals; radical and non-radical mechanism

1. Introduction

Persulfate is a strong oxidizing agent widely used in environmental fields like water purification and soil remediation [1]. In its pure form, persulfate has relatively low reactivity; however, it can be activated, producing sulfate radicals ($SO_4^{\bullet-}$) characterized by high reactivity and participation in degradation reactions of inorganic and organic pollutants [2]. The desirable characteristics of $SO_4^{\bullet-}$ led to the development of a new subcategory of advanced oxidation processes (AOPs), named sulfate radical-based AOPs (SR-AOPs), where $SO_4^{\bullet-}$ are produced either exclusively or together with hydroxyl radicals [2]. When compared to other AOPs, SR-AOPs offer several advantages such as minimized byproducts formation, ease of application and compatibility with existing infrastructure. In addition, besides its well-documented applications in micropollutants removal from aqueous media, some recent studies have focused on diverse sustainability applications of persulfate technologies. These applications encompass a range of areas, including managing solid waste, extracting metals, and facilitating materials circulation [3,4].

Persulfate can be activated through energy input (thermal, radiation, sonochemical/electrochemical), transition metal ions (iron, cobalt, copper, silver), alkali [5], and heterogeneous activators [6]. In addition, today sonochemical activation has sparked increasing interest as an alternative activation method. In that case, the ultrasound's energy is not employed directly by the liquid reactants. Instead, the impact of ultrasound occurs via a phenomenon referred to as cavitation. This process encompasses the creation, expansion, and forceful implosion of voids [7,8]. This leads to the development of hybrid systems, such as cavitation combined with photocatalysis, showing enhanced oxidation efficiency [9]. However, among persulfate activation methods, heterogeneous activation appears to be the most effective method, mostly because it has minimal energy requirements, and provides facile separation and reusability of the catalytic materials [10]. The benefit deriving from this fact is twofold. Firstly, the metal leaching observed in homogeneous activation methods is now minimized, and secondly, the operational costs are reduced due to catalyst recovery. In addition, heterogeneous persulfate activation can significantly enhance the overall efficiency of such degradation systems, as the presence of solid catalysts provides additional active sites for persulfate activation. However, it should be noted that heterogeneous catalysts should be characterized by high stability in order to be recovered and reused, reducing their environmental impact.

Numerous investigations have documented the heterogeneous activation of persulfate, and the activation means can be categorized into two main groups: (i) Carbon-based materials, including well-known materials such as activated carbon [11], carbon nanotubes [12], or even graphene [13]. Such materials are considered promising persulfate activators as they provide a high surface area, facilitating the interaction between persulfate and target pollutants [14]; (ii) Catalytic materials such as metal oxides [15], metal phosphides [16], metal-organic frameworks [17] or monolith catalysts with 3D hierarchical structures like δ-MnO_2 immobilized on 3D nickel foam or 3D α-$Co(OH)_2$ nanosheets developed on robust nickel foam (NF) [18,19]. These catalysts can enhance sulfate radicals production through redox reactions with persulfate, accelerating the oxidation of contaminants [6].

Recently, the development of heterogeneous catalytic materials for persulfate activation has been the focus of research in the water treatment area. Copper, cobalt, and iron oxides have already shown interesting results. For example, elimination of tetracycline in water with the use of a persulfate-activated system was studied by Cao et al., using zero-valent iron [20]. Sulfadiazine, another pharmaceutical degradation, was also successfully achieved using zero-valent iron as a persulfate activator [21]. In addition, nano-zero valent iron immobilized on organo-montmorillonite was synthesized and tested for persulfate activation for sulfamethazine degradation in water [22], while Lei et al. pointed out the advantages deriving from a bimetallic catalytic material, such as CuO-Fe_3O_4, in phenol removal from wastewater [23]. Considering cobalt-based activators, many configurations such as cobalt oxides, cobalt hydroxides, metal-organic frameworks, etc., have been tested and shown promising results [24,25]. However, it is still important to reduce cobalt ion leaching and enhance the stability of such catalytic materials across a wide pH range.

A significant contribution in the field was made by Zhang et al., who studied a CuO/persulfate oxidation system using 2,4-dichlorophenol as the model compound and showed that the peroxydisulfate activation process did not involve $SO_4^{\bullet-}$ generation [26]. In general, the persulfate activation process entails the creation of $SO_4^{\bullet-}$ through electron transfer from the transition metal to persulfate, followed by subsequent reactions between the sulfate radicals and water to form $^{\bullet}OH$. However, in the case of CuO, a non-radical mechanism may take place through reactive complexes formation between persulfate and the activators (such as CuO). A subsequent investigation of the CuO/persulfate system resolved previous observations and indicated that there may be a blend of radical and non-radical mechanisms taking place [27].

However, apart from these well-studied transition metals, the activity of other metal-based catalytic materials remains scarcely reported. Specifically, considering nickel-based catalytic materials that are widely adopted in typical catalytic processes, there are only a very few studies dealing with their use as heterogeneous persulfate activators. In particular, Jiang et al. synthesized nickel stannate–reduced graphene oxide (Ni_2SnO_4–RGO) composites and incorporated them in a heterogeneous activation persulfate system for bisphenol A degradation in water. Experiments carried out with appropriate radical scavengers showed that both $SO_4^{\bullet-}$ and $^{\bullet}OH$) participate in degradation reactions, with the former playing the leading role [28]. Another group studied the efficiency of $Ni_xCo_{3-x}O_4$ for the bisphenol A and tetracycline decomposition. The researcher found that it possesses higher efficiency than NiO and Co_2O_3 [29]. Moreover, nickel was mixed with lanthanum oxide, forming perovskite phases showing appreciable activity and stability. Very recently, Kim et al. demonstrated a non-radical activation mechanism in the case of Ni–NiO nanocomposites [30]. Furthermore, oxygen vacancies-rich NiO could activate persulfate by a non-radical mechanism leading to e fast phenol elimination [31]; this hypothesis was further verified by Xie et al., studying sulfamethoxazole degradation in water using pompon-like NiO microspheres as a heterogeneous persulfate activator [32].

On the other hand, many researchers stated that the persulfate activation mechanism in the case of NiO should include $SO_4^{\bullet-}$ formation, as the Ni^{2+}/Ni^{3+} redox potential allowed Ni^{2+} to react with persulfate generating $SO_4^{\bullet-}$ [33–35].

The above-mentioned contradictory conclusions regarding the use of NiO as a heterogeneous persulfate activator indicate the need for further research. Within this framework, $NiO/SrCO_3$ was chosen as the heterogeneous persulfate activator in this study. Considering strontium-based materials, Gkika et al. investigated sulfamethoxazole degradation in water with the use of lanthanum strontium cobaltite perovskite oxide [36], whereas Miao et al. synthesized $SrCo_{1-x}Ti_xO_{3-\delta}$ perovskite oxides and reported their efficiency towards persulfate activation for phenol degradation [37]. In addition, strontium has been used as a promoter in different catalytic configurations, such as $BiFeO_3$ [38] and $LaFeO_3$ [39].

This is the first time that such a catalytic configuration has been used for persulfate activation. The ability of $NiO/SrCO_3$ to activate SPS was investigated in terms of degrading the antibiotic sulfamethoxazole (SMX) in ultrapure water (UPW). SMX has been detected in concentrations ranging from nanograms per liter to milligrams per liter in treated effluents [40,41]. The persistence of SMX in the environment can lead to potential ecological harm, as it may disrupt the balance of aquatic ecosystems, affecting various organisms, including algae, fish, and other aquatic life. Moreover, it can boost the growth of antibiotic resistance in bacteria, having fatal implications for human health [42].

$NiO/SrCO_3$ was synthesized following a combustion method, while BET, XRD, and TEM techniques were used for the catalyst characterization. The effectiveness of the $NiO/SrCO_3/SPS$ system for SMX removal was reported not only in ultrapure water (UPW) but also in real and synthetic water matrices, such as treated effluent (WW) and bottled water (BW).

2. Materials and Methods

2.1. Chemicals and Aqueous Media

Table 1 summarizes the chemicals (Sigma Aldrich, Merck KGaA, Darmstadt, Germany and/or its affiliates) used in this work.

Table 1. Chemicals employed in this work.

Name	Formula	CAS Number
Sulfamethoxazole, SMX	$C_{10}H_{11}N_3O_3S$	723-46-6
Sodium persulfate, SPS	$Na_2S_2O_8$	7775-27-1
Methanol	CH_3OH	67-56-1
Sulfuric acid	H_2SO_4	7664-93-9
Sodium hydroxide	NaOH	1310-73-2
Sodium bicarbonate	$NaHCO_3$	144-55-8
Sodium chloride	NaCl	7647-14-5

Table 1. *Cont.*

Name	Formula	CAS Number
Humic acid	-	1415-93-6
Tert-butanol	$(CH_3)_3COH$	75-65-0
Strontium nitrate, 99.995% trace metals basis	$Sr(NO_3)_2$	10042-76-9
Nickel (II) nitrate hexahydrate, 99.999% trace metals basis	$Ni(NO_3)_2 \cdot 6H_2O$	13478-00-7
Citric acid	$C_6H_8O_7$	77-92-9
Ammonium nitrate	NH_4NO_3	6484-52-2
Ammonia	NH_3	7664-41-7

In the majority of the experiments presented in this study, ultrapure water (UPW, Millipore Milli-Q Gradient A10) was used as the water matrix. However, in some experiments, treated effluent (WW) from the wastewater treatment plant of the University of Patras and commercially available bottled water (BW) were examined as environmental matrices. The physicochemical characterization of aqueous matrices is shown in Table 2.

Table 2. Water matrices' characteristics.

Parameter	WW	BB	UPW
pH	8	7.5	6
Conductivity [μS/cm]	0.33	0.39	0.012
Alkalinity [mg/L]	190	152	-
Sulfates [mg/L]	33	15	-
Chlorides [mg/L]	69	9.8	-
Bicarbonates [mg/L]	190	209	-
TOC [mg/L]	7	-	-
COD [mg/L]	21	-	-

2.2. *NiO/SrCO$_3$ Synthesis*

The preparation of $NiO/SrCO_3$ took place by the in situ ignition method [43]. First, the precursor compounds $Sr(NO_3)_2$ and $Ni(NO_3)_2$ were diluted under stirring to triple distilled water. Then, a solution of citric acid, $C_6H_8O_7$, and ammonium nitrate, NH_4NO_3, was added and the solution was continuously stirred. Appropriate amounts of NH_3 were added (pH = 9) in order to neutralize the excess of citric acid, followed by heating through the hot plate under stirring.

When the water had evaporated, the container was heated vertically and circumferentially using a heat gun from a distance of 5 cm.

The mixture ignited into a thin crust and crumbled into a fine powder. Finally, it was calcined in air for 5 h at 700 °C [44]. The obtained wt.% of NiO was equal to ca. 33.

2.3. *NiO/SrCO$_3$ Physicochemical Characterization Techniques*

The crystal structure of $NiO/SrCO_3$ was examined using X-ray diffraction (XRD) through an A Brucker D8 Advance device (Billerica, MA, USA) equipped with a Cu Kα source (wavelength λ = 1.5496 Å). For the purpose of capturing transmission electron microscopy (TEM) images, a JEOL JEM-2100 system (Tokyo, Japan) at 200 kV was employed, offering a resolution of 0.23 nm for point measurements and 0.14 nm for lattice measurements. The specific surface area (SSA) was estimated using the Brunauer–Emmett–Teller (BET) method, employing a Gemini III 2375 instrument from Micromeritics [45].

2.4. Analytical Methods and Experimental Procedures

The SMX oxidation was investigated within a 250 mL cylindrical Pyrex reaction container exposed to the atmosphere. At first, the vessel was loaded with a solution (volume 120 mL) with the selected concentration of SMX, followed by the addition of premeasured SPS and NiO/$SrCO_3$ loading. The degradation tests were performed at 25 °C without regulating the solution's pH, and continuous magnetic stirring was maintained throughout the process.

To analyze the samples, 1.2 mL portions were withdrawn from the vessel at specific times. These samples were mixed with methanol to stop the reaction, filtered, and subjected to analysis using high-performance liquid chromatography (HPLC). The HPLC system utilized for this purpose consisted of an Alliance 2695 HPLC system connected with a Waters 2996 photodiode array detector. Separation was achieved using a Kinetex column (C18 100A, 150 mm × 3 mm; 2.6 μm particle size, Phenomenex (Torrance, CA, USA) thermostated at 45 °C.

For the investigation of the combined activation of persulfate by NiO/$SrCO_3$ and simulated solar light irradiation, additional tests were performed using an Oriel, LCS-100 solar simulator equipped with a 100 W Xenon arc lamp.

2.5. Electrochemical Characterization

2.5.1. Electrochemical Measurements

Electrochemical analysis was carried out implementing an PGSTAT128N Auto lab potentiostat (Utrecht, The Netherlands) at room temperature (ca. 25 °C) in a three-electrode setup using 0.1 M Na_2SO_4 as the electrolyte, while a Ag/AgCl (3 M KCl) purchased from (Metrohm A.G., Herisau, Switzerland) served as the reference electrode and a Pt wire served as the counter electrode. The working anodic electrode was the examined NiO/$SrCO_3$ material deposited on FTO (NiO/$SrCO_3$/FTO), whose preparation is described in detail in the following section. Electrochemical impedance spectroscopy (EIS) and linear sweep voltammetry (LSV) measurements were performed in order to examine the potential effect of an electron transfer mechanism to SMX elimination [46,47]. LSV and EIS experiments were performed using the electrolyte solution of only the examined catalyst (NiO/$SrCO_3$/FTO), with the presence of the catalyst and amount of SPS as well as when in the presence of the catalyst and amounts of SPS and SMX. LSV curves were obtained by scanning linearly the applied potential from the open circuit potential (OCP) value (−0.2 V vs. Ag/AgCl) to 1.5 V vs. Ag/AgCl applying a potential scan rate equal to 50 mV s^{-1} whereas EIS data were recorded under OCP conditions in the frequency range 0.1 to 100 kHz and with an amplitude perturbation of 0.01 V.

2.5.2. Synthesis of NiO/$SrCO_3$/FTO Working Electrode

The NiO/$SrCO_3$ was immobilized as a thin film on a coated glass of high conductive fluorine—doped tin oxide. The deposition was performed employing a suspension of 100 mg NiO/$SrCO_3$, 1.2 mL ethanol and 0.4 mL of 5 wt.% Nafion® ionomer solution (Sigma Aldrich, St. Louis, MO, USA). The mixture was ultrasonicated for 30 min in a sonochemical bath (Bandelin electronic GmbH & Co. KG, Berlin, Germany) in order to obtain the desired homogeneity [48]. The Nafion solution was used for electrode stability enhancement and maintenance of the thin film uniformity. The deposition of the aforementioned suspension on the FTO surface was performed via drop-casting and subsequent drying overnight at 75 °C. The prepared electrode surface area and the resulting areal density were ca. 1 cm^2 and 10 mg/cm^2, respectively.

3. Results and Discussion

3.1. Physicochemical Characterization

Figure 1 shows thee X-ray diffraction spectrum of the NiO/$SrCO_3$ sample. It is observed that the spectrum consists of peaks corresponding to the $SrCO_3$ orthorhombic structure (JCPDS No 05-0418). Furthermore, the peaks at 37.4° and 43.5° correspond to NiO (JCPDS No. 1-1239), verifying the successful formation of the composite material. The SSA of NiO/$SrCO_3$ was estimated at ca. 5 m^2/g.

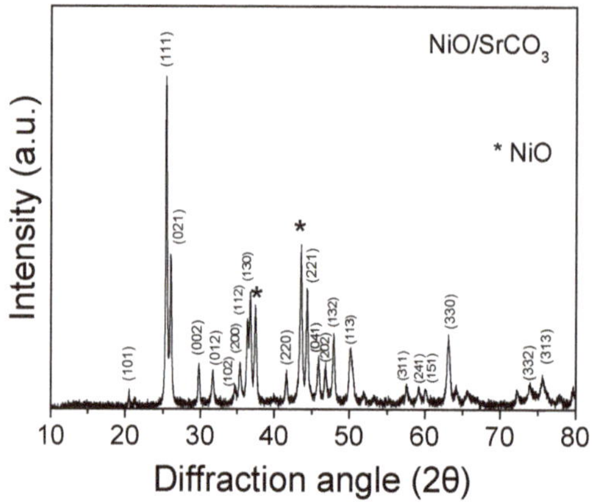

Figure 1. XRD pattern of NiO/$SrCO_3$.

The morphology together with the particle size of the as prepared material was further investigated by TEM images (Figure 2). It was observed that NiO/$SrCO_3$ is made up of spherical-like nanoparticles with an estimated average diameter of approximately 31 nm (Figure 2C). Moreover, it could be stated that the particle size distribution (Figure 2C) is rather broad, including particles from 10 to ca. 65 nm.

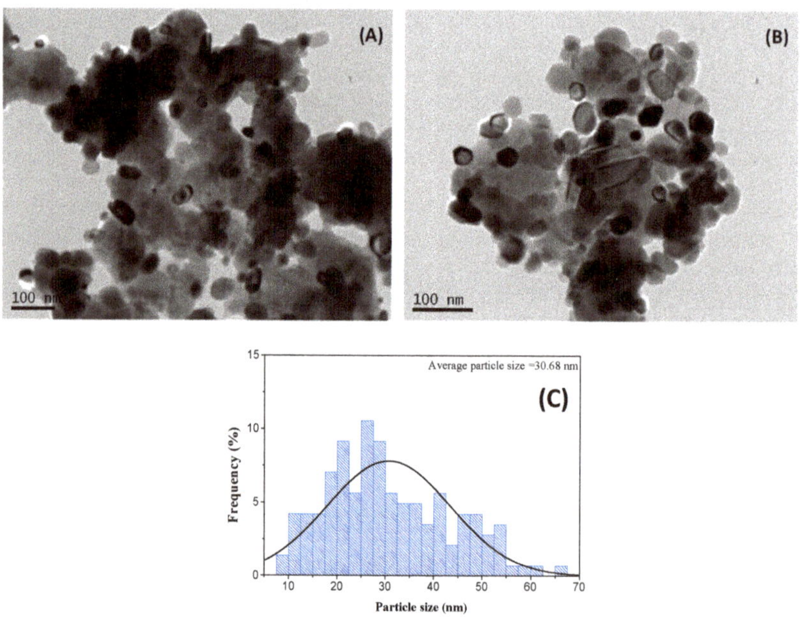

Figure 2. (**A**,**B**) TEM images and (**C**) particle size distribution of NiO/$SrCO_3$.

3.2. *Effect of Experimental Conditions on SMX Removal in the NiO/$SrCO_3$/SPS System*

In the domain of SR-AOPs, the ability of nickel oxide (NiO) to activate sodium persulfate (SPS) efficiently has exhibited inherent limitations, thereby compelling researchers to seek diverse modifications to enhance its efficacy [30,49]. Acknowledging this significant challenge, the present scientific study endeavors to address this limitation by fabricating a novel composite material that combines nickel oxide with strontium carbonate (NiO/$SrCO_3$). The primary objective of this synthesis is to achieve successful activation of SPS for the SMX elimination. Because the latter serves as a representative antibiotic compound, the proposed system could offer a promising approach to environmental remediation.

In the preliminary stage of our research effort, a thorough investigation was conducted to scrutinize the impact of crucial operating parameters, like catalyst loading, SPS concentration, SMX concentration, and the solution initial pH. It is important to highlight that according to the literature, the reaction is governed by pseudo-first-order kinetics [50–52].

3.2.1. Impact of SPS Concentration on the Degradation Reaction

To explore the influence of SPS loading on the oxidation process and effectively eliminate 500 μg/L of SMX, a series of experiments was conducted, gradually increasing the SPS dosage from 0 to 100 mg/L. Throughout these experiments, 100 mg/L of catalyst was employed. The reactions were carried out in ultra-pure water (UPW) at room temperature.

Initially, the adsorption of SMX on the surface of the composite catalyst in the absence of SPS was quantified, and it was found to be negligible, causing only a minor decrease in SMX concentration (less than 5%). This finding could be correlated with the diminutive specific surface area of $NiO/SrCO_3$ (5 m^2/g). However, as shown in Figure 3a, upon increasing the SPS dosage to 50 mg/L, a remarkable enhancement in the degradation reaction rate was observed. In particular, complete elimination of SMX was achieved within 45 min, while the apparent reaction rate constant (k_{app}) presented a notable threefold increase (from 0.032 min^{-1} to 0.092 min^{-1}). Interestingly, when the loading of SPS was further increased to 100 mg/L, SMX degradation was not favored, coinciding with a slight decrease in k_{app} value from 0.092 to 0.068 min^{-1}. This observation can be attributed to the self-quenching, or the scavenging of the reactive species from SPS. Specifically, the generated sulfate radicals ($SO_4^{\bullet -}$) may undergo self-recombination or react with SPS molecules instead of effectively reacting with the organic compounds according to Equations (1) and (2) [53,54]:

$$SO_4^{\bullet -} + SO_4^{\bullet -} \rightarrow S_2O_8^{2-} \quad (1)$$

$$SO_4^{\bullet -} + S_2O_8^{2-} \rightarrow SO_4^{2-} + S_2O_8^{\bullet -} \quad (2)$$

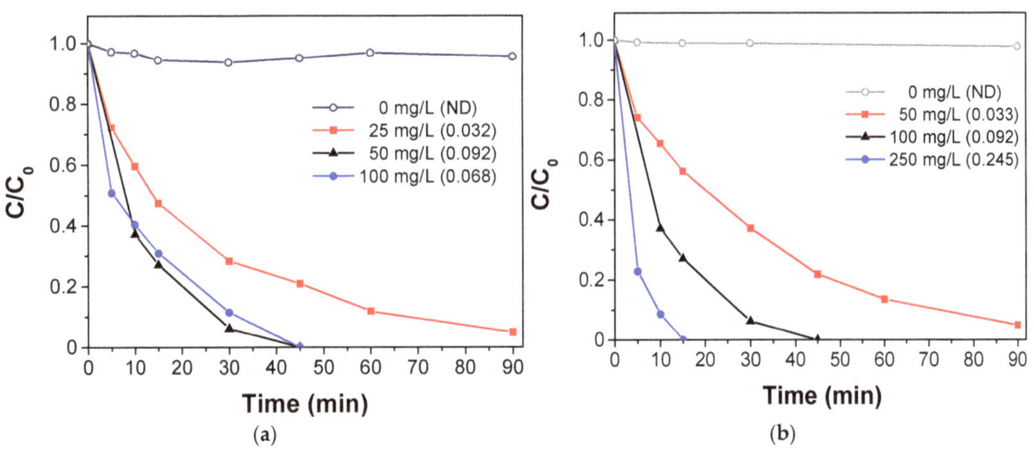

Figure 3. Effect of (**a**) SPS concentration (**b**) catalyst dosage on the removal of 0.5 mg/L SMX.

The above results are in the same order of magnitude as others obtained in similar systems for SMX degradation [4]. Based on the results obtained in Figure 3a, a [SPS] = 50 mg/L was chosen for the following tests.

3.2.2. Effect of NiO/$SrCO_3$ Dosage

The effect of the catalyst was evaluated by ranging the loading from 0 to 250 mg/L while maintaining the loading of SPS constant at 50 mg/L. As depicted in Figure 3b, the inherent oxidizing potential of SPS, in the absence of the catalyst, proved to be insignificant, aligning with the findings reported in prior studies [36,55]. The progressive increase in NiO/$SrCO_3$ dosage from 50 to 250 mg/L engendered a noteworthy increase in the degradation rate, as evidenced by the removal efficiencies of SMX after 15 min, which were 44%, 73%, and 100%, respectively. The most probable rationale behind this observation is that the higher NiO/$SrCO_3$ concentration provides a higher number of active sites for SPS activation, resulting in an increased population of reactive species [54–56]. Despite obtaining optimal results at a catalyst dosage of 250 mg/L, subsequent experiments were performed at 100 mg/L, with the aim of conserving the catalyst usage.

3.2.3. Effect of Antibiotic Concentration

The effect of the antibiotic loading was comprehensively assessed through a set of experiments performed at four distinct SMX loadings (0.25–3.00 mg/L), and the corresponding normalized concentration versus time profiles are presented in Figure 4a. The removal process was substantially hindered as the SMX concentration was raised. Specifically, at the 30-min mark, the removal efficiencies in each case were as follows: 100% for 0.25 mg/L, 93% for 0.50 mg/L, 70% for 1.00 mg/L, and 20% for 3.00 mg/L. From the perspective of kinetics, the corresponding k_{app} values are 0.116, 0.092, 0.037, and 0.004 min^{-1}, confirming the deviation from a true first-order model, since the kinetic constant is contingent on initial SMX concentration. The reason behind this observation is the fact that under defined operating parameters (e.g., catalyst loading, SPS concentration, solution pH, temperature), the population of the reactive species (mainly free radicals) remains practically constant while the concentration of the organic compound is steadily increasing, demanding longer time periods for the same extent of degradation [12].

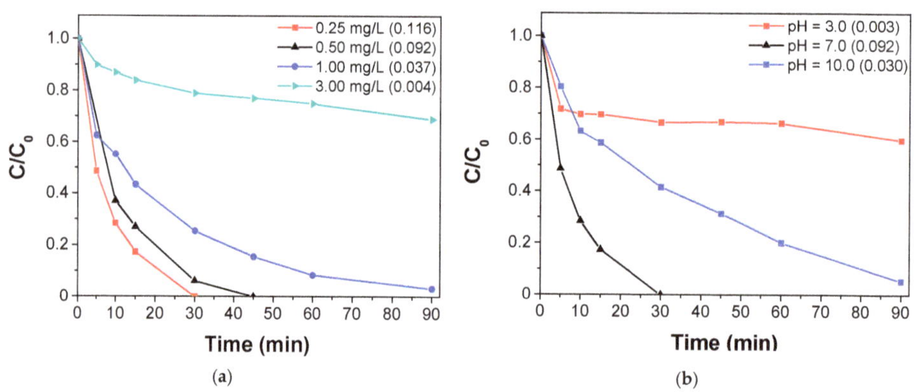

Figure 4. Removal profiles of SMX as a function of (**a**) initial SMX concentration; (**b**) initial solution pH ([SMX] = 0.5 mg/L). Experimental conditions: [Catalyst] = 100 mg/L, [SPS] = 50 mg/L.

3.2.4. Effect of pH

Solution's pH is an essential factor that exerts substantial influence on the removal of organic compounds in a catalyst/SPS system. To thoroughly investigate its impact, a set of experiments was conducted, encompassing a range of pH conditions, spanning from acidic (pH 3) to neutral (pH 7) and alkaline environments (pH 10). According to Figure 4b, a neutral environment proved to be more conducive to the abatement of SMX compared to acidic and alkaline conditions. In the neutral pH setting, complete elimination of SMX was attained within 30 min, whereas the removal efficiencies in the acidic and alkaline cases reached only 30% and 40% in the same time period.

To further elucidate these results, both the catalyst's surface charge and the prevailing chemical form of SMX should be clarified. According to scientific reports, the reported pKa values of the SMX are 1.7 and 5.5 [57]. At a pH value of 3, SMX retains its molecular structure, whereas at pH 7 and 10, it undergoes deprotonation, promoting negatively charged species formation. Having in mind the collected data, it is reasonable to infer that at pH 10, repulsive interactions likely developed between SMX and the $NiO/SrCO_3$ surface. Consequently, SMX encounters hindrances in reaching the catalyst's surface, leading to limited SMX degradation. In contrast, under strong acidic environments, it has been reported that persulfate may be consumed, as described in Equation 3, resulting in the observed slow removal rate [51,58].

$$S_2O_8^{\bullet -} + H^+ \rightarrow HS_2O_8^- \quad (3)$$

3.2.5. Impact of Water Matrix and Additives on Process Efficiency

In the subsequent phase of our study, assays were carried out to gain knowledge into the implementation of the $NiO/SrCO_3/SPS$ system under different water matrices. As practical applications of SPS-based AOPs are still limited, it becomes essential to interpret how the complexity of the water matrices affects the effectiveness of the catalytic system. To explore this aspect, two distinct water sources were utilized: BW and WW. Results are summarized in Figure 5a.

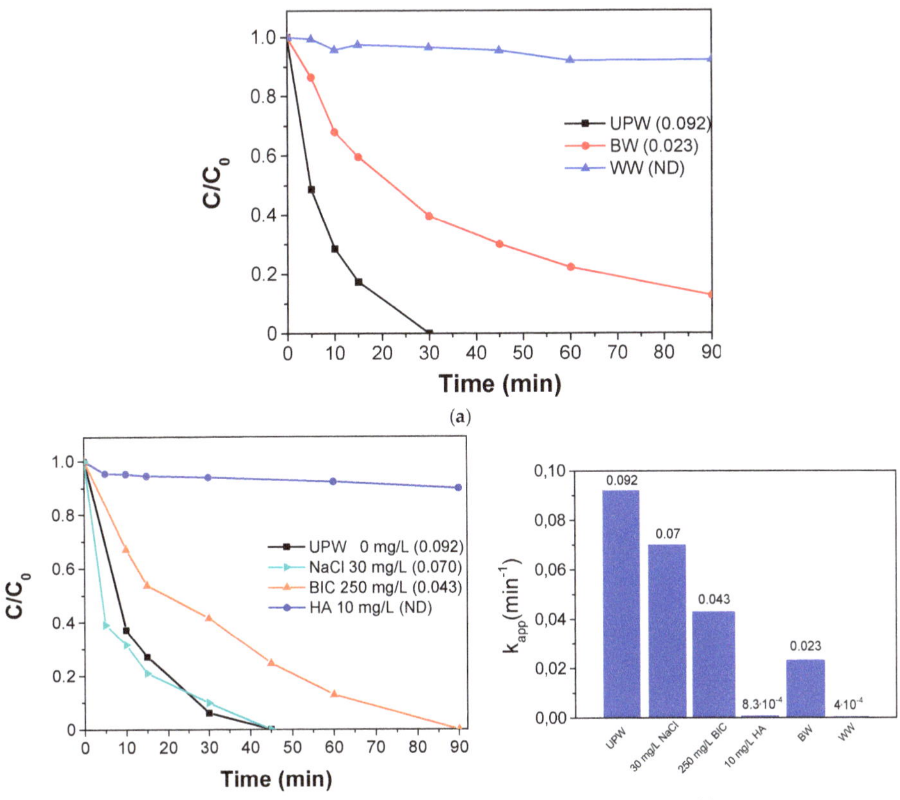

Figure 5. Effect of (**a**) water matrix (**b**) inorganic anions and organic matter on SMX degradation. (**c**) Apparent rate constant (k_{app}) values. Experimental conditions: [Catalyst] = 100 mg/L, [SPS] = 50 mg/L, [SMX] = 0.5 mg/L.

The experimental findings clearly demonstrate that elevating water matrix's complexity exerts a significantly adverse impact on SMX degradation. The SMX removal is 60% in BW and does not exceed 5% in WW after 30 min of reaction. The observed reaction rates are markedly decreased, likely due to the presence of ions which can either act as scavengers, consuming the generated free radicals, or they could potentially be a competitor with the antibiotic, thereby masking the catalyst 's active sites [59,60]. In the case of WW, the inhibitory effects are clearly pronounced, presumably due to the existence of organic matter.

Further tests were conducted in artificial matrices enriched with predetermined amounts of sodium chloride (NaCl 30 mg/L), bicarbonates (HCO_3^- 250 mg/L), and humic acid (HA 10 mg/L). HCO_3^- and Cl^- were specifically selected because they are the dominant components in BW. On the other hand, HA was used as a representative compound accurately imitating the organic moieties detected in treated effluent or environmental samples. Figure 5b reveals that the addition of 30 mg/L NaCl has a minor impact on the oxidation, leading to a slight reduction in the k_{app} value from 0.092 to 0.070 min^{-1}. This phenomenon may be probably attributed to the interaction between the reactive species and the chlorine ions, leading to the formation of chlorine radicals as expressed in Equations (4)–(6) [50,61]:

$$SO_4^{\bullet-} + Cl^- \rightarrow SO_4^{2-} + Cl^\bullet \tag{4}$$

$$HO^\bullet + Cl^- \rightarrow ClOH^{\bullet-} \tag{5}$$

$$ClOH^{\bullet-} + H^+ \rightarrow Cl^\bullet + H_2 \tag{6}$$

Figure 5b highlights the negative effect of bicarbonates on the oxidation of SMX. Specifically, the addition of 250 mg/L HCO_3^- (representative BIC concentration in BW) induced a two-fold decrease in k_{app} values, compared to the experiment conducted in UPW, extending the complete breakdown time of SMX to 90 min. The presence of BIC initiated side reactions, consuming the highly reactive free radicals ($SO_4^{\bullet-}$, $^\bullet OH$) and led to the subsequent creation of secondary species with reduced potential, as described in the following Equations (7)–(11) [16,62]. Additionally, BIC facilitated the development of a strong alkaline environment (pH = 9) which was previously noted to impede the oxidation reaction. In fact, the majority of documented research is conducted using ultrapure water, which overlooks the interplay between organic and inorganic substances such as carbonates and reactive oxygen species. This leads to inaccurately high yields that do not reflect the real outcomes. On a positive note, certain recent investigations have taken into account the matrix impact on the degradation of micropollutants [63].

$$HCO_3^- + SO_4^{\bullet-} \rightarrow SO_4^{2-} + HCO_3^\bullet \tag{7}$$

$$HCO_3^\bullet - \rightarrow H^+ + CO_3^{\bullet-} \tag{8}$$

$$^\bullet OH + HCO_3^- \rightarrow CO_3^{\bullet-} + H_2O \tag{9}$$

$$CO_3^{2-} + SO_4^{\bullet-} \rightarrow SO_4^{2-} + CO_3^{\bullet-} \tag{10}$$

$$CO_3^{2-} + {}^\bullet OH \rightarrow OH^- + CO_3^{\bullet-} \tag{11}$$

Regarding HA, it is evident that it completely suppressed the oxidation of SMX (Figure 5b,c), as SMX removal does not exceed 10% and k_{app} values were nearly identical to those observed in experiments performed in WW. HA competes with SMX, quenching the free radicals through its aromatic groups. Furthermore, it has been reported that HA is able to cover the catalyst's surface, obstructing active sites and thereby impeding the activation of SPS [55,64].

3.3. Degradation Mechanism

3.3.1. Electrochemical Measurements

Figure 6 presents the electrochemical characteristics of the NiO/$SrCO_3$/FTO electrode. The measured current (mA), as demonstrated in Figure 6a, may practically refer to current density (mA cm^{-2}) since the surface area of the working electrode is equal to 1 cm^2. It is obvious from the LSV curves shown in Figure 6a that adding 50 mg L^{-1} SPS in the electrolyte solution increased the current, which could be attributed to the formation of metastable reactive complexes thus facilitating the SPS and active catalytic surface interaction [46]. Specifically, the maximum current density enhanced from approximately 1 mA cm^{-2} when the Ni/$SrCO_3$ was present in the electrochemical system to approximately 1.8 mA cm^{-2} in the case of the Ni/$SrCO_3$/SPS system. The same behavior was observed with the addition of the SMX pollutant (0.5 mg L^{-1}) in the electrolyte solution. The measured current values increased further (above 2 mA cm^{-2}) with the co-presence of the NiO/$SrCO_3$, SPS, and SMX) in the Na_2SO_4 solution, implying an electron transfer mechanism on the interface of NiO/$SrCO_3$/FTO, SPS, and SMX.

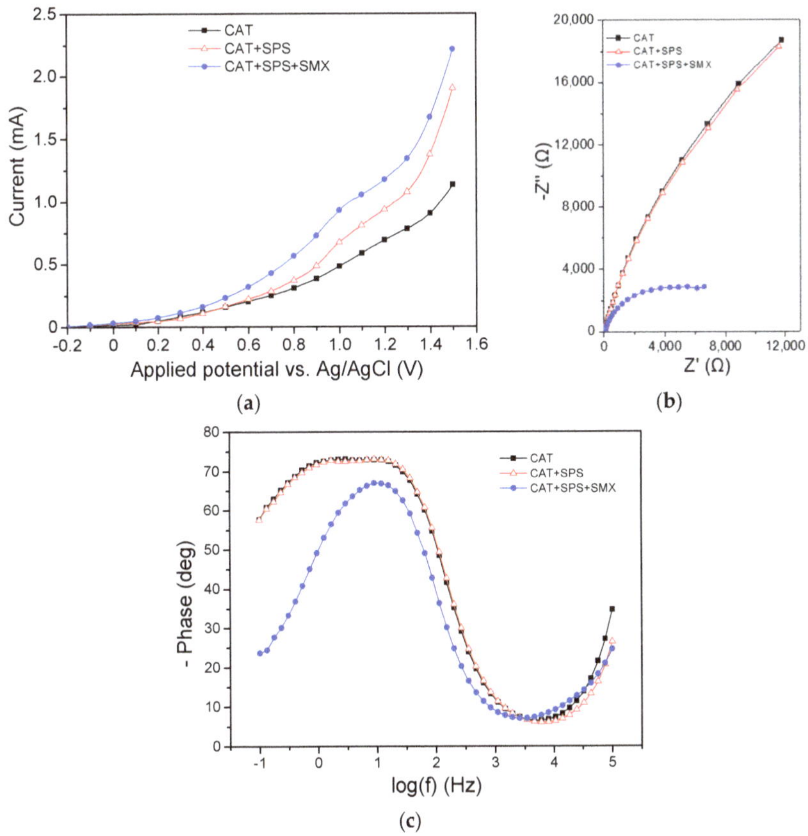

Figure 6. (**a**) Linear sweep voltammograms and EIS characteristics depicted in the form of (**b**) Nyquist plots and (**c**) Bode plots recorded in the presence of catalyst (CAT) and in the co-presence of catalyst (Ni/$SrCO_3$) and SPS (CAT + SPS) as well as catalyst, SPS and SMX (CAT + SPS + SMX); Measurements were conducted in a three-electrode set-up with the Ni/$SrCO_3$/FTO as working electrode, a Ag/AgCl (3M KCl sat.) as reference electrode and a Pt wire as counter electrode. Open circuit potential: −0.2 V vs. Ag/AgCl.

The behavior of the LSVs was reflected in the EIS characteristics as depicted in Figure 6b,c. Figure 6b shows the Nyquist plots as recorded in the aforementioned three scenarios. The calculation of the charge transfer resistance, RCT, cannot be performed since the semicircles are incomplete. However, it is evident that the addition of SPS and SMX decreased the semicircle diameter, which is more evident in the NiO/$SrCO_3$/SPS/SMX system, indicating higher electrical conductivity. This observation is reflected in the shape of the Bode plots as demonstrated in Figure 6c. A reduction in the intensity of the peaks of the Bode plots was observed when comparing the electrochemical system without and in the presence of SMX, which was accompanied by shifting the peak's maximum towards higher frequencies indicating a decline of the characteristic relaxation time of the examined electrochemical system. The appearance of a flattened semicircle in the Nyquist plots and a main peak in the Bode plots implies the realization of a dominant process which could be mainly related to the charge transfer resistance, whereas various subprocesses that may take place simultaneously in the electrochemical system cannot be distinguished. In any case, the appearance of a flattened semicircle implies that the processes that occur are intertwined and overlap with each other, while a dominant process is essentially the one which controls the reaction rate and subsequently the NiO/$SrCO_3$ performance towards SMX degradation.

It can be concluded that the simultaneous presence of NiO/$SrCO_3$/FTO, SPS, and SMX led to an increment in the measured current values when the potential changing and to a RCT decline, implying SMX oxidation via a non-specific pathway. In particular, it can be assumed that the SMX oxidation may occur via electron transfer from the organic compound acting as an electron donor, to SPS acting as an electron acceptor) mediated by the NiO/$SrCO_3$ catalyst [46,65].

3.3.2. Scavengers of Reactive Species

To investigate the reactive species involved in the oxidation of SMX, quenching experiments were conducted utilizing the appropriate scavengers. In this regard, methanol (MeOH, 5 g/L) and tert -butanol (TBA, 5 g/L) were employed to trap $SO_4^{\bullet -}$ ($k = 3.2 \times 10^6\ M^{-1}\ s^{-1}$) and $^{\bullet}OH$ ($k = 3.8–7.6 \times 10^8\ M^{-1}\ s^{-1}$), respectively. It should be mentioned that MeOH also interacts with $^{\bullet}OH$ ($k = 3.0 \times 10^8\ M^{-1}\ s^{-1}$) [66].

The outcomes from Figure 7a clearly indicate the significant contribution of hydroxyl radicals as the dominant radical since the oxidation rate was substantially decreased when TBA was added. Contrarily, the contribution of $SO_4^{\bullet -}$ appears to be minor, as the addition of MeOH leads to a negligible differentiation in k_{app} value compared to the experiments with TBA (Figure 7b).

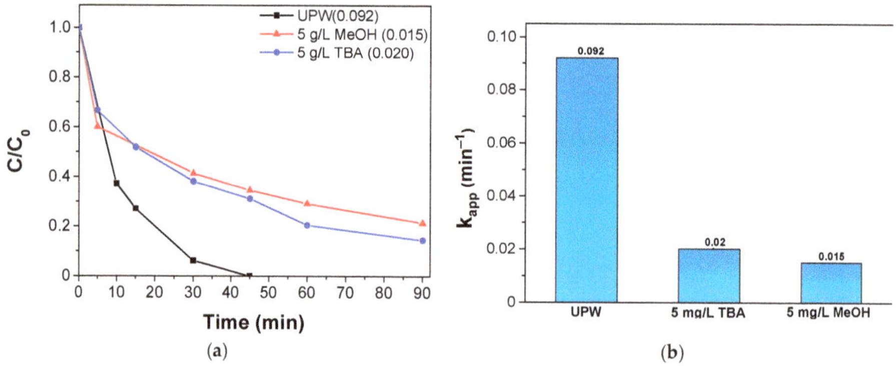

Figure 7. (**a**) Effect of quenching agents on the oxidation of 0.5 mg/L SMX; (**b**) Apparent rate constant (k_{app}) values. Experimental conditions: [Catalyst] = 100 mg/L, [SPS] = 50 mg/L, [TBA] = 5 g/L, [MeOH] = 5 g/L.

Overall, based on the data collected from the electrochemical characterization and trapping experiments, it can be strongly suggested that the successful oxidation of SMX involves a combination of non-radical and radical mechanisms.

4. Conclusions

This study focused on the development of a novel $NiO/SrCO_3$ catalytic material for SMX elimination via SPS activation. This catalytic system was prepared using the combustion method, and its physicochemical characteristics were examined through various methods. The composite material demonstrated exceptional catalytic activity, as complete decomposition of SMX was achieved in short time periods in UPW (45 min). The SMX degradation was described by pseudo first-order kinetics, while the apparent kinetic constant decreased with SMX concentration. SMX degradation was favored at pH 7. A significant inhibition was observed in the test performed in environmental matrices such as secondary effluent or bottled water, possibly due to competition from matrix constituents for reactive species and the surface area, as demonstrated in the experiments conducted in the presence of humic acid or bicarbonates. Notably, chlorides in concentrations up to 30 mg/L did not significantly alter the kinetics. Based on the electrochemical and scavenging measurements, it could be concluded that SMX degradation may occur via both radical and non-radical pathways. Overall, the examined system showed promising results for the decomposition of SMX. However, future studies must assess the catalyst reuse through operation under continuous flow, and further examine the mechanistic aspects of the system, including the study of the transformation by-products and their toxicity.

Author Contributions: Conceptualization, A.P. and Z.F.; Writing. A.P., K.K. and G.B.; writing—review and editing, Z.F.; Investigation, D.J.S., G.B. and K.K.; Methodology, D.J.S. and K.K.; Supervision. Z.F. and A.P. All authors have read and agreed to the published version of the manuscript.

Funding: This work is funded by the project Development of New Innovative Low Carbon Footprint Energy Technologies to Enhance Excellence in the Region of Western Macedonia (MIS 5047197), which is implemented under the Action "Reinforcement of the Research and Innovation Infrastructure", funded by the Operational Program "Competitiveness, Entrepreneurship and Innovation" (NSRF 2014-2020) and co-financed by Greece and the European Union (European Regional Development Fund).

Data Availability Statement: Not applicable.

Acknowledgments: The authors wish to thank M. Kollia staff in the Laboratory of Electron Microscopy and Microanalysis (L.E.M.M.) at the University of Patras for TEM images.

Conflicts of Interest: The authors declare no conflict of interest.

References

1. Domingues, E.; Silva, M.J.; Vaz, T.; Gomes, J.; Martins, R.C. Persulfate Process Activated by Homogeneous and Heterogeneous Catalysts for Synthetic Olive Mill Wastewater Treatment. *Water* **2021**, *13*, 3010. [CrossRef]
2. Giannakis, S.; Lin, K.Y.A.; Ghanbari, F. A Review of the Recent Advances on the Treatment of Industrial Wastewaters by Sulfate Radical-Based Advanced Oxidation Processes (SR-AOPs). *Chem. Eng. J.* **2021**, *406*, 127083. [CrossRef]
3. Lin, D.; Fu, Y.; Li, X.; Wang, L.; Hou, M.; Hu, D.; Li, Q.; Zhang, Z.; Xu, C.; Qiu, S.; et al. Application of Persulfate-Based Oxidation Processes to Address Diverse Sustainability Challenges: A Critical Review. *J. Hazard. Mater.* **2022**, *440*, 129722. [CrossRef] [PubMed]
4. Honarmandrad, Z.; Sun, X.; Wang, Z.; Naushad, M.; Boczkaj, G. Activated Persulfate and Peroxymonosulfate Based Advanced Oxidation Processes (AOPs) for Antibiotics Degradation—A Review. *Water Resour. Ind.* **2023**, *29*, 100194. [CrossRef]
5. Qi, C.; Liu, X.; Ma, J.; Lin, C.; Li, X.; Zhang, H. Activation of Peroxymonosulfate by Base: Implications for the Degradation of Organic Pollutants. *Chemosphere* **2016**, *151*, 280–288. [CrossRef]
6. Li, J.; Liang, Y.; Jin, P.; Zhao, B.; Zhang, Z.; He, X.; Tan, Z.; Wang, L.; Cheng, X. Heterogeneous Metal-Activated Persulfate and Electrochemically Activated Persulfate: A Review. *Catalysts* **2022**, *12*, 1024. [CrossRef]
7. Wang, L.; Luo, D.; Hamdaoui, O.; Vasseghian, Y.; Momotko, M.; Boczkaj, G.; Kyzas, G.Z.; Wang, C. Bibliometric Analysis and Literature Review of Ultrasound-Assisted Degradation of Organic Pollutants. *Sci. Total Environ.* **2023**, *876*, 162551. [CrossRef]
8. Yang, L.; Xue, J.; He, L.; Wu, L.; Ma, Y.; Chen, H.; Li, H.; Peng, P.; Zhang, Z. Review on Ultrasound Assisted Persulfate Degradation of Organic Contaminants in Wastewater: Influences, Mechanisms and Prospective. *Chem. Eng. J.* **2019**, *378*, 122146. [CrossRef]

9. Fedorov, K.; Dinesh, K.; Sun, X.; Darvishi Cheshmeh Soltani, R.; Wang, Z.; Sonawane, S.; Boczkaj, G. Synergistic Effects of Hybrid Advanced Oxidation Processes (AOPs) Based on Hydrodynamic Cavitation Phenomenon—A Review. *Chem. Eng. J.* **2022**, *432*, 134191. [CrossRef]
10. Manos, D.; Papadopoulou, F.; Margellou, A.; Petrakis, D.; Konstantinou, I. Heterogeneous Activation of Persulfate by LaMO3 (M = Co, Fe, Cu, Mn, Ni) Perovskite Catalysts for the Degradation of Organic Compounds. *Catalysts* **2022**, *12*, 187. [CrossRef]
11. Manz, K.E.; Kulaots, I.; Greenley, C.A.; Landry, P.J.; Lakshmi, K.V.; Woodcock, M.J.; Hellerich, L.; Bryant, J.D.; Apfelbaum, M.; Pennell, K.D. Low-Temperature Persulfate Activation by Powdered Activated Carbon for Simultaneous Destruction of Perfluorinated Carboxylic Acids and 1,4-Dioxane. *J. Hazard. Mater.* **2023**, *442*, 129966. [CrossRef] [PubMed]
12. Kemmou, L.; Frontistis, Z.; Vakros, J.; Manariotis, I.D.; Mantzavinos, D. Degradation of Antibiotic Sulfamethoxazole by Biochar-Activated Persulfate: Factors Affecting the Activation and Degradation Processes. *Catal. Today* **2018**, *313*, 128–133. [CrossRef]
13. Bekris, L.; Frontistis, Z.; Trakakis, G.; Sygellou, L.; Galiotis, C.; Mantzavinos, D. Graphene: A New Activator of Sodium Persulfate for the Advanced Oxidation of Parabens in Water. *Water Res.* **2017**, *126*, 111–121. [CrossRef] [PubMed]
14. Huang, W.; Xiao, S.; Zhong, H.; Yan, M.; Yang, X. Activation of Persulfates by Carbonaceous Materials: A Review. *Chem. Eng. J.* **2021**, *418*, 129297. [CrossRef]
15. Wu, Y.; Shi, Y.; Chen, H.; Zhao, J.; Dong, W. Activation of Persulfate by Magnetite: Implications for the Degradation of Low Concentration Sulfamethoxazole. *Process Saf. Environ. Prot.* **2018**, *116*, 468–476. [CrossRef]
16. Alexopoulou, C.; Petala, A.; Frontistis, Z.; Drivas, C.; Kennou, S.; Kondarides, D.I.; Mantzavinos, D. Copper Phosphide and Persulfate Salt: A Novel Catalytic System for the Degradation of Aqueous Phase Micro-Contaminants. *Appl. Catal. B Environ.* **2019**, *244*, 178–187. [CrossRef]
17. Xiong, Z.; Jiang, Y.; Wu, Z.; Yao, G.; Lai, B. Synthesis Strategies and Emerging Mechanisms of Metal-Organic Frameworks for Sulfate Radical-Based Advanced Oxidation Process: A Review. *Chem. Eng. J.* **2021**, *421*, 127863. [CrossRef]
18. Yuan, R.; Jiang, Z.; Wang, Z.; Gao, S.; Liu, Z.; Li, M.; Boczkaj, G. Hierarchical MnO_2 Nanoflowers Blooming on 3D Nickel Foam: A Novel Micro-Macro Catalyst for Peroxymonosulfate Activation. *J. Colloid Interface Sci.* **2020**, *571*, 142–154. [CrossRef]
19. Yuan, R.; Jiang, M.; Gao, S.; Wang, Z.; Wang, H.; Boczkaj, G.; Liu, Z.; Ma, J.; Li, Z. 3D Mesoporous α-$Co(OH)_2$ Nanosheets Electrodeposited on Nickel Foam: A New Generation of Macroscopic Cobalt-Based Hybrid for Peroxymonosulfate Activation. *Chem. Eng. J.* **2020**, *380*, 122447. [CrossRef]
20. Cao, J.; Lai, L.; Lai, B.; Yao, G.; Chen, X.; Song, L. Degradation of Tetracycline by Peroxymonosulfate Activated with Zero-Valent Iron: Performance, Intermediates, Toxicity and Mechanism. *Chem. Eng. J.* **2019**, *364*, 45–56. [CrossRef]
21. Yang, S.; Che, D. Degradation of Aquatic Sulfadiazine by Fe0/Persulfate: Kinetics, Mechanisms, and Degradation Pathway. *RSC Adv.* **2017**, *7*, 42233–42241. [CrossRef]
22. Wu, J.; Wang, B.; Blaney, L.; Peng, G.; Chen, P.; Cui, Y.; Deng, S.; Wang, Y.; Huang, J.; Yu, G. Degradation of Sulfamethazine by Persulfate Activated with Organo-Montmorillonite Supported Nano-Zero Valent Iron. *Chem. Eng. J.* **2019**, *361*, 99–108. [CrossRef]
23. Lei, Y.; Chen, C.-S.; Tu, Y.-J.; Huang, Y.-H.; Zhang, H. Heterogeneous Degradation of Organic Pollutants by Persulfate Activated by CuO-Fe3O4: Mechanism, Stability, and Effects of PH and Bicarbonate Ions. *Environ. Sci. Technol.* **2015**, *49*, 6838–6845. [CrossRef] [PubMed]
24. Zhao, W.; Shen, Q.; Nan, T.; Zhou, M.; Xia, Y.; Hu, G.; Zheng, Q.; Wu, Y.; Bian, T.; Wei, T.; et al. Cobalt-Based Catalysts for Heterogeneous Peroxymonosulfate (PMS) Activation in Degradation of Organic Contaminants: Recent Advances and Perspectives. *J. Alloys Compd.* **2023**, *958*, 170370. [CrossRef]
25. Li, B.; Wang, Y.-F.; Zhang, L.; Xu, H.-Y. Enhancement Strategies for Efficient Activation of Persulfate by Heterogeneous Cobalt-Containing Catalysts: A Review. *Chemosphere* **2022**, *291*, 132954. [CrossRef] [PubMed]
26. Zhang, T.; Chen, Y.; Wang, Y.; Le Roux, J.; Yang, Y.; Croué, J.-P. Efficient Peroxydisulfate Activation Process Not Relying on Sulfate Radical Generation for Water Pollutant Degradation. *Environ. Sci. Technol.* **2014**, *48*, 5868–5875. [CrossRef]
27. Du, X.; Zhang, Y.; Hussain, I.; Huang, S.; Huang, W. Insight into Reactive Oxygen Species in Persulfate Activation with Copper Oxide: Activated Persulfate and Trace Radicals. *Chem. Eng. J.* **2017**, *313*, 1023–1032. [CrossRef]
28. Jiang, L.; Xu, X.; Yuan, J.; Zuo, Y.; Tao, Y.; Yao, D.; He, G.; Chen, H. Heterogeneous Activation of Persulfate for the Degradation of Bisphenol A with Ni_2SnO_4–RGO. *New J. Chem.* **2020**, *44*, 6355–6361. [CrossRef]
29. Wu, Z.; Liang, Y.; Zou, D.; Yuan, X.; Xiao, Z.; Deng, Y.; Zhou, Y.; Jiang, L.; Qin, P. Enhanced Heterogeneous Activation of Persulfate by $NixCo_3$–XO_4 for Oxidative Degradation of Tetracycline and Bisphenol A. *J. Environ. Chem. Eng.* **2020**, *8*, 104451. [CrossRef]
30. Kim, H.H.; Lee, D.; Choi, J.; Lee, H.; Seo, J.; Kim, T.; Lee, K.M.; Pham, A.L.T.; Lee, C. Nickel–Nickel Oxide Nanocomposite as a Magnetically Separable Persulfate Activator for the Nonradical Oxidation of Organic Contaminants. *J. Hazard. Mater.* **2020**, *388*, 121767. [CrossRef]
31. Liu, L.; Liu, Q.; Wang, Y.; Huang, J.; Wang, W.; Duan, L.; Yang, X.; Yu, X.; Han, X.; Liu, N. Nonradical Activation of Peroxydisulfate Promoted by Oxygen Vacancy-Laden NiO for Catalytic Phenol Oxidative Polymerization. *Appl. Catal. B Environ.* **2019**, *254*, 166–173. [CrossRef]
32. Xie, L.; Hao, J.; Xing, S. Enhanced Non-Radical Activation of Persulfate with Pompon-like NiO Microspheres for Removing Sulfamethoxazole in Water. *Environ. Sci. Pollut. Res. Int.* **2023**, *30*, 14455–14463. [CrossRef] [PubMed]
33. Huang, X.; Zhou, X.; Han, S.; Zhou, J.; Qian, G.; Gao, N. Cycle of Ni(II)-Ni(III)-Ni(II) in Ni-Doped Layered Double Hydroxides for Activation of Intercalated Peroxydisulfate. *Chem. Eng. J.* **2020**, *386*, 123937. [CrossRef]

34. Jian, S.; Sun, S.; Zeng, Y.; Liu, Z.; Liu, Y.; Yang, Q.; Ma, G. Highly Efficient Persulfate Oxidation Process Activated with NiO Nanosheets with Dominantly Exposed {1 1 0} Reactive Facets for Degradation of RhB. *Appl. Surf. Sci.* **2020**, *505*, 144318. [CrossRef]
35. Yue, D.; Guo, C.; Yan, X.; Wang, R.; Fang, M.; Wu, Y.; Qian, X.; Zhao, Y. Secondary Battery Inspired NiO Nanosheets with Rich Ni(III) Defects for Enhancing Persulfates Activation in Phenolic Waste Water Degradation. *Chem. Eng. J.* **2019**, *360*, 97–103. [CrossRef]
36. Gkika, C.; Petala, A.; Frontistis, Z.; Bampos, G.; Hela, D.; Konstantinou, I.; Mantzavinos, D. Heterogeneous Activation of Persulfate by Lanthanum Strontium Cobaltite for Sulfamethoxazole Degradation. *Catal. Today* **2021**, *361*, 130–138. [CrossRef]
37. Miao, J.; Sunarso, J.; Su, C.; Zhou, W.; Wang, S.; Shao, Z. $SrCo_{1-x}Ti_xO_{3-\delta}$ Perovskites as Excellent Catalysts for Fast Degradation of Water Contaminants in Neutral and Alkaline Solutions. *Sci. Rep.* **2017**, *7*, 44215. [CrossRef]
38. Wang, C.; Gao, S.; Zhu, J.; Xia, X.; Wang, M.; Xiong, Y. Enhanced Activation of Peroxydisulfate by Strontium Modified $BiFeO_3$Perovskite for Ciprofloxacin Degradation. *J. Environ. Sci.* **2021**, *99*, 249–259. [CrossRef]
39. Cheng, C.; Gao, S.; Zhu, J.; Wang, G.; Wang, L.; Xia, X. Enhanced Performance of LaFeO3 Perovskite for Peroxymonosulfate Activation through Strontium Doping towards 2,4-D Degradation. *Chem. Eng. J.* **2020**, *384*, 123377. [CrossRef]
40. Rodriguez-Mozaz, S.; Chamorro, S.; Marti, E.; Huerta, B.; Gros, M.; Sànchez-Melsió, A.; Borrego, C.M.; Barceló, D.; Balcázar, J.L. Occurrence of Antibiotics and Antibiotic Resistance Genes in Hospital and Urban Wastewaters and Their Impact on the Receiving River. *Water Res.* **2015**, *69*, 234–242. [CrossRef]
41. Song, Z.; Zhang, X.; Ngo, H.H.; Guo, W.; Wen, H.; Li, C. Occurrence, Fate and Health Risk Assessment of 10 Common Antibiotics in Two Drinking Water Plants with Different Treatment Processes. *Sci. Total Environ.* **2019**, *674*, 316–326. [CrossRef] [PubMed]
42. Cycoń, M.; Mrozik, A.; Piotrowska-Seget, Z. Antibiotics in the Soil Environment-Degradation and Their Impact on Microbial Activity and Diversity. *Front. Microbiol.* **2019**, *10*, 338. [CrossRef] [PubMed]
43. da Conceição, L.; Silva, A.M.; Ribeiro, N.F.P.; Souza, M.M.V.M. Combustion Synthesis of $La_{0.7}Sr_{0.3}Co_{0.5}Fe_{0.5}O_3$ (LSCF) Porous Materials for Application as Cathode in IT-SOFC. *Mater. Res. Bull.* **2011**, *46*, 308–314. [CrossRef]
44. Safakas, A.; Bampos, G.; Bebelis, S. Oxygen Reduction Reaction on $La_{0.8}Sr_{0.2}Co_xFe_{1-x}O_{3-\delta}$ Perovskite/Carbon Black Electrocatalysts in Alkaline Medium. *Appl. Catal. B Environ.* **2019**, *244*, 225–232. [CrossRef]
45. Petala, A.; Tsikritzis, D.; Kollia, M.; Ladas, S.; Kennou, S.; Kondarides, D.I. Synthesis and Characterization of N-Doped TiO_2 Photocatalysts with Tunable Response to Solar Radiation. *Appl. Surf. Sci.* **2014**, *305*, 281–291. [CrossRef]
46. Giannakopoulos, S.; Vakros, J.; Frontistis, Z.; Manariotis, I.D.; Venieri, D.; Poulopoulos, S.G.; Mantzavinos, D. Biochar from Lemon Stalks: A Highly Active and Selective Carbocatalyst for the Oxidation of Sulfamethoxazole with Persulfate. *Catalysts* **2023**, *13*, 233. [CrossRef]
47. Li, M.; Bi, Y.-G.; Xiang, L.; Chen, X.-T.; Qin, Y.-J.; Mo, C.-H.; Zhou, S.-Q. Improved Cathodic Oxygen Reduction and Bioelectricity Generation of Electrochemical Reactor Based on Reduced Graphene Oxide Decorated with Titanium-Based Composites. *Bioresour. Technol.* **2020**, *296*, 122319. [CrossRef]
48. Wang, J.; Wei, Y.; Yang, B.; Wang, B.; Chen, J.; Jing, H. In Situ Grown Heterojunction of Bi_2WO_6/BiOCl for Efficient Photoelectrocatalytic CO_2 Reduction. *J. Catal.* **2019**, *377*, 209–217. [CrossRef]
49. Guo, P.-C.; Qiu, H.-B.; Yang, C.-W.; Zhang, X.; Shao, X.-Y.; Lai, Y.-L.; Sheng, G.-P. Highly Efficient Removal and Detoxification of Phenolic Compounds Using Persulfate Activated by MnO(x)@OMC: Synergistic Mechanism and Kinetic Analysis. *J. Hazard. Mater.* **2021**, *402*, 123846. [CrossRef]
50. Ding, S.; Wan, J.; Wang, Y.; Yan, Z.; Ma, Y. Activation of Persulfate by Molecularly Imprinted Fe-MOF-74 @ SiO_2 for the Targeted Degradation of Dimethyl Phthalate: Effects of Operating Parameters and Chlorine. *Chem. Eng. J.* **2021**, *422*, 130406. [CrossRef]
51. Qi, Y.; Zou, M.; Ajarem, J.S.; Allam, A.A.; Wang, Z.; Qu, R.; Zhu, F.; Huo, Z. Catalytic Degradation of Pharmaceutical and Personal Care Products in Aqueous Solution by Persulfate Activated with Nanoscale FeCoNi-Ternary Mixed Metal Oxides. *Sep. Purif. Technol.* **2023**, *314*, 123585. [CrossRef]
52. Sabri, M.; Habibi-Yangjeh, A.; Vadivel, S. Novel ZnO/$Ag_6Si_2O_7$ Nanocomposites for Activation of Persulfate Ions in Photocatalytic Removal of Organic Contaminants under Visible Light. *Mater. Chem. Phys.* **2020**, *239*, 121988. [CrossRef]
53. Liang, C.; Wang, Z.S.; Bruell, C.J. Influence of PH on Persulfate Oxidation of TCE at Ambient Temperatures. *Chemosphere* **2007**, *66*, 106–113. [CrossRef] [PubMed]
54. Matzek, L.W.; Carter, K.E. Activated Persulfate for Organic Chemical Degradation: A Review. *Chemosphere* **2016**, *151*, 178–188. [CrossRef] [PubMed]
55. Petala, A.; Arvaniti, O.S.; Christofili, M.; Safakas, A.; Frontistis, Z.; Mantzavinos, D. Lanthanum Nickel Oxide: An Effective Heterogeneous Activator of Sodium Persulfate for Antibiotics Elimination. *Catalysts* **2020**, *10*, 1373. [CrossRef]
56. Hong, Y.; Peng, J.; Zhao, X.; Yan, Y.; Lai, B.; Yao, G. Efficient Degradation of Atrazine by CoMgAl Layered Double Oxides Catalyzed Peroxymonosulfate: Optimization, Degradation Pathways and Mechanism. *Chem. Eng. J.* **2019**, *370*, 354–363. [CrossRef]
57. Wang, J.; Wang, S. Activation of Persulfate (PS) and Peroxymonosulfate (PMS) and Application for the Degradation of Emerging Contaminants. *Chem. Eng. J.* **2018**, *334*, 1502–1517. [CrossRef]
58. Ji, Y.; Fan, Y.; Liu, K.; Kong, D.; Lu, J. Thermo Activated Persulfate Oxidation of Antibiotic Sulfamethoxazole and Structurally Related Compounds. *Water Res.* **2015**, *87*, 1–9. [CrossRef]
59. Rønn, L.; Gydesen, E.; Bennedsen, L.R.; Muff, J.; Søgaard, E.G. Aalborg Universitet Influence of Chloride and Carbonates on the Reactivity of Activated Persulfate Chemosphere Influence of Chloride and Carbonates on the Reactivity of Activated Persulfate. *Chemosphere* **2012**, *86*, 1092–1097. [CrossRef]

60. Metheniti, M.; Frontistis, Z.; Ribeiro, R.; Silva, A.; Faria, J.; Gomes, H.; Mantzavinos, D. Degradation of Propyl Paraben by Activated Persulfate Using Iron-Containing Magnetic Carbon Xerogels: Investigation of Water Matrix and Process Synergy Effects. *Environ. Sci. Pollut. Res.* **2018**, *25*, 34801–34810. [CrossRef]
61. Oyekunle, D.T.; Cai, J.; Gendy, E.A.; Chen, Z. Chemosphere Impact of Chloride Ions on Activated Persulfates Based Advanced Oxidation Process (AOPs): A Mini Review. *Chemosphere* **2021**, *280*, 130949. [CrossRef] [PubMed]
62. Arvaniti, O.S.; Bairamis, F.; Konstantinou, I.; Mantzavinos, D.; Frontistis, Z. Degradation of Antihypertensive Drug Valsartan in Water Matrices by Heat and Heat/Ultrasound Activated Persulfate: Kinetics, Synergy Effect and Transformation Products. *Chem. Eng. J. Adv.* **2020**, *4*, 100062. [CrossRef]
63. Rayaroth, M.P.; Boczkaj, G.; Aubry, O.; Aravind, U.K.; Aravindakumar, C.T. Advanced Oxidation Processes for Degradation of Water Pollutants—Ambivalent Impact of Carbonate Species: A Review. *Water* **2023**, *15*, 1615. [CrossRef]
64. Feng, Y.; Song, Q.; Lv, W.; Liu, G. Degradation of Ketoprofen by Sulfate Radical-Based Advanced Oxidation Processes: Kinetics, Mechanisms, and Effects of Natural Water Matrices. *Chemosphere* **2017**, *189*, 643–651. [CrossRef]
65. Lee, H.; Kim, H.; Weon, S.; Choi, W.; Hwang, Y.S.; Seo, J.; Lee, C.; Kim, J.-H. Activation of Persulfates by Graphitized Nanodiamonds for Removal of Organic Compounds. *Environ. Sci. Technol.* **2016**, *50*, 10134–10142. [CrossRef]
66. Guo, R.; Xi, B.; Guo, C.; Liu, W.; Lv, N.; Xu, J. Comprehensive Insight into Heterogeneous Persulfate Activation for Environmental Pollutants Degradation: Approaches and Mechanism. *Environ. Funct. Mater.* **2022**, *1*, 239–252. [CrossRef]

environments MDPI

Article

Bacteria and Yeasts Isolated from the Environment in Biodegradation of PS and PVC Microplastics: Screening and Treatment Optimization

Kristina Bule Možar, Martina Miloloža, Viktorija Martinjak, Matija Cvetnić, Vesna Ocelić Bulatović, Vilko Mandić, Arijeta Bafti, Šime Ukić *, Dajana Kučić Grgić and Tomislav Bolanča

Faculty of Chemical Engineering and Technology, University of Zagreb, Trg Marka Marulića 19, 10000 Zagreb, Croatia; kbule@fkit.unizg.hr (K.B.M.); miloloza@fkit.unizg.hr (M.M.); vprevaric@fkit.unizg.hr (V.M.); mcvetnic@fkit.unizg.hr (M.C.); vocelicbulatovic@fkit.unizg.hr (V.O.B.); vmandic@fkit.unizg.hr (V.M.); abafti@fkit.unizg.hr (A.B.); dkucic@fkit.unizg.hr (D.K.G.); tbolanca@fkit.unizg.hr (T.B.)

* Correspondence: sukic@fkit.unizg.hr

Abstract: Biodegradation is the most environmentally friendly and, at the same time, economically acceptable approach to removing various pollutants from the environment. However, its efficiency in removing microplastics (MPs) from the environment is generally low. The successful biodegradation of MPs requires microorganisms capable of producing enzymes that degrade MP polymers into compounds that the microorganisms can use as a source of carbon and energy. Therefore, scientists are screening and characterizing microorganisms that can degrade MPs more efficiently. These microorganisms are often isolated from sites contaminated with MPs because the microorganisms living there are adapted to these pollutants and should be able to better degrade MPs. In this study, five bacterial strains and five yeast strains were isolated from various environmental samples including activated sludge, compost, river sediment, and biowaste. Among them, screening was performed for bacteria and yeasts with the highest potential for the biodegradation of polystyrene (PS) and polyvinyl chloride (PVC) MPs, and the bacterium *Delftia acidovorans* and the yeast *Candida parapsilosis* were identified as the best candidates. Optimization of biodegradation of the selected MPs by each of these two microorganisms was performed, focusing on the influence of cell density, agitation speed and pH of the medium. It was found that within the selected experimental ranges, high values of cell density, low agitation speed, and a slightly basic medium favored the biodegradation of PS and PVC MPs by *Delftia acidovorans*. In the case of *Candida parapsilosis*, favorable conditions also included high cell density followed by a slightly higher, but not maximum, agitation speed and a weakly acidic medium. Broad spectroscopic and imaging methods indicated that *Delftia acidovorans* and *Candida parapsilosis* better adapt to PVC MPs to use it as a carbon and energy source.

Keywords: biodegradation; polystyrene; polyvinyl chloride; microplastics; *Delftia acidovorans*; *Candida parapsilosis*

1. Introduction

Today, it is impossible to imagine life without plastics: plastics have excellent application properties and are used in practically all areas of human life. Accordingly, the annual production of plastics is continuously increasing worldwide. In 2021, around 390.7 million tons were produced, which is 15.2 million tons more than in 2020 and 25.5 million tons more than in 2018 [1]. By type of plastic polymer, polyethylene (PE) with 36%, polypropylene (PP) with 21%, and polyvinyl chloride (PVC) with 12% have the largest share in the total production of plastics, followed by polyethylene terephthalate (PET), polyurethane (PUR), and polystyrene (PS) with shares below 10% [2].

Unfortunately, the extensive production and use of plastics generates large amounts of plastic waste, and a significant portion of this waste ends up in the environment. Thus, plastic waste is found in virtually all parts of the environment, even in deserts, on mountain tops, in the deepest parts of the oceans, and in the Arctic snow [3]. Because plastics decompose very slowly and have a high hazardous potential [4], the presence of plastics in the environment has become a global environmental problem and a serious threat to all organisms on Earth. The special focus is on the hazardous potential of plastic particles smaller than 5 mm, so-called microplastics (MPs), and their fate and behavior in the environment [5,6]. These particles may be intentionally produced in various industries to improve the quality of certain products, or they may be a product of the degradation of larger pieces of plastic due to use or exposure to various environmental conditions [7].

To date, the presence of plastics has been noted in numerous species [8–10]. One of the worrying facts is that MPs have been found in 17% of species on the International Union for Conservation of Nature's Red List of Threatened Species [11]. Humans are also exposed to MPs, primarily through eating and inhalation. A 2019 U.S. study estimated that an adult human ingests 50,000 particles of MP each year through food and drink, and another 55,000 particles through inhalation [12]. Numerous studies have shown that MPs have harmful effects on organisms [11,13–15]. MPs can cause oxidative stress and various metabolic changes, negatively affect reproduction, the immune system, and the nervous system, and eventually lead to cancer and death. In addition to the polymer base, MP particles may contain other harmful substances that increase the overall hazardous potential of the particles. We refer here primarily to various additives that are added to plastic polymers to improve their properties. Commercially available plastics usually contain numerous additives (UV stabilizers, antioxidants, dyes, plasticizers, flame retardants, etc. [14,16]) that are often very dangerous. In most cases, these additives can be easily released from MP particles because they are rarely chemically bonded to the plastic polymer [14,17]. In addition, various hazardous pollutants or even pathogens can be adsorbed to MP particles and thus ingested by organisms along with MPs [18]. Considering all this, it is necessary to find an efficient and environmentally friendly method to remove MP particles from the environment.

Among the various approaches tested so far, biodegradation has proven to be by far the most environmentally friendly and also economically acceptable approach to remove various pollutants from the environment. The organisms most commonly involved in biodegradation processes are microorganisms such as bacteria and fungi [19].

Despite the fact that the most commonly used plastics to date are highly resistant to biodegradation, bioremediation has attracted the attention of researchers as an alternative method of minimizing the presence of MPs in the environment [20]. Indeed, some microorganisms have shown that they can colonize and biodegrade MPs under certain conditions [21,22]. Figure 1 shows the main factors affecting the biodegradation process.

The successful biodegradation of MPs primarily requires microorganisms capable of producing enzymes that can break down plastic polymers into smaller compounds that can pass through the semipermeable membrane into the microorganism's cell and be used as a source of carbon and energy [19]. Therefore, to solve the complex problems of the biodegradation of MPs, it is necessary to perform screening and characterization of efficient microorganisms [20,23,24]. The microorganisms that can potentially degrade MPs are often isolated from contaminated sites because the microorganisms that inhabit such sites are thought to be naturally capable of degrading long-chain fatty acids. Therefore, it is very likely that they are also more efficient in the degradation of plastic polymers [23]. The enzymes known to be involved in the degradation of certain types of plastic polymers and the microorganisms capable of producing these enzymes are listed in Table 1.

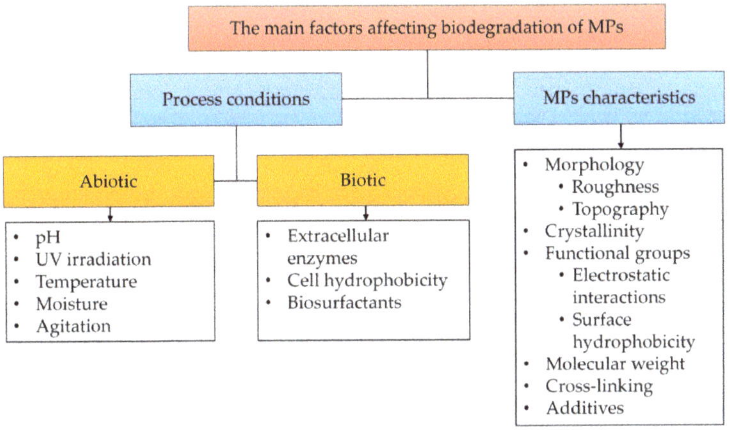

Figure 1. The main factors affecting the biodegradation of MPs.

In addition, the most favorable process conditions (pH, temperature, agitation speed, salinity, cell density, humidity, etc.) are preferred. In biodegradation experiments with aerobic microorganisms, agitation must be continuous to ensure good aeration. The decision on the agitation speed depends on the experimental setup, the organism used, the nutrient source, and other parameters [25]. The pH of the medium is also important for cell activity. For example, most bacteria tolerate a pH between 5 and 8 [26] (i.e., the neutral range), while most fungi grow well in the pH range of 3–7 [27]. Also, the chemistry and morphology of the surface of the MP particles must be such that microorganisms can adhere to them and form a biofilm [19]. One of the most important factors affecting the adhesion of microorganisms to different surfaces is the hydrophobicity of the cells, as hydrophobic cells adhere more strongly to hydrophobic surfaces, while hydrophilic cells adhere more strongly to hydrophilic surfaces [28]. In the case of the biodegradation of MPs, adhesion is complicated by the fact that plastics are a hydrophobic material, while microorganisms are mostly hydrophilic. Bacteria, for example, have hydrophilic cells, with a few exceptions. When we talk about fungi, only the aerial hyphae and conidia are hydrophobic, while yeast cells and vegetative hyphae of filamentous fungi that live in humid environments are generally hydrophilic [29].

Table 1. Enzymes involved in the biodegradation of certain types of plastics.

	Microorganisms	Enzymes	Plastics	Reference
Fungi	*Trametes versicolor*, *Pleurotus ostreatus*, *Pleurotus ostreatus*, and *Trametes pubescen*	Laccase	PE	[21]
	Trichoderma harzianum	Laccase and manganese peroxidase		[30]
	Pleurotus ostreatus	Lignin peroxidase, manganese peroxidase, and laccase		[31]
	Fusarium graminearum	Peroxidase		[32]
	Aspergillus flavus	Laccase and laccase-like multicopper oxidase		[33]
	Bjerkandera adusta	Laccase		[34]
	Phanerocheate chrysosporium	Lignin peroxidase	PVC	[35]
	Cochliobolus sp.	Laccase		[36]

Table 1. *Cont.*

	Microorganisms	Enzymes	Plastics	Reference
	Penicillium citrinum	Polyesterase	PET	[37]
	Candida antarctica	Lipase		[38]
	Lentinus tigrinus	Esterase	PS	[39]
	Candida antarctica	Lipase	PUR	[40]
	Aspergillus flavus	Esterase		[41]
	Aspergillus tubingenesis	Esterase		[42]
Bacteria	*Rhodococcus ruber*	Laccase	PE	[22]
	Pseudomonas sp.	Alkane hydroxylase		[43]
	Pseudomonas aeruginosa	Alkane hydroxylase and rubredoxin reductase		[44]
	Ideonella sakaiensis	PETase	PET	[45]
	Streptomyces scabies	Glycosyl hydrolase and esterase		[46]
	Thermobifida fusca	Cutinase-like hydrolase		[47]
	Delftia acidovorans	Esterase	PUR	[48]

Although it is estimated that more than 400 microorganisms have the potential to biodegrade plastic polymers, there is a general impression that the biodegradation of plastics has only been superficially studied [49,50]. From the literature, it appears that among bacteria, the genera *Pseudomonas* and *Bacillus*, and among fungi, the genera *Aspergillus* and *Penicillium* are most commonly used for the biodegradation of plastics, and that the biodegradation of plastics including MPs is not very efficient [51–57]. Of course, the efficiency of biodegradation largely depends on the duration of treatment, but generally rarely exceeds 10% [58,59]. Thus, *Bacillus* sp. was reported to be able to degrade 10.7% of PE films after 60 days [56], while *Bacillus cereus* and *Sporosarcina globispora* degraded 12% and 11%, respectively, of PP granules after 40 days of exposure [60]. Among the fungi, Sangeetha Devi et al. reported that the mold *Aspergillus flavus* degraded 8.5% of PE films after 30 days [61]. Surprisingly high values for the biodegradation of PE films by the genus *Aspergillus* were reported by Sáenz et al. [62]. The percentage of biodegradation reached 35.3% for *Aspergillus niger* and 22.14% for *Aspergillus terreus*, but the process took 77 days. Chaudhary et al. [63] incubated PS with *Cephalosporium* sp. and *Mucor* sp. for 8 weeks and reported a weight loss of 2.17% and 1.81%, respectively. Some other molds have been studied for the biodegradation of MPs, but the potential of yeasts is almost unexplored in this field.

In this study, the biodegradation of PS and PVC MPs was investigated using different microorganisms isolated from various environmental samples including activated sludge, compost, river sediment, and biowaste. The main objective was to screen the isolated microorganisms to find the bacteria and yeasts that have the greatest potential to utilize PS and PVC MPs as a carbon and energy source for growth and development (i.e., the greatest potential for the biodegradation of PS and PVC MPs). The optimal conditions for biodegradation within the experimental area were determined for the selection of the best bacterial culture and the best yeast culture.

2. Materials and Methods

2.1. Preparation of Microplastics

The plastic material was purchased in the form of granules as DOKI® POLISTIREN 472 (Dioki d.d., Zagreb, Croatia) and GS-28 (Drvoplast d.d., Buzet, Croatia) for PS and PVC, respectively. These materials were ground in a cryo-mill (Retsch, Haan, Germany) and dried for 48 h at room temperature (25.0 ± 0.2 °C). The ground material was sieved using stainless steel sieves (AS 200 jet, Retsch, Hann, Germany) to obtain MPs in the size range of 25–100 μm. Subsequently, the sieved particles were stored in glass bottles. Before the experiments, the MP particles were sterilized in a 100-mL flask containing 70% ethanol for 10 min on a rotary shaker (Incubator 1000 with Unimax 1010 platform shaker, Heidolph Instruments GmbH & Co., Schwabach, Germany) at 160 rpm. Particles were separated from ethanol by vacuum membrane filtration through a sterile 0.45-μm membrane filter made of cellulose nitrate (ReliaDisc^{TM} membrane filter, Ahlstrom, Helsinki, Finland) and additionally washed with sterile deionized water.

2.2. Selection of Bacteria and Yeasts Strains

Five bacterial and five yeast strains were selected as microorganisms for which their potential to biodegrade PS and PVC MPs was tested. For bacteria, *Bacillus cereus* isolated from activated sludge (municipal wastewater treatment plant in Vrgorac, Croatia), *Bacillus subtilis* isolated from compost, and *Pseudomonas alcaligenes*, *Delftia acidovorans*, and *Bacillus licheniformis* isolated from river sediment (Kupa, Croatia) were selected. The yeast strains tested were *Candida parapsilosis* isolated from river sediment (Kupa, Croatia), and *Saccharomyces cerevisiae*, *Rhodotorula glutinis*, *Geotrichum candidum*, and *Trichosporon* sp. isolated from biowaste. The isolation procedure was explained in detail in the paper previously published by Kučić Grgić et al. [64]. Bacteria were cultivated on nutrient agar (Nutrient Broth, Biolife Italiana, Milano, Italy) for 24 h at 37 °C, and yeasts were cultivated on malt agar (Malt Agar, Biolife Italiana, Milano, Italy) for 3–5 days at 28 °C. The cell density of the bacterial or yeast suspensions was measured using a spectrophotometer (DR/2400 Portable Spectrophotometer, Hach, Loveland, CO, USA) at λ = 600 nm, and the CFU (colony forming units) value was determined using the decimal plate method [65].

2.3. Biodegradation Experiments

All biodegradation experiments were performed in 200-mL Erlenmeyer flasks with a working volume of 80 mL for a period of 30 days in a thermostatic rotary shaker (Incubator 1000 with Unimax 1010 platform shaker, Heidolph Instruments GmbH & Co., Schwabach, Germany) at 25.0 ± 0.2 °C. The flasks contained a mineral medium, as described in Table 2, a bacterial or yeast suspension, and PS or PVC microparticles. The mineral medium contained all the nutrients necessary for the growth of the microorganisms, except for a carbon source. The compositions of the mineral media used in our study were chosen in accordance with the reports of Kyaw et al. [53] and Gong et al. [66], for bacteria and yeasts, respectively. Control flasks with the mineral medium and microbial suspension (without addition of MPs) were also used. MP particles in the size range of 25–100 μm and at an amount of 500 mg L^{-1} were used in all experiments. The agitation speed of a rotary shaker was set according to the experimental design. All experiments were performed in duplicate.

During these experiments, the CFU was determined to monitor the growth of the microorganisms used. This is an indirect indicator of the biodegradation of MPs, since all organisms require carbon and energy for growth and the only source of carbon and energy in the system studied was the MP particles. At the end of each experiment, the MPs were separated from the aqueous phase by vacuum membrane filtration and washed in three washing steps to remove the biomass from the surface of the MPs. Each of the washing steps lasted 30 min and was performed in a 100-mL flask on a rotary shaker (Incubator 1000 with Unimax 1010 platform shaker, Heidolph Instruments GmbH & Co., Schwabach, Germany) at 160 rpm. Two percent sodium dodecyl sulfate, 70% ethanol, and sterile deionized water were used as wash solutions during the first, second, and third steps, respectively.

Table 2. Composition of the mineral media used in the biodegradation experiments.

For Bacteria		For Yeasts	
Substance	Conc./g L^{-1}	Substance	Conc./g L^{-1}
K_2HPO_4	12.500	K_2HPO_4	1.000
KH_2PO_4	3.800	KH_2PO_4	1.000
$(NH_4)_2SO_4$	1.000	NH_4NO_3	1.000

Table 2. *Cont.*

For Bacteria		For Yeasts	
Substance	**Conc./g L^{-1}**	**Substance**	**Conc./g L^{-1}**
H_3BO_3	0.232	NaCl	0.500
$ZnSO_4 \cdot 7H_2O$	0.174	$MgSO_4 \cdot 7H_2O$	0.200
$FeSO_4(NH_4)_2SO_4 \cdot 6H_2O$	0.116	$CaCl_2$	0.020
$MgSO_4 \cdot 7H_2O$	0.100	$FeCl_3 \cdot 6H_2O$	0.005
$CoSO_4 \cdot 7H_2O$	0.096		
$(NH_4)_6Mo_7O_{24} \cdot 4H_2O$	0.022		
$CuSO_4 \cdot 5H_2O$	0.008		
$MnSO_4 \cdot 4H_2O$	0.008		

2.4. Modeling of the Optimal Conditions

The optimal conditions for the biodegradation of PS and PVC MPs by the best bacterium and yeast were determined from the results of the main experiments using response surface modeling (RSM), with the final log CFU value (obtained after 30 days of biodegradation) serving as the response parameter. The goodness-of-fit of four regression models described by Equations (1)–(4) was tested. For later discussion, we denote these models by the Roman numerals I through IV.

$$\log \text{CFU} = a_0 + a_1x_1 + a_2x_2 + a_3x_3 \tag{1}$$

$$\log \text{CFU} = a_0 + a_1x_1 + a_2x_2 + a_3x_3 + a_4x_1x_2 + a_5x_1x_3 + a_6x_2x_3 \tag{2}$$

$$\log \text{CFU} = a_0 + a_1x_1 + a_2x_2 + a_3x_3 + a_7x_1^2 + a_8x_2^2 + a_9x_3^2 \tag{3}$$

$$\log \text{CFU} = a_0 + a_1x_1 + a_2x_2 + a_3x_3 + a_4x_1x_2 + a_5x_1x_3 + a_6x_2x_3 + a_7x_1^2 + a_8x_2^2 + a_9x_3^2 \tag{4}$$

The coefficients of the models (Equations (1)–(4)) are denoted by a, while x_{1-3} represent the values of the three tested process parameters: pH, agitation speed, and cell density. To determine the optimal biodegradation conditions for a particular microorganism/MP combination, the model that best described the experimentally obtained data for that combination was used.

All calculations including statistical analyses of the results were performed using Design-Expert 10.0 software (StatEase, Minneapolis, Minnesota, MN, USA).

2.5. Surface Analysis of MP Particles

For additional confirmation of biodegradation, the untreated and treated MP particles were analyzed by attenuated total reflectance-Fourier transform infrared spectroscopy (ATR-FTIR; spectrometer Spectrum One, PerkinElmer, Waltham, MA, USA) and scanning electron microscopy (SEM; electron microscope Tescan Vega Easyprobe 3, Brno, Czech Republic). SEM imaging was performed in secondary electron mode (SE) and backscattered electron mode (BSE) at an accelerating voltage of 10 kV and a working distance less than 10 mm. The powdered samples were attached to the stubs with double-sided adhesive carbon tape. Prior to imaging, gold was sputtered onto the samples using a Quorum Technologies SC7620 sputter coater (Lewes, UK) at 17 mA for 30 s.

3. Results and Discussion

3.1. Screening of the Bacterial and Yeast Strains for Their Potential for PS and PVC Biodegradation

Five bacterial strains (*Bacillus cereus*, *Bacillus subtilis*, *Pseudomonas alcaligenes*, *Delftia acidovorans*, and *Bacillus licheniformis*) and five yeast strains (*Candida parapsilosis*, *Saccharomyces cerevisiae*, *Rhodotorula glutinis*, *Geotrichum candidum*, and *Trichosporon* sp.) were isolated from environmental samples potentially containing PS and PVC MPs and cultivated on agar (Figure 2).

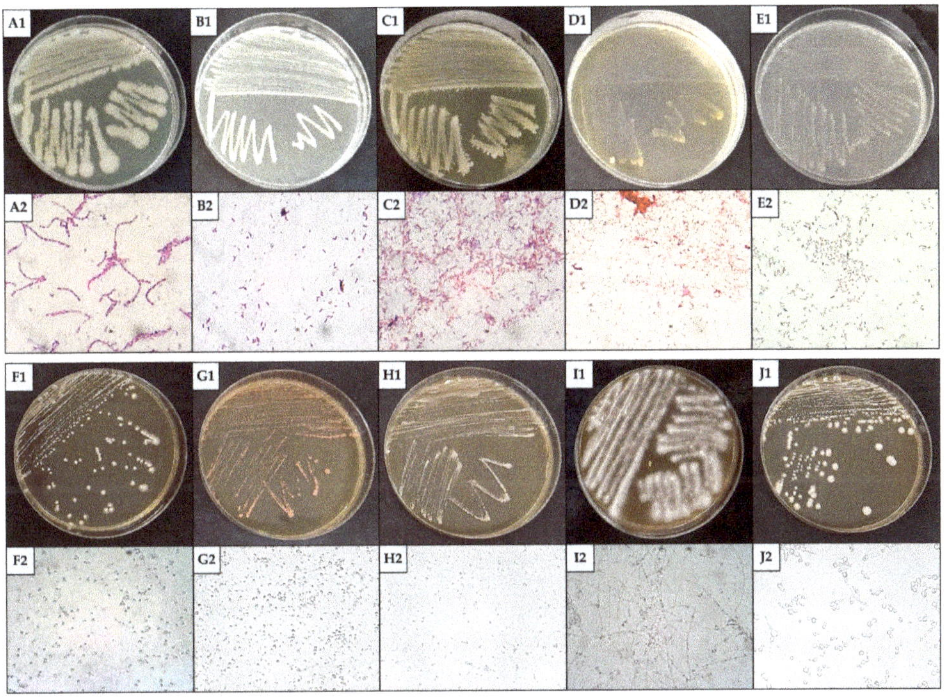

Figure 2. Bacteria *Bacillus cereus* (**A1**), *Bacillus subtilis* (**B1**), *Bacillus licheniformis* (**C1**), *Pseudomonas alcaligenes* (**D1**), and *Delftia acidovorans* (**E1**) cultivated on nutrient agar by the streak plate method, with associated microphotographs of bacterial Gram-stained smears ((**A2–E2**); 1000× magnification). Yeasts *Saccharomyces cerevisiae* (**F1**), *Rhodotorula glutinis* (**G1**), *Candida parapsilosis* (**H1**), *Geotrichum candidum* (**I1**), and *Trichosporon* sp. (**J1**) cultivated on malt agar by the streak plate method, with associated microphotographs of cultivated yeasts ((**F2–J2**); 400× magnification).

After cultivation, the isolated microorganisms were screened to find the bacterium and yeast with the highest potential for the biodegradation of the two MPs mentioned. The selection was based on the criterion of the highest CFU value achieved. The process conditions for the preliminary experiments including pH, temperature, cell density (*OD*), and agitation speed (*AS*) were selected based on information available in the literature [66–68]; the conditions are listed in Table 3. Since the optical density of the suspensions is an indicator of the cell density in the systems, we refer to the cell density as *OD*.

Table 3. Process conditions used for the preliminary biodegradation experiments of PS and PVC MPs.

Condition	For Bacteria	For Yeasts
pH	7.0 ± 0.2	6.0 ± 0.2
$T/°C$	25.0 ± 0.2	25.0 ± 0.2
OD_{initial}	0.5	1.0
AS/rpm	150	160

Figure 3 shows the change in log CFU values during the biodegradation experiments with isolated bacterial cultures. The relatively high log CFU values observed for the control samples are not unusual, as microorganisms under conditions of nutrient deprivation may exhibit some unusual metabolic pathways that are otherwise suppressed in the cells and contribute to maintaining their minimal growth rate [69]. Metabolic pathways that are activated by nutrient deficiency include those that produce secondary metabolites that are not required for normal growth but can be of great importance when the growth conditions become very restrictive. Secondary metabolites can initiate a phase in which the microorganisms try to adapt to the non-ideal environmental conditions in order to survive. In a batch experiment such as the experiments we performed in this study, this phase is generally characterized by zero net growth. Some cells may die, while the others use the nutrients released by cell lysis to continue growing at a limited rate.

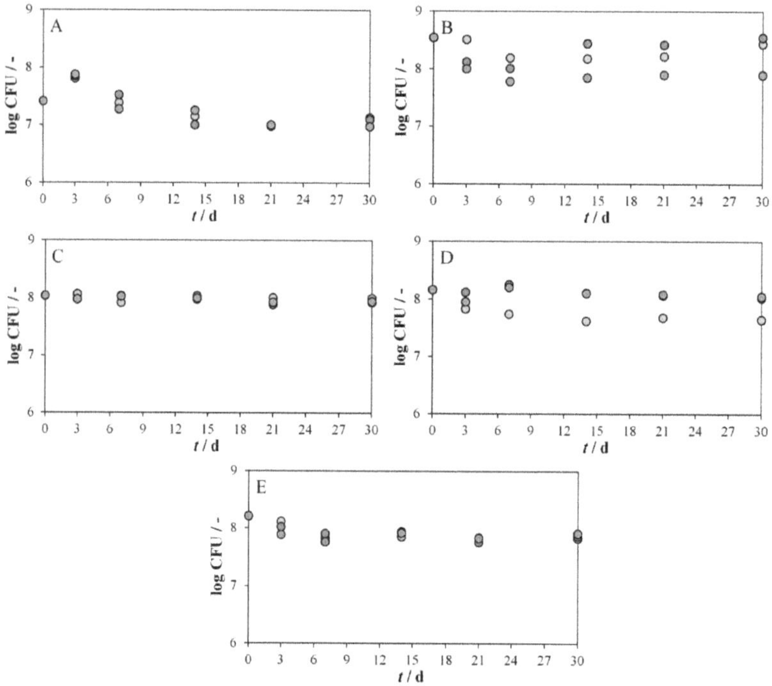

Figure 3. Changes in log CFU values determined during the preliminary biodegradation experiments using bacterial cultures: (**A**) *Bacillus cereus*, (**B**) *Pseudomonas alcaligenes*, (**C**) *Bacillus subtilis*, (**D**) *Delftia acidovorans*, and (**E**) *Bacillus licheniformis*. The blue, orange, and green circles represent PS, PVC, and the control samples, respectively.

For cultures that showed a positive deviation in log CFU values compared to the control sample, we assumed that they were able to use PS or PVC MP as a carbon and energy source for growth and development. All three *Bacillus* strains tested—*Bacillus cereus* (Figure 3A), *Bacillus subtilis* (Figure 3C), and *Bacillus licheniformis* (Figure 3E)—showed similar log CFU values to the control sample throughout the experiment. Compared to the control sample, *Pseudomonas alcaligenes* (Figure 3B) consistently exhibited lower log CFU values in the PVC experiment, while higher log CFU values were observed in the PS experiment starting on day 14. Among the five bacteria tested, *Delftia acidovorans* (Figure 3D) exhibited the highest log CFU value compared to the control sample and was therefore selected for the main experiment as the bacterium with the highest potential for the biodegradation of PS and PVC MPs. This is consistent with some previous reports that bacteria of the genus *Delftia* have high adaptability to different environmental conditions and can be used for the biodegradation of numerous organic compounds including PE [70], PS [71], PET [72], and PUR [73].

Zhang et al. [74] reported several potential pathways for the microbial biodegradation of PS, as the exact degradation pathways and enzymes involved in the microbial biodegradation of PS are still unknown. Among the proposed pathways, in the context of the selection of *Delftia acidovorans*, the pathway focusing on the aromatic ring in the structure of PS is very interesting because it involves the action of aromatic ring hydrolases, and it is known that *Delftia acidovorans* is able to generate such enzymes [75]. Considering the potential of *Delftia acidovorans* for PVC biodegradation, it is interesting to note that *Delftia acidovorans* is capable of producing dehalogenases [76], indicating the chloride atom in the PVC structure as a potential degradation point.

The change in log CFU values during biodegradation with yeasts is shown in Figure 4. For both MPs tested, *Geotrichum candidum* (Figure 4E) had lower log CFU values than the control sample throughout the experiment, while *Trichosporon* sp. (Figure 4A) and *Saccharomyces cerevisiae* (Figure 4D) had very similar values to the control. In the case of *Rhodotorula glutinis* (Figure 4C), the log CFU values for PS were also very similar to the control and lower than the control for PVC. Finally, *Candida parapsilosis* (Figure 4B) showed an adaptation phase to MPs, and after day 14, a positive deviation in log CFU was observed for both the PS and PVC MPs compared to the control. Accordingly, *Candida parapsilosis* was selected for the main experiment as the yeast with the highest potential for the biodegradation of the MPs studied. In general, *Candida* sp. is known to be able to degrade various aliphatic and aromatic hydrocarbons [77] including plastics like PET [78] and PUR [79]. In addition, *Candida parapsilosis* is known to be capable of producing esterases [80], which are considered important enzymes for the biodegradation of PS (Table 1).

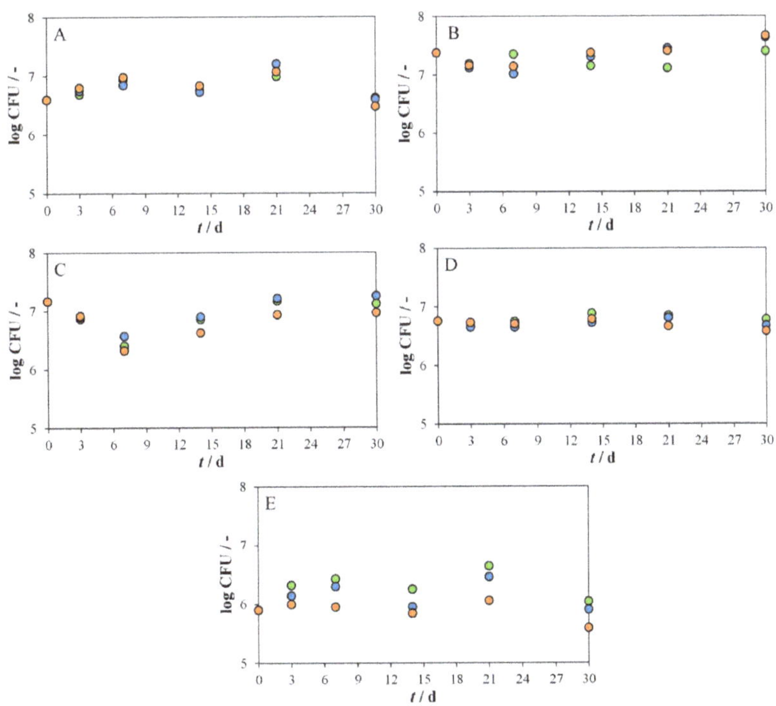

Figure 4. Changes in log CFU values determined during the preliminary biodegradation experiments using yeast cultures: (**A**) *Trichosporon* sp., (**B**) *Candida parapsilosis*, (**C**) *Rhodotorula glutinis*, (**D**) *Saccharomyces cerevisiae*, and (**E**) *Geotrichum candidum*. The blue, orange, and green circles represent PS, PVC, and the control samples, respectively.

3.2. Determination of Optimal Conditions for the Biodegradation of PS and PVC MPs by *Delftia acidovorans* and *Candida parapsilosis*

It is well-known that process conditions have a strong influence on biodegradability [81,82]. In cases where biotreatment is performed on-site, there are numerous factors that may limit microbial activity and thus reduce the efficiency of biodegradation [83]. Therefore, for *Delftia acidovorans* and *Candida parapsilosis*, selected in the previous step as the bacterium and yeast with the highest potential to biodegrade PS and PVC MPs, the optimal biodegradation conditions were determined. These conditions were determined by considering the application of selected microorganisms in bioreactors. The experiments were conducted using a full factorial methodology combining three factors (pH of the mineral medium, agitation speed, and cell density of the bacterial or yeast suspension) at three levels (Supplementary Table S1). The effect of pH on biodegradation by bacteria was tested in the range of 6–8, since bacteria generally tolerate a neutral pH range [26]. Since yeasts generally grow in the acidic to neutral pH range [84], pH conditions in the range of 3–7 were tested for biodegradation by yeasts.

The results of the suitability of four regression models (Equations (1)–(4)) used to describe the behavior of the experimentally obtained data are presented in Table S2 (Supplementary Materials). These models varied in complexity: linear dependence on the parameters studied (model I; Equation (1)), linear dependence with the interaction effect included (model II; Equation (2)), quadratic dependence (model III; Equation (3)), and quadratic dependence with the interaction effect included (model IV; Equation (4)). All four models were found to be significant with 95% confidence (p-value was less than 0.05). However, model IV had the highest values of the adjusted coefficient of determination, R^2_{adj}, for all four microorganism/MP combinations and was thus the best model for determining the optimal conditions for the biodegradation processes tested.

The response surfaces at different agitation speeds are shown in Figures 5 and 6. The red areas indicate the most favorable conditions for biodegradation, while the blue areas represent the least favorable.

In the biodegradation of both PS and PVC MPs by *Delftia acidovorans* (Figure 5), such red areas were observed at the lowest agitation speed (120 rpm) and the highest applied cell densities (Figure 5(A1,B1)). Agitation plays an important role in biodegradation by improving oxygen transfer to the process mixture and providing homogeneous chemical and physical conditions [85]. Since *Delftia acidovorans* is an aerobic microorganism, agitation should be beneficial to its activity. However, when we analyzed our response surfaces at different agitation speeds (Figure 5(A1–A3)), we found that the maximum log CFU values decreased with increasing agitation speed. The cause of this behavior was not investigated in this work, but we can assume that too intense agitation could damage the bacterial cells or perhaps inhibit the adsorption of bacteria on the surface of MPs and the formation of a biofilm. For example, it is known that intensive agitation can change the morphology of the bacterium *Escherichia coli* [86]. In the biodegradation of PVC by *Delftia acidovorans*, the unfavorable effect of the basic pH of the medium was observed at all agitation speeds (Figure 5(B1–B3)), while in the biodegradation of PS, this effect was observed at the highest agitation speed (Figure 5(A3)) and became imperceptible as the agitation speed decreased (Figure 5(A1,A2)). In agreement with what has been written, the optimal process values for the biodegradation of MPs by *Delftia acidovorans* (Table 4) within the selected experimental range included an agitation speed of 120 rpm with high cell densities (0.9 and 1.0 for PS and PVC, respectively) and a slightly basic medium (7.95 and 7.45 for PS and PVC, respectively). The higher log CFU achieved in PVC biodegradation suggests that *Delftia acidovorans* adapts better to PVC MPs than to PS MPs, using these particles more successfully as an energy and carbon source for its development.

Table 4. Optimal conditions for the biodegradation of PS and PVC MPs by *Delftia acidovorans* and *Candida parapsilosis* and the corresponding log CFU values.

Microorganism	Type of MPs	$OD_{initial}$	*AS*/rpm	pH	log CFU *
Delftia acidovorans	PS	0.9	120	7.95	8.10
	PVC	1.0	120	7.45	8.40
Candida parapsilosis	PS	1.0	156	5.67	7.24
	PVC	1.0	136	4.94	7.75

* According to the optimal model.

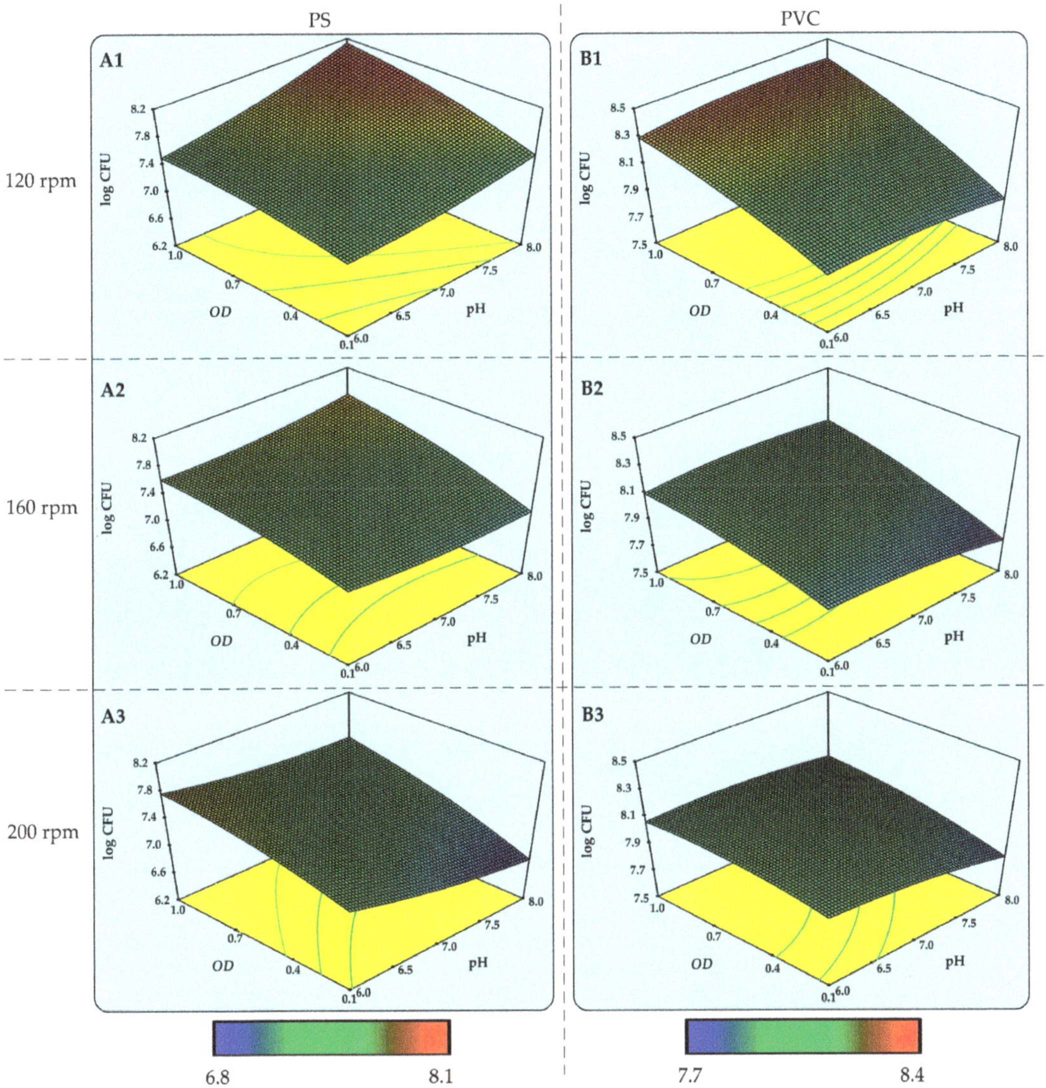

Figure 5. Response surfaces for the biodegradation of the PS (**A1–A3**) and PVC (**B1–B3**) microparticles by *Delftia acidovorans*.

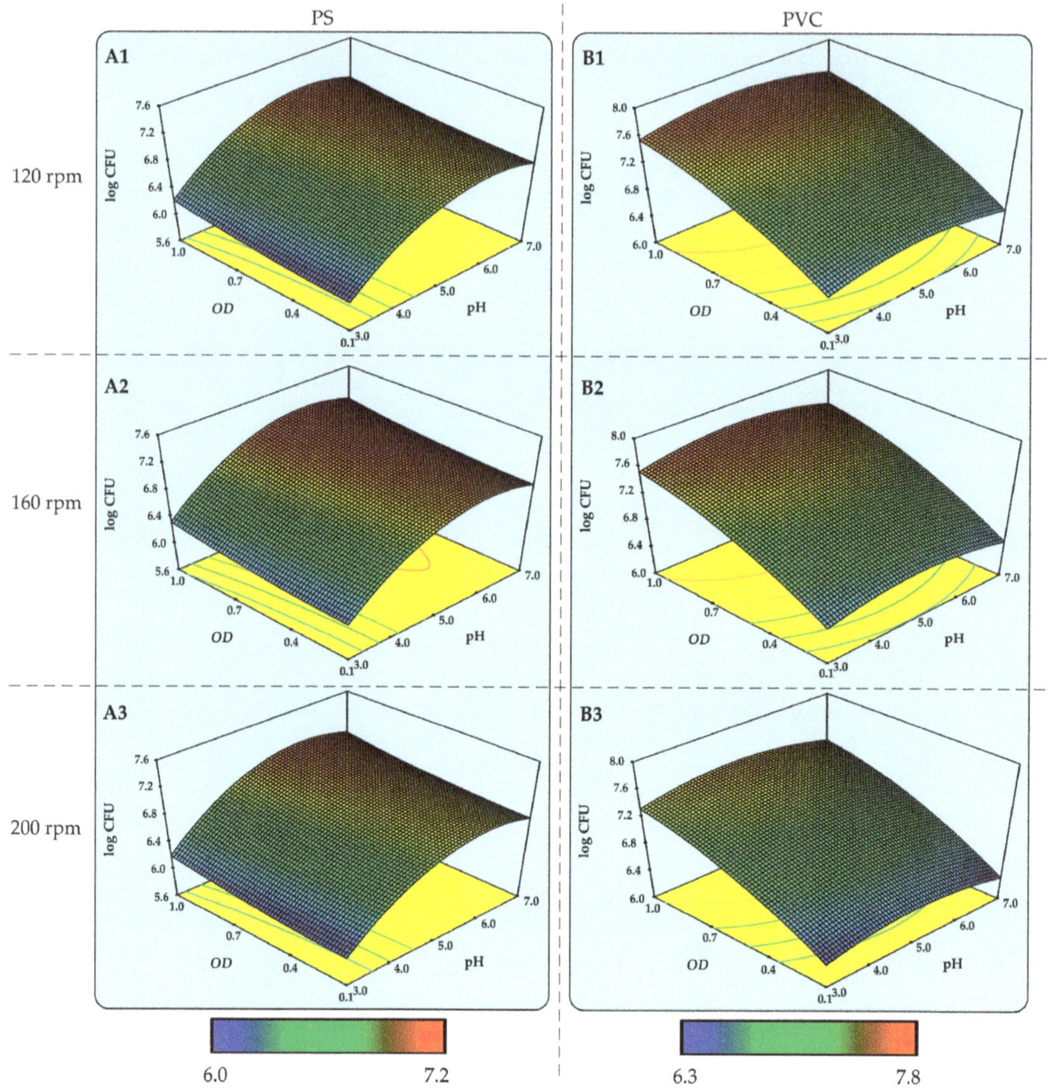

Figure 6. Response surfaces for the biodegradation of the PS (**A1–A3**) and PVC (**B1–B3**) microparticles by *Candida parapsilosis*.

Figure 6 shows the response surfaces for the biodegradation of PS and PVC MPs by the yeast *Candida parapsilosis* at different agitation speeds. As in the case of *Delftia acidovorans*, more efficient biodegradation was associated with higher values of cell density. *Candida parapsilosis* achieved the highest log CFU at pH values of 5.67 and 4.94 for PS and PVC, respectively (Table 4). These results are consistent with reports in the literature that the optimal pH for yeasts should be between 5.5 and 6.0 [84]. The unfavorable effect of more acidic and neutral media on the biodegradation of PS or PVC MPs by *Candida parapsilosis* can be clearly seen in Figure 6. *Candida parapsilosis*, like *Delftia acidovorans*, is an aerobic microorganism, which means that more intensive agitation should have a positive effect on the activity of this yeast. However, it was again found that the maximum agitation speed was not the optimal speed for biodegradation: the results showed an optimal agitation speed of 156 and 136 rpm for PS and PVC, respectively. Agitation speeds above these values were found to be less beneficial, probably for reasons similar to those discussed in the case of *Delftia acidovorans*. We found reports on the negative effects of agitation on the activity of another representative of fungi: molds. Thus, high agitation speed affected the intercellular enzyme activity of *Aspergillus luchuensis* [87] and *Aspergillus niger* [88] and reduced their enzyme production. A comparison of log CFU values at the optimal process conditions (Table 4) showed that *Candida parapsilosis* could also adapt better to PVC MPs than to PS MPs. One of the reasons for this behavior is probably the molecular structure of these two polymers, since PS, unlike PVC, contains an aromatic ring, and it is known that the presence of aromatic rings in the molecular structure increases the toxic potential of a substance [89,90]. Of course, this cannot be reliably asserted without a detailed analysis of the toxicity mechanisms of these two MPs. For example, we recently tested the toxicity of PS and PVC microparticles with different microorganisms and found different orders of toxicity: with applied 200–600 μm particles at an amount of 50–1000 mg L^{-1}, PVC was more toxic to the microalga *Scenedesmus* sp., less toxic to the bacterium *Pseudomonas putida*, and equally toxic to the yeast *Saccharomyces cerevisiae* [91].

3.3. Surface Analysis of the Treated MPs

It is generally known that the biodegradation of synthetic plastics is a very slow process. Therefore, unless the experiments are conducted over an extremely long period of time (much longer than 30 days, as in this experiment), the changes associated with biodegradation mainly occur on the surface of the MP particles, which is insignificant compared to the total volume of the MP sample. Moreover, these effects are generally limited to topographic differences and minor chemical changes. Therefore, to confirm these changes, the untreated and treated MP particles were analyzed by ATR-FTIR and SEM.

FTIR is the perfect method for studying chemical changes at the level of the different bonding of the elemental C, H, and O constituents. The ATR-FTIR spectra recorded before and after treatment are compared in Figure 7. The spectra of the PS and PVC microparticles before the treatment are marked with green and blue lines, respectively. The characteristic peaks of the PS and PVC polymers are marked in the spectra [92,93]. The orange and pink lines represent the spectra of PS and PVC, respectively, recorded after the treatment.

No significant change in the intensities of the characteristic PS peaks was observed (Figure 7A,C). However, two new peaks appeared after both treatments: by bacteria and by yeast. The first was a broad peak in the 2500–3300 cm^{-1} range, corresponding to the stretching of the OH group. The appearance of this peak could indicate that biodegradation has occurred. In addition, a new peak appeared at 1645 cm^{-1}. The peaks in this range mostly relate to the C=C stretching of the alkene [94–97], which could indicate the breaking of the aromatic benzene ring in the PS structure.

In the case of PVC treatment, comparison of the FTIR spectra recorded before and after treatment showed a decrease in the intensity of the characteristic PVC peaks at 616 cm^{-1} (C–Cl stretching), 966 cm^{-1} (CH_2 rocking), 1255 cm^{-1} (CH bending), and 1427 cm^{-1} (CH_2 bending), most likely due to the dehydrochlorination process (Figure 7B,D) [98]. This was much more pronounced in the bacterial treatment of PVC (Figure 7B). The dehydrochlorination of PVC generally leads to the formation of unsaturated C=C bonds [99], which corresponded to the appearance of a new FTIR peak at 1640 cm^{-1}. This peak could be observed after both treatments. Its intensity was much more pronounced in the case of the bacterial treatment (Figure 7B), confirming our assumption that it is closely related to the dehydrochlorination process. It was expected that the biodegradation of PVC should also lead to the formation of the carbonyl group, a group in which the carbon atom is double-bonded to an oxygen atom. The carbonyl group gives a peak in the spectral region 1720–1740 cm^{-1} [100,101]. However, the FTIR spectrum of the untreated PVC already contained such a peak, which obviously represents an additive present in the tested PVC sample, since pure PVC has no carbonyl groups in its structure. The intensity of this peak decreased in both treatments, indicating that *Delftia acidovorans* and *Candida parapsilosis* may have attacked the additive, or that the additive was released from the surface of the MP particles due to biodegradation of the PVC polymer, as later confirmed by LC-MS analysis of the filtrate solution (Section 3.4). This also made it impossible to detect the formation of new carbonyl groups and use this as evidence of biodegradation of the PVC polymer.

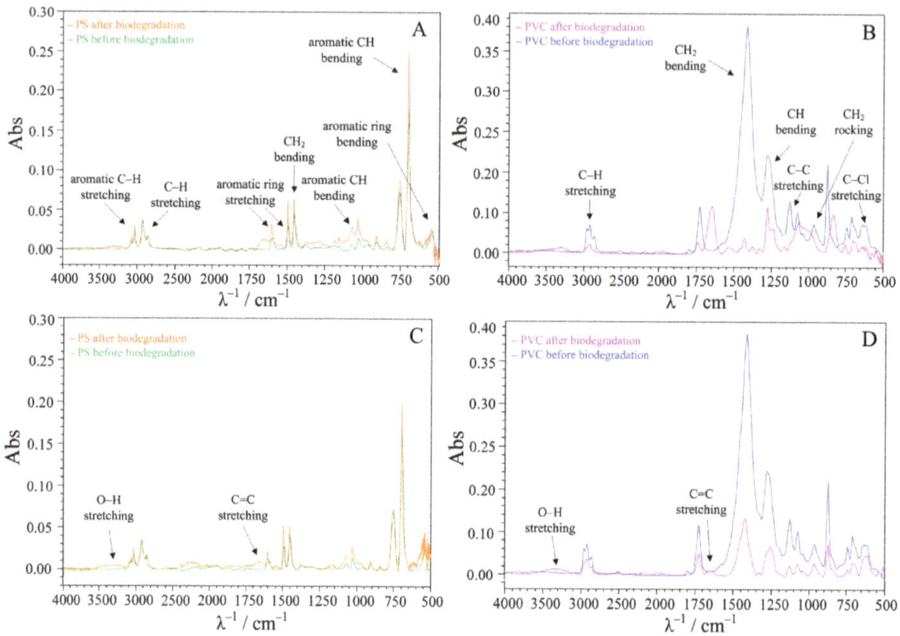

Figure 7. Comparison of the FTIR spectra of the studied MPs before and after biodegradation: (**A**) *Delftia acidovorans* and PS, (**B**) *Delftia acidovorans* and PVC, (**C**) *Candida parapsilosis* and PS, and (**D**) *Candida parapsilosis* and PVC. The biodegradation of the PS MPs was negligible in both treatments. However, for the PVC MPs, there was a significant decrease in the intensity of the characteristic peaks, indicating that both *Delftia acidovorans* and *Candida parapsilosis* initiated biodegradation.

To visualize the degradation effect, the samples were imaged with the SEM in secondary electron mode (SE) and in backscattered electron mode (BSE). SE SEM microscopy provides information about the textural changes on the surface of MPs, which is useful because these changes are the most relevant. BSE SEM is sensitive to chemical composition in terms of atomic mass. Therefore, in the unlikely event that the chemical environment has changed significantly, BSE SEM could visually qualify the areas where the respective changes took place. Untreated PS microparticles (Figure 8a,b) and untreated PVC microparticles (Figure 9a,b) were used as the reference. For all samples scanned, the bulk appearance resembled micrometric polydisperse particles with a wide size distribution in aggregates of several hundreds of microns.

In the case of the untreated PS MPs, the surface topography revealed by SE showed low roughness and a general absence of morphological defects (i.e., the morphological defects were confined to those mechanically induced by grinding plastics into powders) (Figure 8a). BSE imaging also showed that the phase composition was homogeneous in depth beyond the sample surface (Figure 8b).

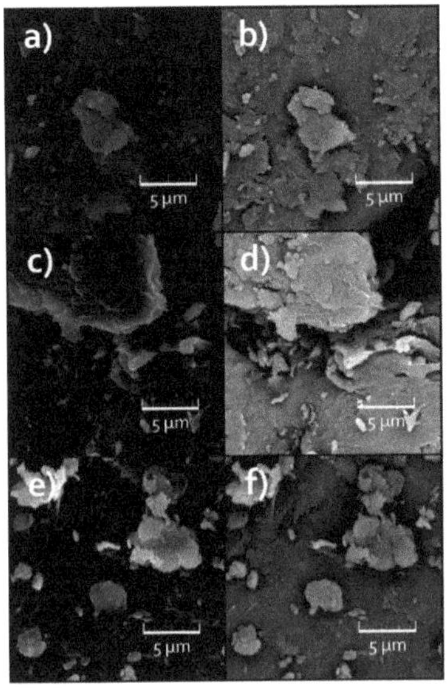

Figure 8. SEM analysis of the PS MPs: (**a**) SE micrograph of the untreated PS MP sample, (**b**) BSE micrograph of the untreated PS MP sample, (**c**) SE micrograph of the PS MP sample treated with the bacterium *Delftia acidovorans*, (**d**) BSE micrograph of the PS MP sample treated with the bacterium *Delftia acidovorans*, (**e**) SE micrograph of the PS MP sample treated with the yeast *Candida parapsilosis*, and (**f**) BSE micrograph of the PS MP sample treated with the yeast *Candida parapsilosis*.

PS microparticles treated with bacteria or yeast showed only a weak difference in surface topography compared to the reference PS sample (i.e., the presence of morphological defects and roughness could be sporadically indicated) (Figure 8c,e). However, it was not possible to quantify any degradation difference between the bacteria and yeast treated PS samples. From the BSE imaging (Figure 8d,f), it was not possible to observe any in depth compositional phase changes for the treated PS samples with respect to the reference. This is consistent with the results of the FTIR analysis discussed previously, which also did not indicate the biodegradation of PS by bacteria or yeast.

Just like in the case of the untreated PS MPs, the surface topography of the untreated PVC microparticles, scanned in SE mode (Figure 9a), showed low roughness and a general absence of morphological defects, while BSE imaging (Figure 9b) indicated a homogeneous phase composition.

However, the consequences of the degradation were more obvious for the case of the treated PVC samples. Specifically, both degraded samples showed the presence of areas characterized with higher surface roughness or higher surface porosity, or an increased presence of morphological defects, or in general, just a more damaged surface. Comparing the PVC MP samples treated with *Delftia acidovorans* with those treated with *Candida parapsilosis*, the apparent presence of degradation phenomena increased from sporadic in the yeast-treated sample to considerable in the bacteria-treated sample, confirming the conclusion from the analysis of the FTIR spectra that PVC MPs are more susceptible to biodegradation by *Delftia acidovorans* (Figure 9c,e). Some faint differences in the BSE images of the treated PVC samples could be observed, pointing out possible in depth compositional phase changes. However, the comparison of BSE between the bacteria and yeast treated PVC samples was not relevant (Figure 9d,f). Figure 9g depicts one of the apparent bacteria degradation areas.

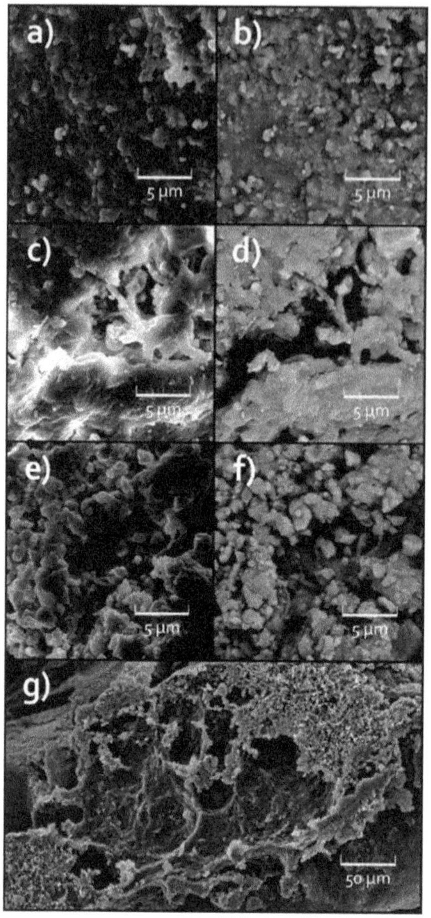

Figure 9. SEM analysis of the PVC MPs: (**a**) SE micrograph of the untreated PVC MP sample, (**b**) BSE micrograph of the untreated PVC MP sample, (**c**) SE micrograph of PVC MP sample treated with the bacterium *Delftia acidovorans*, (**d**) BSE micrograph of PVC MP sample treated with the bacterium *Delftia acidovorans*, (**e**) SE micrograph of the PVC MP sample treated with the yeast *Candida parapsilosis*, (**f**) BSE micrograph of the PVC MP sample treated with the yeast *Candida parapsilosis*, (**g**) SE micrograph of the throughout "surface damage" of the bacteria-treated PVC MP sample.

3.4. Analysis of the Aqueous Phases in Contact with MPs

The MS analyses of the aqueous phases in contact with the MP samples were performed using an LC-MS 2020 (Shimadzu, Japan) to estimate the released additives, as the biodegradation of the surface of the MP particles can lead to additive release [102]. Comparison of the MS spectra obtained before and after the biodegradation experiments of the PS samples showed the appearance of a new peak at m/z ratios of 328 for the experiment with *Delftia acidovorans* (Figure S1, case A2) (i.e., 327 for the experiment with *Candida parapsilosis* (Figure S1, case B2)). This peak could represent tryphenyl phosphate [103], which is commonly used as a flame retardant [104]. In the experiments on the biodegradation of PVC samples, the analysis of the MS spectra showed the appearance of new peaks at m/z ratios of 391 and 447. The peak at 391 could represent bis(2-ethylhexyl)phthalate [105], while the peak at 447 could be associated with diisodecyl phthalate [106]. These two substances are among the most commonly used PVC plasticizers in Europe [107]. In addition, both additives contain a carbonyl group in their structure, so their release from the surface of the PVC microparticles is consistent with the observed decrease in the intensity of the FTIR peak in the range of 1720–1740 cm^{-1} (Figure 7B,D).

4. Conclusions

Although the degradation of microplastics by microorganisms is considered as one of the most acceptable approaches to solving the problem of microplastics in the environment [23], there are still numerous obstacles that need to be overcome before such treatments can be successfully implemented. Foremost among these are the considerably long duration of the treatments and the lack of information on specific enzymes that can successfully degrade some of the plastic polymers including PS and PVC [108].

Five bacterial strains including *Bacillus cereus*, *Bacillus licheniformis*, *Bacillus subtilis*, *Delftia acidovorans*, and *Pseudomonas alcaligenes* and five yeast strains, namely *Candida parapsilosis*, *Geotrichum candidum*, *Rhodotorula glutinis*, *Saccharomyces cerevisiae*, and *Trichosporon* sp. were isolated from environmental samples. Among them, *Delftia acidovorans* was identified as the bacterium with the highest potential to biodegrade PS and PVC MPs, while *Candida parapsilosis* showed the highest potential among the yeasts. The two microorganisms adapted better to the PVC MPs than to the PS MPs to use it as a carbon and energy source. A positive effect of increased cell density and a negative effect of high agitation speed on the biodegradation process were observed.

Based on the results obtained, it would be desirable to study the metabolic processes of *Delftia acidovorans* and *Candida parapsilosis* in the hope of identifying a new enzyme that efficiently degrades PS or PVC MPs.

Supplementary Materials: The following supporting information can be downloaded at: https://www.mdpi.com/article/10.3390/environments10120207/s1, Figure S1: The MS spectra of solutions in contact with MPs for biodegradation experiments with *Delftia acidovorans* (A) and *Candida parapsilosis* (B). The following cases are presented: the solutions at the beginning of the biodegradation experiments (A1 and B1), the solutions after 30 days of exposure to PS (A2 and B2), and the solutions after 30 days of exposure to PVC (A3 and B3); Table S1: Experimental design to determine the optimal conditions for the biodegradation of PS and PVC MPs by bacterium and yeast selected as the best in preliminary experiments. The design included the pH value of the medium, agitation speed (*AS*), and cell density (*OD*); Table S2: The suitability of four regression models (Equations (1)–(4)) to fit the experimental data of log CFU for the biodegradation of PS and PVC MPs by *Delftia acidovorans* and *Candida parapsilosis*. The influence of three factors: medium pH, agitation speed (*AS*) and cell density (*OD*) was considered.

Author Contributions: Conceptualization, V.O.B., Š.U., D.K.G. and T.B.; Methodology, M.M., V.M. (Viktorija Martinjak) and K.B.M.; Formal analysis, K.B.M., V.M. (Vilko Mandić), A.B. and M.M.; Investigation, K.B.M., M.M., Š.U., M.C., V.O.B. and D.K.G.; Writing—original draft preparation, K.B.M. and M.M.; Writing—review and editing, Š.U., V.O.B. and D.K.G.; Visualization, K.B.M. and Š.U. All authors have read and agreed to the published version of the manuscript.

Funding: The authors would like to acknowledge the financial support of the Croatian Science Foundation through the project Advanced Water Treatment Technologies for Microplastics Removal (AdWaTMiR; IP-2019-04-9661).

Data Availability Statement: Data are contained within the article.

Conflicts of Interest: The authors declare no conflict of interest.

References

1. Plastics—The Facts. 2022. Available online: https://plasticseurope.org/wp-content/uploads/2022/10/PE-PLASTICS-THE-FACTS_V7-Tue_19-10-1.pdf (accessed on 20 March 2023).
2. Geyer, R.; Jambeck, J.R.; Law, K.L. Production, use, and fate of all plastics ever made. *Sci. Adv.* **2017**, *3*, e1700782. [CrossRef]
3. MacLeod, M.; Arp, H.P.H.; Tekman, M.B.; Jahnke, A. The global threat from plastic pollution. *Science* **2021**, *373*, 61–65. [CrossRef]
4. Ostle, C.; Thompson, R.C.; Broughton, D.; Gregory, L.; Wootton, M.; Johns, D.G. The rise in ocean plastics evidenced from a 60-year time series. *Nat. Commun.* **2019**, *10*, 1622–1628. [CrossRef]

5. Chen, J.; Wu, J.; Sherrell, P.C.; Chen, J.; Wang, H.; Zhang, W.-X.; Yang, J. How to Build a Microplastics-Free Environment: Strategies for Microplastics Degradation and Plastics Recycling. *Adv. Sci.* **2022**, *9*, 2103764. [CrossRef]
6. Erceg, M.; Tutman, P.; Vojanić Varezić, D.; Bobanović, A. Karakterizacija mikroplastike u sedimentu plaže Prapratno. *Kem. Ind.* **2020**, *69*, 253–260. [CrossRef]
7. Westphalen, H.; Abdelrasoul, A. Challenges and Treatment of Microplastics in Water. In *Water Challenges of an Urbanizing World*; Glavan, M., Ed.; InTech: Rijeka, Croatia, 2018; pp. 71–84. [CrossRef]
8. Ozturk, R.C.; Altinok, I. Interaction of Plastics with Marine Species. *Turk. J. Fish. Aquat. Sci.* **2020**, *20*, 647–658. [CrossRef]
9. Cho, Y.; Shim, W.J.; Jang, M.; Han, G.M.; Hong, S.H. Abundance and characteristics of microplastics in market bivalves from South Korea. *Environ. Pollut.* **2019**, *245*, 1107–1116. [CrossRef]
10. Wang, L.; Nabi, G.; Yin, L.; Wang, Y.; Li, S.; Hao, Z.; Li, D. Birds and plastic pollution: Recent advances. *Avian Res.* **2021**, *12*, 59. [CrossRef]
11. Hardesty, B.D.; Good, T.P.; Wilcox, C. Novel methods, new results and science-based solutions to tackle marine debris impacts on wildlife. *Ocean Coast. Manag.* **2015**, *115*, 4–9. [CrossRef]
12. Cox, K.D.; Covernton, G.A.; Davies, H.L.; Dower, J.F.; Juanes, F.; Dudas, S.E. Human Consumption of Microplastics. *Environ. Sci. Technol.* **2019**, *53*, 7068–7074. [CrossRef] [PubMed]
13. Yeung, C.W.S.; Teo, J.Y.Q.; Loh, X.J.; Lim, J.Y.C. Polyolefins and Polystyrene as Chemical Resources for a Sustainable Future: Challenges, Advances, and Prospects. *ACS Mater. Lett.* **2021**, *3*, 1660–1676. [CrossRef]
14. Zolotova, N.; Kosyreva, A.; Dzhalilova, D.; Fokichev, N.; Makarova, O. Harmful effects of the microplastic pollution on animal health: A literature review. *PeerJ* **2022**, *10*, e13503. [CrossRef]
15. Bhuyan, M.S. Effects of Microplastics on Fish and in Human Health. *Front. Environ. Sci.* **2022**, *10*, 827289. [CrossRef]
16. Ugwu, K.; Herrera, A.; Gómez, M. Microplastics in marine biota: A review. *Mar. Pollut. Bull.* **2021**, *169*, 112540. [CrossRef]
17. Barrick, A.; Champeau, O.; Chatel, A.; Manier, N.; Northcott, G.; Tremblay, L.A. Plastic additives: Challenges in ecotox hazard assessment. *PeerJ* **2021**, *9*, e11300. [CrossRef]
18. Gkoutselis, G.; Rohrbach, S.; Harjes, J.; Obst, M.; Brachmann, A.; Horn, M.A.; Rambold, G. Microplastics accumulate fungal pathogens in terrestrial ecosystems. *Sci. Rep.* **2021**, *11*, 13214. [CrossRef]
19. Bahl, S.; Dolma, J.; Singh, J.J.; Sehgal, S. Biodegradation of plastics: A state of the art review. *Mater. Today Proc.* **2020**, *39*, 31–34. [CrossRef]
20. Miri, S.; Saini, R.; Mohammadreza Davoodi, S.; Pulicharla, R.; Kaur Brar, S.; Mag-douli, S. Biodegradation of microplastics: Better late than never. *Chemosphere* **2022**, *286*, 131670. [CrossRef]
21. Osma, J.F.; Toca-Herrera, J.L.; Rodríguez-Couto, S. Uses of laccases in the food industry. *Enzyme. Res.* **2010**, *2010*, 918761. [CrossRef] [PubMed]
22. Santo, M.; Weitsman, R.; Sivan, A. The Role of the Copper-Binding Enzyme—Laccase—In the Biodegradation of Polyethylene by the Actinomycete *Rhodococcus Ruber*. *Int. Biodeterior. Biodegrad.* **2013**, *84*, 204–210. [CrossRef]
23. Cai, Z.; Li, M.; Zhu, Z.; Wang, X.; Huang, Y.; Li, T.; Gong, H.; Yan, M. Biological Degradation of Plastics and Microplastics: A Recent Perspective on Associated Mechanisms and Influencing Factors. *Microorganisms* **2023**, *11*, 1661. [CrossRef] [PubMed]
24. Yang, X.-G.; Wen, P.-P.; Yang, Y.-F.; Jia, P.-P.; Li, W.-G.; Pei, D.-S. Plastic biodegradation by in vitro environmental microorganisms and in vivo gut microorganisms of insects. *Front. Microbiol.* **2023**, *13*, 1001750. [CrossRef] [PubMed]
25. Hall, B.G.; Acar, H.; Nandipati, A.; Barlow, M. Growth Rates Made Easy. *Mol. Biol. Evol.* **2013**, *31*, 232–238. [CrossRef] [PubMed]
26. Doelle, H.W.; Fiechter, A.; Schlegel, G.; Sell, D.; Shimizu, S.; Ulber, R.; Yamada, H. Biotechnology, 1. General. *Ullmann's Encycl. Ind. Chem.* **2009**, *5*, 689–712. [CrossRef]
27. Ali, S.R.M.; Fradi, A.J.; Al-araji, A.M. Effect of some physical factors on growth of five fungal species. *Eur. Acad. Res.* **2017**, *2*, 1069–1078.
28. Krasowska, A.; Sigler, K. How microorganisms use hydrophobicity and what does this mean for human needs? *Front. Cell. Infect. Microbiol.* **2014**, *19*, 112. [CrossRef]
29. Bayry, J.; Aimanianda, V.; Guijarro, J.I.; Sunde, M.; Latgé, J.-P. Hydrophobins—Unique Fungal Proteins. *PLoS Pathog.* **2012**, *8*, e1002700. [CrossRef]
30. Sowmya, H.V.; Krishnappa, M.; Thippeswamy, B. Degradation of polyethylene by *Trichoderma harzianum*—SEM, FTIR, and NMR analyses. *Environ. Monit. Assess.* **2014**, *186*, 577–6586. [CrossRef]
31. Gómez-Méndez, L.D.; Moreno-Bayona, D.A.; Poutou-Piñales, R.A.; Salcedo-Reyes, J.C.; Pedroza-Rodríguez, A.M.; Vargas, A.; Bogoya, J.M. Biodeterioration of plasma pretreated LDPE sheets by *Pleurotus ostreatus*. *PLoS ONE* **2018**, *13*, e0203786. [CrossRef]
32. Ganesh, P.; Dineshrajn, D.; Yoganathan, K. Production and screening of depolymerising enzymes by potential bacteria and fungi isolated from plastic waste dump yard sites. *Int. J. Appl. Res.* **2017**, *3*, 693–695.
33. Zhang, J.; Gao, D.; Li, Q.; Zhao, Y.; Li, L.; Lin, H.; Zhao, Y. Biodegradation of polyethylene microplastic particles by the fungus *Aspergillus flavus* from the guts of wax moth *Galleria mellonella*. *Sci. Total Environ.* **2020**, *704*, 135931. [CrossRef]
34. Kang, B.R.; Kim, S.B.; Song, H.A.; Lee, T.K. Accelerating the Biodegradation of High-Density Polyethylene (HDPE) Using *Bjerkandera adusta* TBB-03 and Lignocellulose Substrates. *Microorganisms* **2019**, *7*, 304. [CrossRef]
35. Khatoon, N.; Jamal, A.; Ali, M.I. Lignin peroxidase isoenzyme: A novel approach to biodegrade the toxic synthetic polymer waste. *Environ. Technol.* **2019**, *40*, 1366–1375. [CrossRef]

36. Sumathi, T.; Viswanath, B.; Sri Lakshmi, A.; SaiGopal, D.V.R. Production of Laccase by *Cochliobolus* sp. Isolated from Plastic Dumped Soils and Their Ability to Degrade Low Molecular Weight PVC. *Biochem. Res. Int.* **2016**, *2016*, 9519527. [CrossRef]
37. Liebminger, S.; Eberl, A.; Sousa, F.; Heumann, S.; Fischer-Colbrie, G.; Cavaco-Paulo, A.; Guebitz, G.M. Hydrolysis of PET and bis-(benzoyloxyethyl) terephthalate with a new polyesterase from *Penicillium citrinum*. *Biocatal. Biotransform.* **2007**, *25*, 171–177. [CrossRef]
38. Carniel, A.; Valoni, É.; Nicomedes, J.; da Conceição Gomes, A.; de Castro, A.M. Lipase from *Candida antarctica* (CALB) and cutinase from *Humicola insolens* act synergistically for PET hydrolysis to terephthalic acid. *Process Biochem.* **2017**, *59*, 84–90. [CrossRef]
39. Tahir, L.; Ali, M.I.; Zia, M.; Atiq, N.; Hasan, F.; Ahmed, S. Production and Characterization of Esterase in *Lantinus tigrinus* for Degradation of Polystyrene. *Pol. J. Microbiol.* **2013**, *62*, 101–108. [CrossRef]
40. Takamoto, T.; Shirasaka, H.; Uyama, H.; Kobayashi, S. Lipase-catalyzed hydrolytic degradation of polyurethane in organic solvent. *Chem. Lett.* **2001**, *30*, 492–493. [CrossRef]
41. Mathur, G.; Prasad, R. Degradation of Polyurethane by *Aspergillus flavus* (ITCC 6051) Isolated from Soil. *Appl. Biochem. Biotechnol.* **2012**, *167*, 1595–1602. [CrossRef] [PubMed]
42. Khan, S.; Nadir, S.; Shah, Z.U.; Shah, A.A.; Karunarathna, S.C.; Xu, J.; Khan, A.; Munir, S.; Hasan, F. Biodegradation of polyester polyurethane by *Aspergillus tubingensis*. *Environ. Pollut.* **2017**, *225*, 469–480. [CrossRef] [PubMed]
43. Yoon, M.G.; Jeon, H.J.; Kim, M.N. Biodegradation of polyethylene by a soil bacterium and AlkB cloned recombinant cell. *J. Bioremed. Biodegrad.* **2012**, *3*, 145. [CrossRef]
44. Jeon, H.J.; Kim, M.N. Functional analysis of alkane hydroxylase system derived from *Pseudomonas aeruginosa* E7 for low molecular weight polyethylene biodegradation. *Int. Biodeterior. Biodegrad.* **2015**, *103*, 141–146. [CrossRef]
45. Yoshida, S.; Hiraga, K.; Taniguchi, I.; Oda, K. *Ideonella sakaiensis*, PETase, and MHETase: From identification of microbial PET degradation to enzyme characterization. *Meth. Enzymol.* **2021**, *648*, 187–205. [CrossRef]
46. Jabloune, R.; Khalil, M.; Ben Moussa, I.E.; Simao-Beaunoir, A.-M.; Lerat, S.; Brzezinski, R.; Beaulieu, C. Enzymatic Degradation of *p*-Nitrophenyl Esters, Polyethylene Terephthalate, Cutin, and Suberin by Sub1, a Suberinase Encoded by the Plant Pathogen *Streptomyces scabies*. *Microbes Environ.* **2020**, *35*, ME19086. [CrossRef] [PubMed]
47. Müller, R.-J.; Schrader, H.; Profe, J.; Dresler, K.; Deckwer, W.-D. Enzymatic Degradation of Poly(Ethylene Terephthalate): Rapid Hydrolyse Using a Hydrolase from *T. Fusca*. *Macromol. Rapid Commun.* **2005**, *26*, 1400–1405. [CrossRef]
48. Nakajima-Kambe, T.; Onuma, F.; Kimpara, N.; Nakahara, T. Isolation and Characterization of a Bacterium Which Utilizes Polyester Polyurethane as a Sole Carbon and Nitrogen Source. *FEMS Microbiol. Lett.* **1995**, *129*, 39–42. [CrossRef]
49. Ekanayaka, A.H.; Tibpromma, S.; Dai, D.; Xu, R.; Suwannarach, N.; Stephenson, S.L.; Dao, C.; Karunarathna, S.C. A Review of the Fungi That Degrade Plastic. *J. Fungi* **2022**, *8*, 772. [CrossRef]
50. Lear, G.; Kingsbury, J.; Franchini, S.; Gambarini, V.; Maday, S.; Wallbank, J.; Weaver, L.; Pantos, O. Plastics and the Microbiome: Impacts and Solutions. *Environ. Microbiome* **2021**, *16*, 2. [CrossRef]
51. Ahmed, T.; Shahid, M.; Azeem, F.; Rasul, I.; Shah, A.A.; Noman, M.; Hameed, A.; Manzoor, N.; Manzoor, I.; Muhammad, S. Biodegradation of plastics: Current scenario and future prospects for environmental safety. *Environ. Sci. Pollut. Res.* **2018**, *25*, 7287–7298. [CrossRef]
52. DSouza, G.C.; Sheriff, R.S.; Ullanat, V.; Shrikrishna, A.; Joshi, A.V.; Hiremath, L.; Entoori, K. Fungal biodegradation of low-density polyethylene using consortium of *Aspergillus* species under controlled conditions. *Heliyon* **2021**, *7*, e07008. [CrossRef]
53. Kyaw, B.M.; Champakalakshmi, R.; Kishore Sakharkar, M.; Lim, C.S.; Sakharkar, K.R. Biodegradation of low density poly-ethene (LDPE) by *Pseudomonas* species. *Indian J. Microbiol.* **2012**, *52*, 411–419. [CrossRef] [PubMed]
54. Harshvardhan, K.; Jha, B. Biodegradation of low-density polyethylene by marine bacteria from pelagic waters, Arabian Sea, India. *Mar. Pollut. Bull.* **2013**, *77*, 100–106. [CrossRef] [PubMed]
55. Rogers, K.L.; Carreres-Calabuig, J.A.; Gorokhova, E.; Posth, N.R. Micro-by-micro interactions: How microorganisms influence the fate of marine microplastics. *Limnol. Oceanogr. Lett.* **2020**, *5*, 18–36. [CrossRef]
56. Yang, J.; Yang, Y.; Wu, W.-M.; Zhao, J.; Jiang, L. Evidence of polyethylene biodegradation by bacterial strains from the guts of plastic-eating waxworms. *Environ. Sci. Technol.* **2014**, *48*, 13776–13784. [CrossRef]
57. Tokiwa, Y.; Calabia, B.; Ugwu, C.; Aiba, S. Biodegradability of plastics. *Int. J. Mol. Sci.* **2009**, *10*, 3722–3742. [CrossRef] [PubMed]
58. Lee, H.M.; Kim, H.R.; Jeon, E.; Yu, H.C.; Lee, S.; Li, J.; Kim, D.-H. Evaluation of the biodegradation efficiency of four various types of plastics by *Pseudomonas aeruginosa* isolated from the gut extract of superworms. *Microorganisms* **2020**, *8*, 1341. [CrossRef] [PubMed]
59. Auta, H.S.; Emenike, C.U.; Jayanthi, B.; Fauziah, S.H. Growth kinetics and biodeterioration of polypropylene microplastics by *Bacillus* sp. and *Rhodococcus* sp. isolated from mangrove sediment. *Mar. Pollut. Bull.* **2018**, *127*, 15–21. [CrossRef]
60. Auta, H.S.; Emenike, C.U.; Jayanthi, B.; Fauziah, S.H. Screening for polypropylene degradation potential of bacteria isolated from Mangrove ecosystems in Peninsular Malaysia. *Int. J. Biosci. Biochem. Bioinform.* **2017**, *7*, 245–251. [CrossRef]
61. Sangeetha Devi, R.; Rajesh Kannan, V.; Nivas, D.; Kannan, K.; Chandru, S.; Robert Antony, A. Biodegradation of HDPE by *Aspergillus* spp. from marine ecosystem of Gulf of Mannar, India. *Mar. Pollut. Bull.* **2015**, *96*, 32–40. [CrossRef]
62. Sáenz, M.; Borodulina, T.; Diaz, L.; Banchon, C. Minimal Conditions to Degrade Low Density Polyethylene by *Aspergillus terreus* and *niger*. *J. Ecol. Eng.* **2019**, *20*, 44–51. [CrossRef]

63. Chaudhary, A.K.; Vijayakumar, R.P. Studies on biological degradation of polystyrene by pure fungal cultures. *Environ. Dev. Sustain.* **2020**, *22*, 4495–4508. [CrossRef]
64. Kučić Grgić, D.; Miloloža, M.; Ocelić Bulatović, V.; Ukić, Š.; Slouf, M.; Gajdosova, V. Screening the Efficacy of a Microbial Consortium of Bacteria and Fungi Isolated from Different Environmental Samples for the Degradation of LDPE/TPS Films. *Separations* **2023**, *10*, 79. [CrossRef]
65. Black, J.G. *Microbiology: Principles and Explorations*, 8th ed.; John Wiley & Sons: Hoboken, NJ, USA, 2012; p. 975.
66. Gong, Y.; Ding, P.; Xu, M.-J.; Zhang, C.-M.; Xing, K.; Qin, S. Biodegradation of phenol by a halotolerant versatile yeast *Candida tropicalis* SDP-1 in wastewater and soil under high salinity conditions. *J. Environ. Manag.* **2021**, *289*, 112525. [CrossRef] [PubMed]
67. Auta, H.S.; Emenike, C.U.; Fauziah, S.H. Screening of *Bacillus* strains isolated from mangrove ecosystems in Peninsular Malaysia for microplastic degradation. *Environ. Pollut.* **2017**, *231*, 1552–1559. [CrossRef] [PubMed]
68. Wolski, E.A.; Murialdo, S.E.; Gonzalez, J.F. Effect of pH and inoculum size on pentachlorophenol degradation by *Pseudomonas* sp. *Water SA* **2006**, *32*, 93–98. [CrossRef]
69. Gonzalez, J.M.; Aranda, B. Microbial Growth under Limiting Conditions-Future Perspectives. *Microorganisms* **2023**, *11*, 1641. [CrossRef]
70. Peixoto, J.; Silva, L.P.; Krüger, R.H. Brazilian Cerrado soil reveals an untapped microbial potential for unpretreated polyethylene biodegradation. *J. Hazard. Mater.* **2017**, *324*, 634–644. [CrossRef]
71. Lou, Y.; Li, Y.; Lu, B.; Liu, Q.; Yang, S.-S.; Liu, B.; Ren, N.; Wu, W.-M.; Xiang, D. Response of the yellow mealworm (*Tenebrio molitor*) gut microbiome to diet shifts during polystyrene and polyethylene biodegradation. *J. Hazard. Mater.* **2021**, *416*, 126222. [CrossRef]
72. Liu, J.; Xu, G.; Dong, W.; Xu, N.; Xian, F.; Ma, J.; Fang, Y.; Zhou, J.; Jiang, M. Biodegradation of diethyl terephthalate (DET) and polyethylene terephthalate (PET) by a novel identified degrader *Delftia* sp. WL-3 and its proposed metabolic pathway. *Lett. Appl. Microbiol.* **2018**, *67*, 254–261. [CrossRef]
73. Akutsu, Y.; Nakajima-Kambe, T.; Nomura, N.; Nakahara, T. Purification and Properties of a Polyester Polyurethane-Degrading Enzyme from *Comamonas acidovorans* TB-35. *Appl. Environ. Microbiol.* **1998**, *64*, 62–67. [CrossRef]
74. Zhang, Y.; Pedersen, J.N.; Eser, B.E.; Guo, Z. Biodegradation of polyethylene and polystyrene: From microbial deterioration to enzyme discovery. *Biotechnol. Adv.* **2022**, *60*, 107991. [CrossRef] [PubMed]
75. Vásquez-Piñeros, M.A.; Martínez-Lavanchy, P.M.; Jehmlich, N.; Pieper, D.H.; Rincón, C.A.; Harms, H.; Junca, H.; Heipieper, H.J. *Delftia* sp. LCW, a strain isolated from a constructed wetland shows novel properties for dimethylphenol isomers degradation. *BMC Microbiol.* **2018**, *18*, 108. [CrossRef] [PubMed]
76. Harris, J.D.; Coon, C.M.; Doherty, M.E.; McHugh, E.A.; Warner, M.C.; Walters, C.L.; Orahood, O.M.; Loesch, A.E.; Hatfield, D.C.; Sitko, J.C.; et al. Engineering and characterization of dehalogenase enzymes from Delftia acidovorans in bioremediation of perfluorinated compounds. *Synth. Syst. Biotechnol.* **2022**, *7*, 671–676. [CrossRef] [PubMed]
77. Hashem, M.; Alamri, S.A.; Al-Zomyh, S.S.A.A.; Alrumman, S.A. Biodegradation and detoxification of aliphatic and aromatic hydrocarbons by new yeast strains. *Ecotoxicol. Environ. Saf.* **2018**, *151*, 28–34. [CrossRef] [PubMed]
78. Khoddami, A.; Morshed, M.; Tavani, H. Effects of Enzymatic Hydrolysis on Drawn Polyester Filament Yarns. *Iran. Polym. J.* **2001**, *10*, 363–370.
79. Zafar, U.; Houlden, A.; Robson, G.D. Fungal communities associated with the biodegradation of polyester polyurethane buried under compost at different temperatures. *Appl. Environ. Microbiol.* **2013**, *79*, 7313–7324. [CrossRef] [PubMed]
80. Mroczyńska, M.; Brillowska-Dąbrowska, A. Virulence of Clinical *Candida* Isolates. *Pathogens* **2021**, *10*, 466. [CrossRef] [PubMed]
81. Bher, A.; Mayekar, P.C.; Auras, R.A.; Schvezov, C.E. Biodegradation of Biodegradable Polymers in Mesophilic Aerobic Environments. *Int. J. Mol. Sci.* **2022**, *23*, 12165. [CrossRef]
82. Devi, R.; Kannan, V.; Natarajan, K.; Nivas, D.; Kannan, K.; Chandru, S.; Antony, A. The Role of Microbes in Plastic Degradation. In *Environmental Waste Management*; Chandra, R., Ed.; CRC Press: Boca Raton, FL, USA, 2016; pp. 355–384. [CrossRef]
83. Singh, B.; Christina, E. Chapter 13—Indigenous microorganisms as an effective tool for in situ bioremediation. In *Relationship between Microbes and the Environment for Sustainable Ecosystem Services*; Samuel, J., Kumar, A., Singh, J., Eds.; Elsevier: Amsterdam, The Netherlands, 2022; Volume 2, pp. 273–295. [CrossRef]
84. Magan, N. Fungi in Extreme Environments. In *Environmental and Microbial Relationships*; Kubicek, C.P., Druzhinina, I.S., Eds.; Springer: Berlin/Heidelberg, Germany, 2007; pp. 85–103. [CrossRef]
85. Zhou, Y.; Han, L.R.; He, H.W.; Sang, B.; Yu, D.L.; Feng, J.T.; Zhang, X. Effects of Agitation, Aeration and Temperature on Production of a Novel Glycoprotein GP-1 by *Streptomyces kanasenisi* ZX01 and Scale-Up Based on Volumetric Oxygen Transfer Coefficient. *Molecules* **2018**, *23*, 125. [CrossRef]
86. Valdez-Cruz, N.A.; Reynoso-Cereceda, G.I.; Pérez-Rodriguez, S.; Restrepo-Pineda, S.; González-Santana, J.; Olvera, A.; Zavala, G.; Alagón, A.; Trujillo-Roldán, M.A. Production of a recombinant phospholipase A2 in *Escherichia coli* using resonant acoustic mixing that improves oxygen transfer in shake flasks. *Microb. Cell Fact.* **2017**, *16*, 129. [CrossRef]
87. Seth, M.; Chand, S. Biosynthesis of tannase and hydrolysis of tannins to gallic acid by *Aspergillus awamori*—Optimisation of process parameters. *Process Biochem.* **2000**, *36*, 39–44. [CrossRef]
88. Ibrahim, D.; Weloosamy, H.; Lim, S.-H. Effect of agitation speed on the morphology of *Aspergillus niger* HFD5A-1 hyphae and its pectinase production in submerged fermentation. *World. J. Biol. Chem.* **2015**, *6*, 265–271. [CrossRef] [PubMed]
89. Al-Sahhaf, Z.Y. Hematological and biochemical effects of an air freshener in rabbits. *J. Am. Sci.* **2012**, *8*, 70–73.

90. Philibert, D.; Parketon, T.; Marteinson, S.; de Jourdan, B. Assessing the Toxicity of Individual Aromatic Compounds and Mixtures to American Lobster (*Homarus americanus*) Larvae Using a Passive Dosing System. *Environ. Toxicol. Chem.* **2021**, *40*, 1379–1388. [CrossRef]
91. Miloloža, M.; Bule, K.; Prevarić, V.; Cvetnić, M.; Ukić, Š.; Bolanča, T.; Kučić Grgić, D. Assessment of the Influence of Size and Concentration on the Ecotoxicity of Microplastics to Microalgae *Scenedesmus* sp., Bacterium *Pseudomonas putida* and Yeast *Saccharomyces cerevisiae*. *Polymers* **2022**, *14*, 1246. [CrossRef]
92. Jung, M.R.; Horgen, F.D.; Orski, S.V.; Rodriguez, V.; Beers, K.L.; Balazs, G.H.; Jones, T.T.; Work, T.M.; Brignac, K.C.; Royer, S.-J.; et al. Validation of ATR FT-IR to identify polymers of plastic marine debris, including those ingested by marine organisms. *Mar. Pollut. Bull.* **2008**, *127*, 704–716. [CrossRef]
93. Ul-Hamid, A.; Soufi, K.Y.; Al-Hadhrami, L.M.; Shemsi, A.M. Failure investigation of an underground low voltage XLPE insulated cable. *Anti-Corros. Methods Mater.* **2015**, *62*, 281–287. [CrossRef]
94. Smith, B.C. The Infrared Spectroscopy of Alkenes. *Spectroscopy* **2016**, *31*, 28–34. Available online: https://www.spectroscopyonline.com/view/infrared-spectroscopy-alkenes (accessed on 5 June 2023).
95. Smith, B.C. The Infrared Spectra of Polymers IV: Rubbers. *Spectroscopy* **2022**, *37*, 8–12. [CrossRef]
96. IR Spectrum Table & Chart. Available online: https://www.sigmaaldrich.com/HR/en/technical-documents/technical-article/analytical-chemistry/photometry-and-reflectometry/ir-spectrum-table (accessed on 5 June 2023).
97. IR Spectroscopy Tutorial: Alkenes. Available online: https://orgchemboulder.com/Spectroscopy/irtutor/alkenesir.shtml (accessed on 5 June 2023).
98. Amar, Z.H.; Chabira, S.F.; Sebaa, M.; Ahmed, B. Structural changes undergone during thermal aging and/or processing of Unstabilized, dry-blend and rigid PVC, investigated by FTIR-ATR and curve fitting. *Ann. Chim. Sci. Mater.* **2019**, *43*, 59–68. [CrossRef]
99. Ma, D.; Liang, L.; Hu, E.; Chen, H.; Wang, D.; He, C.; Feng, Q. Dechlorination of polyvinyl chloride by hydrothermal treatment with cupric ion. *Process Saf. Environ. Prot.* **2021**, *146*, 108–117. [CrossRef]
100. Kumari, A.; Chaudhary, D.R.; Jha, B. Destabilization of polyethylene and polyvinylchloride structure by marine bacterial strain. *Environ. Sci. Pollut. Res.* **2019**, *26*, 1507–1516. [CrossRef] [PubMed]
101. Subramani, M.; Sepperumal, U. FTIR analysis of bacterial mediated chemical changes in Polystyrene foam. *Ann. Biol. Res.* **2016**, *7*, 55–61.
102. Sutkar, P.R.; Gadewar, R.D.; Dhulap, V.P. Recent trends in degradation of microplastics in the environment: A state-of-the-art review. *J. Hazard. Mater. Adv.* **2023**, *11*, 100343. [CrossRef]
103. MassBank Record: MSBNK-Eawag-EQ363802. Available online: https://massbank.eu/MassBank/RecordDisplay?id=MSBNK-Eawag-EQ363802 (accessed on 23 November 2023).
104. Pawlowski, K.H.; Schartel, B. Flame retardancy mechanisms of triphenyl phosphate, resorcinol bis(diphenyl phosphate) and bisphenol A bis(diphenyl phosphate) in polycarbonate/acrylonitrile–butadiene–styrene blends. *Polym. Int.* **2007**, *56*, 1404–1414. [CrossRef]
105. MassBank Record: MSBNK-BGC_Munich-RP019801. Available online: https://massbank.eu/MassBank/RecordDisplay?id=MSBNK-BGC_Munich-RP019801&dsn=BGC_Munich (accessed on 23 November 2023).
106. MassBank Record: MSBNK-Antwerp_Univ-AN112602. Available online: https://massbank.eu/MassBank/RecordDisplay?id=MSBNK-Antwerp_Univ-AN112602&dsn=Antwerp_Univ (accessed on 23 November 2023).
107. Plasticisers. Available online: http://www.seepvcforum.com/en/content/16-plasticisers (accessed on 23 November 2023).
108. Thew, X.E.C.; Lo, S.C.; Ramanan, R.N.; Tey, B.T.; Huy, N.D.; Wei, O.C. Enhancing plastic biodegradation process: Strategies and opportunities. *Crit. Rev. Biotechnol.* **2023**. *Online ahead of print*. [CrossRef]

environments

MDPI

Article

Magnesium/Silica/Lanthanum@Activated Carbon for the Remediation of As(III) from Water

Athanasia K. Tolkou * and George Z. Kyzas

Hephaestus Laboratory, Department of Chemistry, International Hellenic University, GR-65404 Kavala, Greece; kyzas@chem.ihu.gr

* Correspondence: tolkatha@chem.ihu.gr; Tel.: +30-2510-462-218

Abstract: In this study, activated carbon was suitably modified with Mg/Si/La and its effectiveness in removing As(III) was investigated. The structure of Magnesium/Silica/Lanthamum@Activated Carbon (Mg-Si-La@AC) was fully characterized and several parameters, such the dosage, pH, contact time, and initial As(III) concentration, were studied. Thus, the BET surface area, total pore volume, and micropore volume of Mg-Si-La@AC were measured to be 271.46 m^2/g, 0.006 cm^3/g and 0.52 cm^3/g, respectively. The results showed that the optimal condition for the reduction in As(III) from the initial concentration of 100 μg/L to below 10 μg/L was the addition of 1.5 g/L of adsorbent at pH 7.0. Furthermore, 4 h of contact time showed >90% removal. The Langmuir isotherm model was best fitted to the experimental results, exhibiting a maximum adsorption capacity of 322 μg/g, and the PSO kinetic model was found to be the most applicable according to kinetics. Consecutive regeneration studies were carried out and the results showed that the adsorbent was effectively used up to four cycles.

Keywords: arsenic; adsorption; activated carbon; magnesium; lanthanum; silicate

1. Introduction

Groundwater contamination is a major global environmental concern [1], with many social implications, and can be caused by both human activities and nature. Millions of people in numerous countries, including Bangladesh, Pakistan, Vietnam, Cambodia, India, Argentina, Chile, etc. [2–5], are at risk of high As levels via groundwater consumption. In Pakistan, As was also found, in addition to groundwater [6], in land uses, with dust levels being highest in industrial areas [7]. Arsenic occurs in water as well as in air and soil [1], but it is very toxic in its inorganic form, found as tri-valent (arsenite) or pentavalent (arsenate), especially in its +3 oxidation state (As(III)) [8]. Long-term exposure to arsenic can cause chronic arsenic poisoning, with skin lesions and skin cancer being the most typical consequences [9,10]. In addition, liver or lung cancer, along with a variety of other effects, can also be caused by consuming arsenic via drinking water or food [11]. Consequently, the World Health Organization (WHO) has defined 10 μg/L as the suggested limit for arsenic in drinking water [12]. To maintain the WHO limits, new more effective elimination techniques should be applied, or existing ones should be modified and optimized accordingly. Numerous removal techniques [13] have been established for arsenic elimination from water, such as coagulation [14], adsorption [15], membranes [16,17], precipitation [18,19], ion exchange [20], etc. Comparing these methods, it appears that chemical precipitation and coagulation–flocculation are widely used due to their simplicity. The membrane filtration process is also effective; however, its operating cost is very high and due to membrane fouling, it is disadvantageous. The ion exchange method is a method in which anions and cations are exchanged in solution, but it fails in concentrated metal solutions. These processes are pH-dependent and therefore require the constant monitoring and control of pH. Consequently, the aforementioned procedures have some drawbacks, such as the high cost of both operation and construction, offer the partial removal of certain ions and contribute to high sludge production, thus increasing the operating cost for sludge disposal [21,22]. Adsorption, widely used to remove dyes from aqueous solutions [23–25], is possibly one of the most popular methods for heavy metal and oxyanion elimination, and includes the application of a material that removes the pollutant from the water by binding ions to its surface. A wide range of adsorbents, natural or modified, have been tested successfully for arsenic removal. Among them, there are various minerals such as zeolite [26], silica [27], goethite [28], and calcite [29,30], and also activated carbons [14,31,32], biosorbents [33,34], biochars [35], graphene oxides [36], polyethylenimine-modified materials [37], and metal–organic frameworks (MOFs) [38], etc. In addition, synthetic adsorbents such as CeO_2–ZrO_2 nanospheres [39], molecularly imprinted polymers [40], and the combination of Fe(III)-H_2O_2 [41] and Fe-Mn binary oxides [42], or the addition of graphene oxide/granular ferric hydroxide (GO/GFH) [36], have also been applied in arsenic removal studies.

Among these adsorbents, activated carbon (AC) has been generally used to remove heavy metals from water. A plethora of materials have been produced in several studies by using low-cost origin materials, such as coconut and palm shell, potatoes, fruit peels, bamboo, canola stalks [43–46], etc., as the interest of researchers is high for lower-cost activated carbon sources. Recently, the impregnation of activated carbons with various additives has been widely studied, with the aim of increasing their surface area to make them more suitable for heavy metal and oxyanion removal. The modification of activated carbon offers the chemical groups dispersed within it, resulting in both the optimization of the prevailing activated carbon properties and the enhanced synergy between these chemical groups and the carbon, with the aim of enhancing the removal of various substances (e.g., dyes, heavy metals, etc.) [47,48]. Charged metal cations, which contain high positive charges, can probably recruit several oxyanions such as arsenic. Therefore, several rare earth elements, among them lanthanum, have been used for arsenic removal and exhibited a relative affinity [49]. Consequently, AC has been lately modified with cerium dioxide (PAC–CeO_2), which is a nontoxic rare earth oxide, and found to be highly effective as an adsorbent for the elimination of arsenic from drinking water [50]. Similar studies have combined lanthanum and cerium functionalities to modify activated carbon prepared from avocado kernel seeds [51], focusing on the removal both of fluoride ions and As(V). In addition, a magnesium-activated carbon composite (Mg–AC) was tested for the efficient adsorption of Zn(II) and Cd(II) ions [52]. Moreover, activated carbon modified with $MgSiO_3$ (magnesium silicate) was applied to remove Pb(II) [53]. In recent studes [54,55], the presence of magnesium has shown a significant effect on arsenic removal. Mg^{2+} ions exhibit a corresponding affinity with As that enhances the adsorption effect of the modified activated carbon. In conclusion, the combination of such additives, i.e., La, Mg, and Si, for the alteration of activated carbon derived from low-cost raw materials, with the aim of increasing its surface area, its selectivity to specific pollutants and its effectiveness, has not been used in the recent literature.

Therefore, a hybrid material already synthesized in our previous research [43] and applied with great success in the removal of Cr(VI) from water, combining the benefits of Mg, Si, La and AC (hereafter Mg-Si-La@AC), is examined in this study regarding its potential effectiveness in removing As(III). FTIR and SEM techniques were applied to study both the structure and morphology of Mg-Si-La@AC. Moreover, the effects of numerous parameters (pH, As(III) concentration, dosage of adsorbent and contact time) were studied. Isotherm, kinetic, thermodynamic and regeneration studies were performed to describe and evaluate the adsorption that occurred.

2. Materials and Methods

2.1. Materials

For 100 mg of As(III)/L solution, an appropriate amount of sodium metarsenite ($AsNaO_2$) (Merck) (0.1733 g) was dissolved in 1 L of deionized water and kept in a cool place for further use. For the modification of activated carbon, lanthanum chloride heptahydrate 98% ($LaCl_3$) (Merck), MgO (PMS2 pure magnesia) and silicon dioxide (SiO_2) (Merck) were used. The pH was adjusted accordingly by utilizing a HCl solution of 37% (Panreac) or NaOH pellets ACS $\geq$ 97.0%, (Sigma-Aldrich, St. Louis, MO, USA) (0.01–0.1 M).

Synthesis of Hybrid Mg-Si-La@AC

The Mg-Si-La@AC was synthesized via the modification of activated carbon produced by coconut shells, with magnesium, silicate and lanthanum. The detailed procedure is given in previous studies published by our group [43,44]. In brief, 5 g of AC that was previously activated by 2 M KOH at 25° C for 24 h was added to three beakers of 25 mL of distilled water, each containing 1.8 g of $LaCl_3$, 0.8 g of MgO or 0.6 g of SiO_2, respectively, and mixed for 1 h at 25° C and sonicated for 2 h. After filtering and rinsing with distilled water, the samples produced (i.e., La@AC, Mg@AC and Si@AC, respectively) were dried overnight at 333 K. In order to produce the composite Mg-Si-La@AC, the calcination of the servings of the above samples took place at 773 K for 5 h. Finally, all samples were kept at room temperature until use.

2.2. *Adsorption Experiments*

To carry out the experiments, 10 mL of a constant concentration of arsenic solution was added to 15 mL falcon tubes, followed by the addition of a fixed amount of adsorbent under constant temperature. A Trayster overhead shaker and Loopster rotator were used to stir the suspension at 80 rpm. The experimental parameters considered were the solution pH, which was 4.0–9.0, the initial As(III) concentration, which was 25–500 μg/L, the adsorbent's dose, which varied from 0.5 to 3.0 g/L, and the contact time, which ranged from 10 min up to 360 min for kinetics and up to 24 h when aiming to reach equilibrium. The above-mentioned parameters were measured in the filtrate obtained after filtering the collected samples through a 0.45 μm nylon filter. The following Equation (1) was used to calculate the percentage removal (% R) of As(III):

$$R(\%) = \left(\frac{C_0 - C_f}{C_0}\right) \times 100\% \quad (1)$$

where C_0 is the initial and C_f is the final (residual) As(III) concentration (μg/L).

Moreover, Equation (2) was used for the determination of the adsorbent's adsorption capacity (Q_e) (μg/g):

$$Q_e = \frac{(C_0 - C_e) \times V}{m} \quad (2)$$

where C_e = As(III) concentration (μg/L) at equilibrium, V = volume of solution (L), and m = mass (g) of the adsorbent.

2.2.1. Adsorption Isotherms

For the adsorption isotherms, the results were fitted to the Langmuir and Freundlich isothermal models, which are the commonly used and easiest to apply models.

The Langmuir model is expressed as Equation (3) [56] and determines the relationship between the concentration of the adsorbate in the solid phase (Q_e) and the uptake to the equilibrium liquid concentration:

$$Q_e = \frac{Q_m K_L C_e}{1 + K_L C_e} \quad (3)$$

where Q_m is the maximum adsorption capacity (μg/L), and K_L is the energy of the As(III) (L/μg) adsorption.

Furthermore, according to the Langmuir theory, the adsorption capacity of the adsorbent is partial (Q_m), with the adsorbate forming a monolayer on the surface of the adsorbent; hence, the interface between the adsorbed molecules is almost non-existent [51].

On the other hand, the Freundlich model [57] describes the relationship between the equilibrium As(III) concentrations (μg/L) and the uptake capacities Q_e (μg/g) of the adsorbent; it is stated as Equation (4):

$$Q_e = K_F C_e^{1/n} \tag{4}$$

where K_F (μg/g)(L/μg)$^{1/n}$ is a constant related to the adsorption capacity and *1/n* is a constant related to the intensity of the adsorption or surface heterogeneity:

- *1/n* = 0 means heterogeneity;
- *1/n* < 1 means a normality; and
- *1/n* > 1 designates a cooperativeness.

2.2.2. Kinetics

Two kinetic models were examined to fit the kinetics of As(III)) adsorption; PFO (pseudo-first-order) and PSO (pseudo-second-order) models. These models are the most widely used models in adsorption experiments. Then, the calculated kinetic parameters of adsorption were analyzed in order to evaluate the adsorption rate, as well as to identify the possible reaction mechanism. The PFO model used is expressed as Equation (5):

$$Q_t = Q_e\left(1 - e^{-k_1 t}\right) \tag{5}$$

where Q_e = amount of As(III) adsorbed at equilibrium (μg/g), Q_t = amount of As(III) adsorbed at time *t* (μg/g), k_1 = rate constant of the PFO model (1/min), and *t* = time (min).

The PSO model used for the data analysis is expressed as Equation (6):

$$Q_t = \frac{k_2 Q_e^2 t}{1 + k_2 Q_e t} \tag{6}$$

where Q_e = amount of As(III) adsorbed at equilibrium (μg/g), Q_t = amount of As(III) adsorbed at time *t* (μg/g), k_2 = rate constant of the PSO (g/(μg min), and *t* = time (min).

2.3. Arsenic Determination

After adsorption, water samples were collected from the supernatant of each falcon tube, filtered through a 0.45 μm nylon membrane filter, and saved for immediate determination. The initial and residual arsenic concentrations were then determined using an atomic absorption spectrometer (Varian Zeeman AA240Z with GTA 120; Hansen Way Palo Alto, CA, USA) coupled with a graphite furnace (graphite furnace-AAS). The detection limit of the AAS was 1 μg/L. The values obtained were converted to residual concentrations by corresponding the absorbance to the standard curve of As. To confirm the results and the accuracy of the procedure, the experiments were initially performed in triplicate and each sample obtained was measured three times. Data were expressed as mean trace element values ± standard deviation.

2.4. Thermodynamics

The thermodynamic parameters, such as the change in Gibbs free energy (ΔG^0, kJ/mol), enthalpy (ΔH^0, kJ/mol) and entropy (ΔS^0, kJ/mol·K), were calculated in order to estimate the adsorption process and determine the possible spontaneous nature. Consequently, for this reason, several indicative temperatures (298, 308, 318, and 338 K) were tested and the following equations were used to calculate the thermodynamic parameters [58]:

$$\mathrm{K_c} = \frac{\mathrm{C_s}}{\mathrm{C_e}} \tag{7}$$

$$\Delta \mathrm{G}^0 = -\mathrm{RTln}(\mathrm{K_c}) \tag{8}$$

$$\Delta G^0 = \Delta H^0 - T\Delta S^0 \quad (9)$$

$$\ln(K_c) = \left(-\frac{\Delta H^0}{R}\right) + \frac{\Delta S^0}{R} \quad (10)$$

ΔG^0 was given from Equation (8), and for the calculation of the ΔH^0 and ΔS^0 values, the plot of $\ln(K_c)$ versus $1/T$ (Equation (10)) was applied. The relative slope and intercept resulted in the values.

2.5. Characterization Techniques

To study the surface of the adsorbent, the most common techniques were applied. Therefore, SEM (Scanning Electron Microscopy) (Jeol JSM-6390 LV, Musashino, Akishima, Tokyo 196-8558, Japan, scanning electron microscope), equipped with an EDS (energy-dispersive X-ray) system, was used for surface element analysis; thus, a focused electron beam was used to scan the surface. The results obtained from our electron microscope are given in the form of images of morphological content in relation to the surface and phase distribution, as well as the characteristic X-ray spectra or visualization of the distribution of various data on the surface of the sample. In addition, FT-IR (Fourier-Transform Infrared Spectroscopy) (Perkin Elmer, New York, NY, USA) and BET (Brunauer, Emmett and Teller) analysis software (Quantachrome NovaWin; Data Acquisition and Reduction) for NOVA instruments 1994-2012 , Quantachrome instruments version 11-02, using computer-controlled nitrogen gas adsorption analyzer, were used to determine the specific surface area, including the pore size distribution. BET theory has the effect of determining the molecules, calculated through experimental experience, that are needed to form a single layer, excluding the fact that it is impossible to create one.

3. Results and Discussion

3.1. Characterization of Mg-Si-La@AC

The nitrogen adsorption and desorption isotherms for Mg-Si-La@AC versus the relative pressure (P/P_0) are shown in Figure 1. As depicted, Mg-Si-La@AC shows a Type IV model of isotherm with hysteresis loops [59], according to the IUPAC (International Union of Pure and Applied Chemistry), ascertaining their mesoporous structure. The BET-specific surface area values, median BJH (Barrett–Joyner–Halenda) pore size and pore volume of Mg-Si-La@AC are given in Table 1.

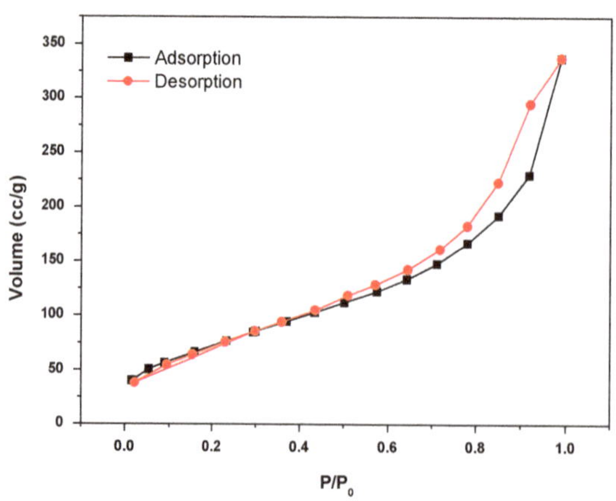

Figure 1. N_2 adsorption/desorption isotherms of Mg-Si-La@AC adsorbent against relative pressure (P/P_0) from 0 to 1.

Table 1. Physical properties of Mg-Si-La@AC.

Parameters	Mg-Si-La@AC
BET Surface area, S_{BET} (m^2/g)	271.46
Micropore volume, V_{micro} (cm^3/g)	0.006
Total pore volume, V_T (cm^3/g)	0.521

According to the determinations, the BET surface area of Mg-Si-La@AC was calculated to be 271.46 m^2/g, the respective total pore volume was 0.006 cm^3/g, and the micropore volume was measured as 0.521 cm^3/g, which show that Mg-Si-La@AC is practically efficient for adsorption, providing a high surface area. The morphology and structure of the modified activated carbon were observed using SEM images. As shown in Figure 2a, Mg-Si-La@AC shows a porous structure with cavities, possibly due to the carbonization stage that takes place during synthesis. Nasri et al. have also observed more developed carbon pore structures, which were synthesized from coconut shell using two-step CO_2 activation [60]. In Figure 2b, it is shown that these pores are filled with As after the adsorption experiments.

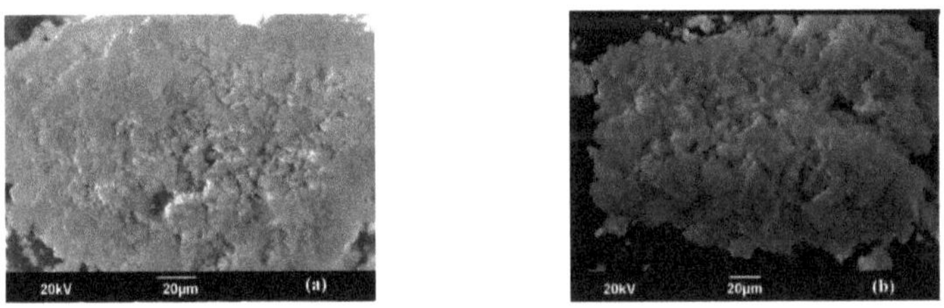

Figure 2. SEM image of Mg-Si-La@AC (**a**) pre and (**b**) post arsenic adsorption.

In addition, an EDS analysis of Mg-Si-La@AC pre and post As(III) adsorption was conducted, as shown in Figure 3 and Table 2. According to the results, Mg, Si and La were detected on the activated carbon surface, confirming its amendment. After the adsorption of As(III), a higher percentage of oxygen content and the appearance of As were observed, which is probably due to the addition of sodium metarsenite ($AsNaO_2$) and its binding. Additionally, a decrease in the Mg, Si and La content was detected, which probably contributed to the removal of As.

Table 2. SEM/EDS analysis of Mg-Si-La@AC pre and post As(III) adsorption.

Element	Mg-Si-La@AC	Mg-Si-La@AC_As
	Content% (*w*/*w*)	
La	3.59	2.46
Mg	6.07	4.79
C	12.64	9.04
Si	8.26	7.56
P	0.79	1.04
O	68.65	74.14
As	0	0.96

The FTIR spectra of Mg-Si-La@AC pre and post As(III) adsorption are shown in Figure 4. As can be seen, the spectra are similar, with a slight change in the intensity and formation of characteristic new bands or the disappearance of others. Specifically, the first broad and stretched peak was obtained at 3395 and 3385 cm^{-1} for the spectra of Mg-Si-La@AC pre and post As(III) adsorption, respectively, and corresponds to the characteristic symmetric and asymmetric stretching vibration of the O–H hydroxyl group, which ranges from 3200–3600 cm^{-1} [61]. Furthermore, the peaks at 2217–1936/2224–1945 cm^{-1}

and 1508–1415/1501–1425 cm^{-1} were related to C–H, –C=C– (carbonyl) and C–C aromatic stretching, corresponding to the IR pre and post adsorption [61]. The ranging from 600–740 cm^{-1} corresponds to MgO stretching vibrations [43], La_2O_3 [61] and to the bending of the O−Si−O [20], confirming the modification of AC with Mg, Si and La. Observing the post-adsorption spectra, there is a new absorption peak that appears at 863 cm^{-1} and can be attributed, according to the literature, to the As–O–La bond [61,62]. In addition, a more intense peak at 738.5 cm^{-1}, post adsorption, corresponds to As–O–Mg vibrations [63]. This proves that As(III) is adsorbed on the Mg-Si-La@AC surface and, indeed, that La and Mg contribute to the binding of arsenic, as was hypothesized earlier in the elemental analysis (Table 2).

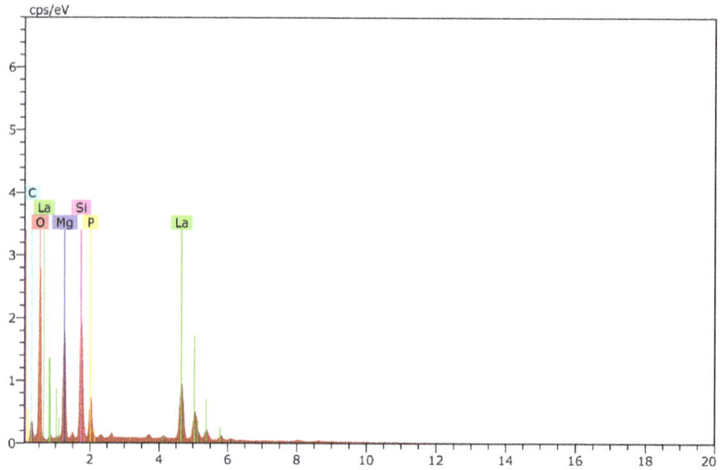

(**a**)

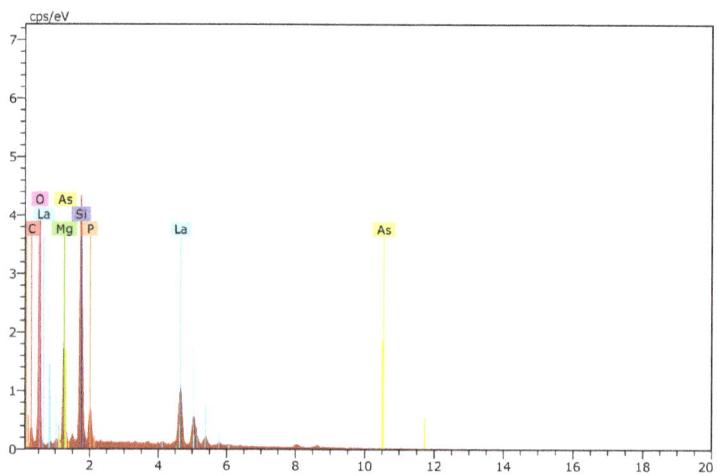

(**b**)

Figure 3. Elemental spectra (**a**) pre and (**b**) post arsenic adsorption on Mg-Si-La@AC, detected using SEM/EDS analysis.

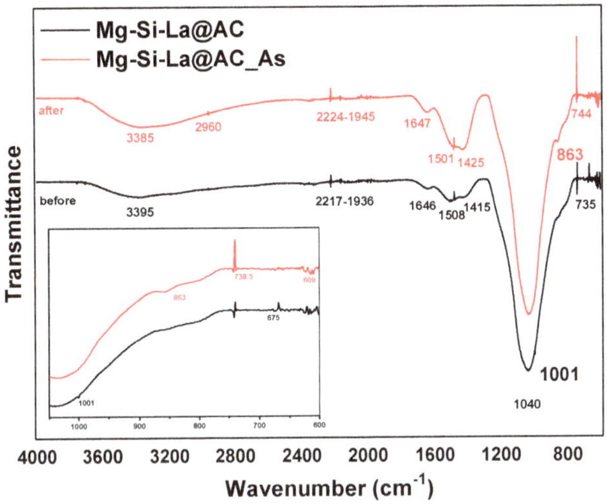

Figure 4. FTIR spectra pre and post arsenic adsorption on Mg-Si-La@AC.

3.2. Batch Adsorption Experiments

Effect of Modification of Activated Carbon

As shown in Figure 5, as the modification of the activated carbon increases, so does its effectiveness in the adsorption of As(III). For instance, when activated carbon is modified with silicon (Si@AC) or lanthanum (La@AC), no significant improvement is observed; however, when modified with magnesium, the performance of this combined material (Mg-Si-La@AC) increases. What follows is that the presence of magnesium contributes significantly to the removal of As(III). This is confirmed both by the elemental analysis presented in Table 2, where the Mg content decreases after As(III) adsorption, and by the appearance of the peak in the FTIR corresponding to the As–O–Mg vibrations (Figure 4). In recent studies [54,55], the presence of magnesium has been shown to have a significant effect on arsenic removal. The Mg^{2+} ions contained in MgO, as demonstrated via EDS analysis and FTIR, are successfully deposited on the surface of the activated carbon and thus exhibit a corresponding affinity with As that enhances the adsorption effect of the modified activated carbon. To conclude, the Mg-Si-La@AC adsorbent is found to be optimal for further study.

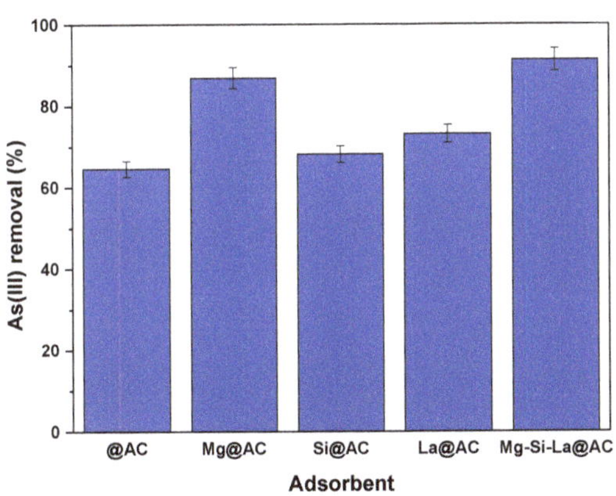

Figure 5. Effect of modification of activated carbon on the removal of As(III). C_0 100 μg/L, dose 1.5 g/L, pH 7.0, 24 h.

3.3. *Effect of Dosage and Initial Solution pH*

The potential of Mg-Si-La@AC to be applied in As(III) removal was evaluated by examining the effect of the adsorbent dosage and initial pH. The pH values tested were in the range of 4.0–9.0. These values were chosen for two reasons. Firstly, the pH of groundwater typically ranges between 6.0–8.5 [15]; secondly, according to the literature [64], As(III) is better removed at pH values of 4.0–6.0, while As(V) is better removed at pH values of 4.0–8.0. As depicted in Figure 6, by increasing the adsorbent's dosage, the removal rate of As(III) is also increased, from 49% (by 0.5 g/L) to 93% (by 1.5 g/L) at pH 7.0 (Figure 6a). Therefore, pH 7.0 was found to be optimal, as for a dose of 1.5 g/L, the residual As(III) concentration (8.82 μg/L) remains below the legislation limit of 10 μg/L (Figure 6b). pH 7.0 was also found to be the most efficient for As(III) removal according to the literature [38,63,65]. Moreover, the pH_{pzc} (point of zero charge) of Mg-Si-La@AC was previously measured [44] using the pH drift method and measured as 7.72. pH_{pzc} is the critical point at which the surface charge of the adsorbent changes from positive to negative, affecting the adsorbent–adsorbed substance interface. Therefore, at pH < pH_{pzc} (i.e., 7 < 7.22), the surface of Mg-Si-La@AC is positively charged, giving it the ability to attract arsenic anions via electrostatic interaction. Furthermore, at pH 7.0, As(III), with a pK_{a1} of 9.22, exists mainly as non-ionic H_3AsO_3 [66]; therefore, at this pH, As(III) adsorption is less sensitive to pH [67] and the adsorption of non-ionized As(III) occurs only through a ligand exchange reaction [68].

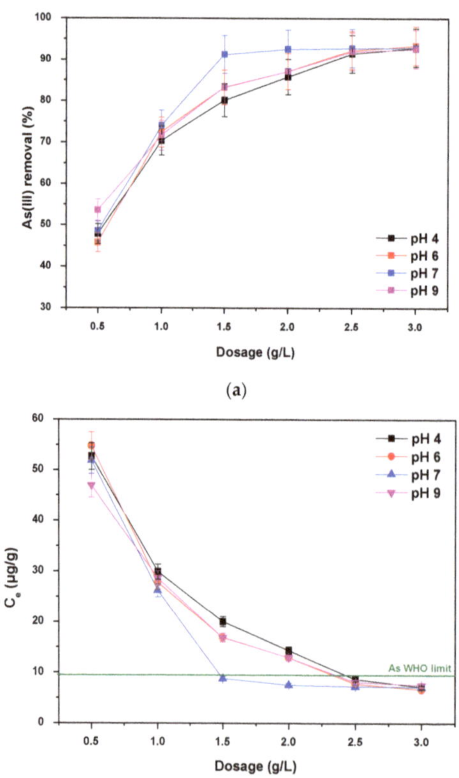

(a)

(b)

Figure 6. Effect of Mg-Si-La@AC dosage and solution pH, in terms of (**a**) removal % and (**b**) C_e (μg/L); C_0 100 μg/L, T = 298 K, 24 h.

3.4. Adsorption Isotherms

Langmuir and Freundlich isotherm models were used in this study to evaluate the adsorption equilibrium data of As(III) on Mg-Si-La@AC at the optimal conditions obtained from the above experiments. These are the most commonly applied models described by two parameters. As presented in Figure 7, the Langmuir isotherm model was found to better fit the adsorption (R^2 = 0.98) and a calculated adsorption capacity (Q_m) of 322 μg/g. On the contrary, the results did not fit well to the Freundlich model (as R^2 = 0.88). Hence, it could be assumed that the monolayer adsorption of As(III) took place on the homogeneous surface of the Mg-Si-La@AC adsorbent (according to the Langmuir model), with the adsorption occurring at a fixed number of sites without lateral interaction between the adsorbent and adsorbate molecule, with the energy of adsorption remaining constant [69]. The comparative fitting parameters (calculated from Equations (3) and (4)) are tabulated in Table 3.

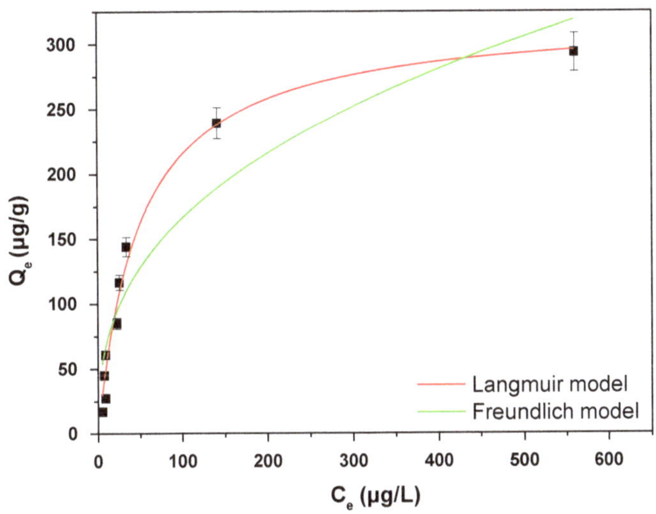

Figure 7. Langmuir and Freundlich isotherm models; dose 1.5 g/L Mg-Si-La@AC (1.5 g/L), pH 7.0, *T* = 298 K, 24 h.

Table 3. Constants of Langmuir and Freundlich isotherm models for the adsorption of As(III) onto Mg-Si-La@AC (1.5 g/L), pH 7.0, *T* = 298 K, 24 h.

	Langmuir Isotherm Model	
Q_m (μg/g)	K_L (L/μg)	R^2
322	0.020	0.9898
	Freundlich isotherm model	
$1/n$	K_F (μg/g)(L/μg)$^{1/n}$	R^2
0.3785	29.04	0.8780

3.5. Effect of Contact Time–Kinetic Models

Figure 8 presents the effect of the contact time varying from 5 to 1440 (24 h). According to the results, to increase the removal rate of As(III) and the relative cost-effectiveness of adsorption, 4 h was considered to be the optimal contact time for the experiments. In addition, adsorption kinetics was used to appreciate the adsorption rate and the possible mechanism controlling the whole process. To interpret the adsorption kinetics of As(III) on AC-Si-Mg-La, the PFO and the PSO kinetics models were applied to evaluate the experimental data (Figure 9). From the results presented in Table 4 it appears that the procedure best followed the PSO model, as the correlation coefficient for the PSO model was calculated as $R^2 = 0.93$ and the relative one for the PFO model was only 0.80. The PSO model was used to define the kinetics of slow adsorption, predicting the long-term adsorption equilibrium. The PSO kinetic model was also found to be the most applicable in recent studies for arsenic removal [14,70].

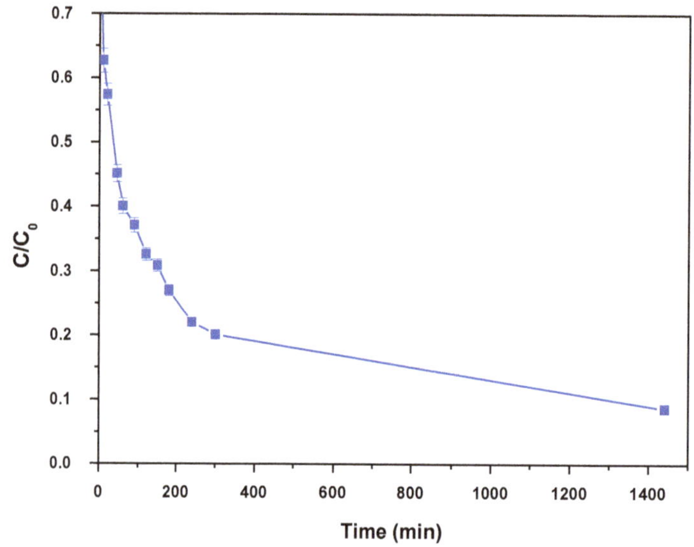

Figure 8. Effect of time on As(III) adsorption onto Mg-Si-La@AC (1.5 g/L), pH 7.0, T = 298 K.

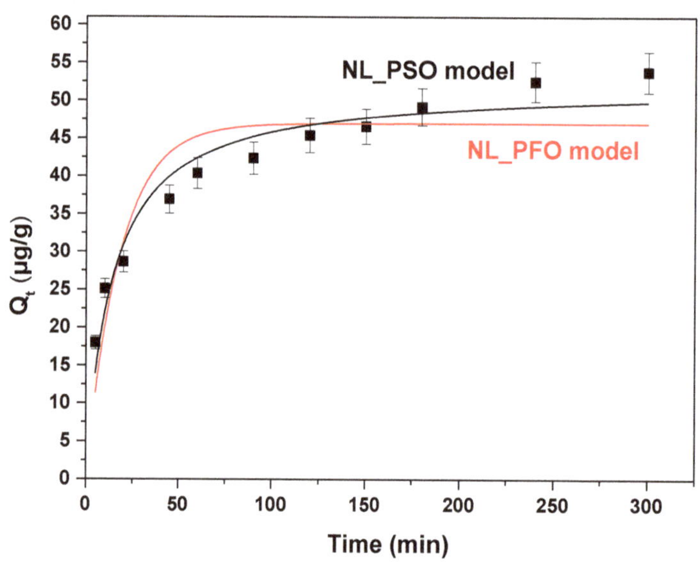

Figure 9. Non-linear kinetic models PFO and PSO for the adsorption of As(III) onto Mg-Si-La@AC (1.5 g/L), pH 7.0, T = 298 K.

Table 4. PFO and PSO kinetic parameters model for the adsorption of As(III) onto Mg-Si-La@AC (1.5 g/L), pH 7.0, T = 298 K.

	Pseudo-First-Order Kinetic Model (PFO)		
$Q_{e \cdot exp}$ (μg/g)	K_1 (L/μg·min)	$Q_{e \cdot cal}$ (μg/g)	R^2
61.47	0.0554	46.96	0.8008
	Pseudo-second order kinetic model (PSO)		
$Q_{e \cdot exp}$ (μg/g)	K_2 (L/μg·min)	$Q_{e \cdot cal}$ (μg/g)	R^2
61.47	0.0014	52.04	0.9318

3.6. Thermodynamics

Temperature is a variable that affects the adsorption process, and its effect on the adsorption of As(III) was observed at different temperatures, such as 298, 308, 318 and 338 K. For the determination of the thermodynamic parameters, the Equations (7–10) were used and the results are presented in Table 5. The values of ΔH^0 and ΔS^0 were obtained by analyzing the plot of $\ln(K_c)$ versus $1/T$ [71], providing a R^2 value of 0.9684. The values of ΔG^0 were calculated to be negative, indicating that the process is spontaneous. The ΔH^0 value for As(III) removal was determined to be 55.114 kJ/mol, demonstrating the endothermic nature of the adsorption process. Moreover, the positive ΔS^0 (0.1916 kJ/mol·K) was due to the increased randomness of adsorbed substances at the interface of the solid and the solute during the adsorption [72].

Table 5. Thermodynamic parameters for the adsorption of As(III) onto Mg-Si-La@AC; C_0 100 μg/L, dose 1.5 g/L, pH 7.0, 3 h.

T (K)	ΔG^0 (kJ/mol)	ΔH^0 (kJ/mol)	ΔS^0 (kJ/mol·K)	R^2
298	−1.969	55.114	0.1916	0.9684
308	−3.884			
318	−5.800			
338	−9.631			

3.7. Regeneration Study

Another important factor in the study of adsorption is the reusability of the adsorbents; thus, recycling studies were carried using similar experimental conditions, which were applied in each cycle (i.e., initial As(III) concentration of 100 μg/L, dose of 1.5 g/L, pH 7.0), using 0.01 M of NaOH. As shown in Figure 10, only a 20% reduction in the adsorption capacity after four regeneration cycles of the process was observed (from 92.8% at the first cycle to 70.6% at the fourth). Therefore, this study showed that the Mg-Si-La@AC adsorbent could be reused for four cycles, with the efficacious regenerate providing an approximately 93% removal of As(III) after the first cycle and approximately 71% removal after the fourth cycle.

3.8. Comparison with Literature

Table 6 provides a comparison between Mg-Si-La@AC and other modified activated carbons found in the literature for removing As(III). A neutral pH (7.0) was found to be optimal in most of the cases analyzed here. The initial As concentration used in most experiments (i.e., 10 mg/L) and the relative dosages of adsorbents (even 50 g/L) were generally much higher than those used in the present study. Regeneration studies showed that the adsorbents could be regenerated and reused up to 4–5 cycles. Finally, the removal rate of As(III) (93%) using Mg-Si-La@AC was high compared to the corresponding literature.

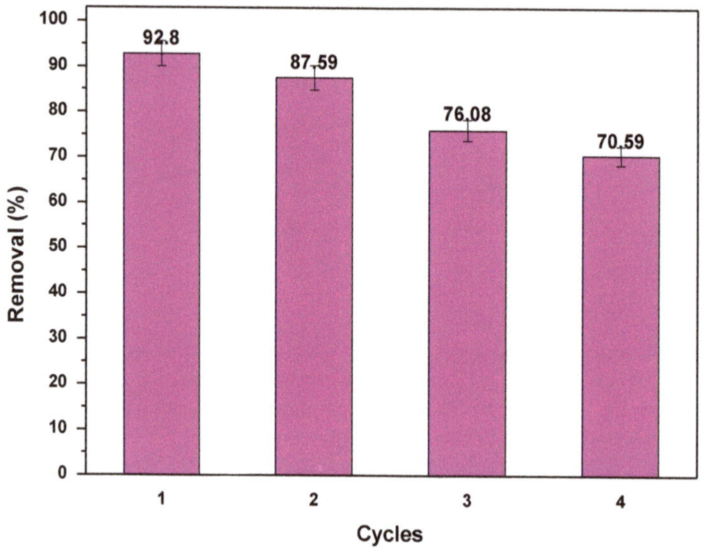

Figure 10. Regeneration study of four adsorption–desorption cycles.

Table 6. Comparison of the adsorption effect of arsenic with other modified activated carbons found in the literature.

Adsorbent	$[As]_0$ (mg/L)	Dosage (g/L)	pH_{init}	Adsorption Capacity (mg/g)	Removal %	Cycles	Ref.
Fe–Zr@AC [1]	1.0	3.0	7.0	1.206	86	5	[73]
AC–Tamarix [2]	10.0	3.0	7.0	37.313	96	-	[74]
$MgO/AC/Fe_3O_4$ [3]	6.0	0.13	7.0	1.166	97	5	[75]
AGSC [4]	0.4	5.0	6.0	0.192	93	5	[76]
ACF [5]	2.0	0.5	6.0	8.651	74	-	[77]
ZrO_2/AC [6]	10.0	50.0	5.0	64.0	90	4	[78]
Mg-Si-La@AC [7]	0.1	1.5	7.0	0.322	93	4	this study

[1] Iron–zirconium-modified activated carbon; [2] Activated carbon adsorbent prepared from Tamarix tree leaves; [3] Fe_3O_4/MgO/Activated carbon composite; [4] Nanoporous activated garlic stem carbon; [5] Fe- and HNO_3-modified activated carbon fiber; [6] Zirconia-modified activated carbon; [7] Magnesium/Silica/Lanthamum@Activated Carbon.

4. Conclusions

The objective of this study was to observe the application of Magnesium/Silica/Lanthamum@Activated Carbon (Mg-Si-La@AC) for the first time in the removal of As(III) from water. BET, FTIR and SEM techniques were used for the characterization of the material. Therefore, the BET surface area, total pore volume, and micropore volume of Mg-Si-La@AC were measured to be 271.46 m^2/g, 0.006 cm^3/g and 0.52 cm^3/g, respectively. According to the results, Mg, Si and La were detected on the activated carbon surface, confirming its amendment. Observing the post-adsorption FTIR spectra, new absorption peaks appeared at 863 cm^{-1} and 738.5 cm^{-1}, and could be attributed to As–O–La and As-O-Mg bonds, indicating that As(III) is adsorbed on the Mg-Si-La@AC surface.

Furthermore, several parameters were examined to evaluate the process. Hence, an optimal dose of 1.5 g/L and pH 7 were selected, resulting in a residual As concentration < 10 μg/L. In addition, 4 h was selected as the optimal time for the experiments and the procedure best followed the PSO kinetic model. The maximum adsorption capacity was determined to be 322 μg/g at pH 7, according to the Langmuir isotherm model. Moreover, thermodynamics studies showed that, by increasing the temperature from 25 to 65 °C, the removal capacity was increased by 80%. Finally, regeneration studies showed that the Mg-Si-La@AC adsorbent could be reused for four cycles with only a 20% decrease in the adsorption capacity.

Author Contributions: Conceptualization, A.K.T. and G.Z.K.; methodology, A.K.T. and G.Z.K.; validation, G.Z.K.; formal analysis, A.K.T. and G.Z.K.; investigation, A.K.T. and G.Z.K.; resources, A.K.T. and G.Z.K.; data curation, A.K.T.; writing—original draft preparation, A.K.T.; writing—review and editing, A.K.T. and G.Z.K.; visualization, A.K.T. and G.Z.K.; supervision, A.K.T. and G.Z.K. All authors have read and agreed to the published version of the manuscript.

Funding: This research received no external funding.

Institutional Review Board Statement: Not applicable.

Informed Consent Statement: Not applicable.

Data Availability Statement: All data analyzed during this study are included in this published article.

Acknowledgments: We acknowledge support of this work by the project "Advanced Nanostructured Materials for Sustainable Growth: Green Energy Production/Storage, Energy Saving and Environmental Remediation" (TAEDR-0535821) which is implemented under the action "Flagship actions in interdisciplinary scientific fields with a special focus on the productive fabric" (ID 16618), Greece 2.0—National Recovery and Resilience Fund and funded by European Union NextGenerationEU. The authors, also thank Olina Makrogianni for taking some characterization measurements.

Conflicts of Interest: The authors declare no conflict of interest.

References

1. Li, P.; Karunanidhi, D.; Subramani, T.; Srinivasamoorthy, K. Sources and Consequences of Groundwater Contamination. *Arch. Environ. Contam. Toxicol.* **2021**, *80*, 1–10. [CrossRef]
2. Bundschuh, J.; Armienta, M.A.; Morales-Simfors, N.; Alam, M.A.; López, D.L.; Delgado Quezada, V.; Dietrich, S.; Schneider, J.; Tapia, J.; Sracek, O.; et al. Arsenic in Latin America: New Findings on Source, Mobilization and Mobility in Human Environments in 20 Countries Based on Decadal Research 2010–2020. *Crit. Rev. Environ. Sci. Technol.* **2020**, *51*, 1727–1865. [CrossRef]
3. Rahaman, M.S.; Mise, N.; Ichihara, S. Arsenic Contamination in Food Chain in Bangladesh: A Review on Health Hazards, Socioeconomic Impacts and Implications. *Hyg. Environ. Health Adv.* **2022**, *2*, 100004. [CrossRef]
4. McKay, G.; Otterburn, M.S.; Sweeney, A.G. The Removal of Colour from Effluent Using Various Adsorbents: Some Preliminary Economic Considerations. *J. Soc. Dye. Colour.* **1978**, *94*, 357–360. [CrossRef]
5. Shaji, E.; Santosh, M.; Sarath, K.V.; Prakash, P.; Deepchand, V.; Divya, B.V. Arsenic Contamination of Groundwater: A Global Synopsis with Focus on the Indian Peninsula. *Geosci. Front.* **2021**, *12*, 101079. [CrossRef]
6. Ur Rehman, H.; Ahmed, S.; Ur Rahman, M.; Mehmood, M.S. Arsenic Contamination, Induced Symptoms, and Health Risk Assessment in Groundwater of Lahore, Pakistan. *Environ. Sci. Pollut. Res.* **2022**, *29*, 49796–49807. [CrossRef]
7. Subhani, M.; Mustafa, I.; Alamdar, A.; Katsoyiannis, I.A.; Ali, N.; Huang, Q.; Peng, S.; Shen, H.; Eqani, S.A.M.A.S. Arsenic Levels from Different Land-Use Settings in Pakistan: Bio-Accumulation and Estimation of Potential Human Health Risk via Dust Exposure. *Ecotoxicol. Environ. Saf.* **2015**, *115*, 187–194. [CrossRef] [PubMed]
8. Pezeshki, H.; Hashemi, M.; Rajabi, S. Removal of Arsenic as a Potentially Toxic Element from Drinking Water by Filtration: A Mini Review of Nanofiltration and Reverse Osmosis Techniques. *Heliyon* **2023**, *9*, e14246. [CrossRef]
9. Mazumder, D.N.G. Health Effects of Chronic Arsenic Toxicity. In *Handbook of Arsenic Toxicology*; Elsevier: Amsterdam, The Netherlands, 2023; pp. 151–192. [CrossRef]
10. Fatoki, J.O.; Badmus, J.A. Arsenic as an Environmental and Human Health Antagonist: A Review of Its Toxicity and Disease Initiation. *J. Hazard. Mater. Adv.* **2022**, *5*, 100052. [CrossRef]
11. Genchi, G.; Lauria, G.; Catalano, A.; Carocci, A.; Sinicropi, M.S. Arsenic: A Review on a Great Health Issue Worldwide. *Appl. Sci.* **2022**, *12*, 6184. [CrossRef]
12. World Health Organization European Standards for Drinking-Water. *Am. J. Med. Sci.* **1970**, *242*, 56.
13. Jadhav, S.V.; Bringas, E.; Yadav, G.D.; Rathod, V.K.; Ortiz, I.; Marathe, K.V. Arsenic and Fluoride Contaminated Groundwaters: A Review of Current Technologies for Contaminants Removal. *J. Environ. Manag.* **2015**, *162*, 306–325. [CrossRef]
14. Katsoyiannis, I.A.; Tzollas, N.M.; Tolkou, A.K.; Mitrakas, M.; Ernst, M.; Zouboulis, A.I. Use of Novel Composite Coagulants for Arsenic Removal from Waters-Experimental Insight for the Application of Polyferric Sulfate (PFS). *Sustainability* **2017**, *9*, 590. [CrossRef]
15. Meez, E.; Tolkou, A.K.; Giannakoudakis, D.A.; Katsoyiannis, I.A.; Kyzas, G.Z. Activated Carbons for Arsenic Removal from Natural Waters and Wastewaters: A Review. *Water* **2021**, *13*, 2982. [CrossRef]

16. Cañas Kurz, E.E.; Hellriegel, U.; Figoli, A.; Gabriele, B.; Bundschuh, J.; Hoinkis, J. Small-Scale Membrane-Based Arsenic Removal for Decentralized Applications–Developing a Conceptual Approach for Future Utilization. *Water Res.* **2021**, *196*, 116978. [CrossRef]
17. Worou, C.N.; Chen, Z.L.; Bacharou, T. Arsenic Removal from Water by Nanofiltration Membrane: Potentials and Limitations. *Water Pract. Technol.* **2021**, *16*, 291–319. [CrossRef]
18. Nur, T.; Loganathan, P.; Ahmed, M.B.; Johir, M.A.H.; Nguyen, T.V.; Vigneswaran, S. Removing Arsenic from Water by Coprecipitation with Iron: Effect of Arsenic and Iron Concentrations and Adsorbent Incorporation. *Chemosphere* **2019**, *226*, 431–438. [CrossRef] [PubMed]
19. Zama, E.F.; Li, G.; Tang, Y.T.; Reid, B.J.; Ngwabie, N.M.; Sun, G.X. The Removal of Arsenic from Solution through Biochar-Enhanced Precipitation of Calcium-Arsenic Derivatives. *Environ. Pollut.* **2022**, *292*, 118241. [CrossRef]
20. Dudek, S.; Kołodyńska, D. Arsenate Removal on the Iron Oxide Ion Exchanger Modified with Neodymium(III) Ions. *J. Environ. Manag.* **2022**, *307*, 114551. [CrossRef]
21. Saleh, T.A.; Mustaqeem, M.; Khaled, M. Water Treatment Technologies in Removing Heavy Metal Ions from Wastewater: A Review. *Environ. Nanotechnol. Monit. Manag.* **2022**, *17*, 100617. [CrossRef]
22. Shankar, S.; Shanker, U. Shikha Arsenic Contamination of Groundwater: A Review of Sources, Prevalence, Health Risks, and Strategies for Mitigation. *Sci. World J.* **2014**, *2014*, 304524. [CrossRef] [PubMed]
23. Pang, X.; Sellaoui, L.; Franco, D.; Dotto, G.L.; Georgin, J.; Bajahzar, A.; Belmabrouk, H.; Ben Lamine, A.; Bonilla-Petriciolet, A.; Li, Z. Adsorption of Crystal Violet on Biomasses from Pecan Nutshell, Para Chestnut Husk, Araucaria Bark and Palm Cactus: Experimental Study and Theoretical Modeling via Monolayer and Double Layer Statistical Physics Models. *Chem. Eng. J.* **2019**, *378*, 122101. [CrossRef]
24. Wang, X.; Xu, Q.; Zhang, L.; Pei, L.; Xue, H.; Li, Z. Adsorption of Methylene Blue and Congo Red from Aqueous Solution on 3D MXene/Carbon Foam Hybrid Aerogels: A Study by Experimental and Statistical Physics Modeling. *J. Environ. Chem. Eng.* **2023**, *11*, 109206. [CrossRef]
25. Wang, X.; Zhang, A.; Chen, M.; Seliem, M.K.; Mobarak, M.; Diao, Z.; Li, Z. Adsorption of Azo Dyes and Naproxen by Few-Layer MXene Immobilized with Dialdehyde Starch Nanoparticles: Adsorption Properties and Statistical Physics Modeling. *Chem. Eng. J.* **2023**, *473*, 145385. [CrossRef]
26. Bilici Baskan, M.; Pala, A. Removal of Arsenic from Drinking Water Using Modified Natural Zeolite. *Desalination* **2011**, *281*, 396–403. [CrossRef]
27. Losev, V.N.; Didukh-Shadrina, S.L.; Orobyeva, A.S.; Metelitsa, S.I.; Borodina, E.V.; Ondar, U.V.; Nesterenko, P.N.; Maznyak, N.V. A New Method for Highly Efficient Separation and Determination of Arsenic Species in Natural Water Using Silica Modified with Polyamines. *Anal. Chim. Acta* **2021**, *1178*, 338824. [CrossRef]
28. Jacobson, A.T.; Fan, M. Evaluation of Natural Goethite on the Removal of Arsenate and Selenite from Water. *J. Environ. Sci.* **2019**, *76*, 133–141. [CrossRef] [PubMed]
29. Ma, M.; Ha, Z.; Lv, C.; Xu, X.; Li, C.; Du, D.; Zhang, T.C.; Chi, R. A Novel Passivator Based on Electrolytic Manganese Residues and Calcite for Arsenic Sorption and Heavy Metal Passivation of Contaminated Soil. *J. Clean. Prod.* **2023**, *414*, 137544. [CrossRef]
30. Wang, Y.; Liu, H.; Wang, S.; Li, X.; Wang, X.; Jia, Y. Simultaneous Removal and Oxidation of Arsenic from Water by δ-MnO2 Modified Activated Carbon. *J. Environ. Sci.* **2020**, *94*, 147–160. [CrossRef]
31. Yao, S.; Liu, Z.; Shi, Z. Arsenic Removal from Aqueous Solutions by Adsorption onto Iron Oxide/Activated Carbon Magnetic Composite. *J. Environ. Health Sci. Eng.* **2014**, *12*, 6–13. [CrossRef] [PubMed]
32. Mondal, M.K.; Garg, R. A Comprehensive Review on Removal of Arsenic Using Activated Carbon Prepared from Easily Available Waste Materials. *Environ. Sci. Pollut. Res.* **2017**, *24*, 13295–13306. [CrossRef] [PubMed]
33. Shakoor, M.B.; Niazi, N.K.; Bibi, I.; Murtaza, G.; Kunhikrishnan, A.; Seshadri, B.; Shahid, M.; Ali, S.; Bolan, N.S.; Ok, Y.S.; et al. Remediation of Arsenic-Contaminated Water Using Agricultural Wastes as Biosorbents. *Crit. Rev. Environ. Sci. Technol.* **2016**, *46*, 467–499. [CrossRef]
34. Nagra, M.A.; Natasha, N.; Bibi, I.; Tariq, T.Z.; Naz, R.; Ansar, S.; Shahid, M.; Murtaza, B.; Imran, M.; Khalid, M.S.; et al. Biowaste-Based Sorbents for Arsenic Removal from Aqueous Medium and Risk Assessment. *Environ. Geochem. Health* **2022**. [CrossRef] [PubMed]
35. Chatzimichailidou, S.; Xanthopoulou, M.; Tolkou, A.K.; Katsoyiannis, I.A. Biochar Derived from Rice By-Products for Arsenic and Chromium Removal by Adsorption: A Review. *J. Compos. Sci.* **2023**, *7*, 59. [CrossRef]
36. Tolkou, A.K.; Rada, E.C.; Torretta, V.; Xanthopoulou, M.; Kyzas, G.Z.; Katsoyiannis, I.A. Removal of Arsenic(III) from Water with a Combination of Graphene Oxide (GO) and Granular Ferric Hydroxide (GFH) at the Optimum Molecular Ratio. *C* **2023**, *9*, 10. [CrossRef]
37. Xanthopoulou, M.; Katsoyiannis, I.A. Enhanced Adsorption of Chromate and Arsenate Ions from Contaminated Water with Emphasis on Polyethylenimine Modified Materials: A Review. *Separations* **2023**, *10*, 441. [CrossRef]
38. Samimi, M.; Zakeri, M.; Alobaid, F.; Aghel, B. A Brief Review of Recent Results in Arsenic Adsorption Process from Aquatic Environments by Metal-Organic Frameworks: Classification Based on Kinetics, Isotherms and Thermodynamics Behaviors. *Nanomaterials* **2023**, *13*, 60. [CrossRef]
39. Xu, W.; Wang, J.; Wang, L.; Sheng, G.; Liu, J.; Yu, H.; Huang, X.J. Enhanced Arsenic Removal from Water by Hierarchically Porous CeO_2-ZrO_2 Nanospheres: Role of Surface- and Structure-Dependent Properties. *J. Hazard. Mater.* **2013**, *260*, 498–507. [CrossRef]

40. Tolkou, A.K.; Kyzas, G.Z.; Katsoyiannis, I.A. Arsenic(III) and Arsenic(V) Removal from Water Sources by Molecularly Imprinted Polymers (MIPs): A Mini Review of Recent Developments. *Sustainability* **2022**, *14*, 5222. [CrossRef]
41. Koutzaris, S.; Xanthopoulou, M.; Laskaridis, A.; Katsoyiannis, I.A. H2O2-Enhanced As(III) Removal from Natural Waters by Fe(III) Coagulation at Neutral PH Values and Comparison with the Conventional Fe(II)-H_2O_2 Fenton Process. *Sustainability* **2022**, *14*, 6306. [CrossRef]
42. Zheng, Q.; Hou, J.; Hartley, W.; Ren, L.; Wang, M.; Tu, S.; Tan, W. As(III) Adsorption on Fe-Mn Binary Oxides: Are Fe and Mn Oxides Synergistic or Antagonistic for Arsenic Removal? *Chem. Eng. J.* **2020**, *389*, 124470. [CrossRef]
43. Tolkou, A.K.; Trikalioti, S.; Makrogianni, O.; Xanthopoulou, M.; Deliyanni, E.A.; Kyzas, G.Z.; Katsoyiannis, I.A. Lanthanum Modified Activated Carbon from Coconut Shells for Chromium (VI) Removal from Water. *Nanomaterials* **2022**, *12*, 1067. [CrossRef] [PubMed]
44. Tolkou, A.K.; Trikalioti, S.; Makrogianni, O.; Katsoyiannis, I.A.; Kyzas, G.Z. Magnesium Modified Activated Carbons Derived from Coconut Shells for the Removal of Fluoride from Water. *Sustain. Chem. Pharm.* **2023**, *31*, 100898. [CrossRef]
45. Sujiono, E.H.; Zabrian, D.; Zharvan, V.; Humairah, N.A. Fabrication and Characterization of Coconut Shell Activated Carbon Using Variation Chemical Activation for Wastewater Treatment Application. *Results Chem.* **2022**, *4*, 100291. [CrossRef]
46. Kyzas, G.Z.; Deliyanni, E.A. Modified Activated Carbons from Potato Peels as Green Environmental-Friendly Adsorbents for the Treatment of Pharmaceutical Effluents. *Chem. Eng. Res. Des.* **2015**, *97*, 135–144. [CrossRef]
47. Heidarinejad, Z.; Dehghani, M.H.; Heidari, M.; Javedan, G.; Ali, I.; Sillanpää, M. Methods for Preparation and Activation of Activated Carbon: A Review. Environ. *Chem. Lett.* **2020**, *18*, 393–415. [CrossRef]
48. Sultana, M.; Rownok, M.H.; Sabrin, M.; Rahaman, M.H.; Alam, S.M.N. A Review on Experimental Chemically Modified Activated Carbon to Enhance Dye and Heavy Metals Adsorption. *Clean. Eng. Technol.* **2022**, *6*, 100382. [CrossRef]
49. Wang, Y.; Liu, Y.; Guo, T.; Liu, H.; Li, J.; Wang, S.; Li, X.; Wang, X.; Jia, Y. Lanthanum Hydroxide: A Highly Efficient and Selective Adsorbent for Arsenate Removal from Aqueous Solution. *Environ. Sci. Pollut. Res.* **2020**, *27*, 42868–42880. [CrossRef] [PubMed]
50. Sawana, R.; Somasundar, Y.; Iyer, V.S.; Baruwati, B. Ceria Modified Activated Carbon: An Efficient Arsenic Removal Adsorbent for Drinking Water Purification. *Appl. Water Sci.* **2017**, *7*, 1223–1230. [CrossRef]
51. Merodio-Morales, E.E.; Reynel-Ávila, H.E.; Mendoza-Castillo, D.I.; Duran-Valle, C.J.; Bonilla-Petriciolet, A. Lanthanum- and Cerium-Based Functionalization of Chars and Activated Carbons for the Adsorption of Fluoride and Arsenic Ions. *Int. J. Environ. Sci. Technol.* **2020**, *17*, 115–128. [CrossRef]
52. Yanagisawa, H.; Matsumoto, Y.; Machida, M. Adsorption of Zn(II) and Cd(II) Ions onto Magnesium and Activated Carbon Composite in Aqueous Solution. *Appl. Surf. Sci.* **2010**, *256*, 1619–1623. [CrossRef]
53. Choong, C.E.; Ibrahim, S.; Yoon, Y.; Jang, M. Removal of Lead and Bisphenol A Using Magnesium Silicate Impregnated Palm-Shell Waste Powdered Activated Carbon: Comparative Studies on Single and Binary Pollutant Adsorption. *Ecotoxicol. Environ. Saf.* **2018**, *148*, 142–151. [CrossRef]
54. Irunde, R.; Ligate, F.J.; Ijumulana, J.; Ahmad, A.; Maity, J.P.; Hamisi, R.; Philip, J.Y.N.; Kilulya, K.F.; Karltun, E.; Mtamba, J.; et al. The Natural Magnesite Efficacy on Arsenic Extraction from Water and Alkaline Influence on Metal Release in Water. *Appl. Geochem.* **2023**, *155*, 105705. [CrossRef]
55. Sugita, H.; Oguma, T.; Hara, J.; Zhang, M.; Kawabe, Y. Removal of Arsenate from Contaminated Water via Combined Addition of Magnesium-Based and Calcium-Based Adsorbents. *Sustainability* **2023**, *15*, 4689. [CrossRef]
56. Swenson, H.; Stadie, N.P. Langmuir's Theory of Adsorption: A Centennial Review. *Langmuir* **2019**, *35*, 5409–5426. [CrossRef]
57. Freundlich, H. Über Die Adsorption in Lösungen. *Z. Phys. Chem.* **1907**, *57U*, 385–470. [CrossRef]
58. Smith, A.H.; Steinmaus, C.M. Health Effects of Arsenic and Chromium in Drinking Water: Recent Human Findings. *Annu. Rev. Public Health* **2009**, *30*, 107–122. [CrossRef]
59. Sotomayor, F.; Quantatec, A.P.; Sotomayor, F.J.; Cychosz, K.A.; Thommes, M. Characterization of Micro/Mesoporous Materials by Physisorption: Concepts and Case Studies. *Acc. Mater. Surf. Res.* **2018**, *3*, 34–50.
60. Nasri, N.S.; Jibril, M.; Zaini, M.A.A.; Mohsin, R.; Dadum, H.U.; Musa, A.M. Synthesis and Characterization of Bio-Based Porous Carbons by Two Step Physical Activation with CO_2. *J. Teknol.* **2014**, *68*, 5–9. [CrossRef]
61. Jais, F.M.; Ibrahim, S.; Yoon, Y.; Jang, M. Enhanced Arsenate Removal by Lanthanum and Nano-Magnetite Composite Incorporated Palm Shell Waste-Based Activated Carbon. *Sep. Purif. Technol.* **2016**, *169*, 93–102. [CrossRef]
62. Dudek, S.; Kołodyńska, D. Arsenic(V) Removal on the Lanthanum-Modified Ion Exchanger with Quaternary Ammonium Groups Based on Iron Oxide. *J. Mol. Liq.* **2022**, *347*, 117985. [CrossRef]
63. Mehanathan, S.; Jaafar, J.; Nasir, A.M.; Ismail, A.F.; Matsuura, T.; Othman, M.H.D.; Rahman, M.A.; Yusof, N. Magnesium Oxide Nanoparticles for the Adsorption of Pentavalent Arsenic from Water: Effects of Calcination. *Membranes* **2023**, *13*, 475. [CrossRef]
64. Mudzielwana, R.; Gitari, M.W. Removal of Fluoride from Groundwater Using MnO_2 Bentonite-Smectite Rich Clay Soils Composite. Groundw. *Sustain. Dev.* **2021**, *14*, 100623. [CrossRef]
65. Tolkou, A.K.; Trikkaliotis, D.G.; Kyzas, G.Z.; Katsoyiannis, I.A.; Deliyanni, E.A. Simultaneous Removal of As(III) and Fluoride Ions from Water Using Manganese Oxide Supported on Graphene Nanostructures (GO-MnO_2). *Sustainability* **2023**, *15*, 1179. [CrossRef]
66. Katsoyiannis, I.A.; Voegelin, A.; Zouboulis, A.I.; Hug, S.J. Enhanced As(III) Oxidation and Removal by Combined Use of Zero Valent Iron and Hydrogen Peroxide in Aerated Waters at Neutral PH Values. *J. Hazard. Mater.* **2015**, *297*, 1–7. [CrossRef] [PubMed]

67. Banerjee, K.; Nour, S.; Selbie, M.; Prevost, M.; Blumenschein, C.D.; Chen, H.; Amy, G.L. *Optimization of Process Parameters for Arsenic Treatment with Granular Ferric Hydroxide*; American Water Works Association: Denver, CO, USA, 2003.
68. Banerjee, S.; Chattopadhyaya, M.C. Adsorption Characteristics for the Removal of a Toxic Dye, Tartrazine from Aqueous Solutions by a Low Cost Agricultural by-Product. *Arab. J. Chem.* **2017**, *10*, S1629–S1638. [CrossRef]
69. Pourhakkak, P.; Taghizadeh, A.; Taghizadeh, M.; Ghaedi, M.; Haghdoust, S. Fundamentals of Adsorption Technology. *Interface Sci. Technol.* **2021**, *33*, 1–70. [CrossRef]
70. Yin, Y.; Zhou, T.; Luo, H.; Geng, J.; Yu, W.; Jiang, Z. Adsorption of Arsenic by Activated Charcoal Coated Zirconium-Manganese Nanocomposite: Performance and Mechanism. *Colloids Surf. A Physicochem. Eng. Asp.* **2019**, *575*, 318–328. [CrossRef]
71. Kyzas, G.Z.; Christodoulou, E.; Bikiaris, D.N. Basic Dye Removal with Sorption onto Low-Cost Natural Textile Fibers. *Processes* **2018**, *6*, 166. [CrossRef]
72. Chen, W.; Luan, J.; Yu, X.; Wang, X.; Ke, X. Preparation of Core-Shell Structured Polystyrene @ Graphene Oxide Composite Microspheres with High Adsorption Capacity and Its Removal of Dye Contaminants. *Environ. Technol.* **2021**, *42*, 3840–3851. [CrossRef]
73. Sahu, N.; Singh, J.; Koduru, J.R. Removal of Arsenic from Aqueous Solution by Novel Iron and Iron–Zirconium Modified Activated Carbon Derived from Chemical Carbonization of Tectona Grandis Sawdust: Isotherm, Kinetic, Thermodynamic and Breakthrough Curve Modelling. *Environ. Res.* **2021**, *200*, 111431. [CrossRef]
74. Koohzad, E.; Jafari, D.; Esmaeili, H. Adsorption of Lead and Arsenic Ions from Aqueous Solution by Activated Carbon Prepared from Tamarix Leaves. *ChemistrySelect* **2019**, *4*, 12356–12367. [CrossRef]
75. Esmaeili, H.; Mousavi, S.M.; Hashemi, S.A.; Chiang, W.H.; Ahmadpour Abnavi, S. Activated Carbon@MgO@Fe3O4 as an Efficient Adsorbent for As (III) Removal. *Carbon Lett.* **2021**, *31*, 851–862. [CrossRef]
76. Prajapati, A.K.; Mondal, M.K. Hazardous As(III) Removal Using Nanoporous Activated Carbon of Waste Garlic Stem as Adsorbent: Kinetic and Mass Transfer Mechanisms. *Korean J. Chem. Eng.* **2019**, *36*, 1900–1914. [CrossRef]
77. Shi, J.; Zhao, Z.; Zhou, J.; Sun, T.; Liang, Z. Enhanced Adsorption of As(III) on Chemically Modified Activated Carbon Fibers. *Appl. Water Sci.* **2019**, *9*, 41. [CrossRef]
78. Chuc, P.N.; Bac, N.Q.; Thao, D.T.P.; Kien, N.T.; Chi, N.T.H.; Noi, N.V.; Nguyen, V.T.; Bich, N.T.H.; Nhiem, D.N.; Khieu, D.Q. Highly Efficient Adsorption of Arsenite from Aqueous by Zirconia Modified Activated Carbon. *Environ. Eng. Res.* **2024**, *29*, 230070–230076. [CrossRef]

environments

MDPI

Article

Assessment of the Impact of War on Concentrations of Pollutants and Heavy Metals and Their Seasonal Variations in Water and Sediments of the Tigris River in Mosul/Iraq

Zena Altahaan [1,*] and Daniel Dobslaw [2,*]

[1] Institute of Sanitary Engineering, Water Quality and Solid Waste Management, University of Stuttgart, D-70569 Stuttgart, Germany

[2] Department of Biological Waste Air Purification, Institute of Sanitary Engineering, Water Quality and Solid Waste Management, University of Stuttgart, Bandtäle 2, D-70569 Stuttgart, Germany

* Correspondence: zenafahkrey@yahoo.com (Z.A.); daniel.dobslaw@iswa.uni-stuttgart.de (D.D.)

Abstract: The war-related contamination of water and sediment of the Tigris River within the urban area of Mosul leads to seasonally independent exceedances of the WHO limit values for Cd, Pb, Cr, and Ni in water and sediments. Furthermore, exceedances consistently occur for conductivity, PO_4^{3-}, and SO_4^{2-}, as well as sporadically for salinity and COD in water samples, and consistently for salinity in sediment samples, highlighting the direct impact of war (ammunition, ignition of sulfur fields), as well as indirect effects (destroyed wastewater infrastructure). Conflict-related emissions from the former conflict zone (S5–S7) are highlighted by the sudden increases in load from S4 to S5, although partially masked by the discharge of highly polluted water from the Khosr River (between S3 and S4). Due to the sorption of sediments and the presumed wind-borne discharge of highly polluted particles into the Tigris River, sediments at S10 on the southern edge of Mosul showed the highest pollutant loads. Significant statistical differences were observed through T-test analyses for E.C., TDS, salinity, COD, PO_4^{3-}, NO_3^-, SO_4^{2-}, Cd, Pb, Zn, Cr, and Ni for water samples, as well as salinity, Cd, Pb, Zn, and Cr for the sediment samples for seasonal comparison. Since the percentage difference of water samples at S4–S7 is smaller than upstream and downstream, contaminant input is not limited to rainwater but also occurs via the year-round infiltration of highly polluted wastewater from the surrounding valleys or suburban areas, as well as presumably polluted groundwater or windblown particulate input.

Keywords: war impact; heavy metals; WHO limit values; contamination; Tigris River; Mosul

1. Introduction

Over the centuries, wars have always brought disasters to humans and the environment, but the environment remains the silent victim [1]. Such environmental damage and destruction can be divided into three categories.

Environmental damage is caused by pre-war activities to extract resources for direct warfare or to finance the conflicts, e.g., oil, sulfur, mineral resources, or timber [2]. Low-tech processing methods are often used by competing armed groups to minimize investment costs and the need for skilled labor. As a result of low-tech processing, heavy metals such as mercury, lead, or cadmium can be released into the environment and pollute water bodies [3].

Environmental damage occurs due to direct acts of war, such as the destruction of agricultural land and specialty crops, as well as the fate of destroyed weapons and munitions remnants in the environment [2].

Anarchic conditions, or the proclamation of martial law, result in a reduced control and monitoring function of state environmental authorities, leading to private companies operating in conflict zones being subject to minimal environmental oversight [4].

In terms of population and area, Mosul is the second largest city in Iraq after Baghdad. The Tigris River is one of the most important rivers in the region, as it is the main source of water for drinking, industry, agriculture, entertainment, and public purposes in the city [5].

ISIS held the city from June 2014 to July 2017. The liberation battles, which lasted from October 2016 to July 2017, resulted in more than 40,000 casualties and the complete destruction of 9518 buildings (of which approximately 80% were residential), according to UN reports. The resulting debris in Mosul is estimated at 11 million tons. This mass is equivalent to three times the Great Pyramid of Giza or four times the Eiffel Tower [6,7].

The most severe pollution in the urban area, in addition to the direct use of weapons and resulting explosions, was caused by the burning of both 18 oil wells by ISIS in Qarrayah and their nine-month fire (soot) and 50,000 tons of pure sulfur in the Mishraq complex (SO_2, SO_3) [8].The release of these contaminants had a direct impact on water, groundwater, soil, and air and led to an accumulation of these contaminants over time. The consequences are clearly evident today in the quality of water and sediment of the Tigris River [5,9], especially in direct comparison with pre-war studies.

In the pre-war years, the water quality of the Tigris River met the limits set by Iraq and the World Health Organization (WHO) for drinking water and domestic use, although a wide variety of tributaries and activities resulted in the discharge of pollutants. For example, the drainage of many valleys to Lake Mosul, the discharges of the tributaries of the Al-Mur Valley, the Northern Aljazeera Irrigation Project, and the sulfur springs of the Badoosh Dam Project led to increased eutrophication and discharge of sulfur compounds. Furthermore, discharges of hazardous industrial wastes (e.g., cement industry), solid buffalo wastes (min 10 tons dry matter/day) in the Badoosh area, and car maintenance, car washing, livestock washing, and livestock treatment facilities located on both sides of the river bank resulted in degradation of the water quality of the Tigris River in terms of the parameters of hardness, salinity, nutrients, organic loads, and algal growth [10].

In 2012, a study [11] investigated the variations in pollutant concentrations in the Tigris River within the Mosul urban area. The water qualities determined were, on average, equivalent to that of clean water, with 17% of the test samples classified as being polluted. Although the river experienced pollutant discharges from many sources such as households, industry, and agriculture, the dilution and self-purification potential of the river was sufficiently high, especially at high flow rates, to sustain good water quality in the Tigris. The WQI average was 83% (good water quality). In a later study by [12], the authors addressed the water quality of the Tigris River upstream and downstream of Mosul city to elucidate the impact of wastewater discharges in Mosul during the period from 2002 to 2012. The detected nutrient and salinity loads were harmless to the growth and reproduction of all aquatic hydrophyte species. However, the organic loads, as well as the rotting of aquatic plants enhanced by eutrophication, resulted in an increased odor, a change in color and taste, and a significant increase in salinity.

Ref. [13] conducted an assessment of the water quality of the Tigris River in the urban area of Mosul for drinking water and domestic use using the CCME WQI based on fieldwork from 2008 to 2014. For the ten selected parameters, namely pH, turbidity, and concentrations of calcium, dissolved oxygen, dissolved solids, nitrate, chloride, phosphate, and sulfate, the Tigris showed water quality during the study period ranging from 66.3 to 93.7%, i.e., from moderate to good.

A study conducted in 2011–2012 by [14] of heavy metal concentrations in water and sediments within the Mosul urban area showed that heavy metal concentrations in both compartments were within the normal concentration ranges of Iraq, with concentrations in sediments always being higher than in water. The concentrations in water followed the descending order of Zn > Cu > Pb > Cd, while in sediment, a concentration order of Cu > Pb > Zn > Cd was observed. The results revealed a significant increase in the concentration of these heavy metals in both the aqueous phase and sediments in the area of Al-Busaif south of Mosul compared with the control in Al-Mushirfa area in the northwestern part of Mosul.

The importance of heavy metal pollution, besides acute toxicity, lies particularly in its ability to exhibit bioaccumulation in living organisms [15]), even though heavy metals and other contaminants lead to a deterioration in water quality and thus in the health status of the population.

The previous study indicated that there is no comprehensive overview of scientific evidence on the impact of armed conflicts on water resources and their management [16], therefore, aimed to quantify the effect of heavy metal inputs resulting from the war on the water and sediment of the Tigris River within the urban area of Mosul, taking into account seasonal variations (test series winter and spring vs. summer and fall) and comparing measurement points upstream and downstream of Mosul, and to compare them with existing global limits. A comparison of the measured values was made considering the percentage difference as well as the *T*-test approach.

2. Materials and Methods

2.1. Investigation Area

The Tigris River basin is located in the city of Mosul between northwest Mosul at Al Kuba and southeast Mosul at Hammam Al leel within the investigation area, with sampling points as shown in Table 1. The total length of the Tigris River in Mosul is about 214 Km, the width is between 50 and 200 m, the depth is around 50 m, and the river gradient is 1:2000. The basin area is 375,000 km^2. The Tigris River annually discharges 250–400 m^3/s. According to [17], the river bed consists of two layers: a surface layer of gravel with an average diameter of 32 mm and a subsurface layer of gravel and sand with an average diameter of 13 mm. The disc shape is dominant among the gravel grains, followed by flat and spherical [18].

Table 1. Coordinates and description of all study sites on the Tigris River in Mosul.

Site No.	Location Names	Latitude N	Longitude E	Describtion
S1	AlKuba	36.398501	43.074622	Residentioal area
S2	Alrashidia	36.39606	43.114278	Agricultural area
S3	Third bridge	36.363435	43.114976	Recreation area
S4	Finfth bridge	36.35485	43.12616	Recreation Area
S5	Old bridge	36.345667	43.136953	Commercial + Residential
S6	Aljumhoria bridge	36.340904	43.145025	Residential area
S7	Fourth bridge	36.332283	43.152415	Residential area
S8	Jarimjah	36.298735	43.176358	Agricultural area
S9	Albosaef	36.277681	43.163921	Agricultural area
S10	Hammam Al aleel	36.160292	43.263505	Residential area

2.2. Sampling Procedure

Thirty water and sediment samples were each taken at ten points on the Tigris (see Table 1) in triplicate during the two test series (series 1: January to March 2022; series 2: July to September 2022) using clean polyethylene bottles with a capacity of 250 mL, resulting in a total sample volume of 120 samples. The sample temperatures during sampling were 0–10 °C during the first test series, and they were between 35–45 °C in the second test series. The water flow rate during the period of the study was about 250–300 m^3/s.

Each sample was directly stored in a refrigerated container at −4 °C after sampling, transported directly to the laboratory, and further analyzed for 13 parameters: pH, electric conductivity (E.C.), % salinity, total dissolved solids (TDS), chemical oxygen demand (COD), SO_4^{2-}, PO_4^{3-}, NO_3^-, Pb, Zn, Cd, Cr, and Ni. The collection and analysis of the samples in the environmental laboratory was carried out according to standard methods [19].

Water samples were taken according to DIN EN ISO 5667-6. During sampling from the river bank, special attention was paid to ensure that sampling was performed from the homogeneous layer of the water body without picking up surface films, dead water, or swirling bottom sediments. Sampling was performed at mid-height between the riverbed and the surface as shown in Figure 1.

Figure 1. Map of lighted sampling locations.

Sediment samples were taken according to DIN ISO 5667-12 from the upper river sediment layer from a sediment depth of about 0–30 cm with a clean plastic scoop and placed in plastic bags. Sampling took place at the same 10 locations along the river as sampling for water samples.

2.3. *Sediment Sample Preparation*

Water samples were directly analyzed according to standard methods after membrane filtration. The sediment samples were dried, and their heavy metal concentrations were determined according to the method described in Jackson [20] in which 0.5 g of dry soil was placed in a glass flask and 5 mL of the digestion solution, consisting of concentrated sulfuric acid (H_2SO_4), concentrated nitric acid (HNO_3), and perchloric acid ($HClO_4$), was added at a ratio of 3:1:1. The samples were then heated on a heater at 90 °C in a hood for two hours. After the samples cooled down, they were diluted to 25 mL through the addition of distilled water, and the respective heavy metal concentrations were quantified in mg/kg dry weight of the soil using an atomic absorption spectrometer.

2.4. *Field Measurement Parameters*

Temperature, pH value, E.C., TDS, and % salinity of water samples were directly analyzed on-site using an Oumefar 5-in-1 digital water quality monitor analyzer of type UPC 886108495111.

2.5. *Laboratory Measurement Parameters*

The parameters described in Section 2.4 were measured for sediment samples by mixing the sample with distilled water (about 10–15 g of sample in 25 mL of distilled water). After 2 to 3 min, until the soil was well loosened, the pH, E.C., and salinity of the solution were measured electrometrically with the Oumefar 5-in-1 analyzer. All of the following parameters were analyzed according to [19].

Phosphate PO_4^{3-} (in mg/L) was photometrical analyzed without digestion at 690 nm as phosphorus molybdenum blue using ammonium molybdate solution according to DIN EN 1189.

Nitrate NO_3^- (in mg/L) was also photometrically analyzed at 324 nm after acidification with HCl of the pre-diluted samples using a colorimetric reaction of nitrate with 2,6-dimethylphenole to 4-nitro-2,6-dimethylphenole according to DIN 38405-D9-2.

Sulfate SO_4^{2-} (in mg/L) was quantified through precipitation from acetic acid solution with barium chloride crystals as barium sulfate according to DIN 38405-D5. The intensity of the turbidity of the final suspension was measured at 420 nm using a turbidity meter.

The COD (in mg O_2/L) was determined according to DIN 38409-H41. Abundant potassium dichromate was titrated with iron(II) ammonium sulfate and ferrous as an indicator until the color changed from blue-green to red-brown.

Heavy metals (Cd, Cr, Ni, Pb, Zn) were analyzed after acidic digestion and were analyzed using a Phoenix-986AAS atomic absorption spectrometer. Absorbance levels were converted into concentrations in ppm through calibrations for each compound.

3. Result and Discussion

3.1. Water Samples

3.1.1. pH Value

The pH analysis at all sampling points revealed pH values between 6.84 and 7.53 (test series 1) and between 6.71 and 7.28 (test series 2), which were within the WHO guidelines [21] Along the flow route of the Tigris River, the pH decreased to 6.8–6.9 within the city passage compared to unpolluted conditions (test site 1; 7.28–7.53) due to the discharge of organic pollutant loads and the biodegradation of these contaminants [22], causing partial acidification due to oxygen-limiting conditions. Due to a combination of higher water temperature and thus lower oxygen solubility, higher biodegradation activity, and lower water flow velocity because of low river levels during dry months, oxygen limitation is more prominent in summer months. However, the drop did not exceed 0.69 pH units because the carbonate hardness of the river water resulted in a high buffering capacity [22] This is consistent with the fact that, if this buffering capacity was not present, a significant fluctuation in pH would be expected with implications for aquatic life [23] especially as lead levels are strongly affected by the presence of both orthophosphate (also a typical wartime contaminant) and organic contaminants in a pH range of 7.2–7.8 [24].

3.1.2. Electric Conductivity, Total Dissolved Solids and Salinity

Electrical conductivity of surface waters is highly regional and can range from 50 to 1500 µS/cm [25]. Studies of US inland fresh waters indicate that streams supporting good mixed fisheries have a range between 150 and 500 µS/cm [25]. Conductivity outside this range may indicate non-suitable water for certain species of fish or bugs. High conductivity of up to 10 µS/cm in inland surface water strongly indicates water contamination by industry or crises. The WHO suggests mid-range conductivity (200 to 1000 µS/cm), and maximal conductivity (1400 µS/cm) is the normal background for most major rivers [26]. Measurements in the Tigris River were within the WHO's target value in both test series, and the values are also within the range of electrical conductivities of North American streams, so this parameter is not an exclusion criterion for the use of Tigris water. However, it can be seen over the flow length that the conductivity increases over all 10 measuring points from originally 444.3 µS/cm to 884.3 µS/cm (test series 1) from 418.3 µS/cm to 623 µS/cm (test series 2). Thus, there is a discharge of saline wastewater or contaminated surface runoff through the urban area. Also, infiltration of saline series water cannot be excluded, since the hydrogeological Aljazeera area is part of the Tigris hydrogeological basin in which water flow converges towards the River Tigris [27–29]. Correspondingly, the detected TDS curve correlates with the electrical conductivity, which also increases continuously between the measuring points S1 to S6 (series 1: 405 mg/L to 787 mg/L; series 2 from 292.3 mg/L to 621 mg/L; mean values in each case) and drops by approximately 10% to S7 in both series and then remains nearly unchanged. The TDS concentrations remain below the WHO specifications, whereby high TDS values are not fundamentally harmful to humans. However, the consumption of water with high TDS values can have an increased negative effect on people with kidney and heart diseases in particular. In general, this water may also have an increased laxative but also constipating effect [30,31].

The same effect was observed for the salinity content, which increased from 0.31% (series 1) or 0.39% (series 2) at the reference point S1 to the measuring point S7 and dropped by 35–71% and decreased by about 15–36% to S10. The WHO limit value was just met at S7 during wintertime. Comparing both series shows that an inconsistent picture is drawn for these three parameters. Due to the lack of precipitation during the summer period, and thus lower levels in the Tigris, an increase in electrical conductivity, TDS value, and salinity due to anthropogenic discharges would be expected. However, a direct comparison of Tables 2 and 3 shows that electrical conductivity outside the urban area remained almost unchanged, while that within the urban area remained 30% lower than during wintertime (series 1).

Table 2. Mean and standard deviation values for the concentration of pollutants and heavy metals for all study sites in the river water (series 1).

S1 Water		pH	E.C. μS/cm	TDS mg/L	Sal. %	COD mg/L	PO_4 mg/L	NO_3 mg/L	SO_4 mg/L	Cd μg/L	Pb μg/L	Zn μg/L	Cr μg/L	Ni μg/L
WHO Std.		**6.5–8.5**	**1400**	**1000**	**1%**	**100**	**0.4**	**50**	**250**	**5**	**10**	**5000**	**50**	**20**
S1	Mean	7.53	444.33	405.00	0.31	29.00	0.28	2.62	90.3	5.1367	30.236	338.2	27.8	26.66
	±Sd	0.31	29.3	21.39	0.09	3.06	0.02	0.62	28.10	1.5	14	19	3	7
S2	Mean	7.51	544.33	488.67	0.59	43.33	0.55	3.00	124.7	9.1333	96.47	292.13	37.98	28.33
	±Sd	0.26	12.10	52.32	0.10	5.51	0.11	0.58	6.24	1.2	18	20	6	7
S3	Mean	6.84	693.67	611.33	0.75	44.67	0.72	3.02	189.3	11.3333	130.53	246.50	56.47	46.67
	±Sd	0.16	22.11	34.59	0.06	4.51	0.05	0.60	24.21	4.5	16	34	13	11
S4	Mean	6.90	702.00	711.00	0.97	96.33	0.88	3.74	307	16.1333	138.73	923.33	57.00	56.67
	±Sd	0.09	19.55	34.59	0.04	7.77	0.06	0.10	6.56	4.7	23	37	12	18
S5	Mean	6.87	812.00	776.00	0.70	112.33	0.98	4.65	385.0	18.3333	143.73	973.33	59.00	69.67
	±Sd	0.22	17.62	48.17	0.07	8.08	0.08	0.06	32.19	5.8	34	52	27	21
S6	Mean	6.86	884.33	787.00	0.40	102.67	0.91	2.67	385.0	28.3667	163.80	1496.33	73.15	94.07
	±Sd	0.21	39.04	63.52	0.06	5.29	0.11	0.45	26.85	8	26	29	29	19
S7	Mean	6.98	829.67	717.00	1.07	90.67	0.94	4.03	493.0	26.0000	159.20	1284.00	78.57	92.33
	±Sd	0.15	53.59	45.65	0.07	9.61	0.06	0.50	130.66	7	32	55	14	17
S8	Mean	7.04	711.00	737.67	0.78	82.33	0.75	3.00	410.0	24.6333	169.00	1087.00	64.11	90.33
	±Sd	0.22	26.69	44.84	0.03	6.56	0.11	0.73	26.91	6	22	58	12	15
S9	Mean	7.20	660.67	748.00	0.49	92.00	0.59	3.04	383.7	18.8333	163.33	1165.33	66.23	85.27
	±Sd	0.23	36.01	71.15	0.09	5.03	0.03	0.45	62.66	4	14	37	16	17
S10	Mean	7.13	616	734	0.68	89	0.522	3.071	334.3	18.3333	148	1811.814	61.51333	83
	±Sd	0.25	28.21	28.75	0.04	5.00	0.06	0.78	51.59	7	12	52	13	18
		Zone 1				Zone 2			Zone 3				Zone 4	

In the case of the TDS values, there was an increase from 0.48% (series 1) or 0.52% (series 2) at reference point S1 to measuring point S5 or S7, which dropped by 6–17% to S10. The WHO limit value is just met at S7 during wintertime, with summer levels also being 35–40% lower. The situation is similar to that of salinity, which is also about 30% lower in summer than in winter but rises by 48% (winter) or 25% (summer) over the city passage. It temporarily increases by a factor of 2–3 within the urban area. The most likely explanation for this scenario is that due to the lack of precipitation in summer and thus the lack of surface runoff, a significantly lower pollutant load is flushed into the Tigris. In addition to anthropogenic discharges, the increase in contaminants in the urban area is due to the presence of valleys loaded with contaminated water that flows from both sides of the river into the Tigris [22,27–29,32].

Table 3. Mean and standard deviation values for the concentration of pollutants and heavy metals for all study sites in the river water (series 2).

S2 Water		pH	E.C. µS/cm	TDS mg/L	Sal. %	COD mg/L	PO_4 mg/L	NO_3 mg/L	SO_4 mg/L	Cd µg/L	Pb µg/L	Zn µg/L	Cr µg/L	Ni µg/L
WHO Std.		6.5–8.5	1000	1000	1%	100	0.4	50	250	5	10	5000	50	20
S1	Mean	7.48	418.33	292.33	0.39	24.67	0.32	1.82	57.7	3.901	15.85	178.53	16.67	16.67
	±Sd	0.306	29.297	21.385	0.090	3.055	0.020	0.616	28.095	0.54	4.5	18	3.2	5.2
S2	Mean	7.300	524.33	315.33	0.34	28.67	0.31	1.999	107.0	6.98	66.67	196.53	24.40	13.31
	±Sd	0.265	12.097	52.320	0.104	5.508	0.111	0.578	6.245	1.8	11	20	5.7	3..5
S3	Mean	7.163	558.00	381.67	0.51	32.67	0.38	2.055	140.3	8.33	98.75	146.50	37.44	23.33
	±Sd	0.158	22.113	34.588	0.064	4.509	0.051	0.602	24.214	2.1	18	15	14	8
S4	Mean	7.093	498.33	401.33	0.47	51.67	0.51	2.487	147.0	14.76	122.40	848.17	50.00	57.67
	±Sd	0.090	19.553	34.588	0.038	7.767	0.064	0.100	6.557	1.8	34	37	16	13
S5	Mean	6.980	597.67	469.67	0.58	74.33	0.70	2.766	315.3	23.25	152.20	1391.07	60.47	83.33
	±Sd	0.220	17.616	48.170	0.068	8.083	0.080	0.056	32.192	3.6	31	134	23	19
S6	Mean	6.810	596.00	554.67	0.52	63.00	0.74	2.167	289.0	23.60	138.80	1266.80	63.93	81.
	±Sd	0.209	39.038	63.516	0.061	5.292	0.109	0.452	26.851	8	22	288	19	17
S7	Mean	6.993	610.00	557.00	0.60	65.30	0.81	2.477	232.5	22.53	130.20	1183.	67.53	68.33
	±Sd	0.148	53.594	45.654	0.070	9.609	0.063	0.503	130.664	8.8	35	242	23	18
S8	Mean	7.017	623.00	621.	0.40	59.00	0.56	2.228	293.0	20.04	99.42	853.67	49.67	60.47
	±Sd	0.222	26.690	44.837	0.025	6.557	0.106	0.730	26.907	6.5	21	95	19	17
S9	Mean	7.087	583.00	516.60	0.49	50.33	0.59	2.370	318.3	15.10	88.25	885.60	46.67	56.17
	±Sd	0.228	36.014	71.150	0.093	5.033	0.030	0.450	62.660	4.5	16	167	15	16
S10	Mean	6.833333	593.00	510.33	0.51	58.00	0.52	2.111	320.7	13.84	57.75	1208.03	47.42	55.63
	±Sd	0.254	28.213	28.746	0.040	5.0	0.055	0.776	51.588	4.80	21	214	13	14
		Zone 1			Zone 2				Zone 3			Zone 4		

3.1.3. Chemical Oxygen Demand COD

The thesis of increased infiltration of contaminated water can be confirmed using COD data. In both series, an increase in the COD value (about 73 mg O_2/L in series 1 and about 50 mg/L in series 2), respectively, occurred across the urban area. Within the city, even higher COD values of 112.3mg O_2/L (S5, series 1) and 74.33 mg O_2/L (S5, series 2) were temporarily achieved, detectable at nearly identical levels at S10 during the winter period, while a decrease of about 35% due to biodegradation processes was observed during the summer period, because of the discharge of polluted wastewater or contaminated surface runoff through the urban area to the river during the winter season and the infiltration of polluter groundwater, and from the valleys that, loaded with pollution, flowed into the river [10,33]. The increased direct or indirect discharge of sanitary wastewater and thus increased organic loads into the river due to war-related damages to the sewage system have been previously described [5].

3.1.4. Nitrate, Phosphate, and Sulfate

There are a variety of potential emission sources for nitrate, phosphate, and sulfate. For example, these anions can be released during the exploration of minerals, the agricultural use of fertilizer, industrial effluents, and general leakages in the destroyed municipal sewage network, but also through the direct use of weapons and ammunition. In the latter case, nitrate is directly used in munitions as ammonium nitrate.

Conflict-related phosphate emissions in the urban area of Mosul are particularly known from the use of phosphorus bombs during the liberation in the summer of 2017 [34]. However, this type of bomb was also used by U.S. forces during the second Gulf War [35]. Sulfate emissions can result from the combustion of antimony sulfides when munitions are fired [36,37]. However, in the case of Mosul, the burning of the Mishraq industrial complex (SO_2, SO_3) [8,38], 45 km south of Mosul in October 2016, in particular, played a substantial role, where the resulting gases entered the river through wet deposition into the soil and surface drainage [5], and the sulfur springs extend from Mosul in the north to Qayara in the south for 60 km along the Tigris River, where the River Tigris is also a zone of discharge [27].

Accordingly, the analyses of the water samples for phosphate and sulfate, apart from S1 as the most northwestern measuring point, showed, in some cases, significant exceedance of the WHO limits. Thus, phosphate concentrations increased by a factor of 1.8–1.9 through the urban passage in both series, regardless of the season. In some cases, detected phosphate levels of 0.81 mg P/L (series 2) and 0.94 mg P/L (series 1) exceeded the WHO limit by a factor of 2–3 within the urban region. Phosphate levels exceeded limits from S2 to S10 in series 1 and from S4 to S10 in series 2.

Sulfate levels exceeded limits from S2 to S10 in series 1 and from S5 to S10 in series 2, at the highest value of 495 mg/L (S7 in Series 1). Again, a key aspect of lower sulfate levels, but also other anion levels, was the lack of surface drainage in combination with valleys contaminated with water infiltration to Tigris. However, biodegradation represents a second, possibly even dominant factor, where both sulfate and nitrate may be used as electron acceptors instead of oxygen. This aspect could be easily verified due to the severe odor emissions of H_2S gas near the valleys. Direct drainage into the river has been described previously [22,32,39]. In addition to the above factors, the detected sulfate levels were further increased by the dissolution of gypsum rock due to the impoundment of Tigris water in Mosul Lake [11].

In the case of nitrate, the course over the river flow showed only a small increase at S7, which is mainly due to agriculture upstream but remains at an almost constant level over the course of the river. As microbiological H_2S formation, and thus inevitably microbiological nitrate respiration as a sink of nitrate, occurred, a continuous discharge of nitrate over the urban area took place. The discharge was caused by defects in the sewage system and infiltration processes as a consequence. Nitrate concentration was clearly below the WHO limit value at all locations, irrespective of the season.

3.1.5. Heavy Metals

Due to international contracts prohibiting the use of expanding bullets, as well as the higher cost of hollow-point ammunition, full metal jacket bullets are the most used cartridges by soldiers as well as militias [40]. They consist of a soft core, usually lead, and a harder jacket metal alloy such as gilding metal (CuZn), CuNi, or steel encasing the core [36]. The purpose of this design is to provide improved trajectory as well as greater penetration of the projectiles into soft tissue [40,41]. Supplementary galvanic cadmium coatings are applied to improve some properties like corrosion resistance, improved tribological properties, and chemical stability [37]. Chromatic layers are applied as supplementary passivation [37]. Therefore, the use of weapons in conflict zones leads to an increase in heavy metal concentrations.

Concentrations of Cd, Pb, Cr, and Ni exceeded WHO limits even before entering the city (S1), while increasing further, especially through the Old City passage as a direct conflict zone (S5–S7) in both series.as shown in Figure 2.

The maximum concentrations (in µg/L) were 28.37 Cd, 169 Pb, 78.5Cr, and 94 Ni in series 1 and 23.6 Cd, 152.2 Pb, 67.5 Cr, and 83.3 Ni in series 2, and the WHO limits were exceeded here in part by a factor of 4.7 for Cd and 32 for Pb and 5.6 for Cr and 14.8 for Ni, as follows: Pb > Ni >> Cr > Cd > Zn. For Zn, a comparable concentration pattern occurred over the urban passage, but with maximum values of 1811.8 µg/L (series 1) and 1391 µg/L (series 2), the limits were not exceeded.

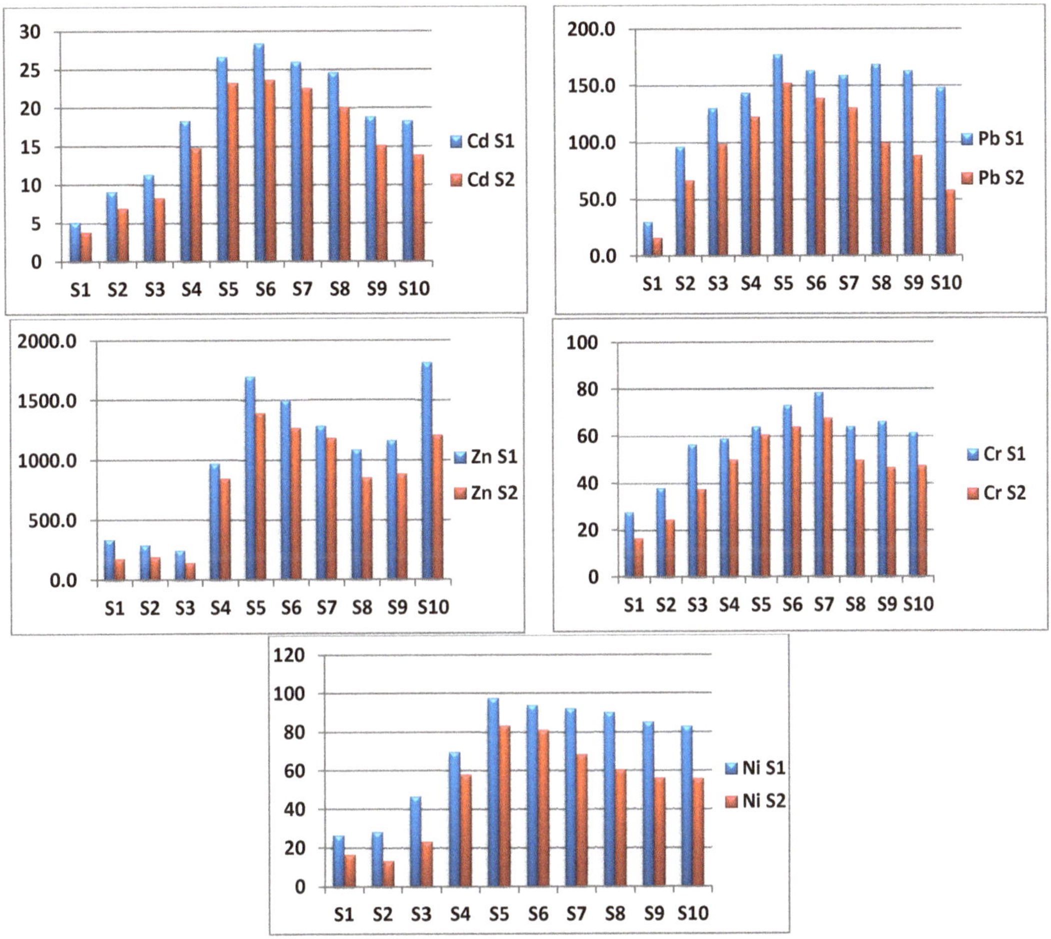

Figure 2. Average values of heavy metal concentration in the first and second series in river water in all sites.

3.1.6. Comparison of Water Samples in Conflict Zone vs. Urban Zone

The pollution of the Tigris River can be divided into four zones over its flow length in the urban area. Zone 1 (S1), as a reference before entering the densely populated urban area, illustrates the back pollution of the river. Zone 2 (S2–S4) essentially represents contamination from agricultural drainage. Since the zone is otherwise characterized by recreation areas, the influence of municipal drains is still low there. Between sampling points S4 and S5, the Al Khosr valley drains into the Tigris, which carries a heavy pollutant load on its flow route from the northeastern valleys through densely populated areas [42] and therefore leads to a strong increase in the parameters E.C., TDS, salinity, COD, nitrate, phosphate, and sulfate at S5.

The military operations to liberate Mosul focused on the Old City of Mosul, where sampling points S5–S7 (zone 3) are located. Therefore, within this zone, an increase in heavy metal concentrations is expected and has been observed as an indicator. Th destruction of the drainage infrastructure also leads to a further increase in the above parameters. zone 4, represented by S8–S10, again represents an area with strong agricultural use and residential development outside the direct conflict zone, so the discharge situation here is limited to agricultural and municipal parameters, and the heavy metals decrease in concentration due to (bio)chemical processes such as adsorption, precipitation, uptake in tissues, and dilution effects.

The differences in concentration between summer and winter periods caused by the lack of surface drains and existing groundwater infiltration have already been pointed out. In addition, the role of entrained sediments should be mentioned, which will be further discussed in Section 3.2.

3.1.7. Comparison of Water Samples in Current Study with Previous Study

The comparison of the annual average values of heavy metals (Cd, Pb, and Zn) in zone 1, zone 3, and zone 4 in the current study in 2022 with a previous study of [14] which was conducted before the occupation and liberation of the city between 2011 and 2012, showed that, in S1, the values of Cd increased by a factor of 4.0, in zone 3 by a factor of 6.25, and in zone 4 by a factor of 3.69 and Pb values increased in zone 1 by a factor of 4.75, in zone 3 by a factor of 6.14, and in zone 4 by a factor of 3.17. Furthermore, the values of Zn in zone 1 decreased by a factor of −1.45, in zone 3 increased by a factor of −1.15, and decreased by a factor of −1.23 in zone 4 as shown in Table 4.

Table 4. Comparison of annual values of heavy metal values in water samples in the current study with previous study [14].

Sites	Element	2013	2022	Factor
Zone 1	Cd	1	4.0	4.0
	Pb	4	19	4.75
	Zn	469	258.4	−1.45
	Cr	ND	22.2	ND
	Ni	ND	21.7	ND
Zone 3	Cd	4	25.0	6.25
	Pb	27	165.1	6.14
	Zn	1339	1544.5	1.15
	Cr	ND	62.2	ND
	Ni	ND	90.5	ND
Zone 4	Cd	5	18.46	4.0
	Pb	33	102.9	3.17
	Zn	1961	1509.9	−1.23
	Cr	ND	54.5	ND
	Ni	ND	69.3	ND

The results indicated that zone 3 represents the most polluted site, especially with Cd and Pb. The increase was significant compared with the previous year, while the variant between the two studies in Zn values was not significant, as shown in Table 4. Even so, a significant increase in zinc concentration was observed within the present study compared to the previous year, and no significant differences were observed in comparison with the study by [14]. Sampling within zone 2 was not carried out by the latter.

3.1.8. Statistical Analysis

An adequate approach for the statistical analysis of deviations in analytical data and their significance is the *t*-test. The results for E.C., TDS, COD, nitrate, phosphate, sulfate, Cd, Pb, Zn, Cr, and Ni showed levels below the critical value ($p < 0.05$), which is why they are considered statistically significant differences. Thus, based on the null hypothesis that the sample mean and population mean are statistically different at the 0.05 significance level due to the time lag between the two analytical series, a significant statistical difference results between these two series ($p < 0.05$, see Table 5).

Table 5. t-test value between first and second series for all sites in river water.

Parameters	t-Test	Df *	p	Status $p < 0.05$
PH	1.745881	18	0.048938	1
EC	2.648184	18	0.008176	1
TDS	3.868649	18	0.000563	1
Salinity%	2.669917	18	0.007808	1
COD	2.586038	18	0.009319	1
PO_4	1.826824	18	0.042179	1
NO_3	5.394127	18	1.99×10^{-5}	1
SO_4	1.813528	18	0.021327	1
Cd	1.798236	18	0.021757	1
Pb	2.008098	18	0.029941	1
Zn	1.668797	18	0.025606	1
Cr	1.6898	18	0.054156	1
Ni	1.484068	18	0.077546	1

* Df: degree of freedom.

The percentage difference in heavy metal analyses between the two series, S1 and S2, at all sites showed that the values of Cd, Pb, Zn, Cr, and Ni in S1, S2, S3, S8, S9, and S10 were higher than the values in S4, S5, S6, and S7, although the concentrations of pollutants in these sites were higher compared to other sites, which is due to the fact that the pollutants at these sites (S4, S5, S6, and S7) is discharged throughout the year via the drains from the valleys and is not only discharged via rainwater. Rather, surface drains from rainwater additionally enter the Tigris in addition to existing drains, which significantly increases the percent of pollution in the river at these sites, especially in winter. According to the topography and the slopes of the Old City, the passage of the water of the runways and their movement from the highest areas and their flow to the low area that represent the highest levels of drainage systems, or the main valleys, that eventually flow into the Tigris River [43].

The lowest percentage differences between both runs were seen at S5 and S6 for Cd (12.8% vs. 16.8%), Pb (14.5% vs. 18.2%), Ni (13.9% vs. 26%), Zn (7.8% vs. 18.2%), and Cr (5.5% vs. 14%), as shown in Table 6 and Figure 3. Differences at other locations (S1, S2, S3, S8, S9, S10) were higher in this case since the main pollution source of the Tigris River is the winter rainwater drain; therefore, the difference in heavy metal concentrations between both series is even more evident.

Table 6. The percentage of difference between the first and second series in all water sites.

Sites	Cd %	Pb %	Zn %	Cr %	Ni %
S1	24.0	47.6	47.2	40.0	37.5
S2	23.6	30.9	32.7	35.8	53.0
S3	26.5	24.3	40.6	33.7	50.0
S4	19.5	14.8	12.9	15.3	17.2
S5	12.8	14.5	18.1	5.5	14.7
S6	16.8	15.3	15.3	12.6	13.9
S7	13.3	18.2	7.8	14.1	26.0
S8	18.6	41.2	21.5	22.5	33.1
S9	19.8	46.0	24.0	29.5	34.1
S10	24.5	61.0	33.3	22.9	33.0

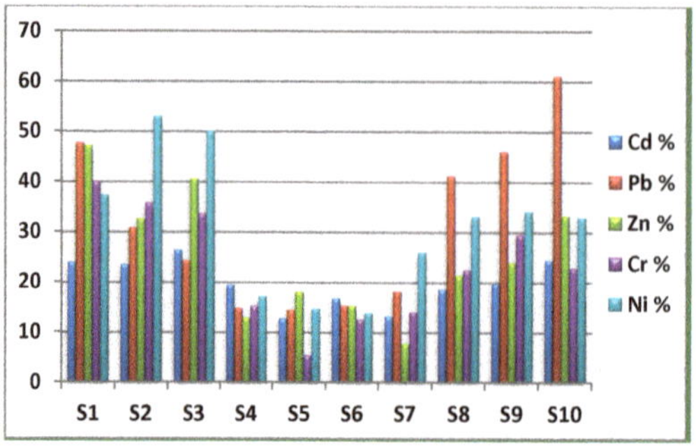

Figure 3. Percentage of difference of concentrations between the first and second series for heavy metals in all sites.

Table 7 and Figure 4 show that the ranking in the percent differences in site contaminants between upstream S1 and downstream S10 in both runs was as follows: Zn > Pb > Cd > Ni > Cr and for other parameters SO_4 > COD > Salinity > TDS >> PO_4 > >E.C. > NO_3. Except for pH values, the percentage increase in all parameters between upstream and downstream was higher in (series 1) (winter, spring) than in (series 2) (summer, Autumn) because the increased biodegradation of organics in summer combined with oxygen limitation led to an enhanced decrease in pH, which is also indicated by [44].

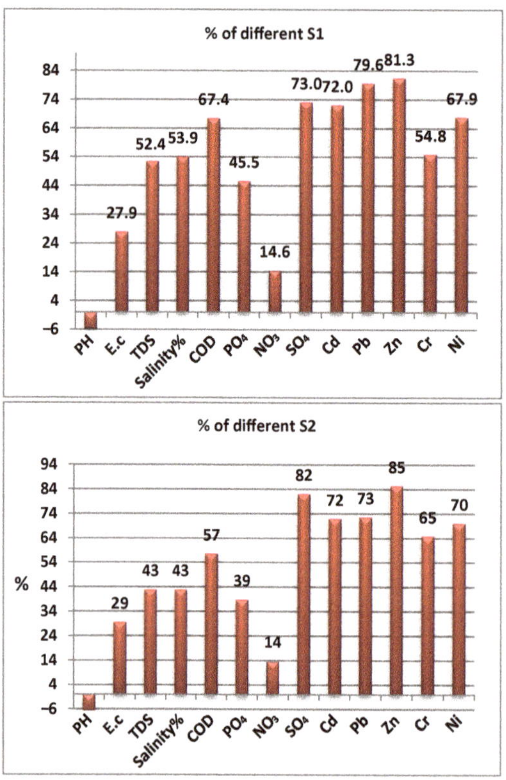

Figure 4. Percentage of difference of concentrations between Upstream (S1) and Downstream (S10) in both.

Table 7. Percentage of difference of concentrations between Upstream (S1) and Downstream (S10) in both.

	Series 1				Series 2			
	Upstream	Downstream	Df	% DF	Upstream	Downstream	Df	%Df
PH	7.5	7.1	−0.4	−5.6	7.18	6.73	−0.45	−6.63
E.C.	444.3	616.0	171.7	27.9	418.33	593.00	174.67	29.45
TDS	405.0	850.0	445.0	52.4	292.33	510.33	218.00	42.72
Salinity%	0.3	0.7	0.4	53.9	0.29	0.51	0.22	42.76
COD	29.0	89.0	60.0	67.4	24.67	58.00	33.33	57.47
PO_4	0.3	0.5	0.2	45.5	0.32	0.52	0.20	38.70
NO_3	2.6	3.1	0.4	14.6	1.82	2.11	0.29	13.78
SO_4	90.3	334.3	244.0	73.0	57.67	320.67	263.00	82.02
Cd	5.1	18.3	13.2	72.0	3.90	13.85	9.95	71.83
Pb	30.2	148.0	117.8	79.6	15.85	57.75	41.90	72.55
Zn	338.2	1811.8	1473.6	81.3	178.53	1208.03	1029.49	85.22
Cr	27.8	61.5	33.7	54.8	16.67	47.42	30.75	64.85
Ni	26.7	83.0	56.3	67.9	16.67	55.63	38.97	70.04

3.2. Sediment Samples

3.2.1. Chemical Parameters and Heavy Metals

Sediments are an important sink for pollutants, but they can also lead to the contamination of the water body through descriptive processes. In particular, if the sediment is primarily of a mineral nature, ions (e.g., heavy metals) may be adsorbed [45]. However, the interaction strength of heavy metals and geochemical fractions of sediments strongly depends on exchangeable, reducible, oxidizable, and residual fractions of the sediment [46,47]. The exchangeable and, therefore, weakly bound geochemical sediment fractions include the fixation of heavy metals to the solid surface via weak electrostatic interactions [48,49]. Hence, changes in environmental parameters like pH, E.C., or salinity can promote the desorption of heavy metals via the breakdown of these weak electrostatic interactions [47,50,51]. If the sediment is primarily of organic nature, organic contaminants (e.g., pharmaceuticals, flame retardants) may be adsorbed [52,53]. In addition to the complexity of the sediment composition, the interpretation of occurring distribution equilibrium between the sediment phase and the liquid phase is particularly complicated by the downstream transport of sediment or small-particle and thus suspended silt fractions. Hence, it is to be expected that even outside the direct conflict zone or the urban area, increased pollutant concentrations may occur in the water phase, as well as in the river sediment, due to sediment transport and subsequent desorption.

In fact, the sediment analysis confirmed a continuous increase in heavy metal concentrations over the flow distance, with the highest concentrations occurring at the southern urban border (S10) due to hydrogeological formation and slope of the Tigris River's bed. A similar result was already presented by [14]. The detected heavy metal concentrations had a mean of 6.30 ppm Cd, 45.33 ppm Pb, 21.33 ppm Zn, 81.33 ppm Cr, 64.00 ppm Ni (series 1), 9200 ppm Cd, 62,800 ppm Pb, 30 ppm Zn, 123,6 ppm Cr, and 78 ppm Ni (series 2). WHO limits were permanently exceeded at all locations except that of Zn. In the case of S10, the limit values for Cd, Pb, Cr, and Ni were exceeded by a factor of 3.1, 1.3, 3.3, and 3.2 (series 1) and by a factor of 4.6, 1.8, 4.9, and 4.0 (series 2). For zinc, the limits were met despite an increase in concentration over the urban passage (see Tables 8 and 9). However, due to the interaction between dissolved heavy metals in the water phase and sorbed heavy metals in the solid phase, severe local variations in concentration, in contrast to the water phase, were not detectable.

In the case of the parameters pH, E.C., and salinity, no significant change occurred for the pH value during the winter series. During the summer series, however, a moderate decrease in the pH value by 0.77 units in absolute terms was observed over the urban area. This decrease is related to the increased degradation of organic pollutants induced by higher temperatures, with partial oxidation of the components. The pH drop turns out to be comparatively small, since both the morphological sediment composition and the existing lime–carbonic-acid equilibrium in the water phase, as well as the sediment phase, lead to a pH buffering [54].This pH buffering prevents the enhanced pH-induced mobilization of heavy metal ions. This effect has already been described for Cr, Ni, and Zn [46]. The mobility of heavy metals is further influenced by E.C., as well as salinity, with enhanced mobilization at increasing E.C. and decreasing salinity, respectively. Both parameters increased by a factor of 1.6–2.4 over the urban passage and just exceeded the WHO limits by a factor of 6.4 for E.C. and by a factor of 3.3 for salinity in the summer measurement, while in the winter period, only salinity exceeded the WHO limit by a factor of 1.8.

Table 8. Average and standard deviation values for the concentration of pollutants and heavy metals for all study sites in sediment (Series 1).

S1		pH	E.C.	Sal.%	Cd	Pb	Zn	Cr	Ni
WHO Std. Limits		**6.5–8.5**	**1000 μS/cm**	**1%**	**2 PPM**	**35 PPM**	**90 PPM**	**25 PPM**	**20 PPM**
S1	Mean	7.44	710	0.86	2.78	25	12	28.20	27
	±Sd	0.31	50	0.07	0.26	2.33	1.00	6.01	2.00
S2	Mean	7.78	680	0.88	2.87	24.70	10.33	36.00	32.67
	±Sd	0.43	20	0.28	0.32	1.12	1.53	7.94	6.51
S3	Mean	7.57	750	1.34	3.50	29.80	14.67	41.67	32.33
	±Sd	0.29	40	0.17	0.96	0.35	3.51	4.16	7.02
S4	Mean	7.80	800	1.50	4.44	33.70	11.67	40.67	37.00
	±Sd	0.10	40	0.00	0.30	0.30	2.07	8.50	2.65
S5	Mean	7.55	1150	1.17	5.63	35.40	16.50	73.67	49.00
	±Sd	0.19	240	0.12	0.83	1.26	3.97	3.51	6.24
S6	Mean	7.53	1640	1.44	5.48	38.20	18.67	80.00	52.00
	±Sd	0.40	710	0.23	0.30	1.53	3.06	1.00	4.36
S7	Mean	7.64	50	1.34	5.87	38.67	19.33	78.67	57.00
	±Sd	0.49	680	0.13	0.35	3.06	3.51	4.51	6.56
S8	Mean	7.93	20	1.03	6.30	40.00	17.83	77.67	59.67
	±Sd	0.05	750	0.12	0.60	1.00	5.11	6.51	4.16
S9	Mean	7.86	40	1.10	5.47	44.33	20.67	80.33	63.67
	±Sd	0.40	800	0.64	0.21	4.51	6.11	3.06	8.50
S10	Mean	7.17	40	1.82	6.23	45.33	21.33	81.33	64.00
	±Sd	0.32	1150	1.05	0.55	2.50	2.50	7.50	8.14
Zone 1			Zone 2		Zone 3			Zone 4	

Table 9. Average and standard deviation values for the concentration of pollutants and heavy metals for all study sites in sediment (Series 2).

S2		pH	E.C.	Sal.%	Cd	Pb	Zn	Cr	Ni
WHO Std. Limits		**6.5–8.5**	**1400**	**1%**	**2 PPM**	**35 PPM**	**90 PPM**	**25 PPM**	**20 PPM**
S1	Mean	7.62	611	1.387	3.3	31.3	10.16	32.0	30
	±Sd	0.19	88	0.528	0.6	7.87	0.76	9.5	3.0
S2	Mean	7.79	777	1.617	3.3	33.66	12.6	50.3	37.6
	±Sd	0.39	110	0.458	0.8	4.51	1.87	11.4	5.6
S3	Mean	7.27	807	1.343	4.28	35.6	15.1	50.6	39.3
	±Sd	0.26	90	0.151	0.7	3.7	2.9	10.4	6.8
S4	Mean	7.80	920	1.667	5.8	40	17.0	42.66	46.0
	±Sd	0.08	42	0.204	0.8	2.3	3.8	12.3	5.8
S5	Mean	7.35	1146	1.517	6.9	43.1	20.3	79.0	58.3
	±Sd	0.19	215	0.264	0.78	2.7	3.52	11.1	6.9
S6	Mean	7.28	1105	1.650	7.37	46.33	22.0	89.0	58.3
	±Sd	0.30	297	0.310	0.88	4.40	3.5	5.0	9.22
S7	Mean	7.13	1267	1.377	7.79	47.6	25.33	92.3	64.80
	±Sd	0.41	367.5	0.6194	0.87	4.41	9.83	9.35	0.41
S8	Mean	7.03	1062	1.600	7.76	51.6	21.6	97.0	67.90
	±Sd	0.53	36	0.431	0.46	7.67	4.7	11.7	3.34
S9	Mean	6.97	1365	1.857	7.86	49.6	24.3	96.0	71.2
	±Sd	0.55	143	0.478	1.20	6.06	5.9	13.1	6.43
S10	Mean	6.86	1445	3.292	9.20	62.8	30.	123.60	79.0
	±Sd	0.45	112	0.340	0.86	4.44	2.8	16.1	5.6
Zone 1			Zone 2			Zone 3		Zone 4	

3.2.2. Comparison of Sediment Samples in Urban Zone between Both Series

The direct comparison of these parameters between both series showed, with the exception of the pH value (see above) and the E.C. (almost unchanged), significant increases of 1.6–1.8 (salinity), 1.2–1.5 (Cd), 1.2–1.4 (Pb), 1.4 (Zn), 1.2–1.5 (Cr), and 1.1–1.2 (Ni) in the summer period (series 2, see Tables 8 and 9). This increase during dry weather is probably related to the presence of valleys that flow into the Tigris River during the summer as well, which are loaded with polluted water from the residential and agricultural areas it passes through [8], the combination of low flow velocity of the Tigris during the dry weather phase, and thus also an increased deposition of small particles, which either infiltrated directly into the Tigris from the conflict zones or were lifted up by wind and settled down above the water surface, as well as an increased particle-bound heavy metal input during the precipitation phase in winter and spring (series 1; These particles, some even forming conglomerates, are slowly transported downstream as sediment mixtures, resulting in the high concentrations detected throughout the entire sampling stretch of the river (see Figure 5).

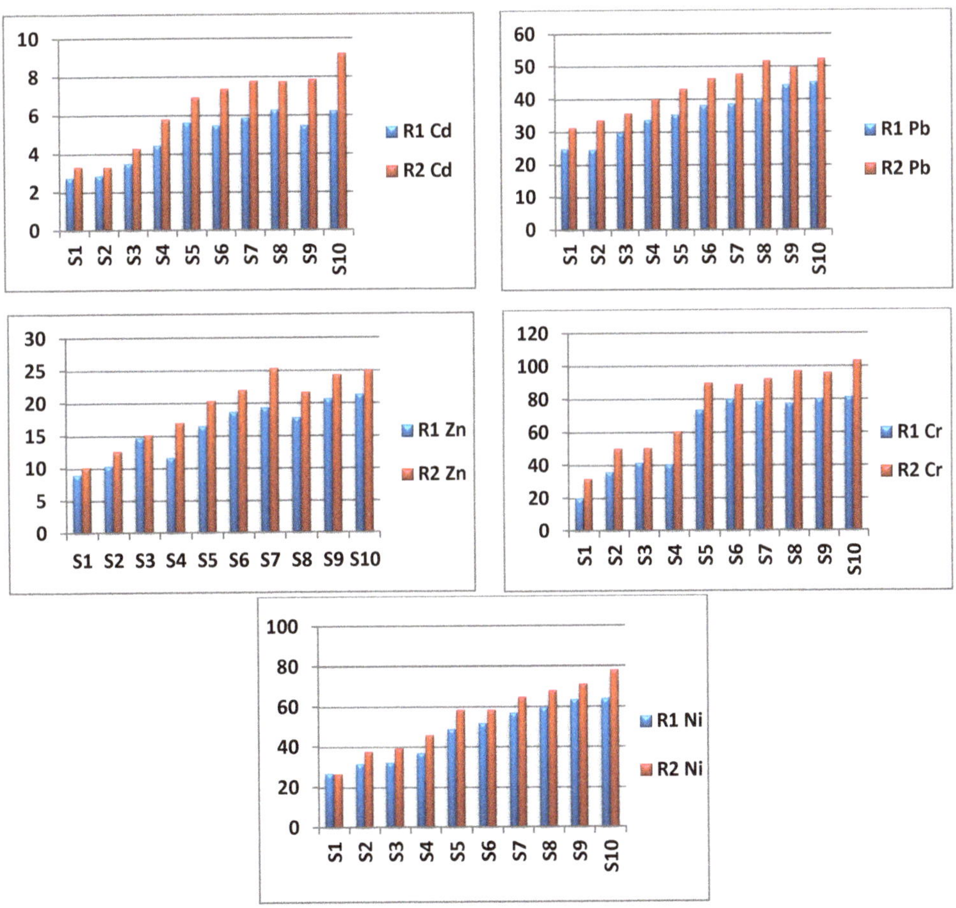

Figure 5. Average values of heavy metal concentration (ppm) between the first and second series in sediment sites.

3.2.3. Comparison of Sediment Samples in the Conflict Zone vs. the Urban Zone

The pollution of the Tigris via the city passage of Mosul can be divided into four zones, whereby zone 1 represents the upstream area before entering the densely populated area and thus serves as a reference, zone 2 describes a predominantly intensively agricultural area, zone 3 comprises the actual conflict zone, and zone 4 includes the southern suburbs of the city. Compared to the reference sampling at S1, the heavy metal concentrations in the sediment increased sharply upon entering Zone 3 and further through the urban area, by a factor of 2.14 (S5) and 2.3 (S10) for cadmium, by a factor of 1.52 (S5) and 1.81 (S10) for lead, and by a factor of 1.3 (S5) and 1.7 (S10) for zinc. The chromium concentrations increased in parallel by a factor of 2.6 (S5) and 2.9 (S10), and the nickel levels increased by a factor of 1.8 (S5) and 2.3 (S10). In the fourth zone in particular, there was a sharp increase in sediment contamination due to the hydrogeological conditions in the river.

The direct comparison of these parameters between both series showed, with the exception of the pH value (see above) and the E.C. (almost unchanged), a significant increase of 1.6–1.8 (salinity), 1.2–1.5 (Cd), 1.2–1.4 (Pb), 1.4 (Zn), 1.2–1.5 (Cr), and 1.1–1.2 (Ni) in the summer period (series 2; see Tables 7 and 8). This increase during dry weather is probably related to a combination of low flow velocity of the Tigris during the dry weather phase and thus also an increased deposition of small particles, which either infiltrated directly into the Tigris from the conflict zones or were lifted up by wind and settled down above the water surface, as well as an increased particle-bound heavy metal input during the precipitation phase in winter and spring (series 1; see Figure 5). These particles, some even forming conglomerates, are slowly transported downstream as sediment mixtures, resulting in the high concentrations detected throughout the entire sampling stretch of the river (see Figure 5).

3.2.4. Comparison of Sediment Samples in Current Study vs. Previous Study

The comparison of the annual average values of heavy metals (Cd, Pb, Zn) in zone1, zone3, and zone4 in the current study 2022 with a previous study by [14] showed that, in zone 1, the values of Cd increased by a factor of 30.5, in zone 3 by a factor of 32.1, and in zone 4 by a factor of 32.1. Furthermore, Pb values increased in zone 1 by a factor of 1.35, in zone 3 by a factor of 1.62, and in zone 4 by a factor of 1.5. The values of Zn in zone 1 increased by a factor of 1.43, in zone 3 increased by a factor of 1.52, and increased by a factor of 1.55 in zone 4. The result indicated that the values of Cd, Pb, and Zn increased in all sites compared with the previous study, and the highest values were in zone 10 due to hydrogeological formation and slope of the Tigris River's bed, as previously mentioned; see Table 10. The highest pollution level was significant in Cd.

Table 10. Comparison of annual values of heavy metal values in sediment samples for the current study with the previous study [14].

Sites	Element	2013	2022	Factor
Zone 1	Pb	14.89	20	1.35
	Cd	0.1	3.05	30.57
	Zn	6.72	9.583	1.43
	Cr	ND	30	-
	Ni	ND	27.8	-
Zone 3	Pb	24.28	39.28	1.62
	Cd	0.197	6.28	32.10
	Zn	12.15	18.42	1.52
	Cr	ND	76.33	-
	Ni	ND	53.6	-

Table 10. *Cont.*

Sites	Element	2013	2022	Factor
Zone 4	Pb	32.45	48.83	1.50
	Cd	0.242	7.72	32.17
	Zn	14.97	23.17	1.55
	Cr	ND	102	-
	Ni	ND	71	-

3.2.5. Statistical Analysis

The percentage difference in sediment analysis data between the locations of both series indicated that there is no regular variation between sites, as the rates of variation between both series were different in all sites, depending on the specific element. This is due to differences in the river's flow rate and the movement of sediments in the river, as well as the different sources of pollution. The highest percent difference in heavy metals was as follows: Cr in S1 (36.8%) > Cd in S10 (32.2%) > Pb in S2 (26.6%) > Zn in S7 (23.6%) > Ni in S1 (23%) (see Table 11).

Table 11. Percentage of the differences in concentration between the first and second series in all sediment sites.

Sites	Cd %	Pb %	Zn %	Cr %	Ni %
S1	16.60	20.12	14.2	36.88	23
S2	13.39	26.63	18.42	28.48	15.93
S3	18.29	16.45	3.08	17.76	17.80
S4	23.51	15.75	31.37	32.97	19.57
S5	18.75	17.99	18.85	18.15	16.00
S6	25.76	17.55	15.15	10.11	10.81
S7	24.69	18.88	23.68	14.80	12.04
S8	18.85	22.58	17.69	19.93	12.13
S9	30.49	10.74	15.07	16.32	10.58
S10	32.25	13.38	14.67	21.29	17.95

There was a general increase in contaminant concentrations between upstream S1 and downstream S10 in both test series, with the highest percentage increase as follows: Cr > Cd > Pb > Zn > Ni > E.C. > salinity (See Table 12).

Table 12. Percentage of difference of concentrations between upstream (S1) and downstream (S10) in sediment in both series.

	S1				S2			
	Upstream	Downstream	Df	% DF	Upstream	Downstream	Df	%Df
pH	7.4	7.2	−0.3	−3.8	7.6	6.9	−0.8	−11.2
E.C.	0.7	1.2	0.5	41.4	0.6	1.4	0.8	57.7
Salinity%	0.9	1.8	1.0	52.5	1.4	3.3	1.9	57.9
Cd	2.8	6.2	3.5	55.4	3.3	9.2	5.9	63.8
Pb	25.0	45.3	20.3	44.9	21.3	62.8	41.5	66.0
Zn	9.0	21.3	12.3	57.8	10.2	30.0	19.8	66.1
Cr	28.0	81.3	53.3	65.6	32.0	123.6	91.6	74.1
Ni	27.0	64.0	37.0	57.8	30.0	79.9	49.9	62.5

The statistical T-test analysis shows that the T-test values for salinity, Cd, Pb, Zn, and Cr were below the critical value of $p < 0.05$. This means that there was a significant statistical difference between series 1 and series 2 ($p < 0.05$). Conversely, the numerical values of Ni and E.C. were greater than 0.05 ($p > 0.05$), which means that there was no significant statistical difference between series 1 and series 2 for these two parameters. This indicates different pollution sources or input sources of nickel to the sediments, while the T-test values for pH were significant, such that the p-value was <0.05 (see Table 13).

The reason for the high concentration of chromium in the sediments at a higher rate than other minerals is due to the waste of the tanning factory, which has been operating since the years before the war, and whose wastewater is thrown into the river, which contains a high percentage of chromium, and accumulated in the sediments over time [14]. However, high levels of cadmium, zinc, lead, and nickel may be mainly caused by military activities. In the case of pH, an enhanced decrease was observed during summer tests (series 2) due to present biodegradation (see Figure 6).

Table 13. *t*-test value between first and second series for all sites in river sediment.

Parameter	*t*-Test	Df *	*p*	Status $p < 0.05$
Ph	2.453	18	0.012	1
E.C.	−0.03433	18	0.486495	0
Salinity	−2.36597	18	0.014703	1
Cd	−2.02299	18	0.02909	1
Pb	−1.79197	18	0.04498	1
Zn	−1.81791	18	0.04392	1
Cr	−1.61075	18	0.04339	1
Ni	−1.17174	18	0.12829	0

* Df: degree of freedom.

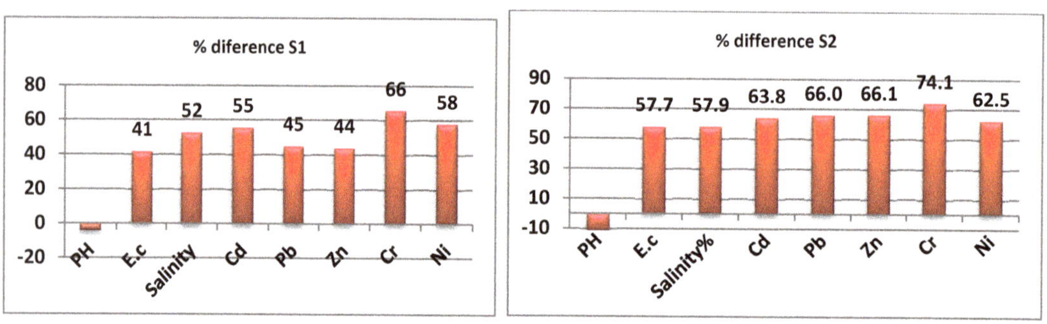

Figure 6. Percentage difference of concentrations between upstream (S1) and downstream (S10) in sediment in both series.

4. Conclusions

The impact of the war was evident due to high contamination levels in water and sediment samples from the Tigris River within the urban area of Mosul. With the exception of Zn, heavy metal loads exceeded WHO limits in a way that was seasonally independent, in both water and sediment. Further exceedances occurred in the water samples for conductivity, phosphate, and sulfate and in individual samples for salinity and COD, regardless of season and location. Salinity was exceeded almost consistently in sediments, while, in contrast, pH, TDS, COD, nitrate, and Zn were (almost) within the limits for all water samples and pH, E.C., and Zn met limits for all sediment samples. The very high loads of heavy metals, but also the exceedances for sulfate, indicated the direct effects of war (ammunition and ignition of sulfur fields). The increased loads of phosphate, but also of COD and TDS, within the urban area indicated indirect effects of war in the form of destroyed wastewater infrastructure. The direct impact of war within the Old City of Mosul as a former conflict zone (S5, S6, S7) is particularly evident in the form of the sudden increase in pollution from S4 to S5. Already between S3 and S4, a clear increase in load occurred due to the discharge of the Khousr valley. The highest load in sediments occurred at S10, which can be attributed to the transport and deposition of polluted sediment due to hydrogeological conditions.

A seasonal comparison showed a decrease in pollutant concentrations in the water samples during the dry season (summer/fall) compared to the wet season (winter/spring), while the pollutant concentration in the sediments increased in contrast during the summer. This effect was caused in the water body by the lack of infiltration of highly contaminated surface drains into the Tigris River and the simultaneous existing infiltration of less contaminated groundwater during the dry season. The increase in pollutants in sediments over the flow distance may have been caused by the deposition of wind-borne, highly contaminated particles and their deposition in the sediment or by low water flow supporting the enhanced deposition of particles transported during the dry season. The hydrogeological conditions of Mosul support the latter scenario.

Apart from pH, the T-test analysis for both series indicated a significant statistical difference between both series for all other parameters in the water samples ($p < 0.05$). For the sediment samples, corresponding significant statistical differences were identified for salinity, Cd, Pb, Zn, and Cr. Since the percentage difference in water samples at S4–S7 is smaller than upstream and downstream, the contaminants are introduced not only via rainwater but also via the year-round infiltration of highly polluted wastewater from the surrounding valleys draining into the river or suburban areas.

It is worth noting that in addition to warfare, a secondary but also relevant source of pollution occurred during the war. This is the use of the river as a disposal system by Mosul's citizens, which was due to the lack and enforcement of environmental laws during and immediately after the war.

A critical comparison of the present data on the water body and the sediment of the Tigris River with studies before the outbreak of the war clearly showed that the water quality deteriorated significantly. Even though the water quality at the entrance to Mosul at an earlier time complied with the WHO limits, the pollution of the river increased sharply during the city passage, and the Tigris required a natural regeneration distance of 40 km to regain acceptable water quality [6]. Due to the effects of the war, as well as the wild discharges by the habitants, the water quality of the river is even more significantly impaired, such that downstream cities such as Hammam Al aleel are still affected by the low water quality.

Author Contributions: Z.A.: Sampling, material preparation, data collection and analysis, writing, literature research. D.D.: Study conception, sampling plan, literature screening, writing. All authors have read and agreed to the published version of the manuscript.

Funding: This research received no external funding.

Data Availability Statement: Data are contained within the article.

Acknowledgments: We are very grateful for the support of the consulting office for laboratory tests in the College of Agriculture and Forestry/Mosul University for analytical support.

Conflicts of Interest: The authors declare no conflict of interest.

References

1. Reuveny, R.; Mihalache-O'Keef, A.S.; Li, Q. The Effect of Warfare on the Environment. *J. Peace Res.* **2010**, *47*. Available online: https://journals.sagepub.com/doi/abs/10.1177/0022343310382069 (accessed on 13 November 2023). [CrossRef]
2. Pearshouse, R. Witnessing the Environmental Impacts of War Environmental Case Studies from Conflict Zones around the World. November 2020. Available online: https://reliefweb.int/report/world/wit (accessed on 13 November 2023).
3. Regional Environmental Center for Central and Eastern Europe(RECEU). Assessment of the Environmental Impact of Military Activities during the Yugoslavia Conflict, European Commission DG XI Environment, Nuclear Safety and CivilProtection. 2015. Available online: https://reliefweb.int/report/albania/assessment-environmental-impact-military-activities-during-yugoslavia-conflict (accessed on 13 November 2023).

4. Bazyan, S. Environmental Impact of War Technology and Prohibition Processes. Master's Thesis, Mid Sweden University, Östersund, Sweden, May 2012. Available online: https://www.diva-portal.org/smash/get/diva2:545221/FULLTEXT01.pdf (accessed on 13 November 2023).
5. UN Habitat for Better Urban (UNHU). City Profile of Mosul, IRAQ, Multi-Sector Assessment of a City under Siege. October 2016. Available online: https://unhabitat.org/city-profile-of-mosul-iraq-multi-sector-assessment-of-a-city-under-siege (accessed on 13 November 2023).
6. UN Report. Environmental Issues in Areas Retaken from ISIS in Mosul, Iraq Rapid Scoping Mission (5_19). July–August 2017. Available online: https://reliefweb.int/report/iraq/technical-note-environmental-issues-areas-retaken-isil-mosul-iraq-rapid-scoping-mission (accessed on 13 November 2023).
7. UNEP. Technical. Mosul Debris Management Assessment (20_55). May 2018. Available online: https://www.unep.org/resources/publication/mosul-debris-management-assessment (accessed on 13 November 2023).
8. UNEO&OCHA. A Rapid Overview of Environmental and Health Risks Related to Chemical Hazards in the Mosul Humanitarian Response. 4 November 2016. Available online: https://reliefweb.int/report/iraq/rapid-overview-environmental-and-health-risks-related-chemical-hazards-mosul (accessed on 13 November 2023).
9. ISPI; Plebani, A. After Mosul, Re-Inventing Iraq. Report Study Submitted for ISPI Executive. 2017. Available online: https://www.ispionline.it/it/pubblicazione/after-mosul-re-inventing-iraq-17080 (accessed on 13 November 2023).
10. Mustafa, M.H.; Muna, H.J. Quality Differences between Two Locations on Tigris River Within Mosul City. *Rafidain J. Sci.* **2007**, *18*, 111–124.
11. Alfadehl, M.; Almasri, N. Pollutant Variation through Tigris River in Mosul City. *Int. J. Interdiscip. Res. Innov.* **2014**, *2*, 38–58.
12. Alfdhel, M. Pollution Investigation on Tigris River within Mosul Area, Iraq. *Plant Arch.* **2020**, *20* (Suppl. S2), 1273–1277.
13. Al-Ahmady, K.; Farhan, A. Assessment of Tigris River Water Quality in Mosul for Drinking and Domestic Use by Applying CCME Water Quality Index. *IOP Conf. Ser. Mater. Sci. Eng.* **2019**, *737*, 012204. Available online: https://iopscience.iop.org/article/10.1088/1757-899X/737/1/012204 (accessed on 13 November 2023).
14. Al-Sarraj, E.S.; Jankeer, M.H.; Al-Rawi, S.M. Estimation of the Concentrations of some Heavy Metals in Water and Sediments of Tigris River in Mosul City. *Rafidain J. Sci.* **2019**, *28*, 1–10. [CrossRef]
15. Aoyoma, I.; Yoshinob, I.; Yoriteru, I. Experimental study on the concentration process of trace element through a food chain from the view point of nutrition ecology. *Water Res.* **1978**, *12*, 831–836. [CrossRef]
16. Schillinger, J.; Özerol, G.; Güven-Griemert, Ş.; Heldeweg, M. Water in war: Understanding the impacts of armed conflict on water resources and their management. *WIREs Water* **2020**, *7*, e1480. [CrossRef]
17. Al-Meshhadani, Y.; Jassim, A. Study of some Characteristics of Tigris River Between Mosul City and Hamam Al-Aleel Province. *Iraq Acad. Sci. J.* **2012**, *23*, 56–67.
18. Othman, K.H.; Bilal, A.; Sulaiman, Y. Morphologic Characteristics of Tigris River with at Mosul City. *Tikrit J. Eng. Sci.* **2012**, *19*, 33–54. [CrossRef]
19. APHA; AWWA; WCPE. *Stand Method for Examination of Wastewater American Public Health Association*, 23rd ed.; Public Health Associated: Washington DC, USA, 2017. Available online: https://www.pdfdrive.com/standard-methods-for-the-examination-of-water-and-wastewater-e11311928.html (accessed on 13 November 2023).
20. Jackson, M.L. Soil Chemical Analysis. Prentice Hall Inc., Englewood Chiffs. 1958, pp. 213–214. Available online: https://www.scirp.org/(S(lz5mqp453edsnp55rrgjct55))/reference/ReferencesPapers.aspx?ReferenceID=2049089I (accessed on 13 November 2023).
21. WHO. *Guidelines for Drinking Water Quality-World Health*, 2nd ed.; All Rights Reserved Surveillance and Control of Community Supplies; World Health Organization: Geneva, Switzerland, 2014; pp. 133–415. Available online: https://apps.who.int/iris/handle/10665/38551 (accessed on 13 November 2023).
22. Al-Saffawi; Almashhadani, M. Application of the Water Quality Index (CWQI) to aquatic life: A case study of the Tigris and Al-Khusser rivers in the city of conductor. In Proceedings of the First and Third International Scientific Conference of the College of Science, Tikrit, Iraq, 17–18 December 2018. Available online: https://www.academia.edu/44468883 (accessed on 13 November 2023).
23. Kevat, D.; Dubey, M.; Saxena, A.K.; Gour, A. Assessment of water quality index of Saank river at Morena, M.P. *Int. J. Sci. Eng. Technol.* **2016**, *5*, 2563–2567. Available online: https://iosrjen.org/Papers/vol9_issue11/Series-3/E0911032331.pdf (accessed on 13 November 2023).
24. Brandt, M.J.; Johnson, K.M.; Elphinston, A.J.; Ratnayaka, D.D. Specialized and Advanced Water Treatment Processes. In *Twort's Water Supply*; Elsevier: Amsterdam, The Netherlands, 2017; pp. 407–473.
25. EPA, US. Available online: https://archive.epa.gov/water/archive/web/html/vms59.html (accessed on 13 November 2023).
26. ENR. Available online: https://www.enr.gov.nt.ca/sites/enr/files/conductivity.pdf (accessed on 13 November 2023).
27. Jassim, S.Z.; Raiswell, R.; Bottrell, S.H. Genesis of the Middle Miocene strata bound sulphur deposits of northern Iraq. *J. Geol. Soc.* **2007**, *156*, 25. Available online: https://www.diva-portal.org/smash/get/diva2:1075532/FULLTEXT01.pdf (accessed on 13 November 2023). [CrossRef]

28. Alsam, S.I.; Jassim, S.Z.; Hanna, F. Water Balance of Iraq, Stage 2, Geological and Hydrogeological Conditions. Unpublished Report, Ministryof Irrigation, Iraq. 1990. Available online: https://www.diva-portal.org/smash/get/diva2:1075532/FULLTEXT01 (accessed on 13 November 2023).
29. Araim, H. Geological Map of Iraq 1:1 000 000 Series, Sheet 5, Hydrogeological Map of Iraq. Publication of the Geological Survey of Iraq. 1991. Available online: https://www.scirp.org/(S(czeh2tfqyw2orz553k1w0r45))/reference/referencespapers.aspx?referenceid=1301794 (accessed on 13 November 2023).
30. Meride, Y.; Ayenew, B. Drinking Water Quality Assessment and Its Effects on Residents Health in Wondo Genet Campus, Ethiopia. *Environ. Syst. Res.* **2016**, *5*, 1. [CrossRef]
31. Sasikaran, S.; Sritharan, K.; Balakumar, S.; Arasaratnam, V. Physical, chemical and microbial analysis of bottled drinking water. *Ceylon Med. J.* **2012**, *57*, 111–116. Available online: https://pubmed.ncbi.nlm.nih.gov/23086026 (accessed on 13 November 2023). [CrossRef] [PubMed]
32. Al-Saffawi; Asaaf, A. An environmental and pehological study of the alternative phosphates of Wadi Al-Danfali and its effect on The flow of the waters of the Tigris River in Al-Sahsal District, Iraq. *J. Educ. Sci.* **2014**.
33. Mustafa, M.H. Wadi Al- Mur a Natural drainagSystem for the Al-Jazzera Northern Irrigation Project. *J. Environ. Res. Evol. Prog. Baghdad* **2002**, *5*, 3–11, (Arabic Version). Available online: https://iraq.un.org/en/184040-rehabilitation-north-al-jazeera-irrigation-scheme-restoring-vital-water-supply-rabea-ninewa (accessed on 13 November 2023).
34. Aljazeera.com. 3 June 2017. Iraq to Probe Use of White Phosphorus in Mosul. abgerufen am. 29 July 2017. Available online: https://www.aljazeera.com/news/2017/6/3/iraq-to-probe-use-of-white-phosphorus-in-mosul (accessed on 13 November 2023).
35. Reuters, A.P.; DPA. Irak: USA Verteidigen Einsatz von Phosphor. stern.de. 17 November 2005. Available online: https://www-stern-de.translate.goog/politik/ausland/irak-usa-verteidigen-einsatz-von-phosphor-3290784.html (accessed on 13 November 2023).
36. Mariussen, E.; Fjellsbø, L.; Frømyr, T.R.; Johnsen, I.V.; Karsrud, T.E.; Voie, Ø.A. Toxic Effects of Gunshot Fumes from Different Ammunitions for Small Arms on Lung Cells Exposed at the Air Liquid Interface. *Toxicol. Vitr.* **2021**, *72*, 105095. [CrossRef]
37. Jovanović, D.; Janković, P.; Radovanović, M.; Đurić, S. The replacement of cadmium coating on parts of the weapon with tungsten-disulphide coating. *Adv. Technol.* **2018**, *7*, 64–68. [CrossRef]
38. BBC. Mosul Battle: Hundreds Treated over Toxic Fumes in Iraq. 22 October 2016. Available online: https://www.bbc.com/news/world-middle-east-37738667 (accessed on 13 November 2023).
39. Al-Sarraj, E.S. Qualitative assessment of water of the Al-Khazer river between Mosul and, Erbil city. *J. Educ. Sci.* **2020**, *29*, 135–848. [CrossRef]
40. Deese 2021—Eindringverhalten von Patronne Full Metal Jacket(FMJ) Ammo vs. Hollow Point Rounds. Miscellaneous by Kyle Deese. 16 February 2021. Available online: https://www.muddybay.com/blog/full-metal-jacketfmj-ammo (accessed on 13 November 2023).
41. Coget, Y.; Novak, J.; Gütter, G.; Demarty, Y.; Rusinek, A. Dynamic testing and simulation of 9 mm full metal jacket ammunition. *Eur. Phys. J. Conf.* **2021**, *250*, 05002. [CrossRef]
42. Al-Saffawi, A.Y.T. Study of the validity of groundwater for the Conseh area/Hamidat area for agricultural purposes. *J. Educ. Sci.* **2007**, *20*, 191–204. [CrossRef]
43. Al-Jawadi, A.S.; Ali, S.H.; Adeeb, H.G.M. Engineering Investigation of the Geological Problems for Old Mosul City by Remote Sensing and GIS. *IOP Conf. Ser. Earth Environ. Sci.* **2022**, *1080*, 012018. Available online: https://iopscience.iop.org/article/10.1088/1755-1315/1080/1/012018/pdf (accessed on 13 November 2023). [CrossRef]
44. Ojo, O.M.; Obiora-Okeke, O.A.; Olabanji, T.O. Seasonal Variation of Physiochemical Properties of River Water Samples in Akure, South-Western Nigeria. *J. Civ. Eng. Urban.* **2022**, *12*, 1–6. [CrossRef]
45. Miranda, L.S.; Ayoko, G.A.; Egodawatta, P.; Goonetilleke, A. Adsorption-desorption behavior of heavy metals in aquatic environments: Influence of sediment, water and metal ionic properties. *J. Hazard. Mater.* **2022**, *421*, 126743. [CrossRef] [PubMed]
46. Buyang, S.; Yi, Q.; Cui, H.; Wan, K.; Zhang, S. Distribution and adsorption of metals on different particle size fractions of sediments in a hydrodynamically disturbed canal. *Sci. Total Environ.* **2019**, *670*, 654–661. [CrossRef] [PubMed]
47. Keshavarzifard, M.; Moore, F.; Sharifi, R. The influence of physicochemical parameters on bioavailability and bioaccessibility of heavy metals in sediments of the intertidal zone of Asaluyeh region, Persian Gulf, Iran. *Geochemistry* **2019**, *79*, 178–187. [CrossRef]
48. Brady, J.P.; Ayoko, G.A.; Martens, W.N.; Goonetilleke, A. Temporal trends and bioavailability assessment of heavy metals in the sediments of Deception Bay, Queensland, Australia. *Mar. Pollut. Bull.* **2014**, *89*, 464–472. [CrossRef]
49. Jayarathne, A.; Egodawatta, P.; Ayoko, G.A.; Goonetilleke, A. Transformation processes of metals associated with urban road dust: A critical review. *Crit. Rev. Environ. Sci. Technol.* **2019**, *49*, 1675–1699. [CrossRef]
50. Chon, H.; Ohandja, D.; Voulvoulis, N. The role of sediments as a source of metalsin river catchments. *Chemosphere* **2012**, *88*, 1250–1256. [CrossRef]
51. Xia, F.; Zhang, C.; Qu, L.; Song, Q.; Ji, X.; Mei, K.; Dahlgren, R.A.; Zhang, M. A comprehensive analysis and source apportionment of metals in riverine sediments of a rural-urban watershed. *J. Hazard. Mater.* **2020**, *381*, 121230. [CrossRef]
52. Dobslaw, D.; Woiski, C.; Kiel, M.; Kuch, B.; Breuer, J. Plant uptake, translocation and metabolism of PBDEs in plants of food and feed industry: A review. *Rev. Envion. Sci. Bio/Technol.* **2021**, *20*, 75–142. [CrossRef]

53. Dobslaw, D.; Kuch, B.; Woiski, C.; Kiel, M. Plant uptake, Translocation and Metabolism of PBDEs in Plants. In *Flame Retardant and Thermally Insulating Polymers*; Xu, Y., Ed.; Intech open German Research Foundation (DFG): Bonn, Germany, 2021; ISBN 978-1-83968-715-0. Available online: https://link.springer.com/article/10.1007/s11157-020-09557-7 (accessed on 13 November 2023).
54. Zarraq, G. Studying the Pollution of Tigris River Sediments Between Al-Qayara-Balad. Research Gate. January 2012. Available online: https://www.iasj.net/iasj/article/64260 (accessed on 13 November 2023).

environments

MDPI

Article

Removal of Cadmium and Lead from Synthetic Wastewater Using *Galdieria sulphuraria*

Hari Lal Kharel [1], Ina Shrestha [1], Melissa Tan [1] and Thinesh Selvaratnam [1,2,*]

[1] Department of Civil and Environmental Engineering, Lamar University, Beaumont, TX 77705, USA; hkharel@lamar.edu (H.L.K.); ishrestha@lamar.edu (I.S.); mtan2@lamar.edu (M.T.)

[2] Center for Advances in Water & Air Quality, College of Engineering, Lamar University, Beaumont, TX 77705, USA

* Correspondence: tselvaratnam@lamar.edu

Abstract: The strain of red microalgae *Galdieria sulphuraria* CCMEE 5587.1 was evaluated in a controlled laboratory environment for its ability to tolerate and remove two heavy metal (HM) ions: cadmium [Cd(II)] and lead [Pb(II)] in aqueous solutions as a single metal species. Various concentrations (0 mg L^{-1} to 5 mg L^{-1}) of Cd and Pb ions were added to the Cyanidium medium in which the chosen microalgae strain *G. sulphuraria* CCMEE 5587.1 was grown at an acidic pH of 2.5. The effectiveness of *G. sulphuraria* CCMEE 5587.1 in tolerating and removing these two metal ions was measured by analyzing its growth profile, growth rate, nutrient removal, and metal ion removal efficiency. The growth of *G. sulphuraria* CCMEE 5587.1 was inhibited during the initial days of incubation, and the growth rate decreased when the HM concentration in the media was increased. Nutrient removal in the HM-containing media is comparable to that in the control media at low metal concentrations but decreases as the metal concentration rises. *G. sulphuraria* CCMEE 5587.1 has the highest removal efficiency for Cd and Pb in a medium containing 2.5 mg L^{-1} of metal ions, which is 49.80% and 25.10%, and the corresponding sorption capacity is 1.45 mg g^{-1} and 0.53 mg g^{-1} of dry biomass, respectively. These findings suggest that *G. sulphuraria* CCMEE 5587.1 holds potential as a viable bioremediation solution for extracting Cd and Pb from wastewater, alongside its capacity to remove nutrients concurrently. The study underscores the dual advantage of *G. sulphuraria* CCMEE 5587.1, making it a promising candidate for addressing heavy metal pollution in wastewater treatment processes.

Keywords: bioremediation; *Galdieria sulphuraria*; heavy metals; wastewater; acidic pH; removal efficiency

1. Introduction

The surge in industrialization and urbanization has given rise to a significant influx of wastewater into the natural water ecosystem. The widespread utilization of metals in industrial processes has resulted in the release of substantial quantities of metal-laden waste into the environment. Industries spanning oils, textiles, fossil fuels, metallurgy, electroplating, galvanization, and others contribute to the discharge of harmful HM ions into the surroundings [1,2]. Given the non-biodegradable nature of HMs and their propensity to accumulate within the food chain, their introduction into the environment carries detrimental implications [3]. The presence of HM pollution bears unfavorable consequences for human health and exerts adverse impacts on ecosystems and aquatic biodiversity. Recognizing the potential severity of the issue, environmental regulations have been instated to safeguard natural water sources against HM contamination. This is especially vital as the pollution of natural water bodies by HMs has emerged as a pressing concern in recent times [4].

Cd and Pb stand as examples of toxic inorganic metals inherent to the Earth's crust. Their exceptionally low tolerance thresholds for toxicity classify both Cd and Pb as non-threshold toxins. Pb, recognized as a hazardous HM, finds its way into wastewater through human-induced activities, such as mining and the combustion of fossil fuels [5]. Similarly, Cd is a very poisonous, invasive HM released into the environment due to mining, battery disposal, and other human activities [6]. Remarkably, these HMs have detrimental implications for the environment and human well-being, even in minuscule quantities. Depending on the extent of exposure, these HMs can elicit both moderate and severe effects on human health. Consequences encompass a spectrum of ailments including, but not limited to, elevated blood pressure, damage to vital organs, such as the kidneys, liver, and lungs, gastrointestinal disorders, cancer, and even fatality [1,7,8]. The pervasiveness of these potential effects emphasizes the urgency of addressing Cd and Pb contamination due to their far-reaching consequences.

An array of physical, chemical, and biological processes, encompassing adsorption, chemical precipitation, ion exchange, and electrochemical treatment, have demonstrated the ability to eliminate metal ions from aqueous solutions [9,10]. However, these technologies often encounter limitations, including heightened energy demands, incomplete removal of HMs, generation of secondary contamination, and the imposition of substantial operation and maintenance costs [11,12]. Furthermore, many of these methods prove ineffective and unsuitable for addressing HM concentrations below 100 mg L^{-1} [8]. In light of these challenges, a pressing necessity arises to establish environmentally sound approaches for extracting HMs from wastewater. This urgency is driven by the need to surmount the shortcomings of existing techniques and develop efficient, cost-effective, and sustainable solutions that tackle varying concentrations of HMs, particularly those at lower levels.

The phycoremediation approach of removing HMs is based on interactions between HMs and microalgae. Microalgae play a vital role in establishing and maintaining healthy ecological relationships and interactions in aquatic environments [13]. The ecological role of microalgae is being investigated for its potential to purify water and wastewater that contains HMs [14]. Microalgae are increasingly recognized for their ability to eliminate both nutrients and HMs from wastewater due to their exceptional photosynthetic efficiency, pronounced binding affinity, abundant binding sites, and expansive surface area [8]. Despite the potential for elevated metal concentrations to induce toxic effects and hinder growth, numerous microalgae have evolved adaptive mechanisms for counteracting excessive metals. These mechanisms include secretion, compartmentalization, and chelation with metal ligands, enabling microalgae to detoxify surplus metals [15]. This dual capability of nutrient removal and HM detoxification underscores the value of microalgae-based strategies in environmental remediation.

G. sulphuraria, CCMEE 5587.1, a unicellular red alga, demonstrates remarkable resilience in challenging environments, such as high temperatures (37–55 °C) and extremely low pH levels (0.0–4.0) [16]. Moreover, its resistance to HMs commonly found in sulfur springs has been documented [17]. Given its affiliation with such demanding conditions, *G. sulphuraria* stands out as a promising subject for investigating its tolerance to HMs and its potential in phycoremediation practices [18]. The versatility of *G. sulphuraria*'s metabolism is notable; it can thrive using over 50 distinct carbon sources, adeptly switching between autotrophic, heterotrophic, and mixotrophic growth modes. This adaptability positions it as a prospective asset in biotechnology applications [19]. Notably, *G. sulphuraria* has showcased its capability to effectively eliminate carbon and nutrients, such as ammoniacal nitrogen (NH_4-N) and phosphate-P (PO_4-P) from various sources, including municipal wastewater, landfill leachate, and produced water [16,20–23]. Research findings reveal that *G. sulphuraria* harnesses these nutrients for its growth, leading to a reduction of these elements within the wastewater effluent. This ability to utilize and mitigate nutrients underscores its potential as a valuable component in wastewater treatment strategies.

Domestic wastewater and landfill leachate are typically characterized by their richness in nutrients and containing trace amounts of HMs. Landfill leachate, for instance, showcases a Pb concentration spanning 2–5 mg L^{-1} [24], whereas Cd content remains below 1 mg L^{-1} [25]. Given the relatively low metal concentrations observed in municipal wastewater and landfill leachate, the implementation of distinct chemical and physical recovery techniques often proves unfeasible or cost-prohibitive. Furthermore, some industries and mining operations, such as acid mine drainage (AMD), release extremely acidic wastewater that includes HM [26,27]. This effluent can then find its way into municipal wastewater treatment systems. Treatment of this kind of effluent may be best accomplished by the use of acid-resistant microalgae, such as *G. sulphuraria* CCMEE 5587.1.

This study embarks on assessing the effectiveness of *G. sulphuraria* CCMEE 5587.1 in the removal of Cd and Pb at low concentrations (<5 mg L^{-1}), simultaneously engaging in nutrient removal at acidic pH. Moreover, the study endeavors to quantitatively gauge the inhibitory impact of Cd and Pb ions on the growth and biomass production of *G. sulphuraria* CCMEE 5587.1. By delving into these aspects, the research aims to ascertain the viability of *G. sulphuraria* CCMEE 5587.1 as a potential solution for mitigating Cd and Pb contamination while also addressing the interplay between metal toxicity and microalgae growth.

2. Materials and Methods

2.1. Algal Strain

This study assessed the feasibility of *G. sulphuraria* CCMEE 5587.1 isolates sourced from the Culture Collection of Microorganisms from Extreme Environments at the University of Oregon. The chosen strain was cultivated within an incubator (Percival, IA, USA) at a temperature of 40 °C, with constant illumination maintained at 4000 lux for a duration of 24 h [23,28]. The cultivation process began with axenic cultures achieved through streaking cultures onto agar plates. Subsequently, single colonies were selected for transfer to progressively larger flasks, ultimately culminating in 4 L Erlenmeyer flasks.

2.2. Cultural Media

In the research, *G. sulphuraria* CCMEE 5587.1 was grown in Cyanidium medium (CM). The following macro and micro level ingredients were utilized in the preparation of CM: $(NH_4)_2SO_4$: 1.32 g L^{-1}; KH_2PO_4: 0.27 g L^{-1}; NaCl: 0.12 g L^{-1}; $MgSO_4{\cdot}7H_2O$: 0.25 g L^{-1}; $CaCl_2{\cdot}2H_2O$: 0.07 g L^{-1}; Nitch's Trace Element Solution: 0.5 mL L^{-1}; $FeCl_3$ (solution = 0.29 g L^{-1}): 1.0 mL L^{-1} [16]. The media's pH was maintained at 2.5 using 10 N H_2SO_4. To ensure the complete sterility of the medium, it underwent autoclaving at 121 °C for a duration of 50 min prior to utilization.

2.3. Synthetic Wastewater Solution Preparation

To simulate actual wastewater concentrations, a solution containing Cd and Pb below 5 mg L^{-1} was prepared by diluting $PbCl_2$ and $CdCl_2$ with CM media that contains the nutrients for algal growth. All the chemicals used were analytical-grade, and all the glassware was autoclaved prior to use. The Cd and Pb stock solution of 5 mg L^{-1} was made first, and then the solutions were diluted to the desired concentrations.

2.4. Exposure of *G. sulphuraria* CCMEE 5587.1 to Cd and Pb Added Media

G. sulphuraria CCMEE 5587.1 was harvested during the exponential growth phase and centrifuged at 3000 rpm for 10 min at 4 °C using Centrifuge 5920R (Eppendorf AG 22331 Hamberg, Germany). After the centrifugation, the supernatant was discarded, and the biomass was re-suspended in Cd and Pb-added media.

2.5. Metal and Nutrient Removal Experiment

The experiments were carried out in 125 mL Erlenmeyer flasks holding 50 mL of the experimental sample. The experiment contained three metal concentrations of 5 mg L^{-1}, 2.5 mg L^{-1}, and 1.25 mg L^{-1}. For each concentration, two controls (one without *G. sulphuraria* CCMEE 5587.1 and one with *G. sulphuraria* CCMEE 5587.1 with no HM ions) were included to verify the presence of metal contamination and the impact of environmental circumstances and factors on the removal process. All flasks were placed on a New Brunswick

Innova 2050 platform shaker (Eppendorf, Edison, NJ, USA) and rotated at 120 rpm for 7 days.

2.6. Measurement of Biomass Density

The biomass density was determined by measuring the optical density (OD) at 750 nm with a HACH DR 3900 spectrophotometer (HACH, Loveland, CO, USA). Daily measurements of OD were taken to determine the maximum possible growth rate. Ash-free dry weight (g AFDW L^{-1}) was used to quantify biomass density, and this value was shown to be proportional to OD at 750 nm using the formula [22]:

$$Y = 0.4775 \times X - 0.0163,\ R^2 = 0.9967,\ \text{n} = 12;\ r^2 = 0.997 \quad (1)$$

Y = Ash-free dry wt. (AFDW)
X = OD value at 750 nm

2.7. Measurements of Nutrients

Nutrient (ammoniacal nitrogen and phosphorus) measurements were carried out at the start and end of each experiment. The measurements were performed with a HACH DR 3900 (HACH, Loveland, CO, USA) spectrophotometer, and the related methods were Salicylate TNT Method 10,031 and Phosver 3Method 8048 [22].

2.8. Measurement of Cd and Pb Concentration

After seven days of incubation, samples were collected and centrifuged at 6000 rpm for 10 min with a AccuSpin 400 centrifuge, (Fisher Scientific, D-77520 Osterode, Germany) to separate the supernatant fraction from the cell fraction. The concentration of Cd and Pb in the supernatant was measured by inductively coupled plasma atomic emission spectroscopy (ICP-AES) (SHIMADZU ICPE 9820 simultaneous ICP atomic emission spectroscopy). The Cd and Pb ion removal efficiency was calculated by using the following equation:

$$\text{Removal } (\%) = \frac{\left(c_i - c_f\right)}{c_i} \quad (2)$$

c_i and c_f represent the metal ion concentrations in the supernatant on Day 7 in the media without algal cells and with algal cells, respectively.

2.9. Statistical Analysis and Graph Plotting

All experimental results were gathered from three independent biological replications and provided as a mean of replicates ± standard deviation. Microsoft Excel was used to determine the mean and standard deviation of the sample. ORIGINPRO 2023 was used to generate all the featured graphs.

3. Results

3.1. Growth of *G. sulphuraria* CCMEE 5587.1 in Different Concentrations of Cd and Pb

Exposure of algal cells to HMs is known to induce notable morphological and metabolic transformations, consequently impacting growth. To investigate the effects of Cd and Pb ions on growth, *G. sulphuraria* CCMEE 5587.1 was cultivated both in the absence and presence of Cd and Pb across concentrations spanning from 0 to 5 mg L^{-1}. Results from the growth profiles demonstrate that when *G. sulphuraria* CCMEE 5587.1 was exposed to varied concentrations of the tested metal ions, there were clear changes in the biomass density of algal cells between the control and treated groups. The growth curves for *G. Sulphuraria* CCMEE 5587.1 subjected to various concentrations of Cd and Pb are shown in Figure 1a,b. *G. sulphuraria* CCMEE 5587.1 went through a lag, exponential, and stationary growth phase during the seven-day incubation period. For Days 1 and 2, after being exposed to Cd and Pb added media, *G. sulphuraria* CCMEE 5587.1's growth was inhibited and decreased. In the Pb-added media, there was a more significant inhibition in the growth of *G. sulphuraria* CCMEE 5587.1 on the initial day, and the exponential growth phase was delayed.

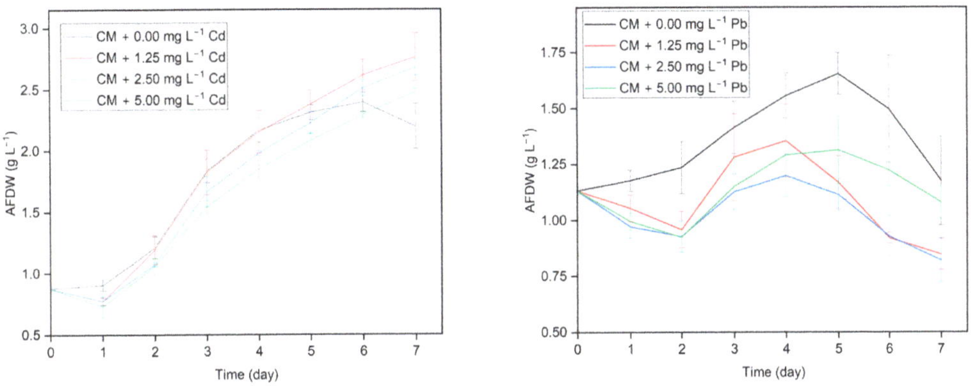

(a) (b)

Figure 1. Growth profile of *G. sulphuraria* CCMEE 5587.1 in the media with different concentrations of (**a**) Cd and (**b**) Pb.

3.2. Growth Rate

The growth rate was estimated based on the growth in the exponential phase. Figure 2a,b illustrates the effects of varying concentrations of Cd and Pb on the growth rate of *G. sulphuraria* CCMEE 5587.1. The first two days after being exposed to Cd and Pb, *G. sulphuraria* CCMEE 5587.1 grew slower than the control group, but after two days, it grew faster than the control group. This could be attributed to the fact that *G. sulphuraria* CCMEE 5587.1 requires a period to acclimate to environments contaminated with HMs, which has not been previously exposed.

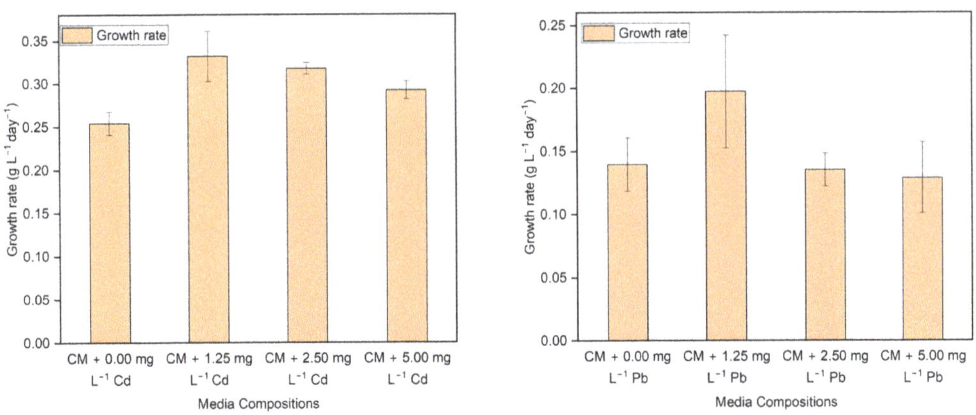

(a) (b)

Figure 2. The growth rate of *G. sulphuraria* CCMEE 5587.1 in the media with different concentrations of (**a**) Cd and (**b**) Pb.

The growth rate of *G. sulphuraria* CCMEE 5587.1 exposed to Cd-added media was greater than that of the control media for every Cd concentration despite a brief initial lag phase on the first day. Exposed to 1.25, 2.5, and 5 mg L^{-1} of Cd, the growth rates were

0.331 ± 0.029, 0.317 ± 0.006, and 0.292 ± 0.010 g L^{-1} day^{-1}, correspondingly, whereas the growth rate was 0.253 ± 0.013 g L^{-1} day^{-1} in the control media. In the case of Pb-added media, the growth rate of *G. sulphuraria* CCMEE 5587.1 is higher than the control media in the media having a Pb concentration of 1.25 mg L^{-1}. However, the growth rate is less than the control media in the media having a Pb concentration of 2.5 and 5 mg L^{-1}. In detail, the growth rate was 0.197 ± 0.044, 0.135 ± 0.013, 0.128 ± 0.028 g L^{-1} day^{-1} when exposed to 1.25, 2.5, and 5 mg L^{-1} of Pb added media, respectively, whereas the growth rate is 0.139 ± 0.021 g L^{-1} day^{-1} in the control media.

3.3. Bioremoval of Cd and Pb

G. sulphuraria CCMEE 5587.1 was able to remove Cd and Pb from the media after seven days of treatment with living cells. The results of the ICP_AES analysis performed on the initial and final samples are shown in Figure 3a,b. Based on the results, *G. sulphuraria* CCMEE 5587.1 exhibits varying levels of effectiveness in the removal of Cd and Pb.

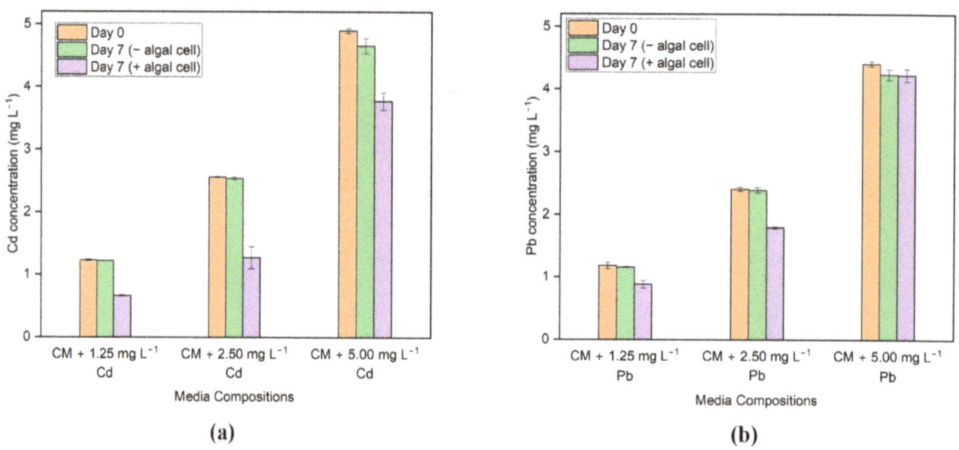

Figure 3. HM concentration before and after the treatment in the media with different concentrations of (**a**) Cd and (**b**) Pb.

In the case of Cd, the highest Cd removal efficiency by *G. sulphuraria* CCMEE 5587.1 occurs in the media containing 2.5 mg L^{-1} Cd, which is 49.80%. However, the removal efficiency was only 19.09% in the media that included a Cd concentration of 5 mg L^{-1}. In addition, the removal effectiveness was 45.90% in the media when the Cd content was 1.25 mg L^{-1}. Moreover, the sorption capacity is highest in the media containing 2.5 mg L^{-1} Cd, which is 1.45 mg g^{-1} of dry biomass, and the lowest value is 0.64 mg g^{-1} of dry biomass in the media containing 1.25 mg L^{-1} Cd.

In the case of Pb, the highest Pb removal by *G. sulphuraria* CCMEE 5587.1 occurs in the media containing 2.5 mg L^{-1} of Pb, which is 25.10%. In addition, the removal efficiency was only 23.47% in the media, which included a Pb concentration of 1.25 mg L^{-1}. However, no removal was observed in the media with a Pb concentration of 5 mg L^{-1}. Moreover, the sorption capacity is highest in the media containing 2.5 mg L^{-1} Pb, which is 0.53 mg g^{-1} of dry biomass, and the lowest value is 0.23 mg g^{-1} of dry mass in the media containing 1.25 mg L^{-1} Pb.

3.4. Nutrient Removal

In addition to Cd and Pb concentrations, nitrogen and phosphorus concentrations in the media were monitored to infer the efficacy of *G. Sulphuraria* CCMEE 5587.1 for the removal of nutrients in the presence of these toxic HM. The nitrogen content is measured as ammoniacal nitrogen (NH_4-N) in the media. Figure 4a,b shows the NH_4-N concentration present in Cd and Pb-added media before and after the incubation. On Day 0, the medium contains around 290 mg L^{-1} NH_4-N. On Day 7, media with 0, 1.25, 2.5, and 5 mg L^{-1} Cd show NH_4-N removal of 78.9%, 76.47%, 67.11%, and 62.10%, respectively. Similarly, on Day 7, NH_4-N removal is 83.6% in media with no Pb, 80.74% in media with 1.25 mg L^{-1} Pb, 75.05% in media with 2.5 mg L^{-1} Pb, and 69.89% in media with 5 mg L^{-1} Pb. The removal of NH_4-N is nearly comparable to the control media when adding lower metal ion concentrations in the media. However, as the metal ion concentration rises, the efficiency of removal decreases.

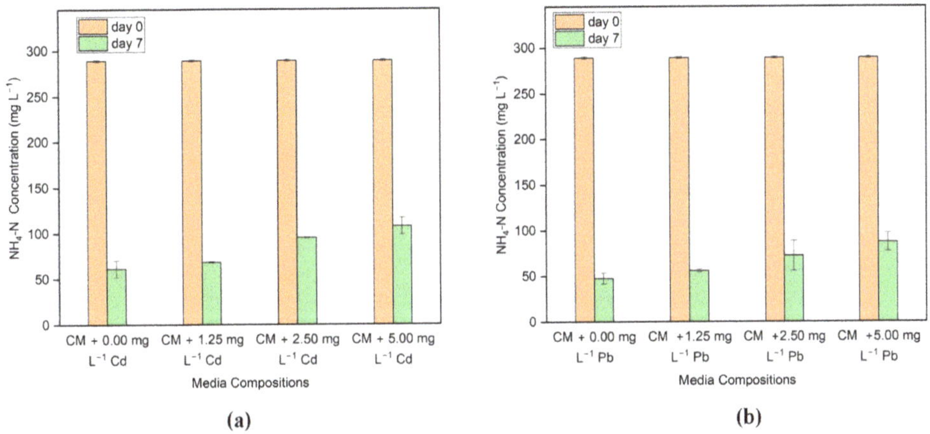

Figure 4. Ammoniacal nitrogen concentration before and after the treatment in the media with different concentrations of (**a**) Cd and (**b**) Pb.

Phosphorus content is measured as phosphate-P (PO_4-P) in the media. Figure 5a,b shows comparisons of the PO_4-P content in Cd and Pb-added media before and after incubation. The initial concentration of PO_4-P in the medium is about 67 mg L^{-1}. Graphs show that on Day 7, PO_4-P removal is nearly the same as the control media in the media having Cd and Pb concentrations. From the nutrient removal results, it can be said that there is no significant effect on PO_4-P removal in the presence of low concentrations of Cd and Pb.

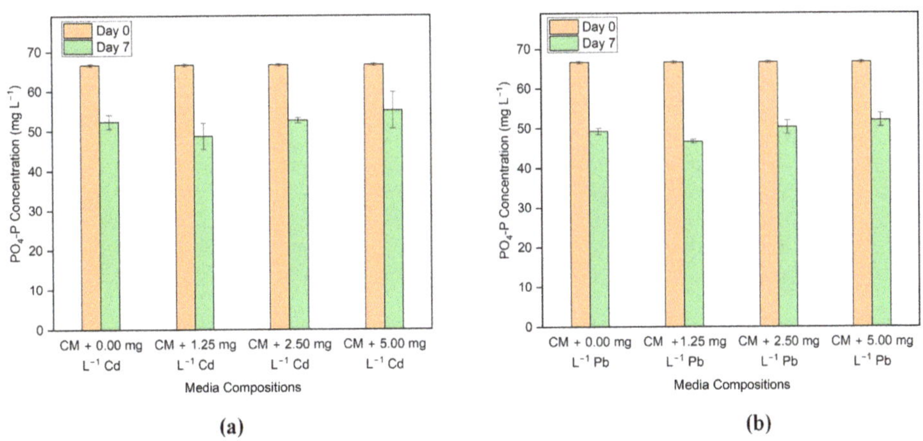

Figure 5. Phosphorus concentration before and after the treatment in the media with different concentrations of (**a**) Cd and (**b**) Pb.

4. Discussions

Phycoremediation represents an innovative strategy for water and wastewater treatment, leveraging the inherent biological capabilities of microalgae to facilitate the removal of contaminants. Current research extensively explores the bioremoval of Cd and Pb from wastewater using diverse green and red algae strains. The primary aim of this study was to evaluate *G. sulphuraria* CCMEE 5587.1's removal potential for Cd and Pb within a seven-day incubation period. It was found that *G. sulphuraria* CCMEE 5587.1 can remove Cd and Pb from the contaminated media with varied effectiveness.

The highest Cd removal efficiency and sorption capacity for *G. sulphuraria* CCMEE 5587.1 for the given concentrations were 49.80% and 1.45 mg g^{-1}, respectively. Previous investigations have explored Cd bioremoval with red algal strains. Isachsen et al. [29] investigated *Cyanidioschyzon merolae* strains MS1 and 10D, finding that strain 10D achieved a 31.55% removal efficiency after a seven-day treatment, while strain MS1 achieved 1.16%. Another study by Ostroumov et al. [30] employed the *G. sulphuraria* IPPAS P-513 strain, effectively removing Cd, Cu, Pb, and Ni over 30 days of incubation, resulting in a 24% reduction in average Cd content. The literature indicates that most of the research for Cd bioremoval has been conducted with green algae strains. According to the research conducted by Abinandan et al. [6], acid-tolerant microalgae *Heterochlorella* sp. MAS3 and *Desmodesmus* sp. MAS1 were used to remove 2 mg L^{-1} Cd from the solution and achieved the results with approximately 80% and 60% removal efficiency and 0.36 and 0.77 mg g^{-1} sorption capacity, respectively. Plohn et al. [3] investigated Cd-tolerant microalgal strains in Nordic environments where *Chlorella vulgaris* and *Coelastrella* sp. demonstrated substantial removal abilities, eliminating 72% and 82% of Cd, respectively, with maximum sorption capacities of 49 mg g^{-1} and 65 mg g^{-1}. Under heterotrophic conditions, *Chlorella minutissima* UTEX2341 exhibited impressive Cd removal efficacy, achieving a 74.34% removal efficiency and a maximum sorption capacity of 35.65 mg g^{-1} [31]. Additional successes include the use of living seaweed *Ulva lactuca* [32], achieving 56% Cd removal efficiency, *Pseudochlorococcum typicum* [33] achieving 86%, *Chlorella vulgaris* [4] reaching 95.20%, and *Oedogonium westti* [34] with a range of 55–95% Cd removal efficiencies. Table 1 demonstrates the comparisons between the results of this study and other literature studies on the removal of Cd from wastewater using various algal species at different concentrations of Cd and pH conditions.

In this research, *G. sulphuraria* CCMEE 5587.1 exhibited a maximum Pb removal efficiency of 25.10% and a sorption capacity of 0.53 mg g^{-1} across the tested concentrations. Studies have primarily focused on green algal strains for Pb bioremoval, with fewer investigations on red algal strains. Cho et al. [17] explored Pb biosorption in three red algal strains: *Cyanidium caldarium* NIES 551, *Cyanidioschyzon merolae* NIES 3377, and *Galdieria maximum*, at a pH of 5.0. Among these, *Cyanidium caldarium* demonstrated the highest maximum Pb sorption capacity of 298.4 mg g^{-1}, while *Galdieria maxima* showed the lowest at 38.2 mg g^{-1}. Similarly, *Cyanidioschyzon merolae* displayed a maximum sorption capacity of 214.0 mg g^{-1}. In the study conducted by Ostroumov et al. [30], after 30 days of incubation, the average Pb concentration was reduced by 84% using *G. sulphuraria* IPPAS P-513. However, in a separate study by Ostroumov et al. [35], both stripping voltammetry and vitrified mortmass techniques failed to detect Pb in the biomass. Investigation into Pb removal efficiency using the green microalga *Chlamydomonas reinhardtii* was conducted by Flouty et al. [36], yielding an 8% removal efficiency. Additionally, living seaweed *Ulva lactuca* [32] achieved 87% Pb removal efficiency, *Pseudochlorococcum typicum* [33] achieved 70%, *Chlorella vulgaris* [37] reached 98.70%, and *Oedogonium westti* [34] exhibited a range of 61–96% removal efficiencies. In the context of literature studies, most investigations involving green algal strains were conducted around neutral pH conditions, simplifying the underlying chemistry. It is worth noting that our investigation employed initial metal concentrations of up to 5 mg L^{-1}, whereas many Pb removal studies used concentrations of 1 mg L^{-1} or lower. For a comprehensive overview, refer to Table 2, which presents comparisons between the outcomes of this study and other literature investigations focusing on

Pb removal from wastewater, highlighting different algal species, Pb concentrations, and pH conditions.

Table 1. Removal efficiency of Cd using live algae from this study vs. the literature review.

Algal Strain	Initial Cd Concentration (mg L^{-1})	pH	Sorption Capacity (mg g^{-1})	Removal Efficiency (%)	References
		Green Algae			
Chlorella vulgaris	2.50	7.2	49.00	72.00	[3]
Chlorella vulgaris	100.00	-	16.34	95.20	[4]
Coelastrella sp.	2.50	7.2	65.00	82.00	[3]
Oedogonium westii	0.50–2.00	5.0	0.974	55.00–95.00	[34]
Ulva lactuca	0.20	7.8	0.018	56.00	[32]
Desmodesmus sp. MAS1	1.00–2.00	3.5	0.37–0.77	60.00	[6]
Heterochlorella sp. MAS3	1.00–2.00	3.5	0.16–0.36	80.00	[6]
Chlorella minutissima UTEX2341	67.48	6.0	35.36	74.34	[31]
Pseudochlorococcum typicum	10.00	7.0	6.26	86.24	[33]
		Red Algae			
Cyanidioschyzon merolae 10D	1.00	1.75	-	31.55	[29]
Cyanidioschyzon merolae MS1	5.00	1.75	-	1.16	[29]
Galdieria sulphuraria IPPAS P-513	5.00	2.7	-	24.00	[30]
Galdieria sulphuraria CCMEE 5587.1	5.00	2.5	1.02	19.09	Existing Study
	2.50		1.45	49.80	
	1.25		0.64	45.90	

The maximum allowable levels of Cd and Pb in drinking water are 0.005 and 0.015 mg L^{-1}, respectively, as set by the United States Environmental Protection Agency (EPA) [38]. However, the regulation limit of Cd and Pb in wastewater effluent is determined by several variables, including the nature and purpose of the industry, the state of the art in terms of feasible control technology, and the method of discharge (direct or indirect). For instance, Cd and Pb levels in centralized waste treatment effluent cannot exceed 0.163 and 0.283 mg L^{-1}, respectively, as stated by the US EPA in 40 CFR part 437 [39]. Under the conditions of this investigation, the final concentration of Cd and Pb is greater than the EPA's allowable concentration in drinking water and wastewater effluent. Future studies will focus on optimizing this system and figuring out how to combine this technology with others to obtain the targeted final concentration.

Table 2. Removal efficiency of Pb using live algae from this study vs. the literature review.

Algal Strain	Initial Pb Concentration (mg L^{-1})	pH	Sorption Capacity (mg g^{-1})	Removal Efficiency (%)	References
		Green Algae			
Chlamydomonas reinhardtii	0.10	6.0	0.042	8.00	[36]
Chlorella vulgaris	1.95–4.83	6.2	-	71.81–98.70	[37]

Table 2. *Cont.*

Algal Strain	Initial Pb Concentration (mg L^{-1})	pH	Sorption Capacity (mg g^{-1})	Removal Efficiency (%)	References
		Green Algae			
Oedogonium westii	0.10–0.80	5.0	0.261	61.00–96.00	[34]
Ulva lactuca	1.00	7.8	0.145	87.00	[32]
Dunaliella salina **AL-1**	5.00–10.00	-	-	95.00	[40]
Scenedesmus **sp.**	0.05–10.00	-	13.60	70.00	[41]
Pseudochlorococcum typicum	10.00	7.0	5.11	70.06	[33]
		Red Algae			
Galdieria sulphuraria	0.10	2.4	-	0.00	[35]
Galdieria sulphuraria **IPPAS P-513**	5.00	2.7	-	84.00	[30]
Galdieria maxima	0.00–500.00	5.0	38.20	-	[17]
Cyanidioschyzon merolae **NIES 3377**	0.00–500.00	5.0	214.00	-	[17]
*Cyanidium caldarium***NIES 551**	0.00–500.00	5.0	298.80	-	[17]
Galdieria sulphuraria **CCMEE 5587.1**	5.00	2.5	0.00	0.00	Existing Study
	2.50		0.5	25.10	
	1.25		0.225	23.47	

Our study's findings highlight the efficacy of *G. sulphuraria* CCMEE 5587.1 in removing Cd and Pb, particularly at lower metal concentrations. However, as the metal concentration increases, the removal efficiency diminishes, aligning with conclusions from other published studies. Pham et al. [41] employed the green algal strain *Scenedesmus* sp. to address various Pb concentrations spanning 0.05–10 mg g^{-1}. Their results revealed the highest removal efficiency at an initial Pb concentration of 2 mg L^{-1}, while the lowest was observed at 10 mg L^{-1}, indicating a decline in efficiency as Pb concentration rises. Similarly, Regaldo et al. [37] employed *Chlorella vulgaris* cells to tackle different Pb concentrations within the range of 1.95–4.83 mg g^{-1}. Their outcomes demonstrated that removal efficiency decreases with higher Pb concentrations but increases with extended exposure time. The observed removal efficiencies ranged from 71.81% to 98.70%. Furthermore, the green macroalgae *Oedogonium westti* exhibited Cd removal efficiencies ranging from 55% to 95% and Pb removal efficiencies ranging from 61% to 96% within a different initial metal concentration [34]. Notably, the highest removal efficiency was achieved at lower initial metal concentrations. This trend can be attributed to the phenomenon where, at lower concentrations, all metal ions in the solution interact with binding sites, optimizing adsorption. Conversely, higher concentrations can saturate these binding sites, resulting in unabsorbed ions [42].

Another noteworthy finding from our study is the initial inhibition and subsequent reduction in the growth of *G. sulphuraria* CCMEE 5587.1 during the experiment's early stages. These findings resonate with the research of Bajguz et al. [43], who worked with *Chlorella vulgaris* and observed a decrease in growth and chemical composition (chlorophyll, monosaccharides, and protein content) during the initial 48 h of cultivation under HM ion stress. Furthermore, the current study found a direct correlation between growth inhibition and the amount of metal concentration at which microalgae were exposed, and similar conclusions were drawn by Monteiro et al. [44] and Yang et al. [31]. Interestingly, our study also underscores that Pb ions have more pronounced inhibitory effects on growth compared to Cd ions. This observation aligns with the results presented by Alharbi et al. [45], wherein Pb ion demonstrated greater inhibiting effects among Pb, Cd, and Zn ion on *Chroococcus minutus* and *Chlorococcum aegyptiacum*. The findings collectively suggest that initial growth inhibition is a common response to HM exposure in various microalgae species, and the extent of inhibition correlates with the concentration of the metal. Furthermore, the differential inhibitory effects of various HMs, such as Pb and Cd, on microalgal growth are consistent with prior research.

Although this study did not investigate the actual mechanisms involved in the removal of Cd and Pb by using *G. sulphuraria* CCMEE 5587.1, an attempt has been made to discuss the possible defense mechanisms involved in microalgae in order to mitigate the toxic effects of HM on microalgal cells. The biosorption process of HMs by microalgal cells involves two distinct stages. Initially, metal ions are adsorbed onto the cell surface. This step may or may not involve metabolic activities. In the subsequent phase, metal ions traverse through the internal pathways of living cells and accumulate within the cytoplasm. This process occurs at a relatively gradual pace and necessitates metabolic processes, as described by Flouty et al. [36] and Priyadarshini et al. [2]. Active uptake of metal ions is a common occurrence in living cells, often driven by the need for microbial growth or intracellular accumulation. However, the reactivity of many HMs makes them hazardous to living cells. The toxicity of HMs primarily stems from their oxidative effects on the components within cells. Within the algal cell, antioxidants, such as glutathione, play a crucial role in safeguarding against oxidative stress induced by metals and other contaminants, as noted by Pinto et al. [46]. When intracellular HM levels rise, they can hinder the effectiveness of antioxidant defenses. This interference can lead to damage in the chloroplast, subsequently slowing down photosynthesis and overall biomass production. Moreover, the presence of HMs can also disrupt essential cellular functions. For instance, the competition between HMs and crucial ions, such as calcium and sodium, at binding sites has been shown to inhibit growth, a phenomenon highlighted by Yang et al. [31]. Additionally, the functional groups of biologically essential molecules, such as enzymes and transport mechanisms for vital nutrients, can be obstructed by HMs, resulting in toxicity within the organism, an observation also made by Napan et al. [47].

The toxicity of HMs is counteracted by a diverse array of extracellular and intracellular removal mechanisms present in various organisms. In the case of living microalgae, a range of defense strategies is employed to counteract HM toxicity. These mechanisms include the biosorption of HMs on the cell wall and extracellular polymeric substances, bioaccumulation of HMs within intracellular compartments, and the transformation of HMs by algal cells, as highlighted in research by Danouche et al. [14]. A prevalent detoxification strategy involves sequestering HMs within internal organelles such as vacuoles, chloroplasts, and mitochondria. Shanab et al. [33] conducted research using *Pseudochlorococcum typicum* and observed an electron-dense layer on the surfaces of cells treated with both Cd and Pb. Additionally, Pb-treated cells exhibited spherical electron-dense bodies within the cell. These observations might be attributed to metal ions bonding with various functional groups on the cell surface, while the presence of spherical electron-dense bodies in Pb treated cells suggests the bioaccumulation of metal ions within the cell. Elleuch et al. [40] investigated metal removal mechanisms using marine microalgae *Dunaliella salina* and found that both intracellular and extracellular removal mechanisms contribute to the removal process. Yang et al. [31] conducted research on *Chlorella minutissima* UTEX2341 and found that Cd removal primarily occurs through intracellular accumulation supplemented by extracellular immobilization. The carboxylic moieties present in the cell walls were identified as responsible for Cd removal in *Chlorella vulgaris* and *Coelastrella sp.*, as observed in the study by Plohn et al. [3]. Notably, *Euglena gracilis* responds to Cd exposure by forming Cd complexes within the chloroplast as a removal mechanism, as indicated by research conducted by Mendoza et al. [48]. These diverse mechanisms collectively underscore the adaptability of microalgae in countering HM toxicity.

5. Conclusions

This research examines the various studies of *G. sulphuraria* CCMEE 5587.1, including daily biomass density, growth rate, nutrient removal, and metal ion removal, to determine its biosorption efficiency for the removal of Cd and Pb from wastewater. The study's findings unveil that both Cd and Pb ions manifest inhibitory effects on the initial growth of *G. sulphuraria* CCMEE 5587.1, contingent on the prevailing concentrations. Nevertheless, by the conclusion of the culture period, effective elimination of Cd and Pb from the aqueous solution is attained, accompanied by notable nutrient removal. While this study contributes valuable insights, there remain unexplored avenues that warrant further investigation. Among these are the intricate mechanisms underpinning the removal of Cd and Pb, the feasibility of scaling up these processes for large-scale applications, and the system's performance when confronted with real wastewater effluents. Notably, *G. sulphuraria* CCMEE 5587.1's ability to thrive within an environment containing toxic HMs, such as Cd and Pb, coupled with its capacity to extract substantial quantities of metal ions from the system, presents an exciting prospect. This discovery might pave the way for the incorporation of *G. sulphuraria* CCMEE 5587.1 into phycoremediation strategies. As the research continues to evolve, it holds the potential to not only enhance our comprehension of microalgal interactions with HMs but also to offer practical solutions for wastewater treatment through the utilization of *G. sulphuraria* CCMEE 5587.1.

Author Contributions: Conceptualization, T.S. and H.L.K.; methodology, T.S., H.L.K. and I.S.; software, T.S. and H.L.K.; validation, T.S.; formal analysis, T.S. and H.L.K.; investigation, T.S., H.L.K., M.T. and I.S.; resources, T.S.; data curation, T.S. and H.L.K.; writing—original draft preparation, T..S., H.L.K. and I.S; writing—review and editing, T.S., H.L.K., M.T. and I.S.; visualization, T.S. and H.L.K.; supervision, T.S.; project administration, T.S.; funding acquisition, T.S. All authors have read and agreed to the published version of the manuscript.

Funding: This study was supported with funds from the State of Texas as part of a program of the Texas Hazardous Waste Research Center (THWRC). The contents do not necessarily reflect the views and policies of the sponsor, nor does the mention of trade names or commercial products constitute endorsement or recommendation for use.

Data Availability Statement: The data presented in this study are available on request from the corresponding authors. The data are not publicly available due to the continuation of a follow-up study by the authors.

Acknowledgments: The support provided by the Center for Advances in Water and Air Quality (CAWAQ) at Lamar University and The Office of Undergraduate Research at Lamar University in the undergraduate author's SURF fellowship are acknowledged.

Conflicts of Interest: The authors declare no conflict of interest.

References

1. Abdi, O.; Kazemi, M. A review study of biosorption of heavy metals and comparison between different biosorbents. *J. Mater. Environ. Sci.* **2015**, *6*, 1386–1399.
2. Priyadarshini, E.; Priyadarshini, S.S.; Pradhan, N. Heavy metal resistance in algae and its application for metal nanoparticle synthesis. *Appl. Microbiol. Biotechnol.* **2019**, *103*, 3297–3316. [CrossRef]
3. Plöhn, M.; Escudero-Oñate, C.; Funk, C. Biosorption of Cd(II) by Nordic microalgae: Tolerance, kinetics and equilibrium studies. *Algal Res.* **2021**, *59*, 102471. [CrossRef]
4. Cheng, J.; Yin, W.; Chang, Z.; Lundholm, N.; Jiang, Z. Biosorption capacity and kinetics of cadmium(II) on live and dead Chlorella vulgaris. *J. Appl. Phycol.* **2016**, *29*, 211–221. [CrossRef]
5. Purushanahalli Shivagangaiah, C.; Sanyal, D.; Dasgupta, S.; Banik, A. Phycoremediation and photosynthetic toxicity assessment of lead by two freshwater microalgae Scenedesmus acutus and Chlorella pyrenoidosa. *Physiol. Plant.* **2021**, *173*, 246–258. [CrossRef]
6. Abinandan, S.; Subashchandrabose, S.R.; Venkateswarlu, K.; Perera, I.A.; Megharaj, M. Acid-tolerant microalgae can withstand higher concentrations of invasive cadmium and produce sustainable biomass and biodiesel at pH 3.5. *Bioresour. Technol.* **2019**, *281*, 469–473. [CrossRef]
7. Yan, C.; Qu, Z.; Wang, J.; Cao, L.; Han, Q. Microalgal bioremediation of heavy metal pollution in water: Recent advances, challenges, and prospects. *Chemosphere* **2022**, *286*, 131870. [CrossRef]
8. Leong, Y.K.; Chang, J.-S. Bioremediation of heavy metals using microalgae: Recent advances and mechanisms. *Bioresour. Technol.* **2020**, *303*, 122886. [CrossRef]
9. Wang, J.; Chen, C. Biosorbents for heavy metals removal and their future. *Biotechnol. Adv.* **2009**, *27*, 195–226. [CrossRef] [PubMed]

10. Mehta, S.; Gaur, J. Use of algae for removing heavy metal ions from wastewater: Progress and prospects. *Crit. Rev. Biotechnol.* **2005**, *25*, 113–152. [CrossRef] [PubMed]
11. Goswami, R.; Agrawal, K.; Shah, M.; Verma, P. Bioremediation of heavy metals from wastewater: A current perspective on microalgae-based future. *Lett. Appl. Microbiol.* **2022**, *75*, 701–717. [CrossRef] [PubMed]
12. Shrestha, R.; Ban, S.; Devkota, S.; Sharma, S.; Joshi, R.; Tiwari, A.P.; Kim, H.Y.; Joshi, M.K. Technological trends in heavy metals removal from industrial wastewater: A review. *J. Environ. Chem. Eng.* **2021**, *9*, 105688. [CrossRef]
13. Afkar, E.; Ababna, H.; Fathi, A. Toxicological response of the green alga Chlorella vulgaris, to some heavy metals. *Am. J. Environ. Sci.* **2010**, *6*, 230. [CrossRef]
14. Danouche, M.; El Ghachtouli, N.; El Arroussi, H. Phycoremediation mechanisms of heavy metals using living green microalgae: Physicochemical and molecular approaches for enhancing selectivity and removal capacity. *Heliyon* **2021**, *7*, e07609. [CrossRef]
15. Hall, J.L. Cellular mechanisms for heavy metal detoxification and tolerance. *J. Exp. Bot.* **2002**, *53*, 1–11. [CrossRef]
16. Selvaratnam, T.; Pan, S.; Rahman, A.; Tan, M.; Kharel, H.L.; Agrawal, S.; Nawaz, T. Bioremediation of Raw Landfill Leachate Using Galdieria sulphuraria: An Algal-Based System for Landfill Leachate Treatment. *Water* **2022**, *14*, 2389. [CrossRef]
17. Cho, Y.-L.; Lee, Y.-C.; Hsu, L.-C.; Wang, C.-C.; Chen, P.-C.; Liu, S.-L.; Teah, H.-Y.; Liu, Y.-T.; Tzou, Y.-M. Molecular mechanisms for Pb removal by Cyanidiales: A potential biomaterial applied in thermo-acidic conditions. *Chem. Eng. J.* **2020**, *401*, 125828. [CrossRef]
18. Kharel, H.L.; Shrestha, I.; Tan, M.; Nikookar, M.; Saraei, N.; Selvaratnam, T. Cyanidiales-Based Bioremediation of Heavy Metals. *BioTech* **2023**, *12*, 29. [CrossRef]
19. Oesterhelt, C.; Schmälzlin, E.; Schmitt, J.M.; Lokstein, H. Regulation of photosynthesis in the unicellular acidophilic red alga Galdieria sulphuraria. *Plant J.* **2007**, *51*, 500–511. [CrossRef]
20. Selvaratnam, T.; Pegallapati, A.; Montelya, F.; Rodriguez, G.; Nirmalakhandan, N.; Van Voorhies, W.; Lammers, P. Evaluation of a thermo-tolerant acidophilic alga, Galdieria sulphuraria, for nutrient removal from urban wastewaters. *Bioresour. Technol.* **2014**, *156*, 395–399. [CrossRef]
21. Selvaratnam, T.; Pegallapati, A.; Reddy, H.; Kanapathipillai, N.; Nirmalakhandan, N.; Deng, S.; Lammers, P. Algal biofuels from urban wastewaters: Maximizing biomass yield using nutrients recycled from hydrothermal processing of biomass. *Bioresour. Technol.* **2015**, *182*, 232–238. [CrossRef] [PubMed]
22. Pan, S.; Dixon, K.L.; Nawaz, T.; Rahman, A.; Selvaratnam, T. Evaluation of Galdieria sulphuraria for nitrogen removal and biomass production from raw landfill leachate. *Algal Res.* **2021**, *54*, 102183. [CrossRef]
23. Rahman, A.; Pan, S.; Houston, C.; Selvaratnam, T. Evaluation of Galdieria sulphuraria and Chlorella vulgaris for the Bioremediation of Produced Water. *Water* **2021**, *13*, 1183. [CrossRef]
24. Gherasim, C.-V.; Křivčík, J.; Mikulášek, P. Investigation of batch electrodialysis process for removal of lead ions from aqueous solutions. *Chem. Eng. J.* **2014**, *256*, 324–334. [CrossRef]
25. Kumar, M.; Kushwaha, A.; Goswami, L.; Singh, A.K.; Sikandar, M. A review on advances and mechanism for the phycoremediation of cadmium contaminated wastewater. *Clean. Eng. Technol.* **2021**, *5*, 100288. [CrossRef]
26. Cui, M.; Jang, M.; Cho, S.-H.; Khim, J.; Cannon, F.S. A continuous pilot-scale system using coal-mine drainage sludge to treat acid mine drainage contaminated with high concentrations of Pb, Zn, and other heavy metals. *J. Hazard. Mater.* **2012**, *215*, 122–128. [CrossRef]
27. Palma, H.; Killoran, E.; Sheehan, M.; Berner, F.; Heimann, K. Assessment of microalga biofilms for simultaneous remediation and biofuel generation in mine tailings water. *Bioresour. Technol.* **2017**, *234*, 327–335. [CrossRef]
28. Toplin, J.A.; Norris, T.B.; Lehr, C.R.; McDermott, T.R.; Castenholz, R.W. Biogeographic and phylogenetic diversity of thermoacidophilic cyanidiales in Yellowstone National Park, Japan, and New Zealand. *Appl. Environ. Microbiol.* **2008**, *74*, 2822–2833. [CrossRef]
29. Isachsen, I. *Cadmium Tolerance in the Thermo-Acidophilic Red Alga C. merolae, Possible Mechanisms and Implications for Bioremediation*; Arizona State University: Tempe, AZ, USA, 2022.
30. Ostroumov, S.; Tropin, I.; Kiryushin, A. Removal of cadmium and other toxic metals from water: Thermophiles and new biotechnologies. *Russ. J. Gen. Chem.* **2018**, *88*, 2962–2966. [CrossRef]
31. Yang, J.; Cao, J.; Xing, G.; Yuan, H. Lipid production combined with biosorption and bioaccumulation of cadmium, copper, manganese and zinc by oleaginous microalgae Chlorella minutissima UTEX2341. *Bioresour. Technol.* **2015**, *175*, 537–544. [CrossRef]
32. Henriques, B.; Teixeira, A.; Figueira, P.; Reis, A.T.; Almeida, J.; Vale, C.; Pereira, E. Simultaneous removal of trace elements from contaminated waters by living Ulva lactuca. *Sci. Total Environ.* **2019**, *652*, 880–888. [CrossRef] [PubMed]
33. Shanab, S.; Essa, A.; Shalaby, E. Bioremoval capacity of three heavy metals by some microalgae species (Egyptian Isolates). *Plant Signal. Behav.* **2012**, *7*, 392–399. [CrossRef] [PubMed]
34. Shamshad, I.; Khan, S.; Waqas, M.; Asma, M.; Nawab, J.; Gul, N.; Raiz, A.; Li, G. Heavy metal uptake capacity of fresh water algae (Oedogonium westti) from aqueous solution: A mesocosm research. *Int. J. Phytoremediation* **2016**, *18*, 393–398. [CrossRef] [PubMed]
35. Ostroumov, S.; Shestakova, T.; Tropin, I. Biosorption of copper by biomass of extremophilic algae. *Russ. J. Gen. Chem.* **2015**, *85*, 2961–2964. [CrossRef]
36. Flouty, R.; Estephane, G. Bioaccumulation and biosorption of copper and lead by a unicellular algae Chlamydomonas reinhardtii in single and binary metal systems: A comparative study. *J. Environ. Manag.* **2012**, *111*, 106–114. [CrossRef] [PubMed]

37. Regaldo, L.M.; Gervasio, S.G.; Troiani, H.E.; Gagneten, A.M. Bioaccumulation and Toxicity of Copper and Lead in Chlorella vulgaris. *J. Algal Biomass Utln.* **2013**, *4*, 59–66.
38. Kinuthia, G.K.; Ngure, V.; Beti, D.; Lugalia, R.; Wangila, A.; Kamau, L. Levels of heavy metals in wastewater and soil samples from open drainage channels in Nairobi, Kenya: Community health implication. *Sci. Rep.* **2020**, *10*, 8434. [CrossRef]
39. USEPA, Effluent Guidelines Database. Available online: https://www.epa.gov/eg/effluent-guidelines-database (accessed on 20 September 2023).
40. Elleuch, J.; Hmani, R.; Drira, M.; Michaud, P.; Fendri, I.; Abdelkafi, S. Potential of three local marine microalgae from Tunisian coasts for cadmium, lead and chromium removals. *Sci. Total Environ.* **2021**, *799*, 149464. [CrossRef]
41. Pham, T.-L.; Dao, T.-S.; Bui, H.N.; Pham, T.K.N.; Ngo, T.T.H.; Bui, H.M. Lipid production combined with removal and bioaccumulation of Pb by Scenedesmus sp. Green Alga. *Pol. J. Environ. Stud.* **2020**, *29*, 1785–1791. [CrossRef]
42. Kumar, J.N.; Oommen, C. Removal of heavy metals by biosorption using freshwater alga Spirogyra hyalina. *J. Environ. Biol.* **2012**, *33*, 27.
43. Bajguz, A. Suppression of Chlorella vulgaris growth by cadmium, lead, and copper stress and its restoration by endogenous brassinolide. *Arch. Environ. Contam. Toxicol.* **2011**, *60*, 406–416. [CrossRef] [PubMed]
44. Monteiro, C.M.; Fonseca, S.C.; Castro, P.M.; Malcata, F.X. Toxicity of cadmium and zinc on two microalgae, Scenedesmus obliquus and Desmodesmus pleiomorphus, from Northern Portugal. *J. Appl. Phycol.* **2011**, *23*, 97–103. [CrossRef]
45. Alharbi, R.M. Toxicity and bioaccumulation of lead, cadmium and zinc in Chroococcus minutus and Chlorococcum aegyptiacum. *Int. J. Pharm. Res. Allied Sci.* **2017**, *6*, 290–300.
46. Pinto, E.; Sigaud-kutner, T.C.; Leitao, M.A.; Okamoto, O.K.; Morse, D.; Colepicolo, P. Heavy metal–induced oxidative stress in algae 1. *J. Phycol.* **2003**, *39*, 1008–1018. [CrossRef]
47. Napan, K.; Teng, L.; Quinn, J.C.; Wood, B.D. Impact of heavy metals from flue gas integration with microalgae production. *Algal Res.* **2015**, *8*, 83–88. [CrossRef]
48. Mendoza-Cózatl, D.G.; Moreno-Sánchez, R. Cd2+ transport and storage in the chloroplast of Euglena gracilis. *Biochim. Biophys. Acta (BBA)-Bioenerg.* **2005**, *1706*, 88–97. [CrossRef]

environments

MDPI

Article

BioWin Modeling of CalPrex Phosphorus Recovery from Wastewater Predicts Substantial Nuisance Struvite Reduction

Donald Vineyard [1,*], K.G. Karthikeyan [2] and Phillip Barak [1]

1 Madison Department of Soil Science, University of Wisconsin-Madison, Madison, WI 53706, USA; phillip.barak@wisc.edu

2 Madison Department of Biological Systems Engineering, University of Wisconsin-Madison, Madison, WI 53706, USA; kkarthikeyan@wisc.edu

* Correspondence: dlvineyardphd@gmail.com

Abstract: The wastewater treatment industry could benefit from new technologies for the removal and recovery of phosphorus (P). The CalPrex precipitation reactor has the potential to recover P in a readily land-applicable form by treating organic acid digestate with calcium hydroxide to produce brushite. Using data from a pilot-scale reactor at the local Nine Springs Wastewater Treatment Plant in Madison, WI, we modified the plant's BioWin configuration using BioWin 6.2 to model the CalPrex technology and estimate performance under a variety of conditions. We produced dose/response curves for a range of possible lime dosages to estimate the impact of reagent dosage on the quantity and composition of precipitate produced by the CalPrex reactor and characterize the effects on downstream anaerobic digester performance. CalPrex was found to capture 46% of the plant's influent P, reducing nuisance struvite precipitates by 57% and biosolid sludge production by 14%. The CalPrex module was also tested in two predesigned plant configurations in the BioWin cabinet with the intention of testing applicability to other configurations and searching for the impacts of CalPrex on treatment train performance. This is the first work simulating a full-scale implementation of CalPrex and the first to model interactions of CalPrex with other treatment processes.

Keywords: nutrient recovery; nutrient management; brushite; precipitates; multi phase digestion; acidogenic digest; two phase digestion

1. Introduction

The United States (U.S.) consumes more than 23 million short tons of nutrient fertilizer each year, 4.7 million short tons of which is phosphorus (P) [1]. While the bulk of U.S. nutrient pollution originates from the use of agricultural fertilizers, manure application, and eroded runoff, a sizeable portion stems from the disposal of treated wastewater effluents [2]. Conventional treatment methods can reduce plant effluent from 4–10 mg P/L to 1–2 mg P/L [3], but most US treatment plants have no nutrient effluent standards [4]. Nutrient pollution resulting in eutrophication has a variety of associated externalities, including tourism and recreation loss, commercial fishing loss, reduced property values, increased drinking water costs, and human sickness [2].

The nutrients in wastewater are derived from, among other things, biological waste and household and industrial chemicals [2]. Environmental policy has shifted to discourage nutrient pollution of waterways [5] as there is a greater recognition of the deleterious effects of eutrophication, and effective nutrient removal technologies are increasingly commercially viable. In the U.S., the Environmental Protection Agency (USEPA) provides guidance and technical support for states willing to enact nutrient-based water quality guidelines. Phosphorus is a non-renewable resource essential for agriculture [6], and eventually, technology will need to shift towards conservation and recovery of P.

Point sources such as municipal wastewater utilities are relatively easy and effective targets for regulation and control. Most emphasis to date is on P reduction in treated effluent water, usually by enhanced biological phosphorus removal (EBPR) and/or chemical precipitation, resulting in enrichment in biosolids, with only a few nascent technologies for P recovery like the Ostara process [2,7]. In a wastewater treatment plant (WWTP) incorporating EBPR processes, upwards of 90% of influent P can be expected to be captured in the plant biosolids [8], which are typically landfilled, incinerated, or land applied.

A side effect of EBPR processes is the accumulation of nuisance precipitates such as struvite [9]. Struvite, an ammonium magnesium phosphate mineral of the formula $NH_4MgPO_4 \cdot 6H_2O$, is frequently a nuisance precipitate that typically forms when ammonium, magnesium (Mg), and phosphate exceed their solubility under basic conditions caused by the off-gassing of carbon dioxide in anaerobic digesters. Struvite can accumulate in and on digesters, pipes, heat exchangers, and pumps. These precipitates cause operational difficulties in the form of reduced reactor volume, clogged piping, reduced heat exchange efficiency, and increased pumping costs and often require expensive remediation [10,11]. However, struvite can also be precipitated in a controlled manner to prevent impediments to WWTP function and recover P as a potential fertilizer [7].

Land application of biosolids returns P to agricultural soil, though high water content limits the distance biosolids can be economically transported from the plant. In biosolids-applied fields, the soil P levels rapidly exceed those required for crop growth, and the biosolids cease to function as a P fertilizer replacement. Typical P management, as currently practiced, cannot be considered effective from a P recovery standpoint, and therefore, extractive nutrient recovery must constitute an important part of future sustainable nutrient management plans [12]. Many wastewater utilities are exploring novel methods of removing nutrients from their effluent, including technologies such as MagPrex, AirPrex, Prisa, Seaborne, and others [12–25].

The CalPrex process for precipitation of soluble P as brushite, a calcium phosphate mineral of the formula $CaHPO_4 \cdot 2H_2O$, is a novel P recovery technology [26] that has been successfully tested at the pilot stage [27] at the Nine Springs Wastewater Treatment Plant in Madison, WI (Figure 1). CalPrex acts on the effluent of an organic acid digester, typically from a two-phase anaerobic digestion system, that is fed waste-activated sludge (WAS) from an EBPR process. The result is that organic acid digestate has a high concentration of soluble orthophosphate and a chemical composition conducive to phosphate precipitation [28]. In CalPrex operation, sludge from the organic acid digester is first dewatered, and the mildly acidic, phosphate-rich liquid is reacted with calcium hydroxide in a clarifier to a pH of ~6.5–7 to precipitate brushite from the solution. The brushite is settled in the clarifier with the aid of polymer, then dewatered and dried. All reject water from the clarifier and solids removed from the acid digestate return to the intercession point, typically moving on to a methanogenic anaerobic digester.

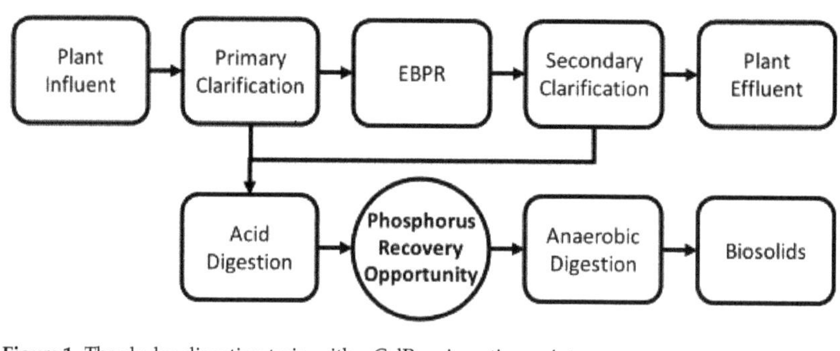

Figure 1. The sludge digestion train with a CalPrex insertion point.

The brushite produced by the CalPrex process can be readily used as a slow-release phosphorus fertilizer [7]. Brushite is a naturally occurring soil phosphate and is rapidly produced when phosphorus is applied to high-calcium soils [29,30]. Pure brushite is 18%

phosphorus by weight, comparable to triple superphosphate. While most phosphorus fertilizers have a mild acidifying effect on soil, brushite has a negative potential acidity and can be considered a mild liming agent [29–31]. Brushite also has a low salt index, reducing runoff and leaching and allowing higher volume, lower frequency application [29–31].

The CalPrex process has not yet been evaluated at full scale nor commercially modeled as part of a full plant configuration. The grand thrust of this work is to evaluate a technology that may remove P from wastewater treatment plants, especially where P precipitates impact essential functions, remove P from land-applied biosolids in fields for which added P might be an environmental concern, and generate a recycled P fertilizer product that will conserve limited P resources. Our primary goal is to make the first prediction of the performance and plantwide impacts of a full-scale CalPrex implementation using a whole-WWTP configuration. The specific objectives of our study are to predict the performance of a full-scale implementation of CalPrex at the Nine Springs Wastewater Treatment plant (Madison, WI), using BioWin 6.2 with modifications to the acid digester process proposed by Vineyard et al. [32,33], and compare results to those of the pilot-scale project [27]. We assess the applicability of these results to other treatment plants and treatment trains using BioWin cabinet models. The resultant configuration will be used in future work to compare phosphorus control technologies, individually and in concert, based on reagent costs, phosphorus recovery, precipitate control, and nutrient emissions. Our results are intended for wastewater engineers considering struvite remediation or nutrient recovery technologies.

2. Materials and Methods

As pilot-scale projects can be prohibitively expensive, it is beneficial to use modeling software to predict the performance of a full-scale installation. BioWin [34] is a wastewater treatment plant modeling program commonly used by engineers to predict the behavior of WWTPs and guide their operations. It can be used to model a variety of wastewater treatment processes and create treatment trains meant to resemble full-scale WWTPs. The software uses a series of hydraulic, chemical, and biological models to predict changes in the chemical and microbial composition of wastewater as it passes through different reactors.

The Nine Springs Wastewater Treatment Plant in Madison, Wisconsin, is a mid-size treatment plant operated by the Madison Metropolitan Sewerage District (MMSD) serving a population of 360,000 over a 180 sq. mile (466 km^2) area, with a design capacity of 50 million gallons per day (50 MGD; 189,300 m^3/day) [35]. This plant contains an EBPR treatment train, generating WAS with a high P content. A 50:50 mixture of primary sludge and WAS is anaerobically digested in a series of mesophilic methanogenic digesters generating biogas, and then the sludge is dewatered to produce biosolids [35]. The first of these digesters is an acid digester, which generates volatile fatty acids (VFAs) from the breakdown of complex biological compounds for the purpose of easier conversion to biogas in downstream digesters.

A portion of this struvite precipitation is observed on-site to form a nuisance scale on heat exchangers and in pipes; another portion gradually accumulates as grit in digesters and reduces reactor volume, and a third portion follows the biosolids out of the plant. At present, BioWin can only model precipitates as suspended, not forming as scale or settling as grit. The first two forms (scale/grit) necessitate expensive treatment stoppages for remediation; struvite reduction was the primary motivation for the subsequent installation of the WASStrip and Ostara systems at Nine Springs. The CalPrex process was piloted in a Water Research Foundation-funded project at the Nine Springs wastewater treatment plant conducted during the spring, summer, and fall of 2018 [27]. The pilot unit handled ~5% of the throughput of the acid digester at Nine Springs. This project [27] yielded a performance report containing details on chemical concentrations, mass balances, and lime usage, which will serve as the primary data source for building our BioWin configuration or from mass balances provided by the Nine Springs engineers. Madison Metropolitan Sewerage District requested that the BioWin model for the Nine Springs Wastewater Treatment Plant not be fully shared for security reasons.

The Nine Springs wastewater treatment plant provided a BioWin configuration, which we assessed in BioWin 6.2. The WASStrip and Ostara reactors were removed from the treatment train, creating a model essentially equivalent to the plant before the incorporation of P control renovations in 2013. The configuration received was from a legacy version of BioWin and did not attempt to model an acid digester, necessitating that we modify it to match plant mass balances. BioWin does not natively model acidogenic digesters, so we created one using the full suite of modifications described in Vineyard et al. [32,33]. These modifications involve the adjustment of multiple default kinetic parameters to replicate the specialized behavior of these digesters and generate predictions in line with sampling data [32,33]. The effluent profile of the resultant acid digester process (Table 1) closely resembled the specifications and chemical analyses of the Nine Springs acid digester [27].

Table 1. Acid Digester Process Effluent Profile.

Parameter	Unit	Value
Flow	m^3/d	830
pH		5.14
Acetate	kg COD/d	2937
Propionate	kg COD/d	7814
Total P	kg P/d	906
Soluble P	kg P/d	635
Soluble Ca	kg Ca/d	207
Ammonia	kg N/d	1104
Total Solids	kg/d	20,800
Volatile Solids	kg/d	17,500

We developed the following two BioWin configurations from the plant data, one intended to serve as a baseline and the other to test for CalPrex impacts:

- The 2013 Plant (baseline configuration): The Nine Springs configuration has no tertiary P removal processes in place, similar to its state before the installation of the Ostara unit in 2014.
- Base configuration plus the CalPrex unit: The 2013 plant with a CalPrex system installed, including its solid/liquid separation, polymer and lime additions, and reactor clarification. Intended as an analog of a treatment plant installing CalPrex as its only P recovery technology.

2.1. CalPrex Modeling

We modified the BioWin reference model by adding a CalPrex system—consisting of polymer dosing, centrifuge, lime dosing, and clarifier—between the acid digester and the anaerobic digesters to the baseline configuration. The CalPrex pilot operation was placed between the acid digester and the subsequent anaerobic digester in the configuration and consisted of (i) dosing the acid digester effluent stream with a polymer to promote solid-liquid separation, (ii) dewatering the acid digestate by centrifugation, (iii) $Ca(OH)_2$ dosing of the centrate in the CalPrex precipitator, (iv) settling the brushite precipitate in a clarifier, (v) centrifuging the settled brushite, and (vi) recombining the acid digestate solids with the liquids from the brushite centrifuge and clarifier before continuing downstream to the anaerobic digestion train. These functions were replicated within the BioWin configuration using a series of unit processes parameterized to match their pilot equivalents in function. The target pH for lime dosing of the CalPrex reactor in our BioWin configurations is 7.0, at which point almost all solubilized P is expected to precipitate out as brushite [28]. Dewatering and clarification parameters were informed by the performance of the pilot-scale equivalents. The CalPrex pilot unit required a large volume of polymer solution, increasing the total liquid volume of the waste stream by 27%.

The pilot reactor [27] added slurried $Ca(OH)_2$ to increase the pH of the acid digestate to pH 6.5–7.5. The primary variable we controlled in our model was the lime dosing, which we controlled using a variable-volume input stream containing 1 kg Ca/L. We experimented with the total lime dosage to examine its effect on pH, P recovery, and reagent efficiency. Marginal return rates for P capture were calculated using the first derivative of second-order Lagrange interpolating polynomials, equivalent to Newton's central difference methods when X-spacing is equal to [36].

To estimate the effects of the CalPrex process on the function of downstream treatment processes, several select metrics of high importance to a treatment plant operator were chosen. This list included material inputs, precipitate quantities, P partitioning, and downstream sludge and biogas generation.

2.2. Uncertainty Analysis

The acid digestate conditions were based in part on the average of 201 pH measurements taken at Nine Springs [27]. Given that WWTPs treat a constantly variable stream of municipal waste, the conditions within the plant are subject to perturbation. The Nine Springs acid digesters did not maintain perfectly constant chemistry over the observation period, as evidenced by the variability in pH measurements taken by plant operators to characterize the conditions in the plant. As both the performance of the BioWin model of an acid digester and the performance of the CalPrex installation itself are sensitive to the pH, including the P solubilization of the acid digester feed, the impacts of this variability are of interest to our project. We use the data collected in the pilot project to predict the effects of acid digestate variability on the modeling of an acid digester and subsequent performance of a CalPrex installation and record the results.

The mean pH of the digester was 5.13, with a standard deviation (SD) of 0.089. 75% of all pH measurements fell within one SD of the mean, and 96% fell within two SD of the mean. From this exercise in uncertainty, measured pH values of 5.13 ± 0.09 (mean $\pm$ standard deviation) require BioWin pH inhibition threshold values of 6.55 ± 0.13 and yield predicted CalPrex P captures of 431 ± 14 kg P/day.

2.3. Cabinet Configuration Demonstration

The modeling effort applied to the Nine Springs Treatment Plant was facilitated by two factors: the well-documented CalPrex pilot conducted at that location [27] and the availability of a comprehensive BioWin configuration of that plant. It is of interest to know how other WWTPs would respond to the application of CalPrex for P recovery. To test the interactions of the CalPrex reactor with other plant reactors, we inserted the CalPrex system into two generic systems in the BioWin cabinet, a repository of prebuilt plant models. We chose full plant models to capture more detailed trains intended to summarize real treatment plants with a complex series of reactors. This would better expose unforeseen complications in treatment that single reactor configurations would not.

The first cabinet configuration, titled "N Removal + Anaerobic Digester + Sidestream Treatment", was a generic plant designed to demonstrate the removal of nitrogen (N) by nitrification/denitrification. The second cabinet configuration, titled "EBPR + Anaerobic Digester", is a generic plant designed to demonstrate the removal of P through EBPR. The two cabinet models were chosen to represent common treatment plant trains in the USA and are the same cabinet models used in Vineyard et al. [32,33] to demonstrate the acid digester modifications. Process layouts for both configurations can be found in the Appendix A.

3. Results

3.1. Baseline Configuration—Nine Springs without Tertiary Phosphorus Control

The 2013 Nine Springs treatment plant configuration models intake of about 950 kg of elemental P per day (Table 2), 94% of which is diverted from plant effluent by the treatment plant's settling tanks, clarifiers, and the EBPR system. The acidogenic digester produces a stream with a P concentration of roughly 1100 mg P/L, of which 706 mg/L are soluble P.

Table 2. Predicted P flows for the Nine Springs Wastewater Treatment Plant, kg P/d.

Process	EBPR-Only Plant	CalPrex Plant
Influent Total P	949	949
Effluent Total P	57.3	54.5
Biosolid Total P	892	457
Biosolid Brushite P	146	206
Biosolid Struvite P	408	83
CalPrex Total P	N/A	438

In the anaerobic digester, destruction of the VFAs by methanogenesis raises the pH to approximately 7.2 under normal operation, favoring the precipitation of P-based precipitates due to the combination of high mineral, ammonia, and soluble P contents. Under these conditions, 554 kg P/d (58% of plant throughput) are projected to precipitate as brushite or struvite within the anaerobic digesters. This equates to roughly 4000 kg of total P precipitate solids per day, approximately three-quarters of which is predicted by the configuration to be struvite. In BioWin, most of these precipitates are captured with the biomass in the biosolids (Table 2) after passing through a gravity belt thickener. The plant is predicted to produce 20 metric tons of dry sludge mass per day, two-thirds of which is volatile suspended solids. Of the non-volatile portion, half is P particulates precipitated in the anaerobic digesters.

3.2. *CalPrex Configuration*

At the target pH of 7.0, CalPrex significantly affects the P flows of the wastewater treatment plant (Table 2) by capturing 438 kg of P per day. The total dry mass of the CalPrex product is 3800 kg/d, of which 2300 kg/d is brushite, resulting in an 11% P purity by dry mass. The reactor captures 1280 kg/d of organic material alongside the brushite. The subsequent methanogenic digester receives 79 kg P/d in uncaptured brushite and precipitates an additional 128 kg P/d for a total flux of 207 kg P/d in brushite flowing through the methanogenic digesters.

Not all the P exiting the acid digester is captured as brushite (Figure 2). First, the process of dewatering the acid digestate leaves 16% of the solubilized P entrained in the solids and, therefore, not accessible for brushite formation. Further, the CalPrex clarifier cannot perfectly capture all the brushite formed, and 17% of brushite formed in the 2018 pilot passed to the anaerobic digesters [27], which is replicated by our model.

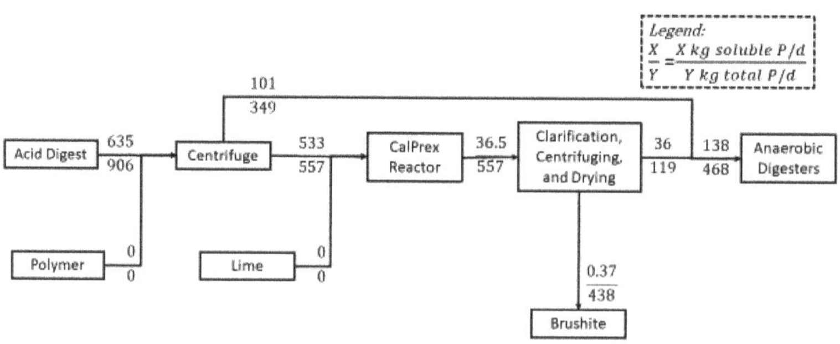

Figure 2. Example P flows within the CalPrex process when dosed with lime to pH 7. Acid digestate is dosed with polymer and centrifuged, diverting most of the organic P and 16% of soluble P to the anaerobic digesters. The CalPrex reactor converts 93% of soluble P to brushite and captures 84% of the precipitates due to imperfections in the subsequent clarification and dewatering processes.

Lime dose–response curves for CalPrex performance were produced by feeding the CalPrex reactor a series of lime doses and calculating a steady-state performance. Dosing the CalPrex reactor with lime caused a predicted pH response curve characteristic of a buffered system with an inflection point at pH 7.6 (Figure 3), here calculated by BioWin but similar to the results calculated using Visual Minteq [28,33]. The titratable acidity of VFAs and soluble phosphates characteristic of acid digestate neutralizes added hydroxides and buffers pH, but once these are neutralized, the pH becomes much more sensitive to lime addition. This sensitivity peaks at the pH 7.6 inflection point, requiring about 1000 kg Ca/day. At higher pH, the solution is buffered by new species such as struvite and ammonia.

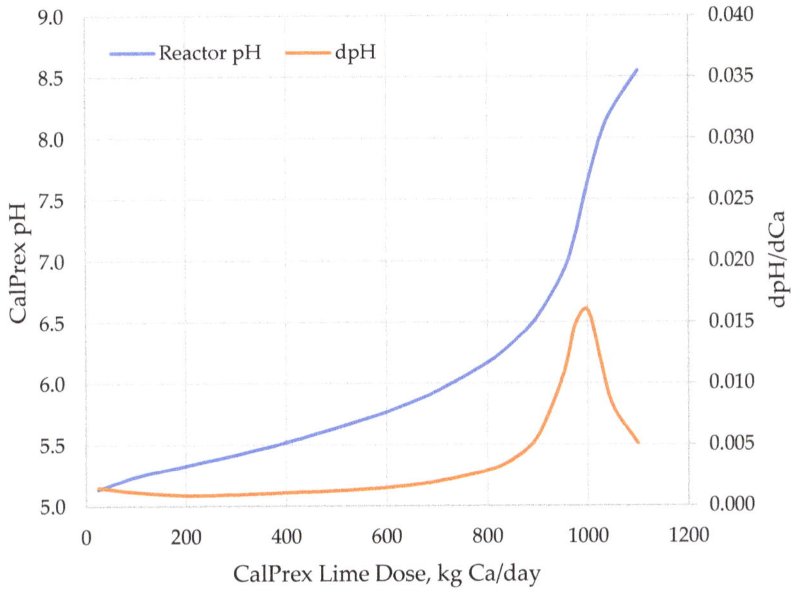

Figure 3. CalPrex reactor pH response to lime dosing.

At low pH, the solubility of brushite is much higher, and the solution was undersaturated with respect to brushite and could not form a precipitate. As a result, small lime doses increase pH but result in no P recovery (Figure 4). Larger doses achieve steady increases in both pH and brushite recovery as acidity is overcome and ionic activities favor precipitation. The response to lime dose is dependent on acid digester stream characteristics. The soluble Ca concentration is more than 50% higher in the pilot stream compared to its BioWin equivalent, which will affect brushite solubility and may need to be considered when interpreting lime dosages; the lower Ca concentration may cause an underprediction of brushite capture or an overprediction of lime needs, but the reaction is limited more by pH than by calcium so this is expected to be minimal. Finding suitable methods to correct the Ca discrepancy or model a range of influent Ca concentrations was outside the scope of this modeling investigation.

During the CalPrex pilot tests [27], some biomass was found to escape the centrifugation and settle in the clarifier with the brushite. Here, even without any lime added, the CalPrex configuration is predicted to capture ~25 kg of non-brushite P per day. Approximately 90% of this P is biomass-P that passes through the centrifuge and settles in the clarifier, and the rest is soluble orthophosphate occluded with the biosolids. The organic P and the associated biomass might be considered either an impurity or an amendment to recovered brushite, depending on the market.

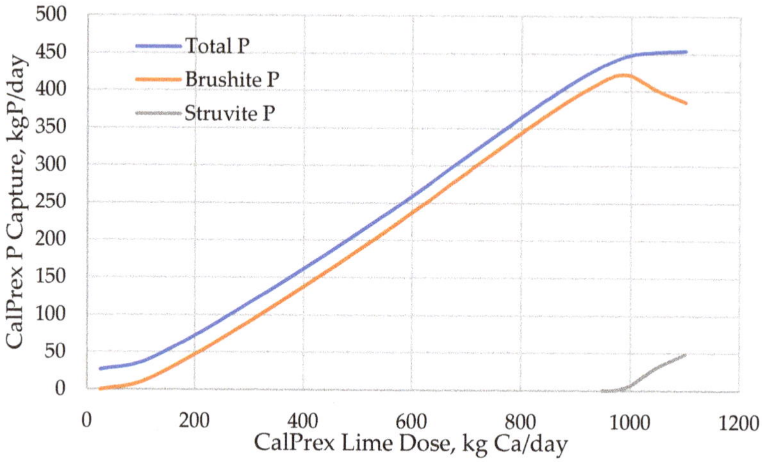

Figure 4. CalPrex reactor P capture response to lime dosing.

Lime addition beyond approximately 800 kg Ca/d showed an acceleration in pH increase as buffering capacity was overcome, and the great majority of the soluble P was removed from the solution as brushite. The target pH of 7 was reached when approximately 960 kg of Ca per day was added, and ~420 kg P/d was captured as brushite. Total predicted brushite capture peaked at roughly 975 kg Ca/d, or pH 7.2, with 82% of soluble-P entering the CalPrex reactor precipitated. Beyond this point, struvite begins to form at the expense of brushite; total P capture continued to increase only slightly with lime addition, and the CalPrex process may be terminated.

These modeling predictions allow calculation of the marginal product, i.e., incremental P capture per lime dose (Figure 5). An ideal system might return brushite P in a 1:1 stoichiometric ratio with added lime-Ca, returning 31 kg P per 40 kg Ca dosed or a 0.775 mass ratio. Instead, the real-world conditions of this acid digestate result in lower yields, reflecting the need for more hydroxide to titrate VFAs and phosphates, to raise pH, drop the solubility of brushite, and precipitate brushite from the digest than for Ca to react with phosphate to form brushite. Each consecutive kg of lime added causes a corresponding quantity of P to precipitate, but this quantity does not remain constant. An operator can use the incremental P capture curve to inform their dosage for tradeoffs between performance and cost-effectiveness. While the lowest doses precipitated no brushite because of undersaturation, by pH 5.7 precipitating 1 kg of soluble P as brushite required less than 2 kg of Ca input; above pH 7.5, additional lime dosage caused a loss of brushite, and the new conditions caused by the added hydroxide favored struvite formation instead. Total P capture per total lime addition peaked around pH 6.4, where 875 kg Ca/d precipitated 400 kg P/d for a 0.460 mass ratio (or a 59% yield). However, the total P capture remained above a 0.4 mass ratio even past pH 8.5, suggesting that operators have a wide range of tolerable lime dosages, at least until the chemical dispersion effects of higher pH become relevant [10]. Marginal total P recovery did not appear to stop until approximately pH 9, requiring 1200 kg Ca/d.

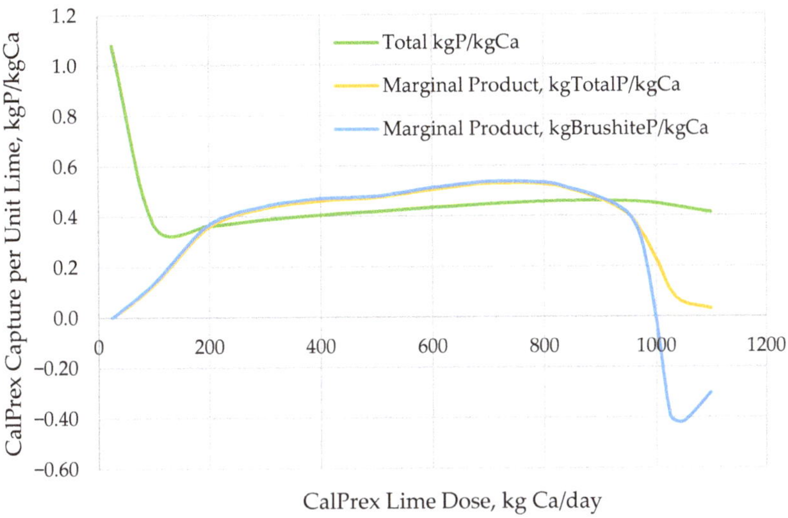

Figure 5. CalPrex reactor reagent efficiency response to lime dosing.

4. Discussion

Comparing plant-wide P distribution between the two scenarios, we find that CalPrex was not predicted to substantially alter the treatment plant effluent P, with the expected effluent P reduced from 57.3 kg P/d in the control state to 54.5 kg P/d after CalPrex installation. Instead, CalPrex implementation greatly reduced P precipitates within the biosolids compared to the control (Table 2). After introducing CalPrex, the total soluble P entering the anaerobic methanogenic digesters is decreased by nearly 250 kg P/d (138 vs. 385 kg/d), and so the total P precipitating in the anaerobic digesters is decreased by over 50% (289 vs. 554 kg/d). The net effect of CalPrex is a 44% reduction in biosolid P content as compared to the reference state with EBPR alone.

A full-scale CalPrex process should reduce the tendency for P-based precipitates to form in pipes and reactors downstream of the anaerobic digester by reducing soluble P. This benefit is potentially shared with other P removal technologies, depending on their location within the treatment stream, though CalPrex directly precedes the anaerobic digester and can specifically reduce the substrate available for struvite precipitation.

During the anaerobic digestion process, precipitate formation shifts from struvite formation to brushite formation. The total struvite production in the anaerobic digester is expected to be reduced from 3.2 tons of dry solids per day in the control state to 0.6 tons of dry solids per day with the CalPrex system in operation. Conversely, the total brushite production increases from 0.81 tons of dry solids per day to 1.1 tons of dry solids per day. The CalPrex process added to the treatment train is, therefore, expected to reduce P precipitation in the anaerobic digester by 55% on a mass basis.

The reduced precipitate formation in the anaerobic digesters impacts the biosolids contents as well, equating to a direct reduction of more than 10% of dry sludge mass. Furthermore, the additional reduction of total P content by almost 50% could have significant dewatering benefits to the treatment plant [10], potentially reducing the total wet mass by more than 10%, as indicated by total solids. Though not reflected in BioWin outputs, this secondary benefit could reduce biosolid removal costs to the operator. The drier sludge with its lower P content might be a more attractive soil amendment to local agriculture. It has better balanced N and P content and lower overall weight, making it easier to land apply in nearby fields.

4.1. Uncertainty Analysis

P solubilization is weakly responsive to the acid digester pH. A pH two SD below the mean solubilizes 0.4% less P than the mean, while a pH two SD above the mean solubilizes 3% less P than the mean. This 3% reduction is a result of inhibited heterotrophs releasing less biological P from degradable organics and, more importantly, the less acidic environment lowering the solubility of brushite and leading to a solution that is saturated in soluble P but at lower P concentrations.

Total capture of P by CalPrex is minimally impacted by low pH conditions (Figure 6), but higher pH conditions are predicted to reduce P capture by 13%. At higher digestate pH values, a higher portion of total P remains in the biosolids and bypasses the CalPrex reactor entirely. Lower pH values are indicative of higher VFA concentrations and require proportionately more lime to reach pH 7 (Figure 7). These two considerations, increased P capture as brushite but at the cost of more lime to overcome VFA acidity, result in a nearly linear net effect of falling brushite returns in terms of P per Ca at pH values below the mean and higher returns about the mean (Figure 8). The range of pH values tested yielded a reagent efficiency of 0.456 $\pm$ 0.11 kg P per kg Ca.

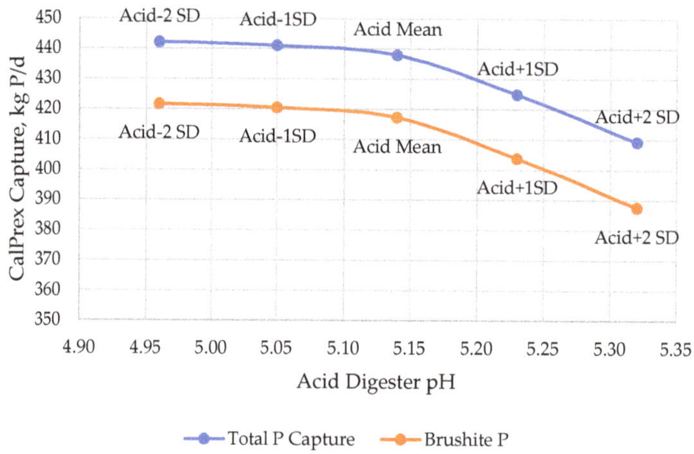

Figure 6. Predicted CalPrex P capture as a function of acid digester pH.

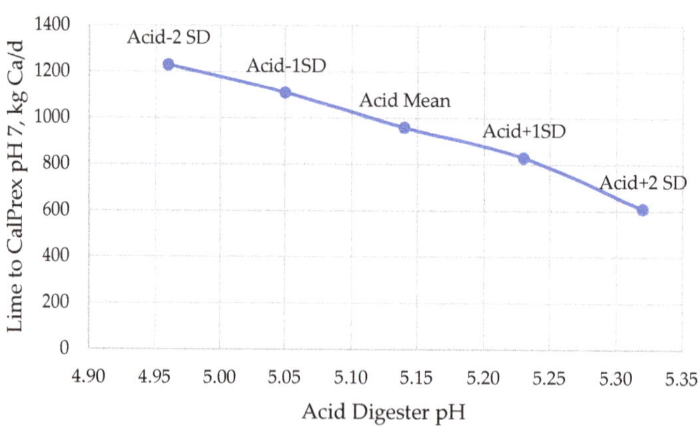

Figure 7. Predicted Ca requirements to reach CalPrex pH 7 as a function of pH values of the acid digest. Higher pH conditions have the beneficial effect of requiring lower lime dosages to capture brushite, potentially reducing costs to the plant as they spend less money neutralizing the buffering capacity of the VFAs.

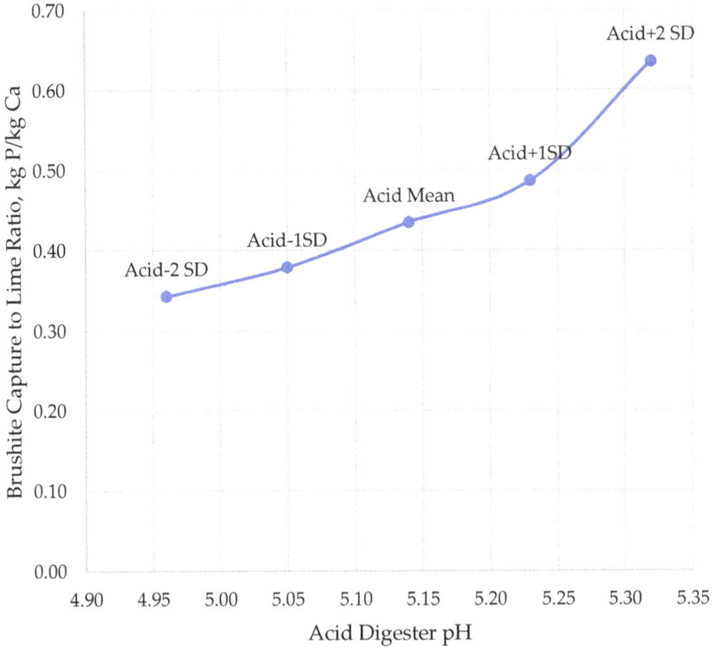

Figure 8. Relationship between the brushite return ratio, kg P per kg Ca, as a function of the acid digester pH used for optimizing the BioWin acid digester kinetics.

4.2. Cabinet Model Demonstrations

4.2.1. Nitrogen Removal Plant Cabinet Model with CalPrex

The N removal cabinet model is comprised of primary and secondary treatment, wherein the sludge is digested in an anaerobic digester. The digester effluent is then dewatered, and the liquids are processed through a nitrification/denitrification step before returning to the headworks. The plant does not intentionally cultivate phosphate-accumulating organisms, so its activated sludge is not particularly rich in P. The 125 m^3/d digester influent stream has only a 106 kg/d P load for a total P concentration of 849 mg P/L (Table 3). Upon splitting the sludge digestion into acid and anaerobic steps as per Vineyard et al. [32,33], the acid digester achieves a 57% P solubilization rate, primarily from hydrolysis of degradable organics from the primary sludge due to the shortage of other readily releasable P entering the acid digester. It has a soluble-P concentration of 483 mg soluble P/L, about 30% less than the expectation from the post-EBPR acid digester of Nine Springs.

Table 3. Cabinet Configuration Plant Acid Digestate Properties.

Acid Digest	Unit	N Plant	EBPR Plant
Flow	m^3/d	125	56
pH		5.21	5.21
Total P	kg P/d	106	61.7
Soluble P	kg P/d	59.9	27.4
Soluble Ca	kg Ca/d	37.9	12.9

The disadvantageous acid digester conditions of the N removal plant are reflected in CalPrex reactor performance. Of the original 106 kg P exiting the acid digester, less than 50% makes it to the CalPrex clarifier as soluble P (Table 4). Due to the lower overall P concentration, it takes a higher lime dose to reach the brushite solubility point, and the proportional P capture is reduced. At pH 7, the reactor has achieved a capture ratio of 0.423 kg P/kg Ca dosage, about 90% of the rate achieved by CalPrex modeled in the Nine Springs configuration. More than 55% of the CalPrex product, by weight, was not brushite but instead organic material that had escaped initial separation and was captured alongside brushite (Table 4).

Table 4. CalPrex Performance in Cabinet Configurations.

CalPrex	Unit	N CalPrex	EBPR CalPrex
Influent Total P	kg P/d	54.3	35.7
Influent Soluble P	kg P/d	50.3	33.8
Total P Capture	kg P/d	42.2	27.3
Brushite P Capture	kg P/d	38.5	25.6
Struvite P Capture	kg P/d	0	0
Lime Dose to pH 7	kg Ca/d	91.0	49.0
Brushite Return Ratio	kg P/kg Ca	0.423	0.522
TSS	kg/d	471	253
TSS Brushite Purity	kg Brushite/kg	45.4%	56.1%

Downstream, the dilution caused by the CalPrex polymer addition reduced the HRT of the anaerobic digesters, which decreased the volatile suspended solids (VSS) destruction and the methane generation by the digesters (Table 5). CalPrex halted struvite generation in the anaerobic digester and thus eliminated it from the biosolids with no increase in brushite contents (Table 6). Overall, the CalPrex process reduced the plant's effluent P by 25%, from 120 to 90 kg effluent-P/d, and its biosolid P by 17% (Table 7), possibly reducing dewatering costs. The addition of CalPrex operations did not noticeably impact the net N content of the effluent.

Table 5. Anaerobic Digestate Properties in Cabinet Configurations with and without CalPrex.

Anaerobic Digest	Unit	N Original	N + CalPrex	EBPR Original	EBPR + CalPrex
Brushite Precipitate	kg P/d	29.0	28.2	10.4	13.4
Struvite Precipitate	kg P/d	8.93	0.0	8.47	3.61
Dry gas flow	m^3/d	130	117	55.9	50.2
VSS Destruction	%	49.0	47.6	45.1	43.5

Table 6. Biosolid properties in Cabinet Configurations with and without CalPrex.

Biosolid	Unit	N Original	N + CalPrex	EBPR Original	EBPR + CalPrex
Total P	kg P/d	72.3	59.9	61.7	34.4
Soluble P	kg P/d	2.6	0.2	27.4	2.28
Struvite P	kg P/d	8.75	0.0	8.47	3.61
Brushite P	kg P/d	28.3	27.5	10.4	13.4
TSS	kg/d	3175	2951	1429	1340

4.2.2. EBPR Plant Cabinet Model with CalPrex

Like the Nitrogen Removal cabinet model above, the cabinet EBPR reactor configuration is composed of primary and secondary treatment but specifically cultivates phosphate accumulating organisms in the sludge. The primary and activated sludges are diverted to a 20-day HRT anaerobic digester process before being partitioned entirely to waste with no solid/liquid separation and, therefore, no recycling of P-bearing liquid from the anaerobic digestate to the headworks.

Table 7. P Distribution Profile of Cabinet Configurations with and without CalPrex.

Whole Plant	Unit	N Original	N + CalPrex	EBPR Original	EBPR + CalPrex
Influent Total P	kg P/d	192	192	80	80
Effluent Total P	kg P/d	120	90.0	18.3	18.3
Biosolid Total P	kg P/d	72.3	59.9	61.7	34.4
CalPrex Total P	kg P/d	0	42.2	0	27.3

After the addition of a CalPrex reactor and an acid digester [32,33], the acid digester is predicted to achieve roughly 64% P solubility (Table 3) in its 56 m^3/d stream for a total soluble P concentration of 719 mg P/L, which is slightly higher than the Nine Springs equivalent. This high soluble P concentration allows the CalPrex clarifier to return 0.522 kg of P as brushite for each kg of Ca as lime dosed and yields a product that is 56% brushite by dry weight (Table 4).

In this configuration, CalPrex increases the brushite precipitation but lowers the struvite precipitation in the anaerobic digester for a net decrease in anaerobic P precipitate formation. The total P in the biosolids is reduced by roughly 44%, and the soluble P in the biosolids is reduced by 92% (Table 6), potentially offering dewatering benefits. The dilution caused by the polymer addition of the CalPrex process reduces both solids destruction and methane generation (Table 5) in the subsequent anaerobic digester because of reduced HRT in a fixed-size reactor. The effect of dilution was greater in this cabinet model than in Nine Springs because of the difference in HRT, 34 days at Nine Springs and 20 days here. Without a recycle stream, there are no effects on the main treatment stream's phosphorus content (Table 7).

Overall, the cabinet model configurations imply that the beneficial effects of CalPrex can be generalized beyond the Nine Springs plant. Any plant with an acidogenic digester may be able to install a CalPrex reactor and achieve benefits such as reduced accumulation of nuisance P precipitates in the anaerobic digesters, reduced P content of plant biosolids, and/or reduced biosolid dewatering costs. Plants with EBPR trains preceding the anaerobic digestion may see higher capture rates and reagent efficiencies, but plants without an EBPR train could reduce the P content of their main effluent stream without needing an overhaul of their treatment train.

4.3. CalPrex Installation and Operation

CalPrex has never been implemented at full scale, so the precise needs for efficient implementation and maintenance are not yet known. CalPrex requires a preceding acidogenic anaerobic digester, so full-scale installations should be considered for treatment plants already using a two-phase anaerobic digestion system. A plant considering transitioning to a two-phase system may see the possibility of CalPrex implementation for struvite control as an additional benefit; similarly, a plant seeking to control struvite may consider the additional cost of adding an acidogenic digester as a worthwhile tradeoff because of the increased digestion performance of a two-phase system.

Experimentation has suggested that a reactor pH above 7.2 can greatly diminish the settling rate of precipitated brushite [10], typically yielding fine granules that form a sludge with a much higher water content than the product of a reactor functioning at pH 7 or below. This dispersion would require both a larger settling tank and additional energy for dewatering to create a viable product. At this point, the buffering capacity of VFAs has been expended, and precipitation of soluble P leaves the pH highly sensitive to input lime dosage, so a target of pH 7.0 is expected to be the highest pH an operator would desire. A full-scale implementation would likely need advanced control systems that continuously monitor solution pH and tightly control the lime dosing.

The post-processing and transportation needs for the brushite produced are also unknown. The brushite is intended to be sold as a fertilizer or soil amendment. In the CalPrex plant, predicted brushite purity was 60% with 11% P by dry weight, which closely matched the pilot test measurements [27] and is comparable to triple and single superphosphates. Increased brushite purity may require more efficient separation of organics out of the liquid phase of acid digester centrate or a preliminary clarifier to further settle the organics before lime dosing and brushite collection. Drying, pelletization, bagging, and marketing needs for the brushite are still being tested but may be comparable to similar struvite-based technologies.

The anaerobic digester was shown to decrease solids destruction and methane generation levels after the installation of the CalPrex precipitator, most likely due to the 27% dilution caused by polymer addition deployed at the pilot-scale plant. Real-world effects of dilution will depend on individual treatment plants and their digester HRTs. A full-scale installation would have better options than a large dilution, however, including replacing the freshwater used for polymer preparation with either CalPrex effluent (already stripped of P) or acid digester centrate (dosed with polymer, solids removed but still high in P). The latter possibility would allow soluble P to enter the CalPrex reactor undiluted, likely eliminating HRT changes in the anaerobic digester and possibly improving brushite recovery by avoiding the reduction of soluble P concentrations upon dilution. Another option would be to seek a polymer that would not require such dilution.

The installation of a CalPrex precipitator uniformly reduced the quantity of digester sludge produced by the anaerobic digester after clarification and/or dewatering, both because of the intended capture of P as brushite and the unintended capture of organics as an impurity in the brushite precipitator. However, the mass of CalPrex product was greater than the reduction in digester sludge produced, so a facility intending to landfill the brushite alongside the biosolids would not experience a net reduction in solids. In all cases, the addition of the CalPrex reactor reduced the daily flow of P to the biosolids, and the P precipitated in those biosolids. This reduction could help alleviate concerns about excess P in biosolids land application and divert that P to more productive use.

4.4. Study Limitations and Future Work

Though based on thorough data, this is ultimately a modeling study. By definition, a model cannot capture all possible mechanisms or sources of variability. This study is limited by BioWin's inability to reliably capture the effects of variable influents in a dynamic simulation. Currently, dynamic simulations using characteristic influent variability of the Nine Springs plant result in static predictions of acid digester function that do not reflect the variability in sampling data [32,33]. This makes it difficult to predict the effects of this variability on the function of nutrient recovery technologies like CalPrex.

While CalPrex has been predicted to greatly reduce nuisance precipitates in the anaerobic digesters and produce a saleable fertilizer product, the precise costs and operational needs of a full-scale implementation are unknown. The technology needs estimates of capital, operation, and maintenance costs, in addition to energy and reagent, which need to be compared with competitor technologies. In addition, the environmental benefits of recovering phosphorus from waste should be compared to the environmental costs of consuming energy and lime. The next step for this research is a detailed life cycle assessment and techno-economic analysis of the CalPrex system. If these prove favorable, additional pilot-scale implementations can increase data reliability and may lead to full-scale installations.

5. Conclusions

The modeling efforts presented in this work yielded the following findings:

- A full-scale CalPrex process was predicted to reduce struvite formation in the anaerobic digesters by 80% and total precipitate formation by 58%.
- The total phosphorus content of the biosolids was decreased by 50%, and the total dry sludge mass was reduced by 12%.
- CalPrex captured 1 kg phosphorus for every 2.1–2.5 kg of calcium as lime added.

- Any plant with an acidogenic digester may be able to install a CalPrex reactor and achieve benefits such as reduced accumulation of nuisance P precipitates in the anaerobic digesters, reduced P content of plant biosolids, and/or reduced biosolid dewatering costs.
- A techno-economic analysis and life cycle assessment is needed to weigh costs and benefits fully.

Author Contributions: Conceptualization, K.K. and P.B.; Methodology, D.V.; Software, D.V.; Formal analysis, D.V.; Investigation, D.V.; Resources, P.B.; Writing—original draft, D.V.; Writing—review & editing, D.V., K.K. and P.B.; Visualization, D.V.; Supervision, K.K. and P.B.; Project administration, K.K. and P.B.; Funding acquisition, K.K. and P.B. All authors have read and agreed to the published version of the manuscript.

Funding: This work was funded by USDA/NIFA INFEWS/T3 Grant (Reference #2017-67003-26055), USDA/NIFA Hatch Grant No. WIS01920 (Accession # 1009816), and USDA/NIFA Hatch Grant No. WIS04092 (Accession # 7003184).

Data Availability Statement: Some MMSD-related data are confidential, but to demonstrate the principles of modeling an acid digester in BioWin, we have made available a modified version of a BioWin cabinet model with full supporting documentation at DOI 10.17605/OSF.IO/8DXM7.

Acknowledgments: We are grateful for the assistance of Matt Seib and the Madison Metropolitan Sewerage District.

Conflicts of Interest: The authors declare the following financial interests/personal relationships, which may be considered as potential competing interests: Phillip Barak reports financial support was provided by the US Department of Agriculture, National Institute of Food and Agriculture, and Hatch Grant. Phillip Barak reports a relationship with Nutrient Recovery and Upcycling, LLC, which includes board membership and equity or stocks. Nutrient Recovery and Upcycling, LLC, is the rights holder to US Patent 8,568,590 (issued 2013), "Phosphate recovery from acid phase digesters".

Appendix A

Table A1. Reference Acid Digester Function.

		BioWin Reference	Tabanpour et al. [27]
Flow	m^3/d	831	844
pH		5.14	5.13
Total P	kg P/d	795	666
Soluble P	kg P/d	434	438
Calcium Concentration	kg Ca/d	213	445
Ammonia Concentration	kg N/d	919	-
Total Solids	%	3.4% (Suspended)	4.75%

Table A2. Reference Anaerobic Digester Function.

HRT	Hours	811
Brushite In	kg ISS/d	0
Brushite P In	kg P/d	0
Brushite Out	kg ISS/d	491
Brushite P Out	kg P/d	88.3
Struvite In	kg ISS/d	0
Struvite P In	kg P/d	0
Struvite Out	kg ISS/d	2624
Struvite P Out	kg P/d	331
Dry gas flow	m3/d	821
Off gas Methane	kg/d	8214
VSS Destruction	%	37.6
Ammonia Concentration	kg N/d	1349

Table A3. Reference Biosolids Contents.

Flow	Unit	Value
Total P	kg P/d	688
Soluble P	kg P/d	27.4
Brushite	kg ISS/d	488
Brushite P	kg P/d	87.8
Struvite	kg ISS/d	2608
Struvite P	kg P/d	329
TSS	kg/d	20,180
VSS	kg/d	13,489

Table A4. Effects of CalPrex on Acid Digester Function.

		Reference	CalPrex, 0 Lime	CalPrex to PH 7
Flow	m^3/d	831	831	831
pH		5.14	5.15	5.15
Total P	kg P/d	795	817	847
Soluble P	kg P/d	434	462	482
Ammonia Concentration	kg N/d	919	886	886
P Solubilization	%	54.50%	56.50%	56.90%

Table A5. Effects of CalPrex on Anaerobic Digester Function.

		Reference	CalPrex, 0 Lime	CalPrex to PH 7
HRT	hours	811	628	628
Brushite In	kg ISS/d	0	0	352
Brushite P In	kg P/d	0	0	63.4
Brushite Out	kg ISS/d	491	628	311
Brushite P Out	kg P/d	88.3	113	56
Struvite In	kg ISS/d	0	0	0
Struvite P In	kg P/d	0	0	0
Struvite Out	kg ISS/d	2624	2353	1381
Struvite P Out	kg P/d	331	297	174
Dry gas flow	m3/d	821	786	749
Off gas Methane	kg/d	8214	7874	7870
VSS Destruction	%	37.6	36.6	36.6
Ammonia Concentration	kg N/d	1349	1283	1337

Table A6. Effects of CalPrex on Biosolid Contents.

		Reference	CalPrex, 0 Lime	CalPrex to PH 7
Total P	kg P/d	688	686	450
Soluble P	kg P/d	27.4	29	5.3
Brushite	kg ISS/d	488	624	309
Brushite P	kg P/d	87.8	112	55.6
Struvite	kg ISS/d	2608	2339	1373
Struvite P	kg P/d	329	295	173
TSS	kg/d	20,180	19,200	17,813
VSS	kg/d	13,489	12,802	12,804

The pH inhibition threshold required to model a given acid digester pH is a nearly linear function of that pH (Figure A1, with a slope of ~1.5, which means that the pH inhibition threshold moves about 1.5x more than the acid digester pH that it is fitting. The mechanistic reason that BioWin causes pH to rise or fall in response to altering the kinetics of the heterotrophs in the acid digester is the production and accumulation of VFAs; the production of both acetic and propionic acids follows the acid digester pH in a nearly linear fashion (Figures A2 and A3), with lower pH associated with higher VFA production.

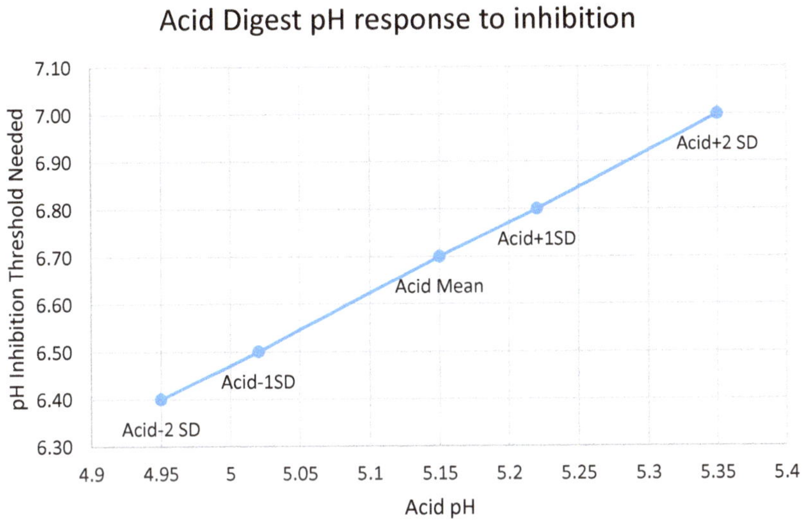

Figure A1. Relationship between the pH of the acid digester and the pH inhibition threshold needed to reach it. Points on the graph are mean and ± one and two standard deviations. Non-uniform spacing of the points is due software difficulty in achieving 0.01 pH precision.

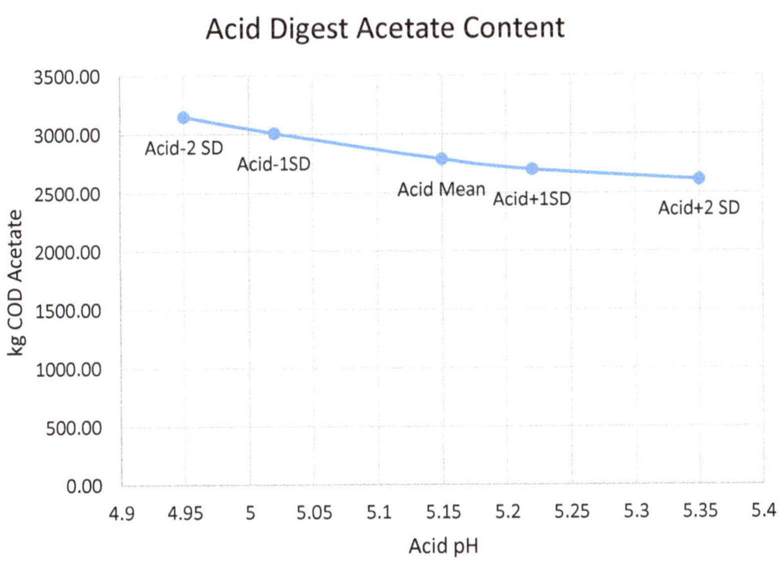

Figure A2. Relationship between the predicted production of acetic acid and the acid digester pH chosen for optimization.

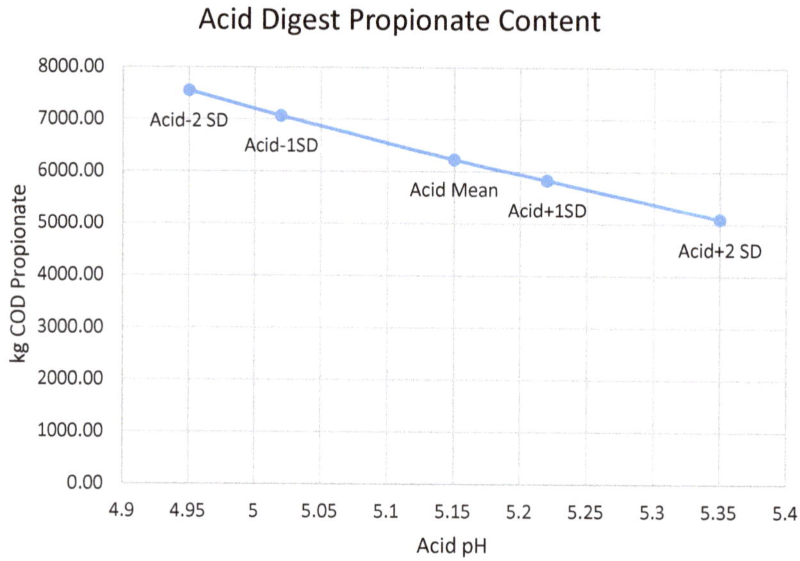

Figure A3. Relationship between the predicted production of propionic acid and the acid digester pH chosen for optimization.

Appendix A.1. Calibration Notes

The reduction in P concentrations suggests that the polymer added a total of 27% to the volume. We match this with an input of unadulterated water (224 m^3/d).

Calibrating the input solids removal

The Calprex system does not react pure acid digester effluent. Instead, it centrifuges a stream and rejects the solids. Under pilot conditions [27], centrifugation reduced total solids in the centrate from 4.75% to 0.41% TS, suggesting 91.4% solids rejection. Centrate was also reported to contain 57.7–59% of total phosphorus and 84% of soluble phosphorus, which suggests the centrate was 84% of total liquid volume. In the BioWin model, we used a dewaterer to reject 16% of liquids and 91.4% of solids directly to the anaerobic digester. In the remainder, the total P concentration is 524 mg P/L and the calcium is 264 mg/L. Our model captures 84% of soluble P but only 48% of the total P due to the lower overall P solubilization (66% observed vs. 52% predicted), but matching P and Ca concentrations is more important for CalPrex calibration purposes.

Calibrating the effluent dewaterer

The CalPrex process ends with a clarifier that collects the precipitate and then centrifuges it to reduce water content. In the pilot plant the clarifier captured 87.1% of total P in the centrate. The clarifier underflow was then centrifuged for another 96.5% capture. We model the double centrifuge process as a single dewaterer with 84.1% solid capture and assume 99% liquid rejection. A total of 49.9% of P from the acid digester stream was captured after that as particulate P.

In the modified configuration, the total quantity of P entering the sludge streams is higher than measured on-site due to differences in settling tank and clarifier function in the primary stream. Our modifications, however, have brought soluble P in the acid digester to within 5% of plant measurements. The configuration acid digester now handles a flow nearly equivalent to the pilot version in volume and pH.

Table A7. CalPrex feed characteristics from site measurements compared to BioWin equivalent.

	CalPrex Pilot	BioWin Plant
Flow m^3/d	50.7	886
Soluble p mg/L	429	421
Total p mg/L	459	460
Soluble Calcium mg/L	361	229
Total solids	0.41%	0.384% (suspended)

The CalPrex precipitator is an unaerated reactor dosed with calcium hydroxide. The precipitation tank is fed the reject water, or solids-reduced, portion from the intake centrifuge while the solids continue to the anaerobic digester. The liquids are dosed with lime and allowed to settle with a retention time of one to two hours before being centrifuged again. These solids are collected for processing while the P-depleted reject water continues to the anaerobic digester.

BioWin Configuration Layouts

The CalPrex unit was modeled using eight BioWin processes (Figure A4). This was used to replace the existing anaerobic digester processes of two BioWin cabinet models (Figures A5 and A6) in order to verify reproducibility.

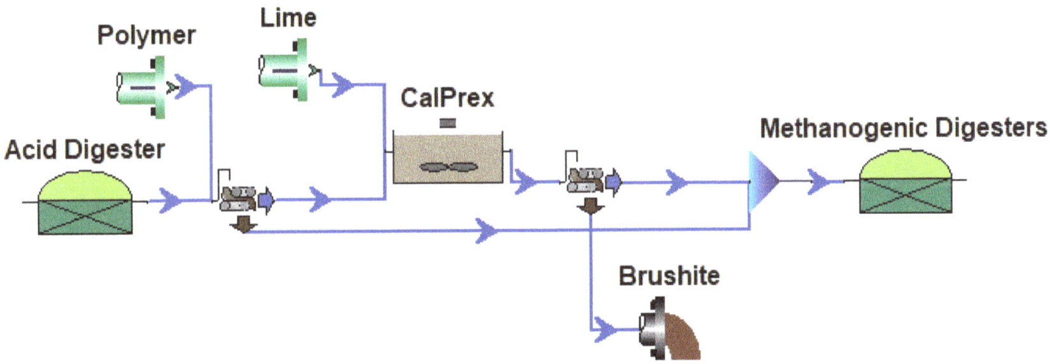

Figure A4. Example BioWin CalPrex Processes.

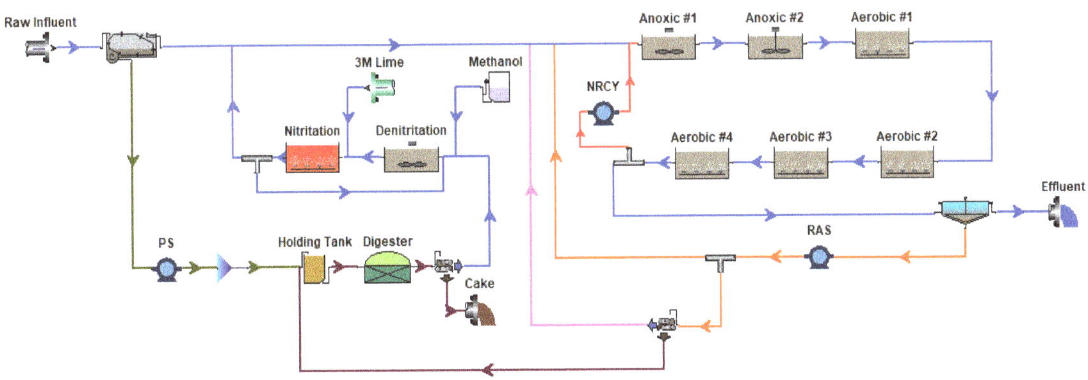

Figure A5. BioWin Cabinet Model: N Removal + Anaerobic Digester + Sidestream Treatment Configuration.

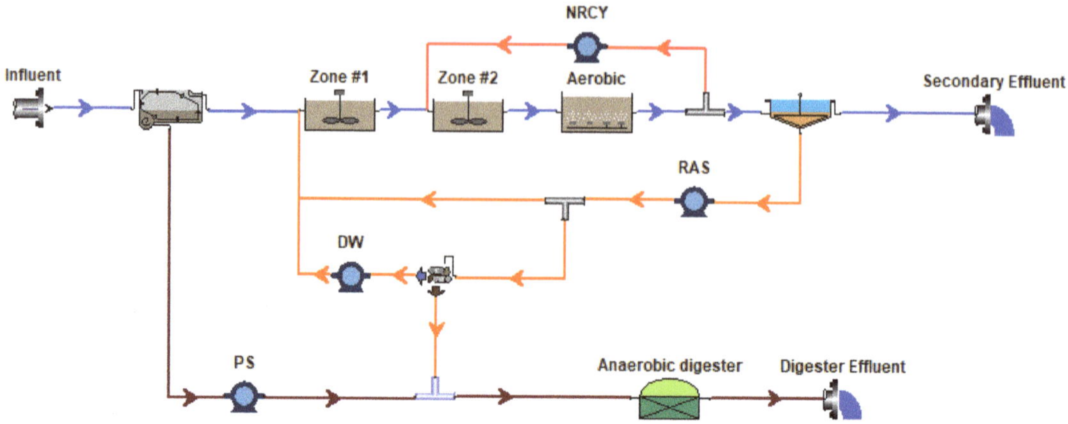

Figure A6. BioWin Cabinet Model: EBPR + Anaerobic Digester Configuration.

References

1. United States Department of Agriculture, Economic Research Service. Fertilizer Use and Price. 2018. Available online: https://www.ers.usda.gov/data-products/fertilizer-use-and-price.aspx (accessed on 25 June 2019).
2. United States Environmental Protection Agency. *A Compilation of Cost Data Associated with the Impacts and Control of Nutrient Pollution*; United States Environmental Protection Agency, Office of Water: Washington, DC, USA, 2015.
3. Carey, R.O.; Migliaccio, K.W. Contribution of Wastewater Treatment Plant Effluents to Nutrient Dynamics in Aquatic Systems: A Review. *Environ. Manag.* **2009**, *44*, 205–217. [CrossRef] [PubMed]
4. United States Environmental Protection Agency. *NPDES Nutrient Data Tables*; United States Environmental Protection Agency: Washington, DC, USA, 2023.
5. United States Environmental Protection Agency. *An Urgent Call to Action—Report of the State-EPA Nutrient Innovations Task Group*; United States Environmental Protection Agency: Washington, DC, USA, 2009.
6. Cordell, D.; White, S. Peak phosphorus: Clarifying the key issues of a vigorous debate about long-term phosphorus security. *Sustainability* **2011**, *3*, 2027–2049. [CrossRef]
7. Kataki, S.; West, H.; Clarke, M.; Baruah, D.C. Phosphorus recovery as struvite: Recent concerns for use of seed, alternative Mg source, nitrogen conservation and fertilizer potential. *Resour. Conserv. Recycl.* **2016**, *107*, 142–156. [CrossRef]
8. Egle, L.; Rechberger, H.; Krampe, J.; Zessner, M. Phosphorus recovery from municipal wastewater: An integrated comparative technological, environmental and economic assessment of P recovery technologies. *Sci. Total Environ.* **2016**, *571*, 522–542. [CrossRef] [PubMed]
9. Achilleos, P.; Roberts, K.R.; Williams, I.D. Struvite precipitation within wastewater treatment: A problem or a circular economy opportunity? *Heliyon* **2022**, *8*, e09862. [CrossRef] [PubMed]
10. Boyce, A.L. Fertilizer-Grade Calcium Phosphate Recovery from Wastewater Treatment Plants. Master's Thesis, University of Wisconsin-Madison, Madison, WI, USA, 2010.
11. Doyle, J.D.; Parsons, S.A. Struvite formation, control and recovery. *Water Res.* **2002**, *36*, 3925–3940. [CrossRef]
12. Jeyanayagam, S.; Khunjar, W.; Mehta, C. Accelerating the Implementation of Extractive Nutrient Recovery as an Integral Component of Sustainable Nutrient Management. *Water E-J.* **2016**, *1*, 1–8. [CrossRef]
13. Jorgensen, T.C.; Weatherley, L.R. Ammonia removal from wastewater by ion exchange in the presence of organic contaminants. *Water Res.* **2003**, *37*, 1723–1728. [CrossRef]
14. Fassbender, A.G. ThermoEnergy Ammonia Recovery Process for municipal and agricultural wastes. *Sci. World J.* **2001**, *1* (Suppl. S2), 908–913. [CrossRef]
15. Gogate, P.R.; Pandit, A.B. A review of imperative technologies for wastewater treatment I: Oxidation technologies at ambient conditions. *Adv. Environ. Res.* **2004**, *8*, 501–551. [CrossRef]
16. Sugiyama, S.; Yokoyama, M.; Ishizuka, H.; Sotowa, K.; Tomida, T.; Shigemoto, N. Removal of aqueous ammonium with magnesium phosphates obtained from the ammonium-elimination of magnesium ammonium phosphate. *J. Colloid. Interface Sci.* **2005**, *292*, 133–138. [CrossRef] [PubMed]
17. Uludag-Demirer, S.; Demirer, G.N.; Chen, S. Ammonia removal from anaerobically digested dairy manure by struvite precipitation. *Process Biochem.* **2005**, *40*, 3667–3674. [CrossRef]
18. El-Bourawi, M.S.; Khayet, M.; Ma, R.; Ding, Z.; Li, Z.; Zhang, X. Application of vacuum membrane distillation for ammonia removal. *J. Membr. Sci.* **2007**, *301*, 200–209. [CrossRef]

19. Tidåker, P.; Sjöberg, C.; Jönsson, H. Local recycling of plant nutrients from small-scale wastewater systems to farmland—A Swedish scenario study. *Resour. Conserv. Recycl.* **2007**, *49*, 388–405. [CrossRef]
20. Turker, M.; Celen, I. Removal of ammonia as struvite from anaerobic digester effluents and recycling of magnesium and phosphate. *Bioresour. Technol.* **2007**, *98*, 1529–1534. [CrossRef] [PubMed]
21. Quan, X.; Wang, F.; Zhao, Q.; Zhao, T.; Xiang, J. Air stripping of ammonia in a water-sparged aerocyclone reactor. *J. Hazard. Mater.* **2009**, *170*, 983–988. [CrossRef] [PubMed]
22. Xie, Z.; Duong, T.; Hoang, M.; Nguyen, C.; Bolto, B. Ammonia removal by sweep gas membrane distillation. *Water Res.* **2009**, *43*, 1693–1699. [CrossRef]
23. Quan, X.; Ye, C.; Xiong, Y.; Xiang, J.; Wang, F. Simultaneous removal of ammonia, P and COD from anaerobically digested piggery wastewater using an integrated process of chemical precipitation and air stripping. *J. Hazard. Mater.* **2010**, *178*, 326–332. [CrossRef]
24. Alitalo, A.; Kyro, A.; Aura, E. Ammonia stripping of biologically treated liquid manure. *J. Environ. Qual.* **2012**, *41*, 273–280. [CrossRef]
25. De Vries, J.W.; Groenestein, C.M.; De Boer, I.J. Environmental consequences of processing manure to produce mineral fertilizer and bio-energy. *J. Environ. Manag.* **2012**, *102*, 173–183. [CrossRef]
26. Barak, P. Phosphate Recovery from Acid Phase Anaerobic Digesters. U.S. Patent US8568590B2, 29 October 2023.
27. Tabanpour, M.; Downing, L.; Kuhnjar, W. Demonstrating the CalPrex System for High Efficiency Phosphorus Recovery. 2021, CNP—Technology Water and Biosolids Corporation and Nutrient Recovery and Upcycling (NRU). In *WEFTEC*; Water Environment Federation: Alexandria, VI, USA, 2021.
28. Barak, P.; Davidson, C.; Minks, A. Chemical composition of organic acid digest from a municipal wastewater treatment plant and chemical modeling of nuisance struvite formation and phosphorus recovery as brushite. *PLoS Water* **2023**, 2, e0000120. [CrossRef]
29. Lehr, J.; Brown, W.E. Calcium phosphate fertilizers: II. A petrographic study of their alteration in soils. *Soil Sci. Soc. Am. J.* **1958**, *22*, 29–32. [CrossRef]
30. Lunt, O.R.; AKofranek, M.; Clark, S.B. Nutrient availability in soil, availability of minerals from magnesium ammonium phosphates. *J. Agric. Food Chem.* **1964**, *12*, 497–504. [CrossRef]
31. Ellis, B.; Foth, H. *Soil Fertility*; CRC Press: Boca Raton, FL, USA, 1996.
32. Vineyard, D.; Karthikeyan, K.G.; Davidson, C.; Barak, P. Modeling an acid-phase digester in BioWin with parameter optimization from site data. *J. Water Process Eng.* **2023**, *54*, 103971. [CrossRef]
33. Vineyard, D.; Karthikeyan, K.; Davidson, C.; Barak, P. BioWin Modeling of Phosphorus Solubilization in an Acid-Phase Digester. *Environments* **2023**, *11*, 31. [CrossRef]
34. Envirosim Associates LTD. BioWin 6 Help Manual. 2020. Available online: http://www.envirosim.com/downloads/BW6 Manual.pdf (accessed on 13 December 2023).
35. AECOM. Chloride Compliance Study: Nine Springs Wastewater Treatment Plant Final Report. Madison Metropolitan Sewerage District. 2015. Available online: https://www.madsewer.org/wp-content/uploads/2021/12/STUDY-2015-District-Chloride-Compliance.pdf (accessed on 23 November 2023).
36. Jeffreys, H.; Jeffreys, B. *Methods of Mathematical Physics*; Cambridge University Press: Cambridge, UK, 1999.

environments MDPI

Article

Improving BioWin Modeling of Phosphorus Solubilization in Acid-Phase Digesters

Donald Vineyard [1,*], K.G. Karthikeyan [2], Christy Davidson [3] and Phillip Barak [1]

1 Department of Soil Science, University of Wisconsin—Madison, Madison, WI 53705, USA; phillip.barak@wisc.edu
2 Department of Biological Systems Engineering, University of Wisconsin—Madison, Madison, WI 53705, USA; kkarthikeyan@wisc.edu
3 Nelson Institute for Environmental Studies, University of Wisconsin—Madison, Madison, WI 53706, USA; cdavidson@wisc.edu
* Correspondence: dlvineyardphd@gmail.com

Abstract: BioWin 6.0 does not accurately predict phosphorus (P) speciation in acidogenic anaerobic digesters under default kinetics characterization and parameterization. The accurate modeling of acid-phase digestion is needed to predict the performance of novel nutrient recovery technologies that act on these digester effluents. The main thrust of this work was to identify and correct the causes of inaccurate P partitioning and precipitation within BioWin models of acid-phase digestion reactors. A BioWin configuration including an organic acid digester was parameterized and recalibrated based on the known traits of acid-phase digestion and then validated against a full-scale digester in a municipal wastewater treatment plant. This digester, with pH 5.14 and 61–74% solubilized P, was predicted by BioWin default parameters to have only 27% soluble P and a net formation of P precipitates. Corrections to the polyphosphate-accumulating organism decay, endogenous product decay, hydrolysis rate, and brushite behavior resulted in 67% solubilization with no precipitate formation. Cabinet configurations showed similar behavior when modified to include an acid-phase digester under default parameters, but predictions were similarly amended by our parameter changes. This improved modeling technique should allow operators to effectively characterize acid digesters for their own treatment trains and allow engineers to predict the performance of novel nutrient recovery technologies acting on acidogenic digest.

Keywords: phosphorus recovery; anaerobic digestion; soluble phosphorus; acidogenic digest; organic acid digester; two-phase digestion

1. Introduction

As the deleterious effects of nutrient pollution have gained importance, municipal wastewater treatment plants have been targeted by regulatory agencies as significant point sources for nutrients [1–4]. In response, treatment plants have explored new technologies for removing nutrients, first from effluent to biosolids, and more recently from treatment to resource recovery. As the circular economy develops, municipal wastewater treatment plants are being increasingly viewed as potential resource recovery facilities to be targeted for fertilizer production [5]. This new wave of experimentation has resulted in different pilot technologies attempting to recycle/recover nutrients.

Anaerobic digestion is a key process in many treatment trains [6]; digesters receive a biomass-rich sludge clarified from other reactors and process it using a microbial community that consumes biomass without the need for oxygen [6,7]. These reactors reduce the pathogen and biomass load while producing a mix of carbon dioxide and methane (i.e., biogas) which provides a potential energy source for the treatment plant [6–8]. Digested effluent is typically separated into liquids and recalcitrant sludge [6] with the liquids returned to the headworks for re-processing while the sludge is disposed of through land application, landfilling, or incineration [6].

The anaerobic digester converts biomass into biogas using four steps: hydrolysis, fermentation, acidogenesis, and methanogenesis. Each process is performed by distinct communities of microorganisms [9–11]. Hydrolysis, fermentation, and acidogenesis sequentially break down complex organic molecules into volatile fatty acids (VFAs) and soluble inorganic nutrients [9,10]; the primary nutrients released include nitrogen (N), phosphorus (P), potassium (K), calcium (Ca), and magnesium (Mg). Subsequent methanogenesis from metabolizing the VFA substrate raises the pH and can precipitate N, P, and Mg out of solution, most commonly as the ammonium magnesium phosphate mineral struvite ($NH_4MgPO_4 \cdot 6H_2O$) [8,12]. Struvite is considered a nuisance precipitate by the wastewater industry because it clogs pipes and reduces the reactor volume, necessitating costly maintenance and repairs; its prevention and removal is an ongoing area of research [8,12–24]. In addition, the sludge generated from processes such as enhanced biological phosphorus removal (EBPR) contains elevated P loads from the digestion of polyphosphate-accumulating organisms, which can exacerbate struvite precipitation issues [8,12,15,25].

The methanogenesis and acidogenesis stages of digestion have an antagonistic relationship. VFAs from acidogenesis dissociate as weak acids and reduce the solution pH, but methanogenesis can be inhibited by pH conditions under 6.5 [9,11]. Because methanogenesis is typically the rate-limiting step, anaerobic digester operators must ensure sufficient methanogenic substrate without creating inhibitory pH conditions through excess acidogenesis [9,11]. A two-phase system can segregate the acidogenic and methanogenic steps into separate reactors to create more favorable conditions for the adverse microorganisms, leading to a net improvement in solids destruction, loading tolerance, methane production, and/or reactor volume needed [1,6,9–12]. The first phase cultivates the hydrolysis, fermentation, and acidogenesis communities using a reactor with a short hydraulic retention time (HRT), referred to as an acid-phase digester. Acid-phase digesters contain high concentrations of VFAs and soluble nutrients; the acidic conditions, typically pH 5–6, inhibit the formation of inorganic precipitates and the growth of methanogens [1,7,9,10,12]. The low-pH, high-nutrient effluent is typically fed to a subsequent anaerobic digester which cultivates methanogens using a longer HRT under near-neutral pH. These methanogens convert VFAs into biogas, but the higher pH favors the formation of nuisance precipitates like struvite [4,9,10,12].

The high-nutrient, low-pH effluent of the acid-phase digester is an ideal target for nutrient recovery technologies. Precipitating P from acid digests in a controlled environment prevents their soluble phosphorus from continuing to the methanogenic digesters, where it may precipitate as struvite [12,26,27]. This offers the potential to reduce treatment stoppages for maintenance due to mineral clogging in subsequent digesters, to reduce the total sludge dry weight and P content, to lower biosolids dewatering costs, and to recover a high-strength fertilizer product [12,26,27]. To predict the potential of emerging technologies in an individual treatment train before a full-scale implementation, wastewater engineers often use computer models [6–8].

BioWin [28] is widely known in the wastewater treatment industry for its comprehensive biological and chemical modeling suite and its ease of use. The modeling of acid-phase digestion is currently not explicitly supported by BioWin and literature examples are rare. Some authors skip the modeling of the acid-phase reactor [29,30] and others note aberrantly low pH conditions [31,32] associated with the acidogenesis steps. Recent work has confirmed that BioWin does not natively make accurate predictions of acid digester function and developed methods to correct the pH predictions [26,33], allowing for more accurate assessments of model performance with respect to the nutrient content. As of writing, there is no literature validating BioWin's modeling of nutrient speciation in acid digesters. Without accurate nutrient speciation, models of nutrient recovery or struvite control technologies would be unable to predict critical performance statistics like reagent requirements, nutrient recovery rates, or struvite mitigation. Improved BioWin methods for modeling acid digesters could allow for the efficient and economical evaluation of new nutrient removal/recovery technologies as well as the troubleshooting of pilot-scale projects.

To develop these methods, we used a BioWin configuration composed by the Nine Springs Wastewater Treatment Plant in Madison, WI, USA, to characterize its treatment train. In contrast to the previous work [33], this evaluation is focused entirely on the speciation of P within the acid digest effluent. We used literature on the behavior of acid digesters to identify gaps in the existing model and corrected them by altering kinetic and stoichiometric parameters. Where possible, we minimized the effects that these parameter alterations have on the performance of other reactors and treatment trains. We compared the model predictions of the Nine Springs configuration using default parameters to predictions using our suggested modified parameters; both model predictions were compared against chemical analyses obtained via grab sampling from the Nine Springs plant. To verify the reproducibility, we replicate the methods using cabinet configurations, which come standard with BioWin for demonstration purposes to enable users to verify our modifications for their specific configurations.

The major goal of this study is to determine whether the BioWin modeling suite can accurately reproduce the conditions of an acid-phase digester with respect to P solubilization. An additional goal is to create a configuration and set of model values for which the predicted acid digester's effluent matches the measured data while being parsimonious about the number of default values altered. This work also seeks to modify the existing model in a way that current users can replicate with available parameter changes without the need for external calculations or additional expertise. This work is intended as groundwork for the future assessment of P removal from acid digester effluent in municipal wastewater treatment plants.

2. Materials and Methods

The primary software used for this study is BioWin 6.2 (EnviroSim Associates, Ltd., Hamilton, ON, Canada), one of the most popular modeling suites for the wastewater treatment industry. BioWin integrates a series of hydraulic, chemical, and biological models to predict changes in the chemical and microbial composition of wastewater as it passes through various reactors. The BioWin software suite is continually updated to meet the needs of the wastewater treatment industry [34], but BioWin has historically not been a tool of choice for modeling P transformations during wastewater treatment [35]. In the past two versions, it has revamped its chemical precipitation models in an attempt to predict formation of P-containing compounds that frequently act as a nuisance; components capturing struvite and brushite formation were incorporated into BioWin version 5 (2016) and version 6 (2019), respectively.

BioWin contains a library of user-adjustable parameters intended to characterize chemical and microbial interactions. BioWin's activated sludge/anaerobic digestion model (ASDM) is a combination of preexisting anaerobic digestion models (ADMs) and activated sludge models (ASMs) intended to integrate the inputs and outputs of each to cover mechanistic gaps [36] and create a seamless method of modeling whole wastewater treatment plants. The BioWin ASDM alone contains more than fifty state variables and over eighty process expressions. The default values are frequently used as is, but users are able to alter parameters to better represent their own systems when appropriate. BioWin's recent updates have introduced new chemical models to predict the precipitation and dissolution of the P-containing minerals brushite and struvite.

The primary simulation model used for our evaluation is a proprietary BioWin configuration for the Nine Springs Wastewater Treatment Plant in Madison, WI, operated by the Madison Metropolitan Sewerage District (MMSD), created by on-site engineers to predict the effects of changes to their treatment train (Table 1) [27,37]. This plant utilizes an EBPR treatment train which generates high-P sludge. The plant anaerobically digests a 50:50 mixture of primary sludge and WAS (waste activated sludge) in a series of mesophilic digesters generating biogas and then dewaters the sludge to produce biosolids for land application [37]. The first of these digesters is an acid-phase digester with a HRT of 1.3 days. The plant has incurred significant costs in the form of treatment stoppages, lost reactor volume, and maintenance expenses due to the buildup of nuisance struvite in its anaerobic digesters. Their BioWin configuration was created, in part, to assist with P remediation efforts, but the inability of BioWin to properly model their acidogenic digester limited its usefulness.

Table 1. Summary statistics for the Nine Springs Wastewater Treatment Plant.

	Unit	Statistic
Population Served	People	365,000
Service Area	km^2	471
Design Capacity	m^3/d	190,000
Average Throughput	m^3/d	155,000
Total Suspended Solids Load	kg/d	35,600
Phosphorus Load	kg P/d	1700

Digester performance data are taken from a Water Environment Federation evaluation of a P recovery technology [12,27], which recorded daily samples for one year from the acid digester effluent at the Nine Springs Wastewater Treatment plant in 2018. Measured pH was generally stable, averaging 5.14 pH (SD 0.20). Percent P solubilization, as reported by Tabanpour et al. [27], was calculated as the concentration of soluble P (determined by ICP-AES on samples passing through a 0.45 µm filter) times the water content of the sludge sample (obtained by drying at 105 °C), divided by the total P of the sludge (determined by dry ashing at 500 °C, dissolution of the ash in acid, and P determination by ICP-AES). Site measurements from the Nine Springs acid digester effluent recorded a consistent P solubilization of 66% $\pm$ 5% and other facilities record solubilities in the 65–85% range [12,27]. Soluble P, as a percentage of the total P in the digest, ranged from 60.7 to 74.0% of P in the fall of 2018 [27].

The BioWin configuration supplied by Nine Springs comprehensively covered the major aspects of treatment, including clarification, a plug flow aeration stage, EBPR, anaerobic digestion, and sludge recycle streams, WASSTRIP biosolids processing, and an Ostara Pearl struvite reactor. However, the anaerobic digestion section of the configuration was underdeveloped as it contained a single anaerobic digester unit with an HRT of 35.5 days rather than the 9-tank system of the real plant. The operators of the Nine Springs plant had previously attempted unsuccessfully to incorporate an acid digester process in the configuration to accurately model the behavior of the P recovery technologies that precipitate struvite (Ostara) or brushite (CalPrex). Default Biowin parameters yielded predictions that were not comparable to site measurements, so the operators compensated using static, user-defined streams. A goal of our study is to replace this inflexible solution with a working acid digester model based on literature figures and the acid digest pH and VFA simulation methods of Vineyard et al. [33].

Our modifications began by splitting the existing 35.5-day HRT anaerobic digestion process into two new processes: a 1.3-day HRT anaerobic digester and a subsequent 34-day HRT anaerobic digester. As BioWin does not natively model acidogenic digesters, the methods detailed in Vineyard et al. [33] were used to achieve appropriate pH and VFA concentrations by inhibiting microbial growth within the acid digester using the pH inhibition thresholds of anaerobic heterotrophs and H_2-utilizing methanogens. In the initial configuration, BioWin predicted a P solubilization of 27%, too low for a meaningful analysis of new P recovery technologies. We found that BioWin had allocated the effluent P primarily to six compartments: soluble, heterotroph biomass, unreleasable polyphosphate (poly-P) within polyphosphate-accumulating organisms (PAOs), endogenous products, degradable organics, and brushite. This distribution suggests that the disparity between model predictions and site observations cannot be explained by a single variable.

Multiple parameter changes were required to bring model behavior closer to process dynamics and biochemical behavior observed on site and documented in existing literature (Table 2). To minimize the effects of our parameterization on plant-wide performance [38,39], parameter modifications were localized only to the modeled organic acid digester. Universal stoichiometric parameters were avoided as these changes cannot be localized to a single reactor and would affect predictions in the configuration's entire treatment train; however, the universal brushite solubility product was altered after validation against an external chemical speciation program, Visual MINTEQ v 3.1 [40]. We validate these results using extensive soluble P measurements collected in 2018.

Table 2. Alteration of parameters to achieve satisfactory predictions of phosphorus solubilization.

Parameter	Unit	Default	Modified	Data Source
Anaerobic decay rate	1/d	0.04	7.0	[9–11,27]
Brushite redissolution rate	L/(mol·d)	10,000	1.0×10^9	[27,28,41]
Brushite solubility product	$(mol/L)^2$	2.49×10^{-7}	3.415×10^{-7}	[12,27,41]
Endogenous product decay rate	1/d	0	1	[27,42]
Hydrolysis rate	1/d	2.1	3	[10,27,43]
Anaerobic hydrolysis factor	-	0.5	1	

Our study considered additional parameters which may be intuitive, but which were found to be unproductive in the pursuit of the major study goals. We tested a range of acid digest HRT values from 0.1 days to 25 days to analyze the effects of HRT on key parameters such as pH, alkalinity, VFA production, and P solubility. We also tested an alternative method of reducing unreleasable poly-P predictions by directly manipulating stoichiometry of the poly-P fractionation. The results of these explorations and their reasons for exclusion from the final parameterization are outlined in the Appendix A. The following changes (Table 2) best influenced the BioWin predictions of P speciation in an acid digester.

Parameter changes and their effects are reported here in sequence for illustrative purposes, but a user can selectively implement any or all of them to better represent their own reactors. The following changes best influenced the BioWin predictions of P speciation in an acid digester:

2.1. Alteration of Local PAO Decay Kinetics

Acid digesters are specialized systems where PAOs are unlikely to survive [9–11]. Tabanpour et al. [27] found no evidence of polyphosphates in their samples, suggesting that the acid digest environment either forces PAOs to rapidly eject their 'unreleasable' P stores or that the P stores are released during destruction of PAOs. In the simulated acid digester, unreleasable stored poly-Pcontained within the PAOs remains a large reservoir of P. Without solubilizing this component, a good agreement between modeling predictions and the soluble P site measurements is not likely.

In BioWin, PAO decay follows the first-order rate at which PAOs die and are broken down. The default kinetics of the BioWin model do not have a unique decay rate for low pH environments or conditions under which PAOs are likely to be predated upon by other organisms and has no mechanism to replicate these effects. As such, there is a negligible decline in the PAO population over the 1.3-day HRT, which directly leads to unrealistic predictions of their P retention within the digester. Raising the local PAO decay rate allows us to simulate the effects of a hostile, predatory environment on the PAO population. We chose an arbitrary 90% reduction in PAO biomass in the acid digest effluent as our target, which at this HRT, corresponded to an increase in the decay rate from 0.04/d to 7/d.

2.2. Altering the Dissolution Kinetics of Brushite

After implementation of the aforementioned step (Section 2.1), stored polyphosphates are reduced and soluble P becomes the largest P compartment, followed by the biomass of ordinary heterotrophs, endogenous products, and precipitated brushite. The predicted

41.2% soluble P (as a fraction of total P) is still considerably less than the 69% average solubility recorded in the Nine Springs Acid Digester. The prediction of brushite within the system is incongruous with expectations. Tabanpour et al. [27] observed no brushite particulates in the acid digest effluent and they would be unlikely to survive in the acidic, mesophilic environment.

Rate of brushite dissolution is dependent on multiple variables including temperature, agitation, and surface area, but BioWin predicts dissolution using a single static rate. As the BioWin manual specifically warns against setting these rates too high due to computational expenses, it is possible that the default dissolution rate was chosen for user experience rather than devoting large amounts of computing power to a single process. This is a known weakness of precipitation and dissolution modeling [41]. Given that brushite precipitates primarily form as small particles on small nucleating surfaces within the wastewater solution, we assume that the brushite will have a high surface area for dissolution and will rapidly dissolve in the acidic environment. A range of higher brushite dissolution rates, in order-of-magnitude increments, were tested to observe the effects of dissolving brushite particulates to the point of near-instantaneous equilibrium of brushite with its ionic products. The kinetic constant for brushite dissolution in BioWin was finally set to 1×10^9 L/(mol·d) to reduce recalcitrant brushite without a noticeable increase in computation needs.

2.3. Modification of the Brushite Solubility Product

Even at high dissolution rates that simulate near-instantaneous equilibrium, BioWin predicted that 33.9 kg brushite P per day would exit the acidic environment of the acid digester. Given that the P solubility (as a fraction of total P) after prior modifications was still significantly underpredicted compared to measurements, the next logical step was to test whether brushite saturation has been correctly simulated in BioWin.

When the pH and solution concentrations of the BioWin-predicted acid digester effluent were input into Visual MINTEQ, there was a good agreement between the two chemical models in terms of charge balance (0.82% difference in the sum of cations and anions). However, Visual MINTEQ reported that the effluent solution was undersaturated with respect to brushite (Figure 1) while BioWin still predicted brushite in the effluent. The ion activity product for brushite calculated by Visual MINTEQ for the acid digester effluent is about one-third less than BioWin's default $K_{sp,brushite}$ value. The $K_{sp,brushite}$ value used by Visual Minteq is nearly identical to that internally set in BioWin (Figure 2), and both BioWin and Visual MINTEQ use the Davies Equation activity model. We hypothesize the reason that BioWin predictions do not agree with Visual MINTEQ is a failure within BioWin to consider ion pairs appropriate for brushite under these conditions [12], leading to an erroneous prediction of saturation.

Chemical speciation calculations by Visual MINTEQ predicted that free Ca^{2+} in the acid digest effluent was only 69% of total soluble Ca, with VFA ion pairs alone neutralizing 21% of total soluble Ca. Visual MINTEQ predicted that free PO_4^{3-} was 93% of the total soluble P, with the primary ion pairs of P species predicted to be with Ca and Mg. Absent consideration of the presence of ion pairs in internal BioWin calculations, particularly those of VFAs, BioWin would overestimate the concentration of free Ca and P and, consequently, overestimate the amount of brushite to be precipitated to relieve supersaturation.

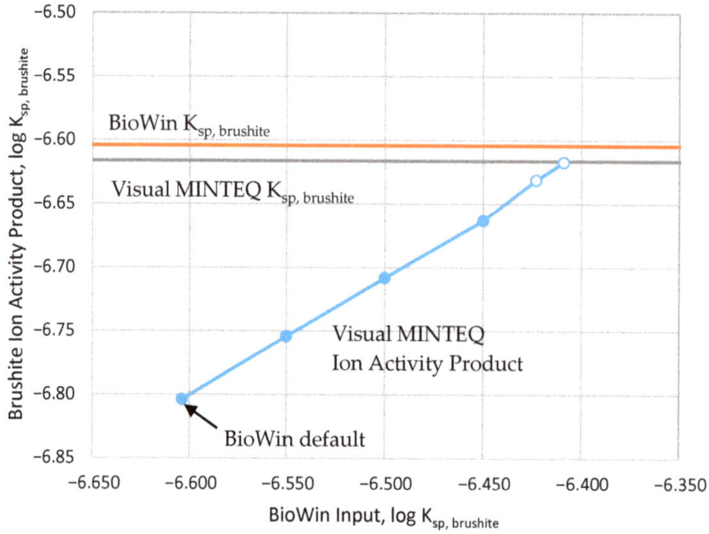

Figure 1. Relationship between value of $K_{sp,brushite}$ input into BioWin and the resulting ion activity product of the effluent as predicted by Visual MINTEQ. Open circles represent BioWin predictions requiring brushite addition to ensure brushite saturation.

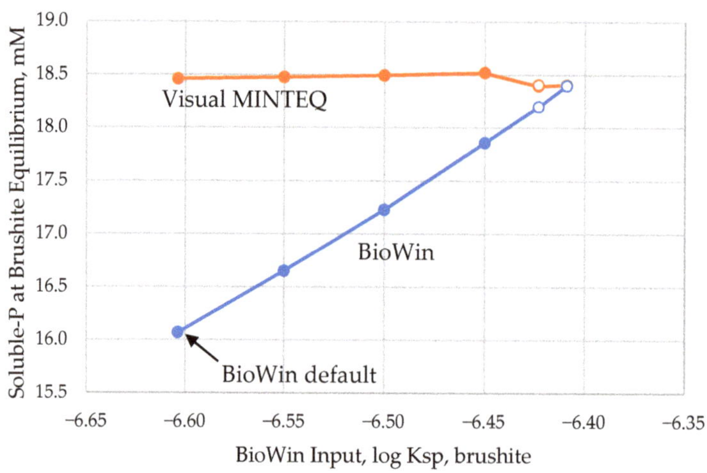

Figure 2. The relationship between value of $K_{sp,brushite}$ input into BioWin and the resulting predicted soluble P concentration in the acid digester effluent compared to the Visual MINTEQ predictions of soluble P at brushite equilibrium. Open circles represent BioWin predictions requiring brushite addition to ensure brushite saturation.

In the absence of an open programming interface within BioWin to correct the chemical modeling of brushite, the ad hoc solution is to alter the $K_{sp,brushite}$ default value used in BioWin to accommodate the expected ion pair formation, essentially creating an apparent or conditional solubility product. We tested a range of brushite solubility products in BioWin and input the resultant effluent concentrations into Visual MINTEQ to determine a $K_{sp,brushite}$ value that indicated the same degree of undersaturation as predicted by Visual MINTEQ.

2.4. Increasing the Endogenous Product Decay Rate

The endogenous products in BioWin are the inert fractions of biomass left over by the microbial processes. Their decay rate is a first order equation that converts their mass to a bioavailable substrate. In BioWin, this parameter is conditionally set to 0 under anaerobic conditions, but a literature review by Menniti et al. [42] reported evidence that endogenous products are degradable under anaerobic conditions and that thermophilic conditions can enhance this degradation. To liberate P from the endogenous products, we increase the local decay rate within the reactor from 0 to 1/d, the maximum allowed by BioWin for this parameter.

2.5. Increasing the Local Hydrolysis Rate

After the previous adjustment step (Section 2.4), degradable organics in the effluent is the largest reservoir of P remaining after heterotroph biomass. As an organic acid digester is a mesophilic system with a high rate of microbial activity, it is specifically engineered to have much higher hydrolysis rates than standard anaerobic digesters [9–11,43]. To more closely represent literature values [10,43], the local hydrolysis rate is increased from 2.1/d to 3/d and the anaerobic hydrolysis factor increased from 0.5 to 1. The acid-representative hydrolysis rate releases P in the organics that would otherwise be erroneously partitioned to the biosolids.

2.6. Validation and Model Testing

Tabanpour et al. [27] sampled eight mass balances over thirty days in the fall of 2018, recording duplicate samples for independent analyses at a local laboratory and the Colorado School of Mines (CSM). The local laboratory reported P solubilities from 60 to 74% while CSM reported from 50 to 85%. Though the averages across all eight mass balances were very similar (66.1% and 64.6%, respectively), the two labs disagreed by an average of ten percentage points on the characterization of any specific sampling date. Using a new influent file comprising 14 days of characteristic variability for the Nine Springs plant from the summer of 2016, we ran two dynamic simulations to record the changes in P solubility predicted by BioWin. The first used default parameters for the acid digester while the second used our modified parameters.

To test for applicability to other datasets, we also replicated these modifications in two new configurations. Because of their ease of access to BioWin users, we selected two preexisting whole-plant configurations in the BioWin cabinet: (i) EBPR with an anaerobic digester, and (ii) N Removal unit with an anaerobic digester. These configurations represent possible EBPR and non-EBPR feeds to acid digesters from realistic treatment trains. Primary modification to the configurations involved pairing a 1.3-day HRT acid digester to the anaerobic digester with appropriate overall HRT adjustment, modifying acid digester performance in accordance with Vineyard et al. [33], and then implementing the above five phosphorus-based parameter changes. We compared the projected acid digester effluent with and without the phosphorus-targeted parameter changes.

3. Results

Five parameter changes were needed to reallocate P to the soluble compartment within a BioWin modeled acid digester for its predictions to match site measurements. For illustrative purposes, these parameters were changed and recorded here one at a time to observe the effects on P speciation within the modeled acid digester effluent. An implementation of these methods can apply them in any order or all at once, depending on the configuration.

3.1. Alteration of the PAO Decay Kinetics

The local decay kinetics for the PAOs under anaerobic conditions were raised from 0.04/d to 7/d to simulate the effects of environments hostile to these organisms and achieve a 90% reduction target. The unreleasable poly-P fraction is reduced in proportion with the biomass, which is then distributed to the soluble P and to the degradable organics fractions (Figure 3). The reduced PAO population in the recycle stream does not significantly affect P reduction in the main treatment train and the effluent P remains within 1% of its value with altered PAO decay rates.

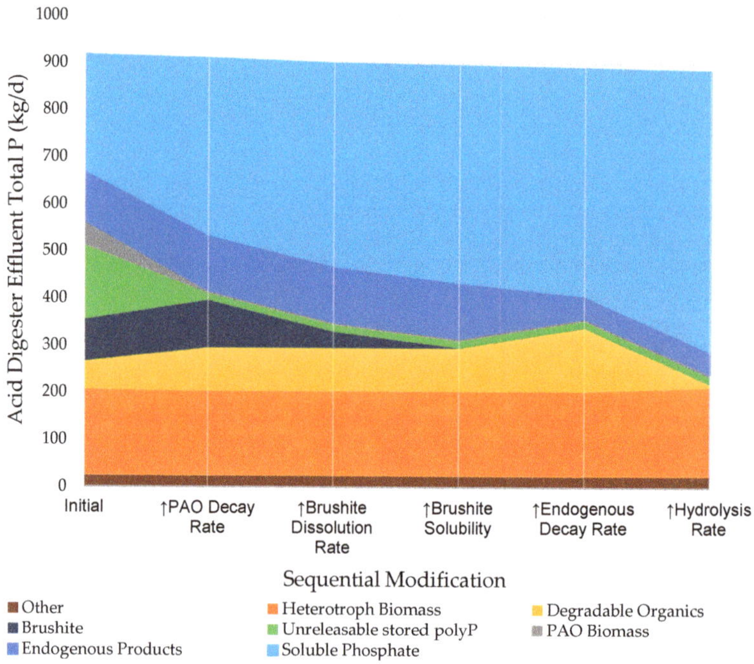

Figure 3. Effects of parameter alteration methods on P speciation in Nine Springs acid digest effluent. Digester set to maintain a pH of ~5.14 across all steps.

3.2. Altering the Dissolution Kinetics of Brushite

In the default BioWin mineral precipitation module, brushite is modeled to precipitate 100 times faster than it dissolves (106 L/mol/d vs. 104 L/mol/d). This parameterization results in the predictions of brushite, once formed in the activated sludge, not dissolving while in the acid digester, though there are no supporting observations at Nine Springs or elsewhere [27]. Progressively higher brushite dissolution rates predict higher soluble P and lower brushite but ultimately yield diminishing returns, seemingly approaching equilibrium in the modeled acid digester effluent (Figure 4). It may be observed in Figure 3 that the soluble P increases are almost identical to the decrease in the mass of the brushite, suggesting that nearly all the dissolved brushite remains as soluble P rather than being reintegrated into another P compartment. An increase in the dissolution rate from the default to 109 L/mol/d achieves an increase in P solubility from 41% to 48%. As secondary effects, an increased brushite dissolution rate is observed to only slightly increase pH, VFA production, and methane generation.

As BioWin does not isolate this precipitation constant to a single reactor, the modified higher brushite dissolution rate applies to every reactor in the treatment train. The performance of other reactors needs to be monitored for unintended side effects when manipulating global variables. Modifying the brushite dissolution showed no significant changes in the brushite content of other streams and the influent brushite to the acid digester compartment was unchanged, suggesting the unintended effects were minimal. We suggest a 109 L/mol/d dissolution rate to simulate instantaneous dissolution while avoiding the computational needs of higher rates.

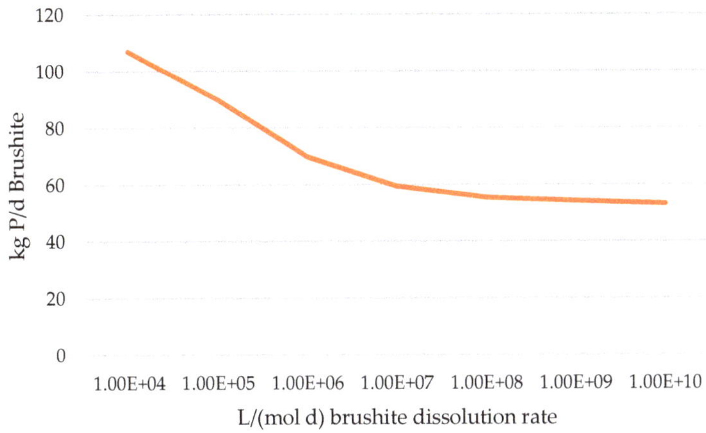

Figure 4. Response of effluent brushite content to increasing dissolution rates shows diminishing returns.

3.3. *Modification of the Brushite Solubility Product*

The BioWin predictions of pH and dissolved ions in the acid digester effluent (NH_4-N, Ca, Mg, P, acetate, propionate, total dissolved CO_2, sulfate, sulfide, "other cations" (represented as Na^+), and "other anions" (represented as Cl^-)) were input into the standalone program Visual MINTEQ v 3.1, a chemical equilibrium model built upon the USEPA's MINTEQA2 v 4.06 software using the NIST Standard Reference Data 46 as its database. We tested the effects of $K_{sp,brushite}$ in BioWin on the modeled acid digester and compared the resulting effluent's brushite ion activity products as calculated by Visual Minteq. A $K_{sp,brushite}$ in BioWin of 3.415×10^{-7} (Figure 5) was found to force the BioWin simulations of the brushite equilibrium to match those of Visual Minteq.

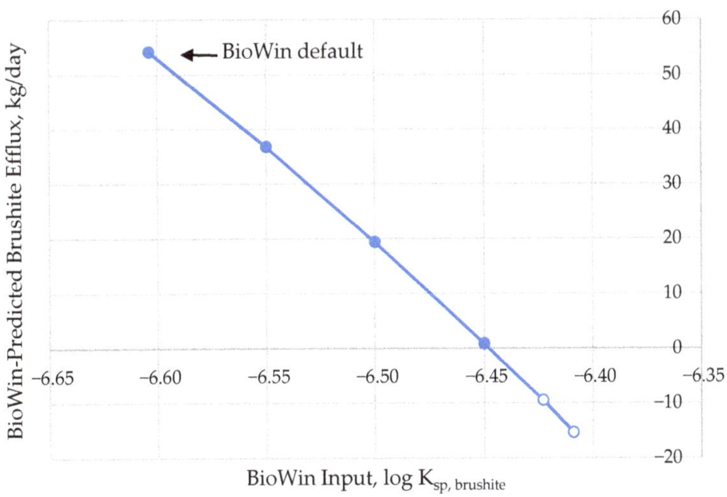

Figure 5. Comparison of saturation indices calculated in Visual Minteq for the BioWin acid digester effluent conditions produced by a range of log $K_{sp,brushite}$ Open circles represent BioWin predictions requiring brushite addition to ensure brushite saturation.

Using the updated conditional $K_{sp,brushite}$ of 3.415×10^{-7}, the brushite persistence and/or formation in the acid digester as predicted by BioWin decreases from 33.9 kg P/d to

0 kg P/d, the soluble P concentration in the acid digester effluent increases from 431 kg P/d to 461 kg P/d, with the acid digester effluent now undersaturated with respect to brushite (Figure 3). With this modification to BioWin, the soluble P is 51.4% of the total P in the acid digester effluent of the Nine Springs model (Figure 3), consistent with some other reactors measured in the 45–49% range [27], but still short of the 69% goal based on the measured performance of the modeled reactor.

The new conditional $K_{sp,brushite}$ determined for the acid digester effluent is applicable only for situations with high concentrations of VFAs and phosphate. For programming reasons, this change must be system-wide in BioWin even though elevated VFAs and phosphate concentrations are typical only of the acid digester. The developers of BioWin could add appropriate ion pairs in later versions and obviate the need for this conditional K_{sp}.

3.4. Increasing the Endogenous Product Decay Rate

After releasing P from PAO stores and biomass, endogenous products remain one of the largest remaining reservoirs of P. Within the modeled acid digester, the endogenous product decay rate was raised from 0 to 1/d, the maximum allowed. This adjustment compensated for the default state of the endogenous products being labeled as indigestible in an anaerobic digester, which may not apply to acid digesters. This modification liberated 60% of the endogenous product P passing through the digester, repartitioning it into degradable organics and soluble P. The total P solubilization increased from 51.4% to 54.2% (Figure 3).

3.5. Increasing the Hydrolysis Rate

By doubling the local hydrolysis rate, the acid digester effluent was predicted to contain lower quantities of degradable organics and higher quantities of VFAs to correspond with the measurements reported in Tabanpour et al. [27]. Nearly tripling the effective hydrolysis rate resulted in a 95% reduction in the degradable organics in the effluent (Figure 3). The degradable organics were the second largest reservoir of P remaining in the effluent, the digestion of which increased the P solubilization from 54.2% to 67.2% (Figure 3), matching the site observations.

4. Discussion

4.1. Benchmark to Full-Scale Grab Sample Data

The dynamic simulations produced a tight range of P solubilizations (Figure 6). Before modification, BioWin predicted an acid digest of 24.2–26.6% P solubilization, with an average of 25.6%. After modification, BioWin predicted an acid digest of 62.3–63.7% P solubilization, with an average of 63.1%. The modifications create a clear improvement in the predictive power of the software. In both dynamic simulations, the total variability of the predicted solubilization was very small compared to the variability in the grab sample data. There are two possible explanations for this discrepancy: either the methods used to acquire and analyze the grab samples were highly uncertain or BioWin's perfect mixing model over-homogenizes the system and does not characterize the full effects of the daily fluctuations in a dynamic simulation. Given the large disagreement between the two laboratories analyzing the same samples, we cannot eliminate the possibility of sampling and/or laboratory error. Further reactor sampling could help determine whether the predicted homogeneity is accurate.

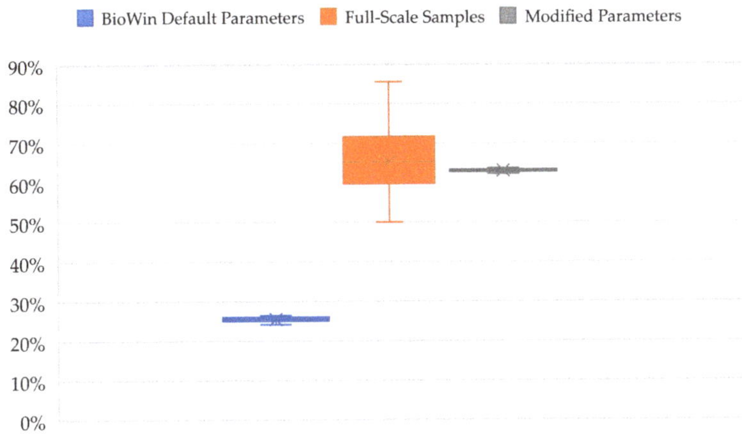

Figure 6. BioWin predictions of acid digester P solubilization using two weeks of summer 2016 plant influent data compared to laboratory analyses of eight acid digest samples from fall 2018.

4.2. Assessment of Plant-Wide Effects

We assessed the impact on the total anaerobic digester train predictions with parameter modifications tailored specifically for the acid digester. A comparison of the anaerobic digester effluent between the original and re-parameterized BioWin configurations revealed some significant differences (Table 3). Both the VSS and TSS decreased after parameter re-calibration due to the enhanced digestion of the endogenous products and PAOs. The phosphorus distribution between the different compartments was also altered. The greatest percentage difference was in the soluble P compartment attributable to the increased destruction of PAOs liberating P from the unreleasable poly-P compartment. In terms of the total mass, the precipitation of struvite was greatly increased due to the extra soluble P driving the equilibrium towards precipitation.

Table 3. Predicted final effluent content of the complete Nine Springs anaerobic digester train.

		Digester Train Effluent		
	Unit	**Default**	**Modified**	**% Change**
VSS	kg/d	16,000	13,600	−15%
TSS	kg/d	23,000	20,700	−10%
pH		7.15	7.19	
Alk	k/d	69.5	77.3	+11%
Total P	kg P/d	813	785	−3.4%
Soluble P	kg P/d	25.1	52.4	+108%
Struvite	kg P/d	277	352	+27%
Brushite	kg P/d	147	130	−12%
Unreleasable Stored PolyP	kg P/d	59.2	6.31	−89%
Calcium	kg/d	100	70.1	−30%

Although all the brushite entering the acid digester was solubilized in the modified (i.e., re-parameterized) reactor, re-calibration had a small effect on the amount of brushite precipitated in the subsequent anaerobic digestion process. Given that the P distribution parameters were altered, their cumulative effect on the P fate can be determined by comparing the P composition in the effluent from the overall configuration before and after the modifications (Table 4). The predicted total P in the plant effluent increased after parameter modification, with changes driven by a shift away from biosolids and towards struvite formation and the treatment plant effluent; the alteration of the brushite solubility and the release of P from endogenous products and PAO biomass, before recirculating to the Ostara reactor, is likely responsible for this change. Modifications to the subsequent anaerobic digester processes to offset these changes may be possible, but it is beyond the scope of this work. Users replicating our methods should be aware of the possible need for rebalancing their treatment train configurations.

Table 4. Predicted distribution of P across all configuration effluents.

	Default (kg P/d)	Modified (kg P/d)	% Change
Plant effluent:			
Brushite	0	0	--
Struvite	0	0	--
Soluble P	8.00	11.7	+46%
Total	67.8	84.3	+24%
Biosolids:			
Brushite	146	129	−11%
Struvite	275	350	+27%
Soluble P	17.0	35.3	+107%
Total	800	764	−4.5%
Recycle to primary:			
Brushite	10	0	−100%
Struvite	0	0	--
Soluble P	27.9	66.3	+138%
Total	104	100	−3.8%

4.3. Application to Other Treatment Plant Models

Both BioWin cabinet configurations were modified by incorporating an acid digester process using the pH correction methods of Vineyard et al. [33]. With default parameters, the acid digesters from both treatment trains projected the P solubility below what is observed in extant acid digesters [27]. After modification using this study's five parameter changes to enhance the P solubilization, both treatment trains increased their soluble P content to be within the observed range of values from acid digesters as recorded by Tabanpour et al. [27] (Table 5). Therefore, a simple 'cut and paste' of new settings into existing configurations could improve the accuracy of the modeled acid digesters. In both cabinet models, it is likely that further calibration using the above methods could bring the predictions of the P distribution closer to the measured/observed values.

Table 5. Responses of an EBPR Train Digester and an N Removal Train Digester to addition of an acid digester, with acid digester effluent properties before and after the generic changes above.

Item	Unit	P Train Digester			N Train Digester		
		Influent	Acid Effluent (Default)	Acid Effluent (Modified)	Influent	Acid Effluent (Default)	Acid Effluent (Modified)
Flow	M^3/d	56	56	56	125	125	125
Total P	kg P/d	61.7	61.7	61.7	106	106	106
Soluble P	kg P/d	0.18	25.4	39.6	0.53	27.8	55.7
Struvite	kg P/d	0	0	0	0	0	0
Brushite	kg P/d	0	0	0	0	0	0
Unreleasable Stored PolyP	kg P/d	4.26	4.05	0.42	0	0	0
Soluble Calcium	kg/d	4.49	9.90	12.85	20.5	29.13	36.6

4.4. Limitations and Future Work

Modeling multiple interactions in a complex treatment train typically encountered in WWTPs is difficult and the assumptions made in the derivation of these estimates may not be valid in certain configurations and circumstances. Our methods sought to increase the accuracy in simulations while making as few changes to the base configuration as possible, but there may be additional complexities that our methods do not capture, which may become significant in alternative treatment trains. As the intended purpose of this work is to better estimate the performance of novel technologies using the acid digester effluent as a starting feed material, future analysts and modelers should be cautious when extending these methods to other applications.

Our work calibrated the acid digester simulation primarily for P solubilization as the groundwork for the evaluation of P recovery by brushite precipitation. This work paves the way for more accurately predicting the performance of nutrient recovery technologies that target soluble P in the acid digester effluent. In the default BioWin settings, such an attempt would predict unrealistically high reagent requirements and unrealistically low recovery rates, leading to inflated estimates of capital and operation costs and low estimates of performance. The consideration of acetate and propionate ion pairs with Ca will be taken up with the programmers at EnviroSim for correction and improvement within future versions of BioWin.

Our re-parameterization process was unable to produce estimates in line with verification data without enacting some plant-wide parameter changes. These new parameter values can still affect the performance of other reactors by altering stream qualities, as recorded in the change in the sludge composition and total plant P distribution despite no changes in the characteristics of the anaerobic digester effluent. A study attempting to predict the plant-wide performance for practical purposes will benefit from accounting for these altered behaviors.

5. Conclusions

The BioWin wastewater treatment plant modeling was unable to match the measured performance of anaerobic acid digesters when using default parameters. It was clear from the data collected from a full-scale facility that a new approach to predicting the P solubility in the model would be required to evaluate the impact of incorporating new nutrient recovery technologies. Through a systematic five-parameter process altering the PAO decay rate, brushite dissolution rate, brushite solubility, endogenous product decay rate, and hydrolysis rate, to align with literature values and account for missing biochemical mechanisms, the BioWin predictions for the soluble P fraction increased from 27% to 69%, matching the site observations. Our novel calibration of the acid digester in BioWin overcomes the limitations of the modeling suite, thus enabling the more rigorous evaluation of novel nutrient treatment/recovery processes.

Author Contributions: Conceptualization, D.V., K.K. and P.B.; methodology, D.V.; validation, D.V.; formal analysis, D.V.; investigation, D.V.; resources, P.B.; data curation, D.V.; writing—original draft preparation, D.V.; writing—review and editing, D.V., K.K., C.D. and P.B.; visualization, D.V. and P.B.; supervision, K.K., C.D. and P.B.; project administration, K.K., C.D. and P.B.; funding acquisition, P.B., K.K. and C.D. All authors have read and agreed to the published version of the manuscript.

Funding: This work was funded by INFEWS/T3 Grant (Reference #2017-67003-26055), USDA/NIFA Hatch Grant No. WIS01920 (Accession # 1009816), and USDA/NIFA Hatch Grant No. WIS04092 (Accession # 7003184).

Data Availability Statement: Some MMSD-related data is confidential but to demonstrate the principles of modeling an acid digester in BioWin, we have made available a modified version of a BioWin cabinet model with full supporting documentation at DOI 10.17605/OSF.IO/8DXM7.

Acknowledgments: We are grateful for the assistance of Matt Seib and the Madison Metropolitan Sewerage District.

Conflicts of Interest: The authors declare the following financial interests/personal relationships which may be considered as potential competing interests: Phillip Barak reports financial support was provided by the US Department of Agriculture, National Institute of Food and Agriculture, Hatch Grant. Phillip Barak reports a relationship with Nutrient Recovery and Upcycling, LLC, that includes board membership and equity or stocks. Nutrient Recovery and Upcycling, LLC, is rights holder to US Patent 8,568,590 (issued 2013), "Phosphate recovery from acid phase digesters".

Appendix A.

Appendix A.1. Alternative Parameters

Appendix A.1.1. Altering Stored Poly-P Distribution

Polyphosphate-accumulating organisms store polyphosphate (poly-P) in their cells, particularly in the EBPR processes, and BioWin considers that some part of that poly-P is not depolymerized in subsequent digestions. In the simulations for the Nine Springs plant, there remains 147 kg/d of P within the unreleasable poly-P compartment, one of the largest remaining destinations for P within the acid digester effluent. The acid conditions of the digester break poly-P bonds and favor its decomposition to an extent other reactors do not, so this high quantity is unexpected. As no direct measurements support the presence of a large fraction of P retained within the PAOs at Nine Springs, we consider that it can function as a source of soluble P with few unintended side effects for the total treatment train performance. This parameter serves as an alternative to altering the PAO decay rate within the acid digester, to be undertaken after the alterations of low pH inhibition points and increasing the brushite dissolution rate.

Altering low poly-P in the PAOs from the default value (94%) to 99% reduces the throughput of the unreleasable poly-P from 147 kg/d to 29 kg/d (Table A1). Of this 118 kg/d difference, 29 kg/d are solubilized and 48 kg/d are redistributed to brushite precipitate despite the use of a higher dissolution rate. Interestingly, the influent P content is reduced by 41 kg/d, 4% of the total P, because the new P distribution reduces the recirculating P fraction and increases the predicted P capture in the biosolids.

Table A1. Acid digester effluent profile predictions in response to an increased low poly-P yield in PAOs.

Item	Unit	0.94 Low Poly-P (Default)	0.97 Low Poly-P	0.99 Low Poly-P
Total P	kg P/d	1050	1026	1009
Soluble P	kg P/d	326	341	355
Brushite	kg P/d	22.7	52.41	71.08
Unreleasable Stored PolyP	kg P/d	147	78.3	29
Calcium	kg/d	260	247	237
VSS Destruction	%	34.06	34.04	33.97
Dry Gas Production	m^3/hr	16.64	16.62	16.68
Gas Methane Content	%	16.76	17.62	17.97

Appendix A.1.2. Impacts of HRT on Acid Digester Outputs

One of the main parameters a reactor operator might use to alter their reactor conditions is the HRT. BioWin operators may also see HRT as a potential tool for solving the discrepancies due to the incorporation of an acid digester. While VFA production is influenced by the HRT, there is no simple way to characterize all the effects of HRT manipulation. Instead, we experimented with a range of HRTs to analyze their effects on key parameters (Table A2). We chose pH, acetate, propionate, alkalinity, total P, soluble P, and VSS as our response variables for analysis. Our configuration is the base configuration after the separation of the acid digester from the anaerobic digester. As the HRT is increased from 0 d, the pH of the modeled acid digester rapidly decreases to the 4.3–4.5 range, below the measured pH values. At the 1.3–1.8 day HRT range, the predictions of the acid digester effluent soluble P content match the range of observed values, but at this HRT, the predicted pH is still lower than the measured values.

Table A2. Effects of HRT on key digester effluent parameters.

HRT	pH	Acetate	Propionate	Alkalinity	Total P	Soluble P	VSS
Days		kg COD/d	kg COD/d	k/d	kg P/d	kg P/d	kg/d
0.1	5.78	556	1336	16.03	1004	124	39,601
0.125	5.46	669	1659	16.35	1004	141	39,302
0.15	5.25	779	1986	16.4	1004	152	39,003
0.2	5.00	998	2640	16.4	1004	173	38,409
0.25	4.86	1216	3291	16.32	1004	191	37,816
0.3	4.77	1433	3939	16.2	1004	208	37,226
0.4	4.65	1865	5223	15.85	1004	240	36,054
0.5	4.57	2293	6491	15.41	1004	268	34,896
0.6	4.52	2713	7737	14.9	1004	294.5	33,755
0.65	4.5	2921	8352	14.65	1004	307	33,192
0.7	4.48	3126	8958	14.39	1004	319	32,635
0.7125	4.47	3177	9108	14.33	1004	322	32,497
0.725	4.47	3228	9258	14.26	1004	325	32,359
0.75	4.46	3323	9556	14.13	1004	331	32,084
0.8	4.45	3528	10,144	13.87	1004	342	31,541
0.9	4.42	3915	11,280	13.37	1004	363	30,482
1	4.41	4280	12,345	12.9	1005	380	29,467
1.2	4.38	4925	14,196	12.13	1005	403	27,618
1.3	4.37	5209	15,002	11.76	1005	413	26,821
1.5	4.36	5725	16,444	10.8	1005	430	25,497
1.8	4.34	6295	17,959	9.43	1005	448	24,218
2.1	4.33	6640	18,731	9.05	1005	461	23,460
2.5	4.34	6962	19,284	8.95	1005	474	22,815
3	4.34	7287	19,647	8.86	1005	487	22,263
3.5	4.34	7592	19,807	8.60	1005	497	21,845
4	4.34	7904	19,819	8.11	1005	507	21,503
4.5	4.34	8230	19,716	7.45	1005	515	21,212
5	4.35	8566	19,530	6.71	1005	522	20,957
5.5	4.35	8901	19,295	5.96	1005	529	20,729
6	4.35	9228	19,038	5.25	1005	535	20,524
6.5	4.35	9540	18,776	4.62	1005	541	20,336
7	4.35	9835	18,521	4.06	1005	546	20,163
10	4.35	11,236	17,286	1.91	1006	571	19,355
15	4.36	12,646	16,124	0.46	1006	598	18,498
20	4.36	13,491	15,945	−0.17	1006	616	17,946
25	4.36	14,060	15,104	−0.52	1006	629	17,557

Using the 1.3-day HRT of the verification system as a baseline, we can track how the key parameters of the effluent are affected by alterations to the acid digester HRT (Figure A2). The predicted pH, VSS, and alkalinity all respond negatively to an increasing HRT; the pH shows a rapid decline due to the logarithmic scaling of the metric, while the alkalinity increases briefly before the onset of the negative trend. The acetate, propionate, and soluble P concentrations all respond positively to an increase in the HRT.

As the HRT is further increased beyond 2.5 days, the reactor in the base configuration shows an interesting behavior. Increasing the HRT by small steps and then allowing the model to iteratively converge to a new steady state will continue to yield acid digester conditions far beyond the HRT range under which those conditions would be expected (Figure A2).

Performing the same HRT analysis on the finalized version of the system, with all the parameter changes made, shows similar trends (Figure A3). The primary differences are that the pH remains much more stable over HRT changes and that the alkalinity shows an inverse trend from the original, increasing with higher HRT. This is likely due to the increased significance of brushite and calcium in the solution.

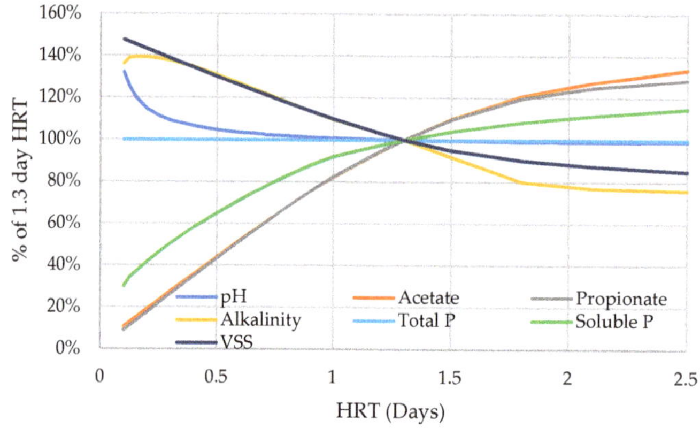

Figure A1. Behavior of various indicators of acid digester function as HRT is adjusted in comparison to the 1.3-day default. Range of 0-2.5 days.

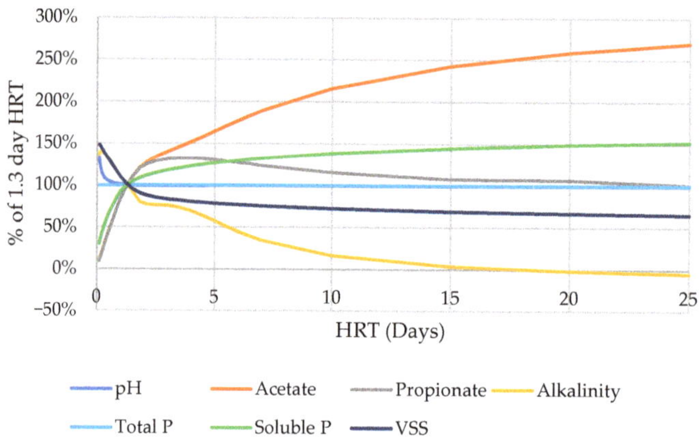

Figure A2. Behavior of various indicators of default acid digester function as HRT is adjusted in comparison to the 1.3 day default. Range of 0–25 days.

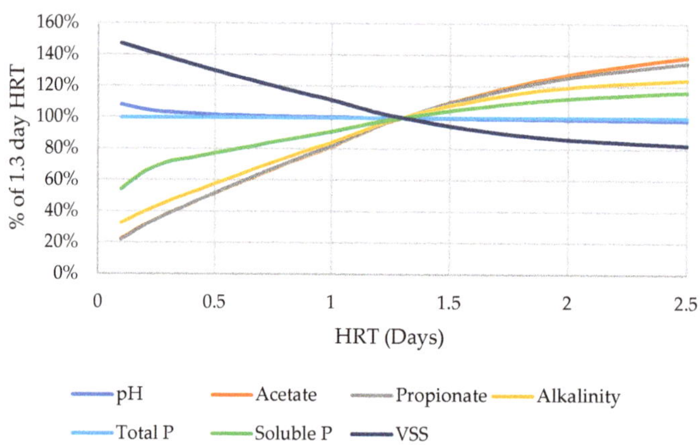

Figure A3. Behavior of various indicators of modified digester function as HRT is adjusted in comparison to the 1.3 day default. Range of 0–2.5 days.

References

1. Jeyanayagam, S.; Khunjar, W.; Mehta, C. Accelerating the Implementation of Extractive Nutrient Recovery as an Integral Component of Sustainable Nutrient Management. *Water e-J.* **2016**, *1*, 1–8. [CrossRef]
2. United States Environmental Protection Agency. *Case Studies on Implementing Low-Cost Modifications to Improve Nutrient Reduction at Wastewater Treatment Plants*; United States Environmental Protection Agency: Washington, DC, USA, 2015; Volume 2.
3. Egle, L.; Rechberger, H.; Krampe, J.; Zessner, M. Phosphorus recovery from municipal wastewater: An integrated comparative technological, environmental and economic assessment of P recovery technologies. *Sci. Total Environ.* **2016**, *571*, 522–542. [CrossRef]
4. Clark, D.L.; Hunt, G.; Kasch, M.S.; Lemonds, P.J.; Moen, G.M.; Neethling, J.B. *WERF Nutrient Management Report*; Montana Department of Environmental Quality: Helena, MT, USA, 2010.
5. Rahman, S.M.; Eckelman, M.J.; Onnis-Hayden, A.; Gu, A.Z. Life-Cycle Assessment of Advanced Nutrient Removal Technologies for Wastewater Treatment. *Environ. Sci. Technol.* **2016**, *50*, 3020–3030. [CrossRef]
6. Vesilind, P.A. *Wastewater Treatment Plant Design*; Water Environment Federation: Alexandria, VA, USA, 2003.
7. Rittmann, B.E.; McCarty, P.L. *Environmental Biotechnology: Principles and Applications*; McGraw-Hill: Singapore, 2001.
8. Water Environment Federation. *Introduction to Water Resource Recovery Facility Design*, 2nd ed.; McGraw-Hill Professional: New York, NY, USA, 2014.
9. Harirchi, S.; Wainaina, S.; Sar, T.; Nojoumi, S.A.; Parchami, M.; Parchami, M.; Varjani, S.; Khanal, S.K.; Wong, J.; Awasthi, M.K. Microbiological insights into anaerobic digestion for biogas, hydrogen or volatile fatty acids (VFAs): A review. *Bioengineered* **2022**, *13*, 6521–6557. [CrossRef]
10. Demirel, B.; Yenigün, O. Two-Phase Anaerobic Digestion Processes: A Review. *J. Chem. Technol. Biotechnol.* **1002**, *77*, 743–755. [CrossRef]
11. Jain, S.; Jain, S.; Wolf, I.T.; Lee, J.; Tong, Y.W. A comprehensive review on operating parameters and different pretreatment methodologies for anaerobic digestion of municipal solid waste. *Renew. Sustain. Energy Rev.* **2015**, *52*, 142–154. [CrossRef]
12. Barak, P.; Davidson, C.; Minks, A. Chemical composition of organic acid digest from a municipal wastewater treatment plant and chemical modeling of nuisance struvite formation and phosphorus recovery as brushite. *PLOS Water* **2023**, *2*, e0000120. [CrossRef]
13. Doyle, J.D.; Parsons, S.A. Struvite formation, control and recovery. *Water Res.* **2002**, *36*, 3925–3940. [CrossRef] [PubMed]
14. Elliot, T.J. *Benefits of Ferric Chloride Addition on Struvite and Sulfide Control*; University of Wisconsin-Madison: Madison, WI, USA, 2002.
15. Fattah, K.P.; Sabrina, N.; Mavinic, D.S.; Koch, F.A. Reducing operating costs for struvite formation with a carbon dioxide stripper. *Water Sci. Technol.* **2008**, *58*, 957–962. [CrossRef] [PubMed]
16. Ghosh, S.; Lobanov, S.; Lo, V.K. An overview of technologies to recover phosphorus as struvite from wastewater: Advantages and shortcomings. *Environ. Sci. Pollut. Res.* **2019**, *26*, 19063–19077. [CrossRef]
17. Hao, X.D.; van Loosdrecht, M.C.M. Model-based evaluation of struvite recovery from an in-line stripper in a BNR process (BCFS). *Water Sci. Technol.* **2006**, *53*, 191–198. [CrossRef]
18. Hao, X.D.; Wang, C.C.; Lan, L.; van Loosdrecht, M.C. Struvite formation, analytical methods and effects of pH and Ca^{2+}. *Water Sci. Technol.* **2008**, *58*, 1687–1692. [CrossRef] [PubMed]
19. Ishii, S.K.; Boyer, T.H. Life cycle comparison of centralized wastewater treatment and urine source separation with struvite precipitation: Focus on urine nutrient management. *Water Res.* **2015**, *79*, 88–103. [CrossRef] [PubMed]
20. Kataki, S.; West, H.; Clarke, M.; Baruah, D.C. Phosphorus recovery as struvite: Recent concerns for use of seed, alternative Mg source, nitrogen conservation and fertilizer potential. *Resour. Conserv. Recycl.* **2016**, *107*, 142–156. [CrossRef]
21. Marti, N.; Bouzas, A.; Seco, A.; Ferrer, J. Struvite precipitation assessment in anaerobic digestion processes. *Chem. Eng. J.* **2008**, *141*, 67–74. [CrossRef]
22. Sharp, R.; Vadiveloo, E.; Fergen, R.; Moncholi, M.; Pitt, P.; Wankmuller, D.; Latimer, R. A theoretical and practical evaluation of struvite control and recovery. *Water Environ. Res.* **2013**, *85*, 675–686. [CrossRef]
23. Song, Y.H.; Qiu, G.L.; Yuan, P.; Cui, X.Y.; Peng, J.F.; Zeng, P.; Duan, L.; Xiang, L.C.; Qian, F. Nutrients removal and recovery from anaerobically digested swine wastewater by struvite crystallization without chemical additions. *J. Hazard. Mater.* **2011**, *190*, 140–149. [CrossRef] [PubMed]
24. Barak, P.; Stafford, A. Struvite: A recovered and recycled phosphorus fertilizer. In Proceedings of the 2006 Wisconsin Fertilizer, Aglime & Pest Management Conference, Madison, WI, USA, 17–19 January 2006; p. 199.
25. Booker, N.; Priestley, A.; Fraser, I. Struvite formation in wastewater treatment plants: Opportunities for nutrient recovery. *Environ. Technol.* **1999**, *20*, 777–782. [CrossRef]
26. Vineyard, D.L., II. Phosphorus Recovery of Brushite at Wastewater Treatment Plants: BioWin Modeling, Economic and Life Cycle Assessment of CalPrex. Ph.D. Thesis, The University of Wisconsin, Madison, WI, USA, 2022.
27. Tabanpour, M.; Downing, L.; Kuhnjar, W. *Demonstrating the CalPrex System for High Efficiency Phosphorus Recovery*; Water Research Foundation: Alexandria, VA, USA, 2021.
28. *BioWin 6 Help Manual*; Envirosim Associates Ltd.: Hamilton, ON, USA, 2020.
29. Pritty, C. Anaerobic Digestion Modeling of a High Rate Acidification Pre-Treatment Technology. Master's Thesis, The University of Western Ontario, Ontario, CA, 2019.

30. Forgacs, G.; Smyth, M.; Law, I.; Arnot, T. BioWin modelling to reduce the Avonmouth digester commissioning programme from 6 months to 6 weeks. In Proceedings of the European Biosolids and Organic Resources Conference, Leeds, UK, 20–21 November 2017.
31. Yang, W.; Young, S.; Munoz, A.; Palmarin, M.J. Dynamic modeling of a full-scale anaerobic mesophilic digester start-up process for the treatment of primary sludge. *J. Environ. Chem. Eng.* **2019**, *7*, 103091. [CrossRef]
32. Nghiem, L.D.; Wickham, R.; Ohandja, D.-G. Enhanced biogas production and performance assessment of a full-scale anaerobic digester with acid phase digestion. *Int. Biodeterior. Biodegrad.* **2017**, *124*, 162–168. [CrossRef]
33. Vineyard, D.; Karthikeyan, K.G.; Davidson, C.; Barak, P. Modeling an acid-phase digester in BioWin with parameter optimization from site data. *J. Water Process Eng.* **2023**, *54*, 103971. [CrossRef]
34. Envirosim. A Brief History of Activated Sludge Process Models and Simulators. 2000. Available online: https://www.envirosim.com/downloads/Model%20history.pdf (accessed on 7 November 2023).
35. Flores-Alsina, X.; Solon, K.; Kazadi Mbamba, C.; Tait, S.; Gernaey, K.V.; Jeppsson, U.; Batstone, D.J. Modelling phosphorus (P), sulfur (S) and iron (Fe) interactions for dynamic simulations of anaerobic digestion processes. *Water Res.* **2016**, *95*, 370–382. [CrossRef] [PubMed]
36. Mo, R.; Guo, W.; Batstone, D.; Makinia, J.; Li, Y. Modifications to the anaerobic digestion model no. 1 (ADM1) for enhanced understanding and application of the anaerobic treatment processes—A comprehensive review. *Water Res.* **2023**, *244*, 120504. [CrossRef] [PubMed]
37. AECOM. Chloride Compliance Study: Nine Springs Wastewater Treatment Plant Final Report. *Madison Metropolitan Sewerage District.* 2015. Available online: https://web.archive.org/web/20160505154642/%20http://www.madsewer.org/Portals/0/ProgramInitiatives/ChlorideReduction/MMSD%20Chloride%20Compliance%20Study%20Report%20-%20Final%206-19-15bookmarks.pdf (accessed on 7 November 2023).
38. Liwarska-Bizukojc, E.; Biernacki, R. Identification of the most sensitive parameters in the activated sludge model implemented in BioWin software. *Bioresour. Technol.* **2010**, *101*, 7278–7285. [CrossRef] [PubMed]
39. Liwarska-Bizukojc, E.; Olejnik, D.; Biernacki, R.; Ledakowicz, S. Calibration of a complex activated sludge model for the full-scale wastewater treatment plant. *Bioprocess Biosyst. Eng.* **2011**, *34*, 659–670. [CrossRef] [PubMed]
40. Gustafsson, J. Visual MINTEQ ver. 3.1. 2013. Available online: https://vminteq.com (accessed on 7 November 2023).
41. Wang, R.; Li, Y.; Chen, W.; Zou, J.; Chen, Y. Phosphate release involving PAOs activity during anaerobic fermentation of EBPR sludge and the extension of ADM1. *Chem. Eng. J.* **2016**, *287*, 436–447. [CrossRef]
42. Menniti, A.; Rieger, L.; Boltz, J.P.; Johnson, B.; Daigger, G.; Habermacher, J.; Derlon, N.; Morgenroth, E. Critical review on modeling of endogenous processes and the degradability of endogenous decay products. In Proceedings of the WWTmod, Mont-Sainte-Anne, QC, Canada, 26–28 February 2012.
43. Eastman, J.A.; Ferguson, J.F. Solubilization of particulate organic carbon during the acid phase of anaerobic digestion. *J. Water Pollut. Control Fed.* **1981**, *53*, 352–366.

environments MDPI

Article

Analysis of Siphonic Roof Drainage Systems with EPANET

Gonzalo López-Patiño, Pedro L. Iglesias-Rey, Francisco Javier Martínez-Solano * and Vicente S. Fuertes-Miquel

Hydraulic Engineering and Environment Department, Universitat Politècnica de València, 46022 Valencia, Spain; glpatin@upv.es (G.L.-P.); piglesia@upv.es (P.L.I.-R.); vfuertes@upv.es (V.S.F.-M.)
* Correspondence: jmsolano@upv.es

Abstract: Analysis and design of siphonic roof drainage systems are usually performed with specific software developed by the system's manufacturers. An experimental study was carried out to verify if a general software for the analysis of hydraulic pressurized pipe networks can be used to analyze a siphonic roof drainage system. A test model was built and tests were conducted to compare the prototype results with simulation runs in EPANET. A new EPANET data model was developed to overcome software limitations for the siphonic drainage systems analysis. Considering the results, less than 5% of average error was observed between the measures in the real test model and the simulation results, which can be attributed to measurement error. To validate the EPANET data model, it was compared with a specific software that analyzes siphonic roof drainage systems. It can be concluded that EPANET is a software that can be used to analyze and, therefore, to design, siphonic roof drainage systems in buildings.

Keywords: rainwater; sump; fully primed pipe; pressurized pipe; software

1. Introduction

Conventional roof drainage systems (CRDSs) have been the most widely used to evacuate the falling rainwater in buildings [1]. These systems are designed in the same way as sewage systems are [2]. The pipes are sized to maintain an annular flow through the downpipes, with the air flowing through a central interior path, maintaining the depression near the atmospheric value to guarantee the water-trap seal depth of the toilets within the building [3]. This design criterion makes perfect sense for the sewage. However, it is not necessary for rainwater drainage systems, as the downpipes are not directly connected to toilets, and seals are not needed [4]

The evacuation capacity of a CRDS is limited by the water level on the roof surface since the entry of air must be guaranteed. This will make the evacuation flow limited. For more than 35 years, siphonic roof drainage systems (SRDS) have been an alternative to conventional systems due to the benefits they offer over CRDSs [5,6].

The performance of the SRDS varies according to the rainwater flow, in such a way that three phases can be distinguished in its behavior. In the first phase, with a low rainwater flow, the system behaves like a traditional system, with the air entering through the sump and filling most of the cross section of the pipes. The drainage capacity is very limited [7,8]. In the second phase, as the rainwater flow increases, the pipes become more and more full. At some points in the system, the pipes may be completely filled, while at others, they remain only partially filled [9]. In the third phase, the rainwater flow is so great that the pipe is fully primed. It is in this situation that the maximum drainage capacity of the system is reached [10].

In a climate change scenario, with increasingly more intense rains, it is necessary to provide a greater evacuation capacity for drainage systems [10,11]. SRDSs always have greater evacuation capacity than an equivalent conventional system when fully primed. So, this should be the design condition of the facility [12].

In a fully primed SRDS, the system works as a pressurized network in a quasi-steady state flow condition. The only energy involved is the gravity energy due to the difference between the water surface elevation over the roof and the elevation at the discharge point. This energy is used in accelerating the flow before the sump and overcoming head losses in hydraulic elements.

All pipes work with a below zero pressure because the pressure drop in the sump is greater than the elevation gained in it. The horizontal layout also produces a low pressure that is maintained, and only in the final downpipe does the pressure grow again. The pipes are designed to be able to transport the maximum drainage flow but to avoid collapse, withstanding the low pressure [13]. To design the system, SRDS manufacturers have developed dedicated software that must be handled by specialists [14–17]. These software are used with semi-empirical data and are applicable only for the commercial models marketed by each manufacturer. The reality is that the equation that explains the flow in the pipes of an SDRS is Bernoulli's equation; it is widely known.

Dependence on the manufacturers for the design of the system reduces the use of the SDRS as the project engineer loses control over the installation, which is left in the hands of someone outside the legal responsibilities of the project. Developers of the software indicate that corrections must be made in the pressure flow equations to adjust them to the real behavior of the system, which is not so clear [18].

The layout of the SDRS, when several inlets are disposed, is a looped hydraulic network. The analysis of the flow through the pipes in this type of network requires applying iterative calculation methods [19].

Arthur [9] is the first reference of a software used to analyze SRDSs. The method of characterization is used to solve both the transient flow from a partially primed phase to a fully primed one and the stationary flow during the fully primed phase of the system. Boundary conditions were established but no correction of the results were made in order to explain any difference between the simulation results and the measured ones. Other software used by some manufacturers do not indicate so clearly the resolution method used, so there is no evidence as to whether the corrections required to adjust the results of the analysis to reality are due to the iterative calculation method used or to other factors. In any case, validation of the results of the analysis of every software and the experimental results of a physical model are necessary to verify the goodness of such a program [20].

EPANET [21] is a widely used software that performs the analysis of pressurized looped hydraulic networks. In most cases, its application has been limited to the analysis of water distribution networks, and there are no references to its use for the analysis of hydraulic drainage systems under pressure by gravity. EPANET's capabilities show that, in theory, it is perfectly possible for the analysis of a fully primed SRDS. It remains to be demonstrated whether the results of the analysis of a real SRDS and the results of the simulation with EPANET are perfectly consistent, or on the opposite side, whether some corrections, such as those made by the manufacturers' software, would have to be made.

The main objective of this document was to verify whether it is possible to use a general pressure hydraulic network analysis program, such as EPANET, for the analysis of the hydraulic performance of an SRDS under fully primed conditions. If this were possible, it would be possible to use some looped network dimensioning techniques, based on the use of EPANET, to dimension the pipes of an SDRS, and the use of many types of pressure pipes would be extended to equip this type of pipework systems. The model proposed in this work will be validated, both by means of an experimental prototype developed at the laboratory scale and by comparing its results with the results of a specific SRDS analysis software developed by a manufacturer.

2. Materials and Methods

2.1. Experimental Equipment

For the development of this research, a small-size prototype was designed and built. It simulates the drainage surface of a single sump on the roof of a building, and the siphonic drainage pipes from the drain itself to the discharge manhole at atmospheric pressure. A better description can be seen in Figure 1.

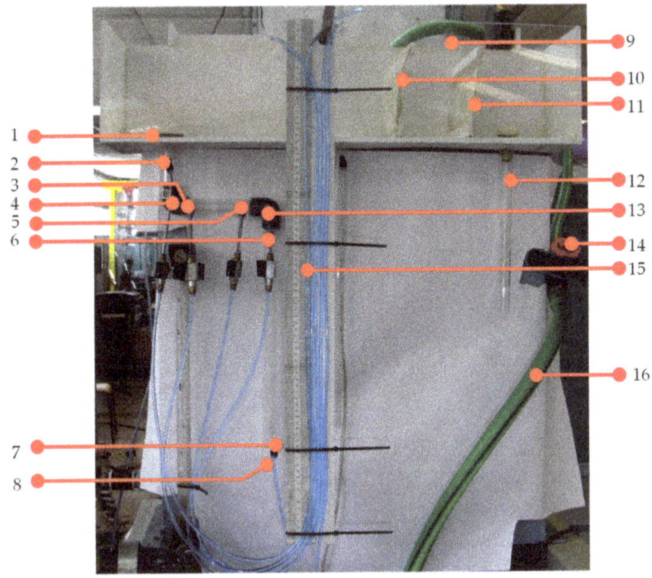

1 . Roof sump
2 . Measure point 1 (M1)
3 . Measure point 2 (M2)
4 . Elbow 1 (E1)
5 . Measure point 3 (M3)
6 . Measure point 4 (M4)
7 . Measure point 5 (M5)
8 . Discharge
9 . Overflow
10. Tranquillizer structure
11 . Tranquillizer structure
12 . Water Level Control sump (LCS)
13 . Elbow 2 (E2)
14 . Control valve
15 . U-manometer
16 . Supply pipe

Figure 1. Laboratory SRDS small-size test model.

The prototype consists of three functional parts: the small-size test model of the siphonic system itself, the system to regulate the intensity of rain to be simulated, and the measurement system.

For the design of the SRDS test model, the prototype must allow the simulation of different rain intensities and the turbulence generated by the drainage water flowing into the sump must be avoided, just as it would in a real gutter.

The SRDS test model consists of a lower reservoir that collects all the water drained by the system. The whole prototype works in a closed circuit. From a lower tank, the water is pumped with a submerged pump through a pressurized pipe (16 in Figure 1) to the first elevated tank and then drained back to the lower tank. From this elevated tank, the water overflows towards a lower surface that acts as a gutter in the roof in the building.

The water overflows on the surface away from the sump (9 in Figure 1). Falling water creates undesirable turbulence. Two flow tranquilizers (10 and 11 in Figure 1) are installed on the surface to eliminate turbulence and ensure that the flow of water to the sump is as calm as possible. The flow passes through the tranquilizers and reaches the sump. The water is drained through the sump and a vertical conduit. Next, a 90° elbow (4 in Figure 1) and a horizontal section were arranged to simulate a real system that collects several sumps. Finally, another 90° elbow (13 in Figure 1) and the vertical discharge conduit were installed (8 in Figure 1).

The levels in both tanks, the one that drains all water and the elevated one, remain constant, as does the flow pumped. A regulating system is necessary to simulate the intensity of the rain, as desired. A control valve was installed in the pumping line (14 in Figure 1).

In addition, to keep the level at the drainage surface constant, a level control sump (12 in Figure 1) was installed before the flow tranquilizers.

Several pressure measurement points were arranged throughout the pipes, before and after every singular element. All these pressure taps are connected with a pipe to a U-shaped manometer, the same for all of them, with a precision of 1 mm.

With these pressure measures, it is possible to calculate the real losses in any singular element of the prototype. All these data are entered in the EPANET model to run the SRDS simulation, so the results obtained with the software analysis are the most accurate to the real measures.

The drainage flow is also measured. For the flow measurement, the volume of water drainage in a controlled period of time is weighed on a precision scale.

2.2. Determination of Local Losses in the Elements of the System

Although the design of the elements that are part of the test model is not an objective of this work, it is necessary to know what their hydraulic behavior is, to insert them in the EPANET simulation model.

The behavior of pipes are well known when their diameter, length, and material are defined. However, there are elements whose local head losses depend on the installation in which they are mounted. This is the case for the drain (1 in Figure 1), the 90° elbows (4 and 13 in Figure 1), and the discharge section (8 in Figure 1).

The value of the local head loss depends on the flow through the installation and cannot be set, but hydraulic resistance can be, which is what is going to be calculated.

The location of the different measurement points was arranged so that, by making an energy balance between two successive measurement points, each local head loss can be estimated and, therefore, every hydraulic resistance, also.

Thus, to calculate the local head loss in the sump, Bernoulli's energy equation must be applied between the free surface of the water over the roof (point FS) and the measurement point M1,

$$\frac{P_{FS}}{\gamma} + z_{FS} + \frac{v_{FS}^2}{2g} = \frac{P_{M1}}{\gamma} + z_{M1} + \frac{v_{M1}^2}{2g} + r_{RD}Q^2 + h_{p,L1} \quad (1)$$

where z_{FS} is the level of the free surface of the water (referred to as the discharge section), P_{FS}/γ is the atmospheric pressure at the free surface of the water and its value is 0, v_{FS} is the velocity at the free surface of the water far from the sump and is neglected, v_{M1} is the velocity at the M1 point, and $h_{p,L1}$ is the head loss distributed through the pipelines between the sump and the measurement point M1.

Head losses distributed through the pipes are calculated from the hydraulic slope of the design of the installation, since all the pipes are of the same material and diameter. In particular, the one between the roof sump and the measurement point M1 is

$$h_{p,L1} = j_d L_{L1} \quad (2)$$

So, the hydraulic resistance of the roof sump, r_{RD}, is

$$r_{RD} = \frac{\left(0 - \frac{P_{M1}}{\gamma}\right) + (z_{FS} - z_{M1}) + \left(0 - \frac{Q^2}{2gS_{M1}^2}\right) - j_d L_{L1}}{Q^2} \quad (3)$$

To calculate the hydraulic slope of the design, the pressure measurements of a pipeline with no local head loss elements between them are used. For example, the pipeline between M4 and M5 in Figure 1.

$$\frac{P_{M4}}{\gamma} + z_{M4} + \frac{Q^2}{2gS_{M4}^2} - j_d L_{M4\text{–}M5} = \frac{P_{M5}}{\gamma} + z_{M5} + \frac{Q^2}{2gS_{M5}^2} \tag{4}$$

So, the hydraulic slope of the design for this installation, j_d, is

$$j_d = \frac{\left(\frac{P_{M4}}{\gamma} - \frac{P_{M5}}{\gamma}\right) + (z_{M4} - z_{M5})}{L_{M4\text{–}M5}} \tag{5}$$

where P_{M4}/γ and P_{M5}/γ are the pressures, with an accuracy of 1 mmwc, at the measuring points M4 and M5; z_{M4}–z_{M5} is the level difference between those points; and $L_{M4\text{–}M5}$ is the pipeline section.

To establish the hydraulic resistance for the first 90° elbow, Bernoulli's energy equation is applied between the measuring points M1 and M2.

$$\frac{P_{M1}}{\gamma} + z_{M1} + \frac{Q^2}{2gS_{M1}^2} - j_d L_{M1-E1} - r_{EBW1} Q^2 - j_d L_{E1-M2} = \frac{P_{M2}}{\gamma} + z_{M2} + \frac{Q^2}{2gS_{M2}^2} \tag{6}$$

where P_{M1}/γ and P_{M2}/γ are the pressures, with an accuracy of 1 mmwc, at the measuring points M1 and M2; $L_{M1\text{–}E1}$ is the pipeline section between the measuring point M1 and the elbow upstream end; and $L_{E1\text{–}M2}$ is the pipeline section between the downstream end of the elbow and the measurement point M2, which in this case, is negligible.

As the cross sections S_{M1} and S_{M2} are equal, the kinetic energy is the same value and cancels out, such that the hydraulic resistance of the first 90° elbow is calculated with

$$r_{EBW1} = \frac{\left(\frac{P_{M1}}{\gamma} - \frac{P_{M2}}{\gamma}\right) + (z_{M1} - z_{M2}) - j_d (L_{M1-E1} + L_{E1-M2})}{Q^2} \tag{7}$$

In the same way, by applying Bernoulli's energy equation between the measuring points M3 and M4, the hydraulic resistance of the second 90° elbow is calculated.

$$r_{EBW2} = \frac{\left(\frac{P_{M3}}{\gamma} - \frac{P_{M4}}{\gamma}\right) + (z_{M3} - z_{M4}) - j_d (L_{M3-E2} + L_{E2-M4})}{Q^2} \tag{8}$$

In this case, P_{M3}/γ and P_{M4}/γ are the pressures, with an accuracy of 1 mmwc, at the measuring points M3 and M4; $L_{M3\text{–}E2}$ is the length between the measuring point M3 and the upstream end of elbow 2; and $L_{E2\text{–}M4}$ is the length between the downstream end of the elbow 2 and the measuring point M4, which are both negligible.

Finally, to calculate the pressure drop at the installation discharge section, Bernoulli's energy equation is applied between M5 measurement point and the discharge section DSC. The hydraulic resistance in the discharge is

$$r_{DSC} = \frac{\left(\frac{P_{M5}}{\gamma} - 0\right) + (z_{M5} - z_{DSC}) - j_d (L_{M5-DSC}) + \left(\frac{Q^2}{2g\Sigma_{M5}^2} - \frac{Q^2}{2g\Sigma_{DSC}^2}\right)}{Q^2} \tag{9}$$

In Equation (9), the kinetic head is considered because there is a small reduction in the cross section at the measure point where the intake is installed, which does not occur in the outlet section. This velocity reduction affects the pressure reading and the goodness of the results.

2.3. System Model in EPANET

In conventional drainage systems, open channel flow equations are used to explain their behavior [22]. EPANET is an open source, free software, developed by the U.S. EPA that performs the analysis of hydraulic behavior within pressurized pipe networks [21]. Obviously, EPANET cannot be used to analyze a conventional drainage system.

In fully primed siphonic drainage systems, the water flow behaves as in a pressurized hydraulic system in which the analysis carried out by EPANET does have a place. This software solves the equations of the pressure network using the node applying the gradient method [23].

However, there are two reasons for EPANET not to be considered, as is, for the SRDS analysis. On one hand, singular elements with local head losses such as elbows, cross section pipe reductions, T-junctions, and, of course, sumps are not represented as EPANET model objects.

On the other hand, kinetic energy in the network is neglected in the EPANET simulation model. For the singular elements that produce local losses, EPANET uses the minor loss coefficient of the pipes to be modeled. Every element can be accounted for by assigning an additional head loss, through a local loss coefficient, to the pipeline friction head loss where the singular element is installed or, in isolation, considering a virtual pipe of negligible length whose local head loss coefficient is the same one of that element.

In the EPANET model of the analyzed installation, three virtual pipes were created to model the singular elements: one for the sump, and another two for each of the 90° elbows.

The local loss coefficients of each of the elements are not known and will be calculated as part of this study.

The kinetic energy of the Euler equation must be considered in SRDS analysis for some reasons. The flow gains velocity along the roof sump. If kinetic energy is not considered, as in EPANET, the real static pressure does not match the EPANET pressure simulation value.

If the pipes' cross section changes throughout the installation, the kinetic energy in the pipes will change too, although to a lesser extent than in the previous case. If the section of the pipes is maintained, the kinetic energy will not change, as happened in the test model analyzed.

In any case, if kinetic energy is not considered, the measured pressure in the test model and the pressure results of the EPANET simulation will not match. A modification has to be made to the EPANET pressure results to be reliable.

Finally, in the discharge section, the pressure is zero and velocity is again important since the kinetic head is not zero.

The discharge section of an SDRS behaves like an EPANET pressure-dependent flow node, in which the demand cannot be set. This node can be modeled as an "emitter". This element models the flow through a free discharge to the atmosphere, changing with the pressure as

$$q = C_E p^{\varphi} \tag{10}$$

where q is the flow rate at the discharge point, C_E is the discharge coefficient, p is the pressure, and φ is the pressure exponent, usually equal to 0.5 for a turbulent flow.

The value of the emitter coefficient, C_E, at the discharge section, adding local head losses at the node due to the discharge itself, and with the units indicated for flow and pressure, can be calculated as

$$C_E = \left[\frac{g\pi^2 D^4}{8(k_{dsc} + k_d)}\right]^{1/2} \tag{11}$$

where D is the diameter of the pipeline in m, k_d is the loss coefficient due to the kinetic head and is equal to 1, and k_{dsc} is the local head losses coefficient in the discharge.

3. Results and Discussion

3.1. Experimental Test Results

Measurements were made with a double purpose. First, the hydraulic characteristics of the elements of the test model installation (sump, 90° elbows, and discharge) must be defined. One of the reasons for the discrepancies between measurements and computed results is the estimation of the hydraulic losses of the elements on the system [24]. Some authors [25] estimated these losses from the literature [26]. For this work, the hydraulics' characteristics were calculated to gain accuracy.

Secondly, readings of the hydraulic variables in the real model must be compared with the EPANET model simulation ones to verify the reliability of the software.

Nine series of measurements were carried out. For each of them, the flow control valve (element 14 in Figure 1) was operated to change the flow pumped and, therefore, the drainage flow rate.

In every series of measurements, the following variables were read: the free surface water level, the U-manometer pressure for each of the five measurement points in Figure 1, and the volume discharged in a timestep. Accuracy of the measurement scale for the readings of the free surface water level and U-manometer pressure readings are 1 mm. The discharge water volume was calculated by measuring the mass with a weighing machine with a precision of 1 g. Time was recorded with a stopwatch with a precision of 1 s.

Readings for the level of the free surface of water in every series of measurements, pressure readings on the U-manometer in each test, for each of the measurement points, and values for drainage flow of the SRDS, calculated with the time and around a water volume of ten liter measurements, are presented in Table 1.

Table 1. Water level, pressure reading, and drain flow for each test.

Test	1	2	3	4	5	6	7	8	9
Water level (mm)	−68	−67	−124	−55	−83	−83	−83	−82	−80
P_{M1} (mmwc)	−417	−415	−431	−414	−407	−410	−405	−405	−417
P_{M2} (mmwc)	−572	−573	−577	−563	−561	−557	−562	−563	−558
P_{M3} (mmwc)	−580	−580	−579	−570	−567	−566	−569	−567	−566
P_{M4} (mmwc)	−724	−724	−723	−722	−722	−722	−722	−723	−724
P_{M5} (mmwc)	−761	−774	−765	−765	−765	−766	−765	−766	−770
Q (l/min)	26.96	27.04	25.92	26.99	26.73	27.26	26.81	26.95	26.38

3.2. Hydraulic Resistance of Elements Calculation

Hydraulic resistances of each singular element of the test model are calculated with the results of the measurements presented in Tables 1–3 and by applying Equations (5), (7), (8) and (10).

Table 2. Coefficients ki of head loss for a D = $19 \cdot 10^{-3}$ m pipe. Dimensionless.

Element	1	2	3	4	5	6	7	8	9	Mean
Roof sump	1.71	1.68	1.57	1.78	1.56	1.48	1.52	1.51	1.73	1.61
Elbow 1	1.17	1.18	1.19	1.12	1.18	1.08	1.20	1.19	1.10	1.17
Elbow 2	1.13	1.12	1.22	1.19	1.23	1.20	1.20	1.22	1.23	1.19
Discharge	0.06	−0.04	0.00	0.03	0.02	0.03	0.02	0.02	−0.03	0.00

Table 3. EPANET analysis results for total pressure, dynamic pressure, and static pressure.

Intake	P_{EPANET} (Pa)	P_d (Pa)	P_s (Pa)
M1	−216	1599	−1814
M2	−932	1599	−2531
M3	−1040	1599	−2639
M4	−1942	1599	−3541
M5	1089	1599	−510

Once the hydraulic resistance of the sump, 90° elbow 1, 90° elbow 2, and discharge are known, the head loss coefficients to be introduced in the EPANET model are calculated by applying Equation (12).

$$K_i = r_i \frac{g\pi^2 D^4}{8} \tag{12}$$

where k_i is the head loss coefficient of any element in the EPANET model, dimensionless; r_i is the hydraulic resistance of every singular element, in SI; and D is the diameter of the virtual pipe where the object is added in the EPANET model, in SI.

The result of the head loss coefficients of the EPANET model for each series of measurements is presented in Table 4.

Table 4. Static pressure from the analysis with EPANET vs. real measure.

Intake	$P_{s,EPANET}$ (Pa)	$P_{s,real}$ (Pa)	Error (%)
M1	−1814	−1864	2.7
M2	−2531	−2589	2.3
M3	−2639	−2639	0.0
M4	−3541	−3650	3.1
M5	−510	−520	2.0

The last column in Table 2 shows the mean value calculated by discarding the values that fall outside the 90% confidence interval of the sample.

3.3. Experimental Results Validation

Considering that the test model tested in the laboratory is well defined when knowing the head loss coefficients of each singular element and the pipes' lengths and diameters, now, it is possible to create the EPANET simulation model of the installation.

As shown in Figure 2, three virtual pipes were created to simulate the behavior of the singular elements: Pipe RD, Pipe Elbow1, and Pipe Elbow2. Each of them has a negligible length and a head loss coefficient according to Table 2.

The water on the roof was modeled as a reservoir with a fixed elevation as the free water-surface level. Each of the measurement points was modeled as a pressure node without demand, located at the elevation that corresponded to it in the real installation.

Finally, the discharge was modeled as the DSC-named emitter, with a head loss coefficient, according to Table 2, equal to zero and a head velocity coefficient equal to 1.

The EPANET simulation model was analyzed considering a free water-surface level, z_{FS}, of 0.665 m above the discharge level, corresponding to the fourth test in Table 1.

The results of the analysis for pressures at the measurement points are those indicated in the P_{EPANET} column in Table 3. This value is the result of the total pressure at the node, the sum of the static pressure plus the velocity head, since EPANET neglected the velocity head in its simulations.

ID Node	Elevation (m)	Demand (m3/d)	Coef. (m3/d; mH2O)	Node type
SMP	0.665	-		Reservoir
Node6	0.518	0		Junction
M1	0.496	0		Junction
M2	0.421	0		Junction
M3	0.419	0		Junction
M4	0.370	0		Junction
M5	0.008	0		Junction
Node7	0.438	0		Junction
DSC	0	0	108.5	Emitter

ID pipe	Node1	Node2	Length (m)	Diameter (mm)	Roughness (mm)	MinorLoss	Element Type
RD	SMP	Node6	0.00001	19	0.0001	1.61	Drain
Line 1	Node6	M1	0.022	19	0.001	0	Pipe
Line 2	M1	Node7	0.047	19	0.001	0	Pipe
Line 3	M2	M3	0.092	19	0.001	0	Pipe
Elbow 1	Node7	M2	0.00001	19	0.0001	1.17	Elbow
Elbow 2	M3	M4	0.00001	19	0.0001	1.19	Elbow
Line 4	M4	M5	0.362	19	0.001	0	Pipe
Line 5	M5	DSC	0.008	19	0.001	0	Pipe

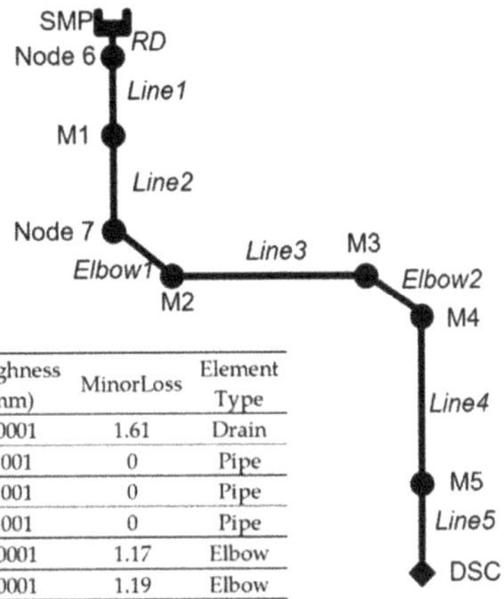

Figure 2. EPANET model.

The real static pressure is calculated by discounting velocity head from the EPANET total pressure results.

$$P_s = p_{EPANET} - p_d = p_{EPANET} - \gamma \frac{Q^2}{2g\Sigma^2} \tag{13}$$

The EPANET pressure results, velocity head, and static real pressure calculated are shown in Table 3.

To verify if the EPANET model is good enough to analyze the SRDS, the results of the simulation were compared with those of the measurements.

The results of the pressure measurements on the U-manometer are available in mmwc, Table 1. As the static pressure resulting from the simulation with EPANET is expressed in Pa, the measurements were calculated in the same units. To obtain them, it must be taken into account that the pressure at the measurement point is referring to the elevation at which the water level is on the roof. Therefore, the static pressure at the measurement point is calculated according to Equation (14).

$$P_s[\mathrm{Pa}] = (p_{UM} - z_{FS})\,[\mathrm{mmwc}]\,\frac{1}{1000}\gamma_{H2O}\left[\frac{N}{m3}\right] \tag{14}$$

where P_s is the static pressure at the measurement point, p_{UM} is the pressure observed on the U-manometer for every measurement point, z_{FS} is the elevation at the free surface of the water on the deck, and γ_{H2O} is the specific weight of the water.

Taking the EPANET results from Table 3 and those calculated from the measurements according to the test 4 set of data from Table 1, and then applying Equation (14), Table 4 is presented. A comparison between the measured results and those simulated for pressure can be observed. A calculation of the error produced at each measurement point was performed. It was verified that the error was less than 5%.

Proceeding in a similar way with the flow measurements and the flow obtained in the simulation with EPANET, a comparison of the results is shown in Figure 3. In this figure, each point corresponds to a test in Table 2. It was verified that the error was below 3%.

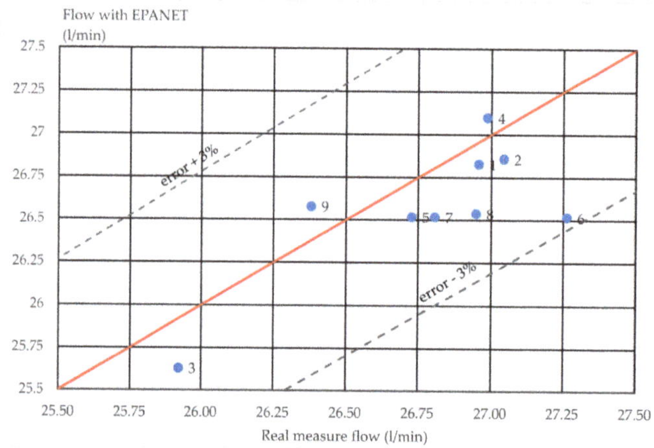

Figure 3. Real flow measure vs. EPANET flow for each series of measurements. In this figure, each blue circle corresponds to a test in Table 2.

The error, both in pressures and flows, is less than that observed in similar studies [23], where an error of 13% was reached.

Considering the results, it can be indicated that the simulation results with EPANET are good enough. It can be stated that, with the appropriate changes to the EPANET data model, it is possible to use it to analyze SRDSs.

These changes in the EPANET model are modeling the discharge point as a pressure dependent demand node, rather than a known demand node. On the other hand, the kinetic term of the energy equation must be taken into account. To do this, the value of the static pressure calculated with EPANET must be corrected by subtracting the kinetic energy at the node.

3.4. *Software Validation*

Once the EPANET model has been experimentally confirmed, we compared its results with the results of a commercial software from a manufacturer to ensure its suitability. We checked with SIPHONITEC®, a commercial software developed by the RMS company for the design and analysis of SRDSs. It uses commercial components whose hydraulic behaviors are well known.

A small installation of an SRDS was analyzed using both software. This system consists, as can be seen in Figure 4, of two siphonic drains that collect rain from two roofs of different size with a single discharge.

The system analysis was performed with both programs. For the EPANET simulation, the individual components (elbows, reductions, unions) were modeled as regulation valves with the same loss coefficient as the real commercial component.

The results of the analysis are shown in Table 5. It can be observed that the flow rate and head values are practically identical in both cases.

It is verified that analysis of complex SRDSs can be performed with EPANET, a free software, with no need to use a specific software developed by manufacturers.

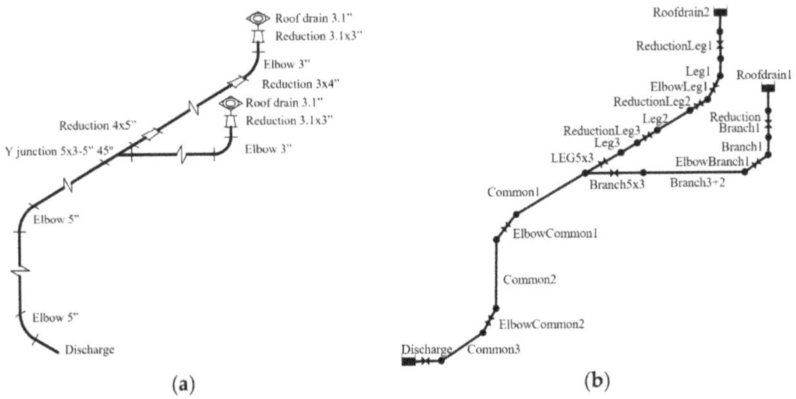

Figure 4. (**a**) SRDS model in SIPHONITEC®. (**b**) SRDS model in EPANET.

Table 5. Results of the simulation for node head.

	SIPHONITEC Results			EPANET Results		
	Flow (cfs)	**Head in (ftwc)**	**Head out (ftwc)**	**Flow (cfs)**	**Head in (ftwc)**	**Head out (ftwc)**
Discharge	1.634	2.342	0.000	1.637	2.350	0.000
Common3	1.634	3.714	2.342	1.637	3.720	2.350
ElbowCommon2	1.634	4.159	3.714	1.637	4.160	3.720
Common2	1.634	6.490	4.159	1.637	6.490	4.160
ElbowCommon1	1.634	6.799	6.490	1.637	6.800	6.490
Common1	1.634	10.776	6.799	1.637	10.770	6.800
LEG5x3	0.690	9.621	10.776	0.693	9.600	10.770
Leg3	0.690	9.646	9.621	0.693	9.630	9.600
ReductionLeg3	0.690	9.926	9.646	0.693	9.910	9.630
Leg2	0.690	12.285	9.926	0.693	12.270	9.910
ReductionLeg2	0.690	13.007	12.285	0.693	13.000	12.270
ElbowLeg1	0.690	13.979	13.007	0.693	13.970	13.000
Leg1	0.690	14.706	13.979	0.693	14.700	13.970
ReductionLeg1	0.690	14.818	14.706	0.693	14.820	14.700
RoofDrain2	0.690	15.267	14.818	0.693	15.270	14.820
Branch5x3	0.944	15.269	10.776	0.944	15.280	10.770
Branch3+2	0.944	16.691	15.269	0.944	16.700	15.280
ElbowBranch1	0.944	18.511	16.691	0.944	18.520	16.700
Branch1	0.944	19.866	18.511	0.944	19.870	18.520
ReductionBranch1	0.944	20.075	19.866	0.944	20.080	19.870
RoofDrain1	0.944	20.916	20.075	0.944	20.920	20.080

4. Conclusions

Considering the results, it may be concluded that it is possible to use the EPANET to hydraulically analyze SRDSs, although it must be adapted so that the calculation can be carried out. In the same way, it would be possible to size the SRDS using simulation techniques with the EPANET program. It has been verified, by means of a test prototype of an SRDS, that EPANET analysis results are accurate enough.

The errors observed between measurements in the laboratory and the results of the simulation with EPANET are less than 5% for pressures and 3% for flows. These errors improve the results obtained by Arthur and Swaffield [24]. Part of the merit arises from the fact that the minor losses were tested in the laboratory.

In order to perform the analysis with EPANET, it is necessary to modify the conventional data model used for pressure hydraulic networks and define a specific data model for SRDSs.

On the one hand, the discharge point had to be modeled as a node whose demand is not constant but dependent on pressure. On the other hand, the kinetic term of the energy equation had to be added to the data model of each conduction, which, in the conventional model, EPANET data is not considered.

It was verified, through comparison with a specific commercial SRDS analysis software, that EPANET results obtained by means of the simulation model are good enough to consider the analysis as valid.

The error observed between the commercial software and EPANET is negligible.

Due to limitations in the size of the test model used, a roof with a single sump was simulated in the prototype. In future developments, a larger test model will be built to simulate the real behavior of a roof with several sumps.

The result of this work is of great use to engineers as it adds an additional free access tool, widely used at a global level, for the analysis, design, and dimensioning of SRDSs.

Author Contributions: All authors contributed extensively to the work presented in this paper. Conceptualization, G.L.-P., P.L.I.-R., F.J.M.-S. and V.S.F.-M.; data curation, G.L.-P. and P.L.I.-R.; formal analysis, G.L.-P., P.L.I.-R., F.J.M.-S. and V.S.F.-M.; investigation, G.L.-P., P.L.I.-R., F.J.M.-S. and V.S.F.-M.; methodology, G.L.-P., P.L.I.-R. and F.J.M.-S.; resources, G.L.-P., P.L.I.-R., F.J.M.-S. and V.S.F.-M.; software, G.L.-P., P.L.I.-R. and F.J.M.-S.; supervision, G.L.-P., P.L.I.-R., F.J.M.-S. and V.S.F.-M.; validation, G.L.-P., P.L.I.-R., F.J.M.-S. and V.S.F.-M.; visualization, G.L.-P., P.L.I.-R., F.J.M.-S. and V.S.F.-M.; writing—original draft, G.L.-P. and P.L.I.-R.; writing—review and editing, G.L.-P., P.L.I.-R., F.J.M.-S. and V.S.F.-M. All authors have read and agreed to the published version of the manuscript.

Funding: This research received no external funding.

Data Availability Statement: Data sharing not applicable. All data used are included in the text. Detailed raw data are available upon reasonable request.

Conflicts of Interest: The authors declare no conflict of interest.

References

1. Gormley, M.; Kelly, D.; Campbell, D.; Xue, Y.; Stewart, C. Building Drainage System Design for Tall Buildings: Current Limitations and Public Health Implications. *Buildings* **2021**, *11*, 70. [CrossRef]
2. May, R.W.P. The Design of Conventional and Siphonic Roof-Drainage Systems. *Water Environ. J.* **1997**, *11*, 56–60. [CrossRef]
3. Arthur, S.; Wright, G.B. Recent and future advances in roof drainage design and performance. *Build. Serv. Eng. Res. Technol.* **2005**, *26*, 337–348. [CrossRef]
4. Fujimura, K.; Sakaue, K. Analysis of pneumatic pressure vibration affected by connecting WCs and discharge load types. *Water* **2017**, *9*, 6. [CrossRef]
5. Arthur, S.; Swaffield, J.A. Siphonic roof drainage: Current understanding. *Urban Water* **2001**, *3*, 43–52. [CrossRef]
6. Escarameia, M.; May, R.W.P. *Performance of Siphonic Drainage Systems for Roof Gutters*; HR Wallingford: Wallingford, UK, 1996.
7. Guan, Y.; Fang, Z.; Tang, Z.; Yuan, J. Influence of the vent pipe diameter on the discharge capacity of a circuit vent building drainage system. *Build. Serv. Eng. Res. Technol.* **2019**, *41*, 5–24. [CrossRef]
8. Banisoltan, S.; Rajaratnam, N.; Zhu, D.Z. Experimental and Theoretical Investigation of Vertical Drains with Radial Inflow. *J. Hydraul. Eng.* **2016**, *143*, 04016103. [CrossRef]
9. Arthur, S.; Swaffield, J. Understanding Siphonic Rainwater Drainage Systems. In Proceedings of the CIB W62 Symposium Edinburgh, Water supply and Drainage for Buildings, Edinburgh, UK, 21–23 September 1999; pp. B1/1–B1/16.
10. Lucke, T.; Beecham, S.; Qu, Y.Y. Estimating Flowrates Through Individual Outlets of Siphonic Roof Drainage Systems. *Build. Res. Inf.* **2016**, *44*, 289–300. [CrossRef]
11. Beattie, R.K.; Jack, L.B. How siphonic roof drainage systems perform in a future climate. In Proceedings of the 2011 Symposium CIB W062, Aveiro, Portugal, 25–28 September 2011; pp. 301–307.
12. Arthur, S.; Wright, G.B. Siphonic roof drainage systems-priming focused design. *Build. Environ.* **2007**, *2*, 2421–2431. [CrossRef]
13. Lucke, T.; Arthur, S. Plastic pipe pressures in siphonic roof drainage systems. *Build. Res. Inf.* **2011**, *39*, 79–92. [CrossRef]
14. Geberit Verwaltungs AG, Geberit Proplanner. Available online: https://www.geberit-global.com/planning-installation/geberit-proplanner/ (accessed on 25 May 2023).
15. R. M. S. (RMS), SiphoniTec®Siphonic Roof Drain Sizing Software Information. Available online: https://www.jrsmith.com/siphonitec-siphonic-roof-drain-sizing-software-information (accessed on 25 May 2023).
16. Siphonix Worldwide Ltd. Roof Drainge System. Available online: http://www.siphonix.com/software.html (accessed on 25 May 2023).

17. Saint Gobain PAM. EPAMS Software. Available online: https://www.pam-drainage-solutions.com/rainwater-epams (accessed on 25 May 2023).
18. Roland, R.E. *Geberit ProPlanner 2020—Roof Drainage Systems;* Training Manual; Geberit International AG: Rapperswil, Switzerland, 2020.
19. Brkicbrkic, D. Iterative Methods for Looped Network Pipeline Calculation. *Water Resour. Manag.* **2011**, *25*, 2951–2987. [CrossRef]
20. Wright, G.B.; Swaffield, J.A.; Arthur, S. Investigation into the performance characteristics of multi-outlet siphonic roof drainage systems. In *Global Solutions for Urban Drainage;* American Society of Civil Engineers: Reston, VA, USA, 2002; pp. 1–16. [CrossRef]
21. Rossman, L. *EPANET 2.0 Users Manual;* U.S. Environmental Protection Agency: Washington, DC, USA, 2000.
22. Chaudhry, M.H. *Open-Channel Flow*, 2nd ed.; Springer: New York, NY, USA, 2008.
23. Todini, E.; Pilati, S. A gradient algorithm for the analysis of pipe networks. In *Computer Applications in Water Supply: Volume 1—Systems Analysis and Simulation;* Coulbeck, B., Orr, C.-H., Eds.; John Wiley & Sons: Hoboken, NJ, USA, 1988; pp. 1–20.
24. Arthur, S.; Swaffield, J.A. Siphonic roof drainage system analysis utilising unsteady flow theory. *Build. Environ.* **2001**, *36*, 939–948. [CrossRef]
25. May, R.W.P. *Design Criteria for Siphonic Roof Drainage Systems;* Report SR654; HR Wallingford: Wallingford, UK, 2004.
26. Idelchik, I.E. *Handbook of Hydraulic Resistance*, 3rd ed.; Begell House Inc.: New York, NY, USA, 1996.

MDPI AG
Grosspeteranlage 5
4052 Basel
Switzerland
Tel.: +41 61 683 77 34

Environments Editorial Office
E-mail: environments@mdpi.com
www.mdpi.com/journal/environments

www.ingramcontent.com/pod-product-compliance
Lightning Source LLC
Chambersburg PA
CBHW061612220926
51850CB00005B/491
* 9 7 8 3 7 2 5 8 5 6 0 9 1 *